Intelligent Systems and Robotics

Intelligent Systems and Robotics

Edited by

George W. Zobrist

and

C. Y. Ho[†]

University of Missouri at Rolla, USA

CRC Press is an imprint of the
Taylor & Francis Group, an **informa** business

CRC Press
Taylor & Francis Group
6000 Broken Sound Parkway NW, Suite 300
Boca Raton, FL 33487-2742

First issued in paperback 2019

ISBN-13: 978-90-5699-665-9 (hbk)
ISBN-13: 978-0-367-39885-9 (pbk)

British Library Cataloguing in Publication Data

Intelligent systems and robotics
1.Intelligent control systems 2.Robotics
I.Zobrist, George W. (George Winston), 1934– II.Ho, C. Y.
(Chung You), 1933–1998
629.8′92

Visit the Taylor & Francis Web site at
http://www.taylorandfrancis.com

and the CRC Press Web site at
http://www.crcpress.com

To *Pete* —
a good friend

CONTENTS

EDITORS' PREFACE

This monograph focuses on new developments in robotics and intelligent systems, provides insight, guidance, and specific techniques vital to those concerned with the design and implementation of robotics and intelligent system applications. The reader will find authoritative presentations on autonomous land vehicle navigation, manipulator reachable workspace problems and the formulation of algorithms for their solution. The next several chapters cover methods for medical applications utilizing expert adaptive control, the integrated piezoelectric sensor/actuator design for distributed identification and control of smart machines, and this includes theory, experiments, finite element formulation and analysis. The monograph continues with a chapter on the automatic repair of aircraft transparencies and geometric modeling utilized in robot task planning. Then the evolution of standard fieldbus networks utilized in the factory environment is presented. Triangle based surface models and their techniques are presented with a representation and conversion of solid modeling and the issues involved.

ACKNOWLEDGMENTS

The publisher's staff, including Frank Cerra, Tara Lynch, Susanne Jones, Lloyd Black, and Matt Uhler, is greatly appreciated. The editors also wish to thank the authors and co-authors for their patience in the publication of this monograph and for their contribution of the material that has been included.

CONTRIBUTORS

D. Brzakovic, Department of Electrical Engineering and Computer Science, Lehigh University, Bethlehem, PA 18015

Silvia Bussi, Elsag bailey, R&D Department, Via Puccini, 2, 16154 Genova, Italy

Gianluca Cena, Dip. di Automatica e Informatica, Politecnico di Torino, Corso Duca degli Abruzzi, 24-10129 Torino, Italy

Leila De Floriani, Information and Computer Science Department, University of Genova, Via Dodecaneso, 35, 16146 Genova, Italy

Luca Durante, Dip. di Automatica e Informatica, Politecnico di Torino, Corso Duca degli Abruzzi, 24-10129 Torino, Italy

L. Hong, Department of Electrical Engineering, Wright University, Dayton, OH 45435

Howard Kaufman, Rensselaer Polytechnic Institute, Electrical, Computer and Systems Engineering Department, Troy, NY

James A. Luckemeyer, Engineering Spectrum, Inc., 6953 Alamo Downs Parkway, San Antonio, TX 78238

Paola Magillo, Information and Computer Science Department, University of Genova, Via Dodecaneso, 35, 16146, Genova, Italy

David S. McFalls, Consulting and Management Services, 603 Williamsport, League City, TX 77573

Claudio Mirolo, Dip. di Matematica e Informatica dell'Universita de Udine, Via delle Science 206, I-33100 Udine, Italy

Gregory W. Neat, Jet Propulsion Laboratory, Pasadena, CA

Enrico Pagello, 1st. LADSEB del CNR, Corso Stati Uniti 4, I-35020 Padova, Italy & Dep. di Elettronica e Informatica, Universita di Padova

Enrico Puppo, Istituto per la Matematica Applicata, Consiglio Nazionale delle Ricerche, Via De Marini, 6 (Torre de Francia), 16149 Genova, Italy

C. I. Tseng, Department of Mechanical Engineering, Feng-Chia University, Taichung, Taiwan, R.O.C.

H. S. Tzou, Department of Mechanical Engineering, Center for Robotics and Manufacturing Systems, University of Kentucky, Lexington, KY 40506-0046

Adriano Valenzano, Centro per l'Elaborzione Numerale dei Segnali

Dennis J. Wenzel, Engineering Spectrum, Inc., 6953 Alamo Downs Parkway, San Antonio, TX 78238

Yugeng Xi, Department of Automatic Control, Shanghai Jiao-Tong University 1954 Hua Shan Road, Shanghai, P.R. China 200030

Zhiyuan Ying, Department of Automatic Control, Shanghai Jiao-Tong University 1954 Hua Shan Road, Shanghai, P.R. China 200030

Zhongjun Zhang, Department of Automatic Control, Shanghai Jiao-Tong University 1954 Hua Shan Road, Shanghai, P.R. China 200030

1

Detection of Low-Contrast Road Edges for Autonomous Land Vehicle Navigation[1]

D. Brzakovic
Department of Electrical Engineering and Computer Science
Lehigh University
Bethlehem, PA 18015

L. Hong
Department of Electrical Engineering,
Wright University, Dayton, OH 45435

1. INTRODUCTION

The problem of autonomous land vehicle navigation has attracted considerable attention in recent years, e.g., [1–15]. Navigation systems have been developed for road scenes of various complexity using different sensors. Some of these systems incorporate object recognition modules and aim at complete scene understanding, e.g., [9]. In many cases attention is primarily focused on determining the road edges, which can then be used for autonomous navigation. Various algorithms that employ properties of road geometry [10, 13] and intensity e.g., [3], have been developed for specific road conditions. Due to the low contrast of images and the required processing speed, road edge detection is, in general, very difficult and still remains an open research problem. The difficulty of road edge detection is somewhat eased by more flexible requirements on the accuracy of edge detection because it is necessary to obtain only a rough estimate of the road edges rather than very accurate positions of individual edge pixels.

In this chapter we propose a hybrid road edge detection method which locates low-contrast road edges in sequences of noisy images acquired by a standard TV camera. Edge detection, when processing the first image frame in the sequence, involves two steps. First, statistically homogeneous regions and intermediate regions wherein road edges lie are identified using low-pass filtering and thresholding. Second, road edges are estimated using a linear recursive filter and

[1] This work has been sponsored in part by Martin Marietta Aerospace, Orlando, Fla.

statistical properties of the homogeneous regions. When processing the subsequent image frames only the second step is executed, and the estimated road edges in the previous frame are used as the initial estimates of the road edges in the current frame. The problem of edge detection is posed in the framework of estimation theory: (1) By formulating mathematical system models based on the assumed geometric edge models; (2) by utilizing statistical properties of the adjacent regions to develop linear measurement models; and (3) by invoking a linear recursive filter (LRF) for edge estimation. The theoretical basis, implementation, and performance evaluation of the proposed road edge detection/description method are presented in this chapter.

The chapter is organized as follows. Section 2 discusses the formulation of the problem of edge estimation in the framework of linear recursive filtering. It details the processing of a single image frame and describes the road edge models and measurement models as well as the necessary preprocessing steps. Section 3 discusses the relationships between the parameters used by the linear recursive filter for different frames, emphasizing initial conditions. This section also discusses the relationship between the filter parameters and the convergence rate. Section 4 discusses the computational requirements and the accuracy of the obtained edges. For completeness, the appendix reviews the relevant properties of spline functions, which are the basis of the road edge model.

2. PROCESSING OF A SINGLE FRAME

The road edge estimation method described in this chapter involves, in general, two steps. The first step identifies statistically homogeneous regions, A and B, and the intermediate region, W, hereafter called the ambiguity zone (Figure 1.1). Typically in road edge detection, one of the regions is the road and the other is the surrounding area. The second step is the actual road edge estimation. In general, it is not necessary to determine accurately the ambiguity zone, but only to provide a rough outline of the region where the road edge lies. Because road edges lie in the ambiguity zone, the first step reduces the problem of estimating a road edge between regions of unknown properties to the problem of the edge estimation in the ambiguity zone between regions of known properties. In this chapter, identification of the statistically homogeneous regions is done only when processing the first frame. The subsequent frames are processed as described in Section 3.

Section 2.1 discusses the preprocessing stage, whereas Section 2.2 formulates the problem of road edge detection in the framework of linear recursive filtering. Section 2.3 describes road edge models and various measurement procedures. Finally, Section 2.4 summarizes various properties that can be utilized by the measurement models.

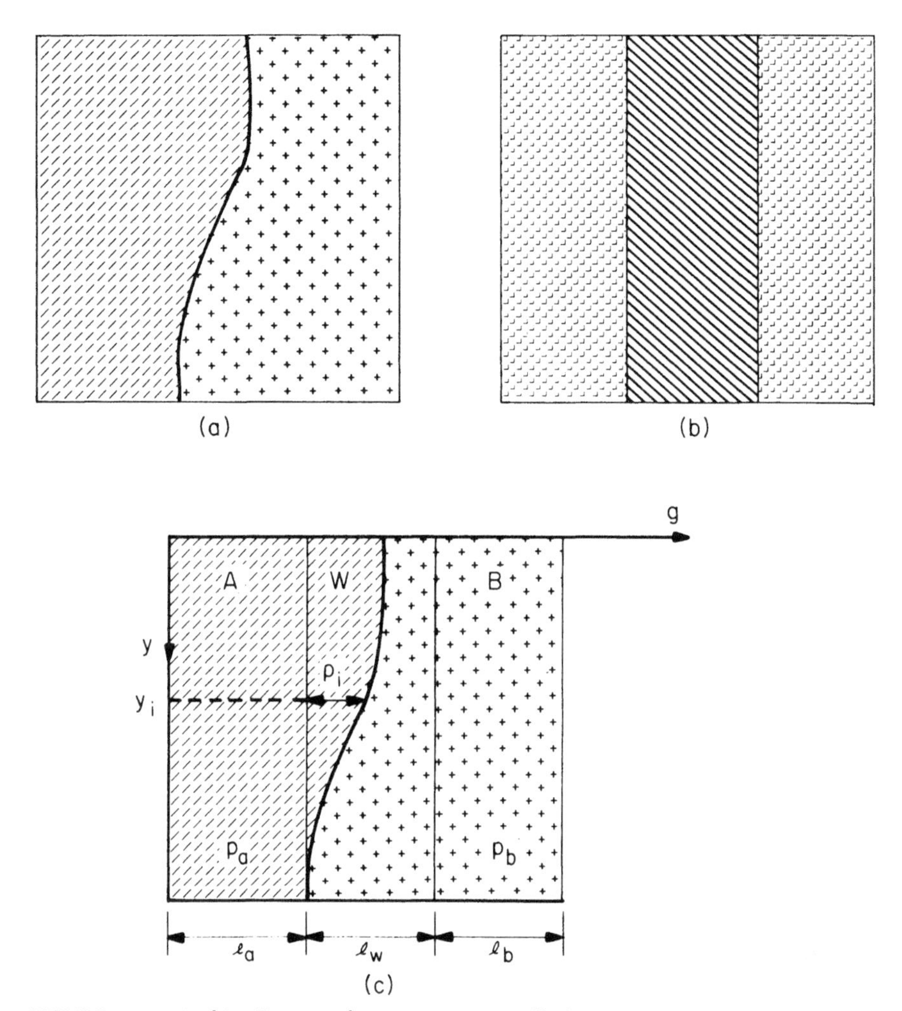

FIGURE 1.1. Ambiguity zone between statistically homogeneous regions: (a) Original image: (b) rectangular ambiguity zone; and (c) notation.

2.1. Preprocessing

Various region-based segmentation methods may be used to identify the statistically homogeneous regions and the ambiguity zones. In this work we employ pixel typicality, T, defined as [16]

$$T(x, y) = \frac{1}{(2P + 1)(2Q + 1)} \sum_{p=-P}^{P} \sum_{q=-Q}^{Q} [I(x, y) - I(x + q, y + p)]^2, \quad (1)$$

where $I(x, y)$ denotes the gray level function (pixel intensity). Generation of the T image highlights regions where significant changes in pixel intensity occur. Consequently, thresholding the T image leads to the extraction of the ambiguity zone. The thresholding method employed in this work is described in [17]. In order to simplify the measurement model equations (Section 2.3), we surround the resulting ambiguity zone by the minimum rectangle with sides parallel to the coordinate axes [Figure 1(b)]. The size of image subregions, $[(2Q + 1) \times (2P + 1)]$, in Equation (1) affects the width of the obtained ambiguity zone. An increase in P and Q results in an increase in the width of the ambiguity zone. In this work we use $P = Q = 9$.

In cases of low signal-noise ratio, the effects of noise are reduced by first subjecting the images to Gaussian low-pass filtering. Gaussian filtering is also very useful when processing texture images because it reduces the effects of texture irregularity (for details see [16]). For computational efficiency the results shown in this chapter were obtained by using an approximation of Gaussian filter by a three-level hierarchical discrete correlation [18].

2.2. Problem Formulation

The objective of the estimator is to determine the road edge location based on measurements (contaminated by noise) performed in the ambiguity zone. In this work, spline functions are used to approximate road edges. By approximating an edge in W by a spline function, in contrast to similar approaches to edge detection, e.g., [19, 20], we formulate system and measurement models of the form

$$\begin{aligned} \underline{x}_{k+1} &= \mathbf{\Phi}_k \underline{x}_k + \underline{w}_k, \qquad \underline{w}_k \sim N(0, \mathbf{Q}_k), \\ \underline{z}_k &= \mathbf{H}_k \underline{x}_k + \underline{v}_k, \qquad \underline{v}_k \sim N(0, \mathbf{R}_k) \end{aligned} \tag{2}$$

where $\underline{x}_k$ denotes the state vector; $\underline{z}_k$ contains the noisy measurements; and $\underline{w}_k$ and $\underline{v}_k$ are system modeling and measurement noise vectors, respectively. Different edge approximation models result in different system matrices, $\mathbf{\Phi}_k$, and measurement matrices, $\mathbf{H}_k$.

With Equation (2) at our disposal, LRF [21] is used to estimate the state vector $\underline{x}_k$. The calculations involved are summarized below.
State estimate extrapolation:

$$\hat{\underline{x}}_k(-) = \mathbf{\Phi}_{k-1} \hat{\underline{x}}_{k-1}(+), \tag{3}$$

Error covariance extrapolation:

$$\mathbf{P}_k(-) = \mathbf{\Phi}_{k-1} \mathbf{P}_{k-1}(+) \mathbf{\Phi}^T_{k-1} + \mathbf{Q}_{k-1}, \tag{4}$$

State estimate update:

$$\hat{\underline{x}}_k(+) = \hat{\underline{x}}_k(-) + \mathbf{K}_k\,[\underline{z}_k - \mathbf{H}_k\,\hat{\underline{x}}_k(-)], \tag{5}$$

Error covariance update:

$$\mathbf{P}_k(+) = [\mathbf{I} - \mathbf{K}_k\,\mathbf{H}_k]\,\mathbf{P}_k(-), \tag{6}$$

Kalman gain matrix:

$$\mathbf{K}_k = \mathbf{P}_k(-)\,\mathbf{H}_k^T\,[\mathbf{H}_k\,\mathbf{P}_k(-)\,\mathbf{H}_k^T + \mathbf{R}_k]^{-1}. \tag{7}$$

The notation used in Equations (2)–(7) is explained in Appendix D and has been adopted from [22]. Details of linear recursive filtering are discussed in [22].

2.3. Formulation of Measurement and System Models

The measurement model relates the values of a property p in the statistically homogeneous regions A and B (known from the first processing stage) to the value of property p in the ambiguity zone, W, at y = const. The selection of an appropriate property p for specific families of images is discussed in Section 2.4. The formulation presented in this section is independent of the particular choice of property p as long as the assumption of linear dependence on the states [Equation (2)] is satisfied. Assuming that p_i, i.e., the value of property p in W at $y = \bar{y}_i$ = const, is a linear combination of p_a and p_b, their relationship (see Figure 1.1) is

$$p_i = \frac{p_a\,\rho_i + (l_w - \rho_i)p_b}{l_w}. \tag{8}$$

Multiple measurements are performed simultaneously in different regions of the ambiguity zone to ensure observability of all states. In general, we consider m simultaneous measurements performed along windows of dimensions $s \times l$ centered in the y-direction at $\bar{y}_j$, $j = 1, 2, \ldots, m$. Using the edge model in two different ways, we formulate two system and measurement models, of the form of Equation (2), which are described in detail in Sections 2.3.1 and 2.3.2. The properties of spline functions used to derive these two models are summarized in Appendixes A, B, and C.

2.3.1. Global Edge Estimator Based on Complete Cubic Splines.

In this formulation, an edge is approximated on the interval $\alpha \leq a \leq \beta$ by an interpolatory cubic spline function with equidistant knots (Figure 1.2). Using

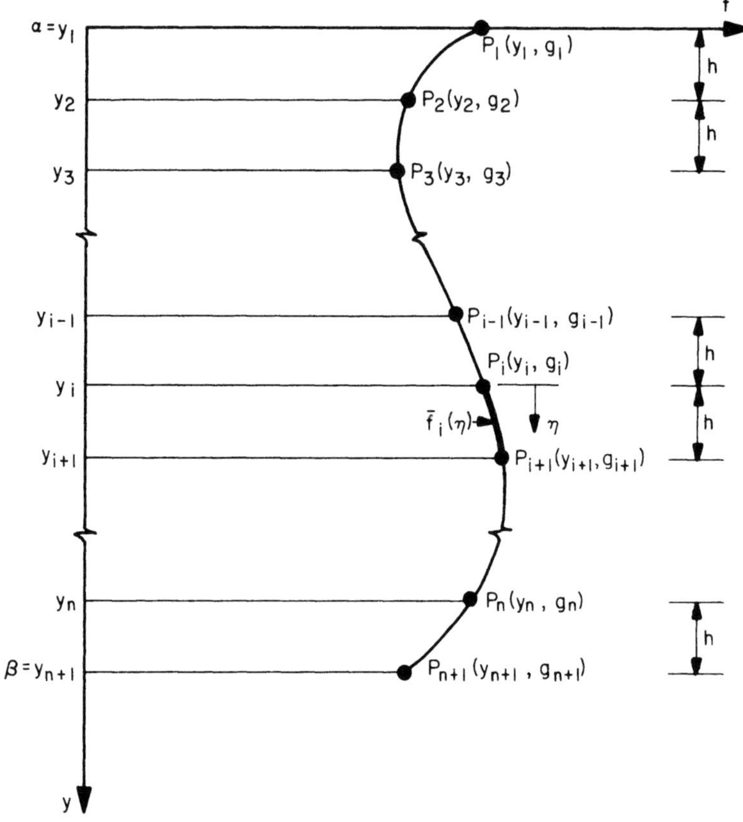

FIGURE 1.2. Complete cubic spline function *f*(*y*) interpolating *g*(*y*) at points $P_i(y_i, g_i)$ $i = 1, 2, \ldots, n + 1$. Equidistant knots.

Equation (34) in Appendix C, the edge is determined in terms of the function values $\{g_i; i = 1, 2, \ldots n + 1\}$ at the fixed knots $\{y_i; i = 1, 2, \ldots, n + 1\}$ and the derivatives g'_1 and g'_{n+1} at the interval ends.

The state vector in this formulation is chosen to be $\underline{x}^T = [g_1, g_2, \ldots, g_{n+1}, g'_1, g'_{n+1}]$. Consequently, the system matrix $\mathbf{\Phi}_k$ becomes the $(n + 3) \times (n + 3)$ identity matrix. At $y = \bar{y}_i =$ const, the quantity $(p_i - p_b)$ is measured. For enhanced observability, n simultaneous measurements are performed. At the kth cycle of the filtering process, measurements of $(p_i - p_b)$ are performed at $\bar{y}_i^k = \alpha + k + (i - 1)h$, $i = 1, 2, \ldots, n$ and arranged into a column vector $\underline{z}_k = [z_1^k, z_2^k, \ldots, z_n^k]^T$ in accordance with the notation of estimation theory. Symbol h denotes the distance between knots (Figure 1.2). Based on Equations (34) and (35) in Appendix C, $(p_i - p_b)$ can be easily expressed as a linear combination of the states, and, consequently, the measurement matrix $\mathbf{H}_k$ takes the form

$$\mathbf{H}_k = \frac{p_a - p_b}{l_w} \begin{bmatrix} a_{11} & a_{12} & \cdots & a_{1(n+1)} & b_{11} & b_{1(n+1)} \\ a_{21} & a_{22} & \cdots & a_{2(n+1)} & b_{21} & b_{2(n+1)} \\ \cdot & \cdot & \cdots & \cdot & \cdot & \cdot \\ a_{n1} & a_{n2} & \cdots & a_{n(n+1)} & b_{n1} & b_{n(n+1)} \end{bmatrix}$$

where a_{ij}, b_{i1}, $b_{i(n+1)}$ are defined in Appendix C. For large values of k, the filter converges to the true state vector. This allows the global analytical description of the edge curve using Equations (34) and (34a).

2.3.2. *Edge Point Estimator.*

The objective of the edge point estimator is to estimate the positions of individual edge points based on measurements performed in the ambiguity zone. In constructing the mathematical model for the edge point estimator, the imposition of interpolation conditions is abandoned. Furthermore, in order to be able to satisfactorily approximate edges with isolated discontinuities in their low-order derivatives, we have focused on spline approximations exhibiting good localization properties [23.] Here, we present the development of system and measurement models for a cubic spline ($\kappa = 3$) with equidistant knots. An alternative edge point estimator can be found in [24]. The range of variable y is the interval $[\alpha, \beta]$. We assume equidistant knots $t_i = t_0 + ih$, $i = 0, 1, \ldots, n$, such that $\alpha = t_0$ and $t_n = \beta$ (Figure 1.3). Introducing six extra knots outside the interval $[\alpha, \beta]$, three at each end, we obtain the augmented sequence of knots $t_i = t_0 + ih$, $i = -3, -2, \ldots, n + 3$. On this knot sequence, we define $(n + 3)$ B-splines using Equation (32) in Appendix B. For the special case of equidistant knots and $\kappa = 3$ we have

$$B_p^3(y) = \frac{1}{24h^4} [(y - t_p)_+^3 - 4(y - t_{p+1})_+^3 + 6(y - t_{p+2})_+^3 - 4(y - t_{p+3})_+^3 + (y - t_{p+4})_+^3], \qquad p = -3, -2, \ldots, n - 1. \tag{9}$$

As described in Appendix B, any element of space L_{3,t_i} can be written in the form

$$f(y) = \sum_{p=-3}^{n-1} \lambda_p B_p^3(y). \tag{10}$$

It can be proved that the spline corresponding to choice

$$\lambda_p = \lambda_p(g) = 4h\left[-\frac{1}{6} g(t_{p+1}) + \frac{4}{3} g(t_{p+2}) - \frac{1}{6} g(t_{p+3}) \right] \tag{11}$$

has good localization properties and yields high-order accuracy when the function to be approximated, g, is sufficiently smooth [25]. It should be noted that the

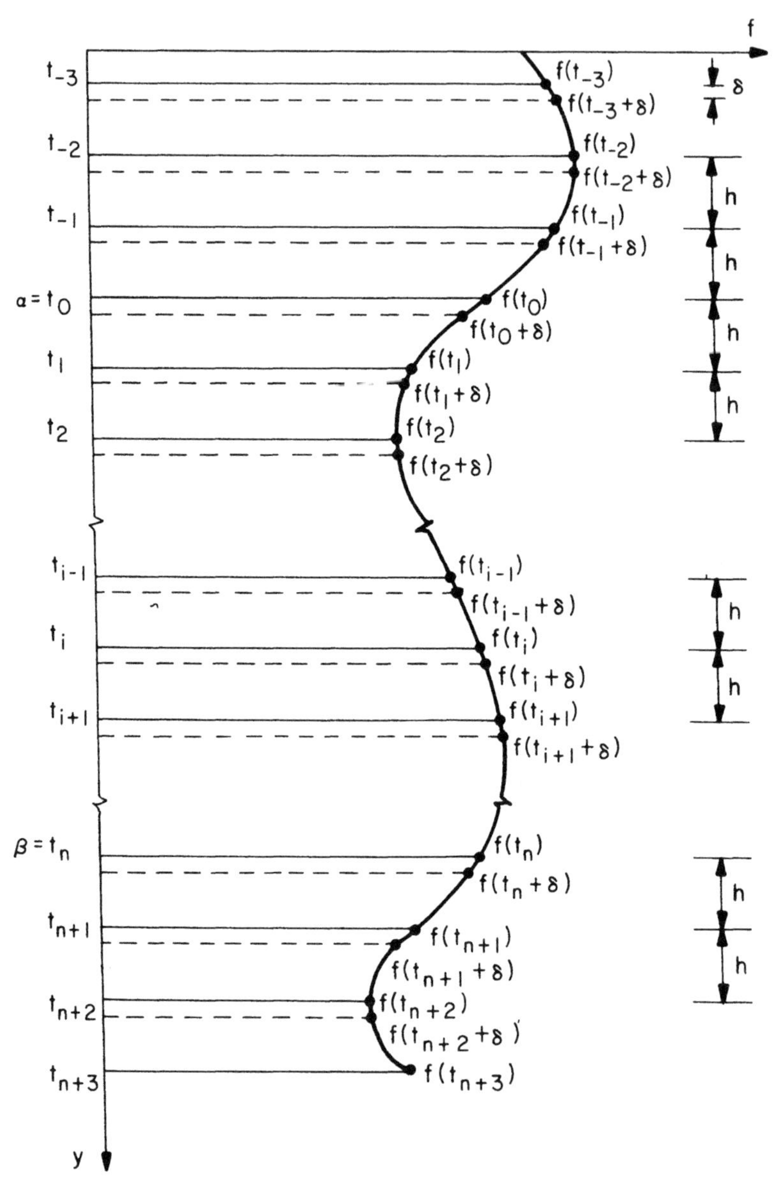

FIGURE 1.3. Knot arrangement for the edge point estimator.

spline defined by Equations (10) and (11) is not interpolatory when g is an arbitrary function.

By forming the vectors

$$\tilde{\underline{f}} = [f(t_0), f(t_1), \ldots, f(t_{n-1})]^T, \tag{12}$$

$$\underline{g} = [g(t_{-2}), g(t_{-1}), \ldots, g(t_{n+2})]^T, \tag{13}$$

and evaluating Equation (9) at $y = t_p$, $p = 0, 1, \ldots n - 1$ we obtain

$$\tilde{\underline{f}} = \tilde{\mathbf{E}}\, \underline{g}, \tag{14}$$

where $\tilde{\mathbf{E}}$ is an $n \times (n + 5)$ constant matrix defined by

$$\tilde{\mathbf{E}} = \begin{bmatrix} a_1 & a_2 & a_3 & a_4 & a_5 & 0 & 0 & 0 & \cdot & \cdot & \cdot & 0 \\ 0 & a_1 & a_2 & a_3 & a_4 & a_5 & 0 & 0 & \cdot & \cdot & \cdot & 0 \\ \cdot & \cdot & \cdot & \cdot & \cdot & \cdot & \cdot & \cdot & \cdot & \cdot & \cdot & \cdot \\ 0 & 0 & \cdot & \cdot & \cdot & 0 & a_1 & a_2 & a_3 & a_4 & a_5 & 0 \end{bmatrix} \tag{15}$$

and $a_1 = -1/36$, $a_2 = 1/9$, $a_3 = 5/6$, $a_4 = 1/9$, $a_5 = -1/36$.

By expanding vector $\tilde{\underline{f}}$ to

$$\underline{f} = [f(t_{-2}), f(t_{-1}), f(t_0), f(t_1), \ldots, f(t_n), f(t_{n+1}), f(t_{n+2})]^T \tag{16}$$

and introducing the square matrix, $\mathbf{E}$, defined by

$$\mathbf{E} = \begin{bmatrix} 1 & & & & & \\ & 1 & & & & \\ & & \tilde{\mathbf{E}} & & & \\ & & & 1 & & \\ & & & & 1 & \\ & & & & & 1 \end{bmatrix}, \tag{17}$$

Equation (14) is rewritten as

$$\underline{f} = \mathbf{E}\, \underline{g}. \tag{18}$$

Considering a set of points displaced by δ pixels (relative to t_i, $i = -2, -1, \ldots, n + 2$; see also Figure 2.3) in the y-direction we can similarly write

$$\underline{f}_\delta = \mathbf{F}\, \underline{g}, \tag{19}$$

where $\mathbf{F}$ is an $(n + 5) \times (n + 5)$ matrix given by

$$\mathbf{F} = \begin{bmatrix} 1 & 0 & 0 & \cdot & \cdot & \cdot & \cdot & \cdot & \cdot & \cdot & \cdot & \cdot & \cdot & 0 \\ 0 & 1 & 0 & \cdot & \cdot & \cdot & \cdot & \cdot & \cdot & \cdot & \cdot & \cdot & \cdot & 0 \\ b_1 & b_2 & b_3 & b_4 & b_5 & b_6 & 0 & 0 & \cdot & \cdot & \cdot & \cdot & \cdot & 0 \\ 0 & b_1 & b_2 & b_3 & b_4 & b_5 & b_6 & 0 & \cdot & \cdot & \cdot & \cdot & \cdot & 0 \\ \cdot & \cdot & \cdot & \cdot & \cdot & \cdot & \cdot & \cdot & \cdot & \cdot & \cdot & \cdot & \cdot & \cdot \\ 0 & 0 & \cdot & \cdot & \cdot & \cdot & \cdot & 0 & b_1 & b_2 & b_3 & b_4 & b_5 & b_6 \\ 0 & 0 & \cdot & \cdot & \cdot & \cdot & \cdot & 0 & 0 & 0 & 0 & 1 & 0 & 0 \\ 0 & 0 & \cdot & \cdot & \cdot & \cdot & \cdot & \cdot & \cdot & \cdot & \cdot & 0 & 1 & 0 \\ 0 & 0 & \cdot & \cdot & \cdot & \cdot & \cdot & \cdot & \cdot & \cdot & \cdot & 0 & 0 & 1 \end{bmatrix} \tag{20}$$

and $\underline{f}_\delta = [f(t_{-2} + \delta), f(t_{-1} + \delta), \ldots, f(t_{n+1} + \delta), f(t_{n+2} + \delta)]^T$. The elements of matrix $\mathbf{F}$ are

$$b_1 = \frac{1}{36h^3} [-(3h + \delta)^3 + 4(2h + \delta)^3 - 6(h + \delta)^3 + 4\delta^3], \tag{21}$$

$$b_2 = \frac{1}{36h^3} [8(3h + \delta)^3 - 33(2h + \delta)^3 + 52(h + \delta)^3 - 38\delta^3], \tag{22}$$

$$b_3 = \frac{1}{36h^3} [-(3h + \delta)^3 + 12(2h + \delta)^3 - 39(h + \delta)^3 + 56\delta^3], \tag{23}$$

$$b_4 = \frac{1}{36h^3} [-(2h + \delta)^3 + 12(h + \delta)^3 - 39\delta^3], \tag{24}$$

$$b_5 = \frac{1}{36h^3} [-(h + \delta)^3 + 12\delta^3], \quad \text{and} \tag{25}$$

$$b_6 = -\frac{1}{36h^3} \delta^3. \tag{26}$$

Combining Equations (18) and (19) we obtain

$$\underline{f}_\delta = \mathbf{\Phi} \underline{f}, \tag{27}$$

where

$$\mathbf{\Phi} = \mathbf{F}\,\mathbf{E}^{-1}. \tag{28}$$

Equation (27) relates two sets of points on the edge separated by distance δ in the y-direction.

The state vector, $\underline{x}$, in this formulation is chosen to be equal to vector $\underline{f}$ defined in Equation (16), i.e., the state vector contains the values of the approximating spline function at the knots t_j, $j = -2, -1, \ldots, n + 2$. At each cycle of the filtering process, the knot sequence t_j, $j = -3, -2, \ldots, n + 3$ is displaced by δ pixels in the direction of increasing y. Then, in view of Equation (27), Equation (2) is satisfied at the kth step of the filtering process with $\underline{x}_k = \underline{f}$, $\underline{x}_{k+1} = \underline{f}_\delta$ and $\mathbf{\Phi}_k = \mathbf{\Phi}$.

For the edge point estimator, the measurable quantity considered at $y = \bar{y}_i =$ const is

$$q(y = \bar{y}_i) = \frac{l_w(p_i - p_b)}{p_a - p_b}. \tag{29}$$

At the kth cycle of the filtering process, measurements of $q(y = t_j)$ are performed for $j = -2, -1, \ldots, n + 2$ and arranged into a column vector $\underline{z}_k = [z^k_{-2}, z^k_{-1}, \ldots, z^k_{n+2}]$, in accordance with the notation of estimation theory. Then, taking into account Equation (8) and the definition of the state vector, $\mathbf{H}_k$ becomes the $(n + 5) \times (n + 5)$ identity matrix. It is noted that the data needed for measurements outside the interval $[\alpha, \beta]$ are generated by first reflecting the image about the lines parallel to the f-axis at locations $(t_0, f(t_0))$ and $(t_n, f(t_n))$ and then reflecting the result about the lines parallel to the y-axis at locations $(t_0, f(t_0))$ and $(t_n, f(t_n))$.

2.4. Property Selection

The choice of property p, Equations (8) and (29), strongly depends on the type of images considered. It has to reflect meaningful characteristics of regions involved and allow differentiation between them. A particular property measurement is considered valid if $\min(p_a, p_b) \leq p_i \leq \max(p_a, p_b)$ where p_a and p_b are properties of regions A and B, respectively; otherwise, it is rejected and a new measurement along the consecutive strip is performed. Two general categories of properties have been studied: statistical measures and texture measures. In an ideal case, properties of an observed subregion approximate corresponding properties of a region as a whole. Such regions exhibit ergodic behavior. Two classes of properties, statistical measures and co-occurrence matrices, suited for representing statistical and structural properties of ergodic regions, are discussed in Sections 2.4.1 and 2.4.2. In addition, an alternative measurement procedure that relies on the detection of individual edge points is described in Section 2.4.3.

2.4.1. Statistical Measures.

Frequently, an image region can adequately be modeled by a collection of random variables governed by Gaussian distribution. Then, regions A and B are completely characterized by their means, μ_a, μ_b, and variances, σ_a^2, σ_b^2. In the case of two regions of uniform intensity, A and B degraded by white noise, only the region means need to be taken into consideration. In this case it can be easily verified that the mean of the intermediate zone satisfies Equation (8).

2.4.2. Co-Occurrence Matrix Approach.

In many cases meaningful regions in an image do not appear as clearly defined regions of contrasting brightness. Instead, they exhibit distinct textural properties. This section focuses on edge estimation between regions characterized by regular textures, i.e., textures are periodic in particular direction(s). Because structural properties are predominant in distinguishing between regular textures, the measurement model must incorporate structural information. Co-occurrence matrices, [26], are particularly useful statistical tools for accomplishing this task.

An element c_{ij} of a co-occurrence matrix $\mathbf{C}(d, \Phi) = [c_{ij}]$ $i, j - 1, 2, \ldots, L$ (L is the number of intensity levels) is the estimated probability of gray levels i and j appearing at an intersample spacing distance d in the angular direction Φ. In the case of two texture regions A and B, where the texture in A is periodic with period π along direction Φ, and the texture in B is periodic with period ω in direction Ψ and $\Phi \neq \Psi$, and/or $\omega \neq K\pi$ $K = 1, 2, \ldots$, matrix $\mathbf{C}_a(\pi, \Phi)$ is diagonal while matrix $\mathbf{C}_b(\pi, \Phi)$ is an arbitrary matrix. Therefore, it is convenient to choose a particular element c_{jj} of co-occurrence matrices $\mathbf{C}_{a/b}(\pi, \Phi)$ to be property $p_{a/b}$. Because an observed subregion does not necessarily contain an integer number of texture periods, bias needs to be added to Equation (8) [16].

2.4.3. Utilization of Individual Edge Points in the Measurement Procedure.

Vector $\underline{z}_k$ [Equation (2)] contains the noisy measurements, i.e., estimates of the locations of individual edge points. Instead of utilizing region measurements, the locations of individual edge points determined by using an edge detection method may be used. However, in this case LRF has clearly only the smoothing role. An example of utilizing edge points is described in [2]. This method employs an edge point search about a reference curve. In the first frame the reference curve consists of the midpoints of the ambiguity zone and in the other image frames the estimated road edges in the previous frame are used as the measurement reference.

The individual edge point at location $f = \beta_i$ is determined by considering the mean and variance of a $1 \times w$ region centered at (α_i, β_i) (Figure 1.4). The measured image property p_i (mean and variance) of the region is compared with the average property p from one of the neighboring regions, e.g., the road. If $|p_i - p| < \epsilon$, where ϵ is a specified tolerance, the search continues. When performing measurements in the region corresponding to the road, the center of the measurement region is shifted to the right for the right road edge and to the left for the left road edge. The amount of shift is proportional to the value w.

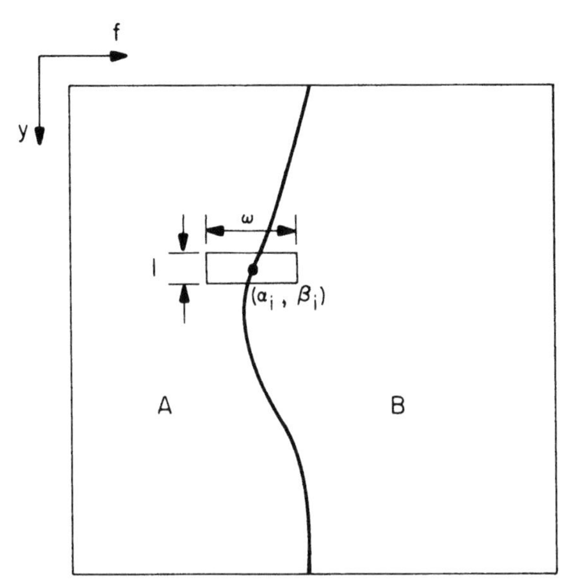

FIGURE 1.4. A measurement region corresponding to a point (α_i, β_i).

After shifting, the measured property p_i is calculated and compared to p again. This procedure stops when $|p_i - p| \geq \epsilon$, and then $z_i = \alpha_i - w/2$ for the right road edge, and $z_i = \alpha_i + w/2$ for the left road edge. Figure 1.5 shows measured edge points when employing region mean and variance as measured image properties. As can be seen from Figure 1.5, the measurements are very noisy. Therefore, LRF is used to estimate true road edges.

It is important to note that the reference need not be accurate. Displacements for 20 or 30 pixels from the true edges are sufficiently accurate because of the nature of the LRF and the measurement process. Even larger displacements can be tolerated by LRF; however, the convergence rate is slower.

3. PROCESSING OF IMAGE SEQUENCES

Under the conditions that the road is relatively smooth and that the road images are acquired sufficiently frequently, it is reasonable to assume that no abrupt changes in road geometry are likely to occur between two consecutive frames. Therefore, only the first frame in the sequence is preprocessed (Section 2.1). Subsequent frames are processed using estimated road edges from the previous frame. Section 3.1 describes the interframe update, and Section 3.2 discusses the sensitivity of LRF to input parameters in road edge detection.

FIGURE 1.5. Edge point search: (a) Original image; and (b) results obtained by applying edge point search, based on region mean and variance, for the right road edge. Curve *B* is the reference while curve *A* contains the measured points. Curves are represented in their local coordinate system.

3.1. Parameter Update

Processing a single frame requires a measurement of statistical properties of the uniform regions, specifications of the ambiguity zone, and LRF parameters. The latter are system modeling error, initial state estimate, measurement error covariance, and number of knots used when approximating a road edge by a spline. Having defined the ambiguity zone as described in Section 2.1 and in Figure 1(b), the width of the ambiguity zone is determined for all frames by the width of the ambiguity zone in the first frame. Moreover, when processing the subsequent frames, the ambiguity zone is positioned so that its center in the g-direction (for each value y) is positioned at the estimated location of the road edge (for the same y) in the previous frame. Also, the statistical properties of the uniform regions are calculated locally at each y.

The LRF parameters, system modeling errors, measurement error covariance, as well as the number of knots, are retained the same for all frames. The initial estimate $\underline{x}_0$ is updated for each frame, and it is the estimate of the road edge in the

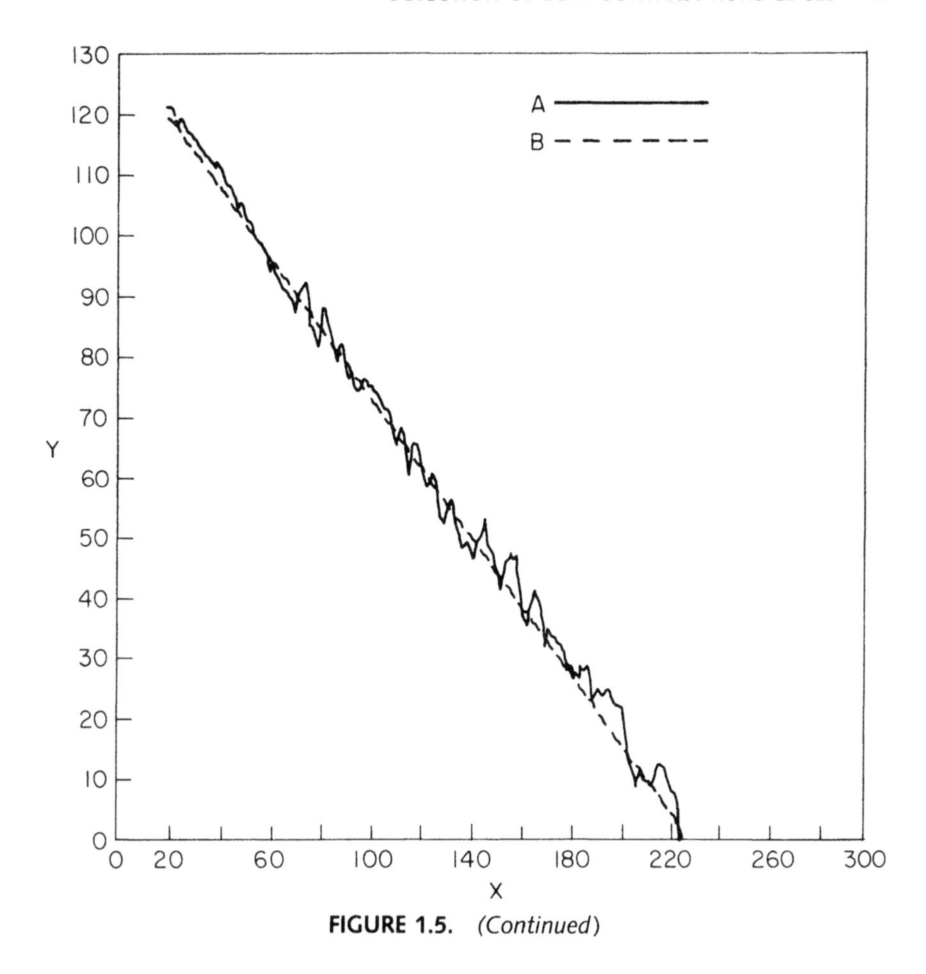

FIGURE 1.5. *(Continued)*

previous frame. In the first frame, the initial estimate consists of the middle points (in the g-direction) of the ambiguity zone.

3.2. Parameter Sensitivity

The sensitivity analysis of the road edge estimators, described in Section 2, reveals that

- The estimators are relatively insensitive to system modeling errors. This is attributed to the flexibility of the spline functions to conform with the local behavior of the boundary curves. Furthermore, both estimators are found to be insensitive to errors in the initial state estimate $\underline{x}_0$.

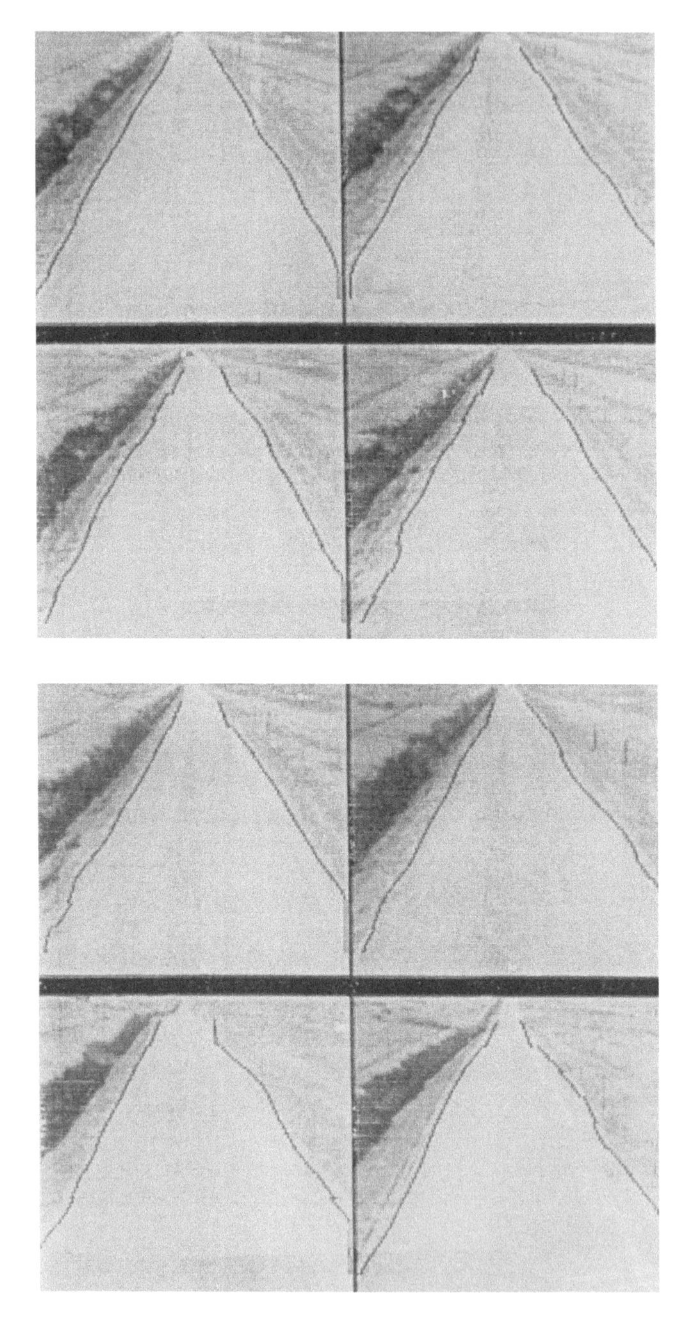

FIGURE 1.6. Road edges determined using region mean and standard deviation and the measurement procedure described in Section 2.4.3 and formulation described in Section 2.3.2. The road edge is approximated with $n = 11$ knots.

FIGURE 1.7. Road edges determined using region mean and standard deviation and the measurement procedure described in Section 2.4.3 and formulation described in Section 2.3.2. The road edge is approximated with $n = 6$ knots.

- When determining smooth boundaries, the results are not significantly affected by the number of knots as long as this number is greater than some minimum required to model a particular boundary. Examples of the effects of number of knots on the obtained results are illustrated in Figures 1.6 and 1.7. The accuracy of the boundary approximation increases as the number of knots increases. However, the number of knots determines the dimension of the recursive filters and thus directly effects the processing time. Therefore, it is possible to determine an optimal number of knots for an application by considering a cost function that appropriately weighs the accuracy against the processing time. It should be noted that the placement of knots becomes crucial in cases when the road edge has intervals of high gradients as well as intervals of flat behavior. Efficient algorithms for the optimal placement of nonequidistant knots are under development. These algorithms use low-order splines and splines under tension [25].
- The state estimates are sensitive to the measurement error covariance matrix $\mathbf{R}_k$. When considering uniform regions with noise of known standard deviation, the diagonal elements of the $\mathbf{R}_k$ matrix can be calculated; otherwise, proper values for $\mathbf{R}_k$ need to be determined by using appropriate test images.

4. RESULTS AND CONCLUSIONS

The proposed boundary detection models have been applied to various road images and the results obtained are very good. Examples of typical results are shown in Figures 1.6 and 1.7. In general, the global boundary estimator, which provides a means of obtaining an analytic description of road edges, is useful in cases when the road edges are smooth. In contrast, the point estimator very accurately detects individual road edge points. Consequently, both estimators yield good results when estimating road edges, since they are, in general, smooth.

The correct choice of property p, Equations (8) and (29), is crucial for the success of the segmentation schemes described in Sections 2 and 3. It is necessary that the selected property allows differentiation between the regions involved. The results shown in this chapter are obtained using region mean gray level. In this case, $p_a = \mu_a$, $p_b = \mu_b$ and $p_i = \mu_w$, where μ_a and μ_b denote region means for regions A and B, respectively, and μ_w is the mean of the window $s \times l$ centered (in the y-direction) at $y = y_i$ in W. In this chapter we use $s = l_w$ and $l = 1$.

The proposed estimators aim at estimation of relatively long boundaries between statistically homogeneous regions. The estimators are capable of performing very well in the presence of noise. Compared to the alternative methods of road edge detection, the proposed estimators offer two major advantages: (1) They provide an insight into the uncertainty of the obtained road edge through the estimation of the error covariance matrix; and (2) they are well suited for real-time applications and cases where available memory is limited.

APPENDIX. SPLINE FUNCTIONS

In this section we briefly review some properties of spline functions that are employed in our boundary model and form the basis for the construction of matrices $\mathbf{\Phi}_k$ and $\mathbf{H}_k$ appearing in Equation (2). A complete discussion of these properties can be found in [23, 25].

A function f is called a spline function of degree κ if it is in space $C^{\kappa-1}[\alpha, \beta]$ and if there exist points $\{t_j; j = 0, 1, 2, \ldots, n\}$ satisfying the conditions $\alpha = t_0 < t_1 < t_2 < \ldots < t_{n-1} < t_n = \beta$ such that f is a polynomial of degree at most κ on each of the intervals $\{[t_{j-1}, t_j]; j = 1, 2, \ldots, n\}$ [23]. The points $\{t_j; j = 0, 1, \ldots, n\}$ are called knots. For a given knot sequence $\{t_j;_j = 0, 1, \ldots, n\}$ the spline functions of degree κ form a linear space denoted by $L_{\kappa;t_j}$. The dimension of the space is $(n + \kappa)$.

A. Truncated Power Function Representation of Spline Functions

Each spline function in space $L_{\kappa;t_j}$ may be represented on $[\alpha, \beta]$ [23] by

$$f(y) = \sum_{m=0}^{\kappa} c_m y^m + \frac{1}{\kappa!} \sum_{j=1}^{n-1} d_j (y - t_j)_+^{\kappa}, \tag{30}$$

where c_m and d_j are constants and $(y - t_j)_+^{\kappa}$ denotes the truncated power function defined as

$$(y - t_j)_+^{\kappa} + \begin{cases} (y - t_j)^{\kappa} & \text{if } y \leq t_j \\ 0 & \text{if } y < t_j. \end{cases} \tag{31}$$

For a fixed knot t_j and $\kappa \geq 0$, the truncated power function is a piecewise polynomial that has $(\kappa - 1)$ continuous derivatives. Its κth derivative has a jump discontinuity at t_j of size $\kappa!$.

The evaluation of spline functions based on Equation (30) may suffer a severe loss of accuracy under the circumstances described in [23, 25]. A more accurate and efficient spline function evaluation is obtained when using B-spline functions described in the next section.

B. B-Spline Representation of Spline Functions

For a knot sequence $\{t_j; j = 0, 1, \ldots, n\}$, the corresponding sequence of B-splines of degree κ are defined as [23]

$$B_p^\kappa(y) = \sum_{j=p}^{p+\kappa+1} \left[\sum_{\substack{i=p \\ i \neq j}}^{p+\kappa+1} \frac{1}{(t_i - t_j)} \right] (y - t_j)_+^\kappa, \qquad p = -\kappa, -\kappa + 1, \ldots, n - 1. \tag{32}$$

The following properties of B-splines are needed for the development of system and measurement models (Section 4):

Property 1. The support of B_j^κ is $[t_j, t_{j+\kappa+1}]$, i.e., $B_j^\kappa(y) = 0$ if $y \notin [t_j, t_{j+\kappa+1}]$.
Property 2. Among the B-splines B_j^κ $j = 0, 1, 2, \ldots$ only the $\kappa + 1$ splines $B_{j-\kappa}^\kappa(y), B_{j-\kappa+1}^\kappa(y), \ldots, B_j^\kappa(y)$ are nonzero in the interval $t_j \leq y \leq t_{j+1}$.
Property 3. If we normalize the B_j^κ splines by defining $N_j^\kappa = (t_{j+\kappa+1} - t_j)B_j^\kappa$, then for any $y \in (t_j, t_{j+1})$ $\Sigma_j N_j^\kappa(y) = 1$.
Property 4. If the knot sequence t_j is augmented by adding the knots $t_{-\kappa} < t_{-\kappa+1} < \ldots < t_{-1} < t_0 = \alpha$ and $\beta = t_n < t_{n+1} < t_{n+2} \ldots < t_{n+\kappa}$, any spline function in space $L_{\kappa;t_j}$ can be represented on $[\alpha, \beta]$ by

$$f(y) = \sum_{j=-\kappa}^{n-1} \lambda_j B_j^\kappa(y). \tag{33}$$

The $(n + \kappa)$ coefficients λ_j define a member of the linear space L_{κ,t_j}. The most important advantage of selecting the B-splines of degree κ as the basis for the vector space of the splines of degree κ and knot sequence t_j is their minimal support. This property leads to significant computational advantages when B-splines are used [23, 25].

C. Complete Cubic Splines

The $(n + \kappa)$ unknowns c_m and d_j, appearing in Equation (30), may be determined by any $(n + \kappa)$ linearly independent conditions that one may choose to impose. A very useful choice that leads to interpolatory splines requires the spline function $f(y)$ to take assigned values g_i at the knots. Additional conditions have to be imposed in order to close the system.

The only interpolatory splines used in the present work are complete cubic splines with equidistant knots, and, consequently, we restrict our presentation to this special class of interpolatory spline functions. However, the extension to spline functions of different degrees and/or arbitrarily placed knots is straightforward.

Let $g(y)$ denote the function to be approximated by a complete cubic spline on the interval $\alpha \leq y \leq \beta$. For the sequence of equidistant knots $y_{i+1} = \alpha + ih$, $i =$

$0, 1, \ldots, n$ and $h = \beta - \alpha/n$, denote the values of function g and its first derivative with respect to y at the jth knot by g_j and g_j', respectively. Then, the cubic polynomial

$$\bar{f}_i(\eta) = \sum_{j=1}^{n+1} a_{ij}(\eta)g_j + b_{i1}(\eta)g_1' + b_{i(n+1)}(\eta)g_{n+1}', \qquad i = 1, 2, \ldots, n \tag{34}$$

approximates g in the ith interval. In Equation (34), η is a local coordinate for the ith interval (see Figure 1.2) defined as $\eta = y - y_i$ $(0 \leq \eta \leq h)$,

$$a_{ij}(\eta) = \frac{(h-\eta)^3 - h^2(h-\eta)}{6h}\omega_{ij} - \frac{\eta(h^2-\eta^2)}{6h}\omega_{(i+1)j} + \delta_{ij}\frac{h-\eta}{h} + \delta_{(i+1)j}\frac{\eta}{h},$$

$$b_{i1}(\eta) = \frac{(h-\eta)^3 - h^2(h-\eta)}{6h}\sigma_{i1} - \frac{\eta(h^2-\eta^2)}{6h}\sigma_{(i+1)1},$$

and

$$b_{i(n+1)}(\eta) = \frac{(h-\eta)^3 - h^2(h-\eta)}{6h}\sigma_{i(n+1)} - \frac{\eta(h^2-\eta^2)}{6h}\sigma_{(i+1)(n+1)},$$

where $[\omega_{ij}] = [\mathbf{C}]^{-1}[\mathbf{D}]$, $[\sigma_{ij}] = [\mathbf{C}]^{-1}$,

$$\mathbf{C} = \begin{bmatrix} 1/3 & 1/6 & 0 & 0 & \cdots & 0 & 0 \\ 1/6 & 2/3 & 1/6 & 0 & \cdots & 0 & 0 \\ 0 & 1/6 & 2/3 & 1/6 & \cdots & 0 & 0 \\ \cdot & \cdot & \cdot & \cdot & \cdots & \cdot & \cdot \\ 0 & 0 & 0 & 0 & \cdots & 1/6 & 1/3 \end{bmatrix} h,$$

$$\mathbf{D} = \begin{bmatrix} -1 & 1 & 0 & 0 & \cdots & 0 & 0 \\ 1 & -2 & 1 & 0 & \cdots & 0 & 0 \\ 0 & 1 & -2 & 1 & \cdots & 0 & 0 \\ \cdot & \cdot & \cdot & \cdot & \cdots & \cdot & \cdot \\ 0 & 0 & 0 & 0 & \cdots & 1 & -1 \end{bmatrix} 1/h$$

and δ_{ij} is the Kronecker's delta. It is noted that the cubic polynomial $\bar{f}_i(\eta)$ depends linearly on all function values g_i and the slopes g_1' and g_{n+1}' at the extreme knots. The resulting complete cubic spline, $f(y)$, approximating the function $g(y)$ on $[\alpha, \beta]$, is given by

$$f(y) = \sum_{i=1}^{n} \delta_i \bar{f}_i(\eta), \tag{34a}$$

where

$$\delta_i = \begin{cases} 1 & \text{if } y_i < y < y_{i+1} \\ 0 & \text{otherwise.} \end{cases}$$

D. Explanation of Symbols

$a_{ij}, b_{i1}, b_{i(n+1)}$ = cubic polynomials in η (Appendix C).
b_i = coefficients [Eqs. (21)–(26)].
c_{ij} = an element of the co-occurrence matrix.
c_m, d_j = spline parameters [Eq. (30)].
d = distance at which co-occurrence of gray levels is measured.
$\bar{f}_k$ = vector (Section 2.3.2).
$\bar{f}_i(\eta)$ = cubic polynomial approximating g in the ith interval [Eq. (34)].
f = spline function approximating function g.
f_i = $f(y_i)$.
g = function defining the edge curve.
g_i = $g(y_i)$.
h = distance between knots in equidistant knot arrangement.
l_a, l_b = widths of the homogeneous regions at $y = y_i$.
l_w = width of the ambiguity zone at $y = y_i$.
n = number of segments in $[\alpha, \beta]$.
p_a, p_b = values of property p in regions A and B, respectively.
p_i = value of property p in the ambiguity zone, W, at $y = y_i$.
q = measurable quantity used by the edge point estimator.
$t_j, j = 0, 1, \ldots, n$ = knot sequence.
v_k = measurement noise vector at step k.
w_k = system noise vector at step k.
$\underline{x}_k$ = state vector at step k.
$\underline{x}_k(-)$ = extrapolated state vector at step k.
$\underline{x}_k(+)$ = updated state vector at step k.
$y_j, j = 1, 2, \ldots, n+1$ = knot sequence.
$\bar{y}_j, j = 1, \ldots, n$ = locations of simultaneous measurements.
$\bar{y}_i^k, i = 1, 2, \ldots, n$ = positions of simultaneous measurements of global estimator.
z_k = measurement vector at step k.
A, B = homogeneous regions.

B_j^κ = jth B-spline of degree κ.
C, D = matrices (Appendix C).
C^n = space of n times continuously differentiable functions.
C = co-occurrence matrix.
$\tilde{\mathbf{E}}$, **E**, **F** = matrices (Section 2.3.2).
$\mathbf{H}_k$ = measurement matrix at step k.
I = unity matrix.
$I(x, y)$ = gray level function, i.e., image element (pixel) at location (x, y).
K_k = Kalman gain at step k.
$L_{\kappa;t_j}$ = space of spline functions of degree κ with knots at t_j, $j = 0, 1, \ldots$.
$N(0, G)$ = normal distribution with zero mean and covariance matrix **G**.
$\mathbf{P}_k$ = estimation error covariance matrix at step k.
$\mathbf{Q}_k$ = system noise covariance matrix at step k.
$\mathbf{R}_k$ = measurement noise covariance matrix at step k.
$T(x, y)$ = pixel typicality of image element $I(x, y)$.
W = ambiguity zone.
$[\alpha, \beta]$ = closed interval on which function $g(y)$ is approximated by $f(y)$.
δ = displacement (in the y-direction) between an edge point and t_i.
δ_{ij} = Kronecker's delta.
η = local coordinate, $\eta = y - y_i$, $0 \leq \eta \leq h$.
κ = degree of spline function.
λ_j = spline coefficient [Eq. (33)].
μ_a, μ_b = region means for regions A and B.
ρ_i = local (in W) g edge coordinate at $y = y_i$ (Figure 1.1).
σ = standard deviation of noise.
σ_a, σ_b = variance of regions A and B.
ω_{ij}, σ_{ij} = parameters (Appendix C).
π, ω = texture periodicity.
$\mathbf{\Phi}_k$ = system matrix at step k.
Φ, Ψ = angular direction for co-occurrence matrix.

ACKNOWLEDGMENT

The authors would like to thank to Dr. Antonios Liakopoulos for his many comments and indepth discussions that have led to this work.

REFERENCES

1. B. Bhanu, P. Symosek, J. Ming, W. Burger, H. Nasr, and J. Kim, "Qualitative target motion detection and tracking," in *Proc. Image Understanding Workshop,* May 1989, pp. 370–398.
2. D. Brzakovic and L. Hong, "Road edge detection for mobile robot navigation," in *Proc. 1989 IEEE Int. Conf. on Robotics and Automation,* May 1989, pp. 1143–1147.
3. L.S. Davis and T.R. Kushner, "Vision-based navigation: A status report," in *Proc. of Image Understanding Workshop,* pp. 153–170, February 1987.
4. B.A. Draper, R.T. Collins, and J. Brolio, "Tools and experiments in the knowledge-directed interpretation of road scenes," in *Proc. of Image Understanding Workshop,* pp. 178–193, February 1987.
5. H. Frohn and W. von Seelen, "VISOCAR: An autonomous industrial transport vehicle by visual navigation," in *Proc. 1989 IEEE Int. Conf. on Robotics and Automation,* May 1989, pp. 1155–1159.
6. M. Hebert, "Building and navigation maps of road scenes using active sensor," in *Proc. 1989 IEEE Int. Conf. on Robotics and Automation,* May 1989, pp. 1136–1142.
7. R.M. Inigo, E.C. McVey, B.J. Berger, and M.J. Wirtz, "Machine vision applied to vehicle guidance," *IEEE Trans. Pattern Analysis and Machine Intelligence,* vol. 6, pp. 820–826, November 1984.
8. K. Kluge and C. Thorpe, "Explicit models for robot road following," in *Proc. 1989 IEEE Int. Conf. on Robotics and Automation,* May 1989, pp. 1148–1154.
9. D.T. Lawton, T.S. Levitt, C. McConnell, P.C. Nelson, and J. Glicksman, "Environmental modeling and recognition for an autonomous land vehicle," in *Proc. of Image Understanding Workshop,* pp. 107–121, February 1987.
10. S. Liou and R. Jain, "Road following using vanishing points," *Computer Vision, Graphics, and Image Processing,* vol. 39, pp. 116–130, 1987.
11. C.D. McGillem and T.S. Rappaport, "Infra-red location system for navigation of autonomous vehicles," in *Proc. 1988 IEEE Int. Conf. on Robotics and Automation,* April 1988, pp. 1236–1238.
12. H. Nasr and B. Bhanu, "Landmark recognition for mobile robots," in *Proc. 1988 IEEE Int. Conf. on Robotics and Automation,* April 1988, pp. 1218–1223.
13. C. Thorpe, S. Shafer, T. Kanade, "Vision and navigation for the Carnegie Mellon Navlab," in Proc. of Image Understanding Workshop, February 1987, pp. 143–152.
14. C. Thorpe and T. Kanade, "Carnegie Mellon Navlab vision," in Proc. of Image Understanding Workshop, May 1989, pp. 273–282.
15. C.M. Wang, "Location estimation and uncertainty analysis for mobile robots," in *Proc. 1988 IEEE Int. Conf. on Robotics and Automation,* April 1988, pp. 1230–1235.
16. D. Brzakovic and A. Liakopoulos, "Estimation theory based segmentation of texture images," in *Advances in Image Processing and Pattern Recognition,* V. Cappellini, and R. Marconi, Eds. Amsterdam: North-Holland, pp. 234–238, 1986.
17. N. Otsu, "A threshold selection method for gray-level histograms," *IEEE Trans. on Systems, Man and Cybernetics,* vol. SMC-9, January, pp. 62–66, 1979.
18. P.J. Burt and E.H. Adelson, "The Laplacian pyramid as a compact image code," *IEEE Transactions on Communications,* vol. 31, pp. 532–540, October 1983.

19. M. Basseville, B. Espiau, and J. Gasnier, "Edge detection using sequential methods for change in level–Part I: A sequential edge detection algorithm," *IEEE Trans. Acoustics, Speech, and Signal Processing,* vol. ASSP 29, January, pp. 24–31, 1981.
20. M. Basseville, "Edge detection using sequential methods for change in level–Part II: Sequential detection of change in mean," *IEEE Trans. on Acoustics, Speech, and Signal Processing,* vol. ASSP 29, January, pp. 32–50, 1981.
21. R.E. Kalman, "A new approach to linear filtering and prediction problems," *Journal of Basic Engineering, Trans. ASME,*Series D, vol. 82, pp. 35–45, January 1960.
22. A. Gelb, *Applied Optimal Estimation.* Cambridge, MA: MIT Press, 1974.
23. M.J.D. Powell, *Approximation Theory and Methods.* Cambridge, MA: Cambridge University Press, 1981.
24. L. Hong, D. Brzakovic, and A. Liakopoulos, "Boundary detection in digital images based on spline functions and estimation theory," in *Proc. of 26th IEEE Conference on Decision and Control,* December 1987, pp. 1048–1049.
25. C. De Boor, *A Practical Guide to Splines.* New York: Springer Verlag, 1978.
26. R.M. Haralick, "Statistical and structural approaches to texture," *Proc. of the IEEE,* vol. 67, pp. 786–803, May 1979.

2

Formulation and Algorithms for Solving the Manipulator Reachable Work Space Problems

Zhiyuan Ying, Yugeng Xi, and Zhongjun Zhang
Department of Automatic Control,
Shanghai Jiao-Tong University
1954 Hua Shan Road, Shanghai,
P.R. China 200030

1. INTRODUCTION

The reachable work space [16] is one of the most important aspects of manipulator kinematics. It may be concerned in many significant problems such as:

1. Feasibility test of the planned trajectory for a robot manipulator,
2. Reachability test of a robot to an object or a region,
3. Evaluation and display of the manipulator reachable work space, etc.

In the past several years, the above problems have been investigated independently by many authors. Tsai and Soni [1] investigated the accessible region of planner two-three link manipulators with pin joints. They also presented an algorithm for an n-R manipulator to determine the work space on an arbitrary plane [2]. Gupta and Roth [3] proposed the ideas of holes and voids. Gupta [4] discussed the quantitative and qualitative evaluation of the robot work space. Lee and Yang [5, 6] presented in algorithm for detecting the boundary of work space and gave the criteria for detecting the holes and voids in the work space. Sugimoto and Duffy [7] developed an algorithm for calculating the extreme distances reached by a robot hand. Kumar and Patel [8] gave a real-time algorithm for computing and displaying the work space of a generalized manipulator. Rastegar and Deravi [9] presented a general method to determine the work space and its subspace with different numbers of configurations. Hsu and Kohli [10] provided a method of conducting the work space analysis to determine the boundaries of the work space and to determine the voids and holes inside the workspace. Wu and Young [11] developed a procedure for determining the feasibility of a planned Cartesian trajectory.

In this chapter, a general formulation for solving the reachable work-space–

related problems is proposed, and based on it some numerical algorithms for different problems are developed. The main idea here is to transform the related problems into a constrained optimization problem and to solve it by using the nonlinear programming method. The chapter is divided into six sections. In Section 2, the basic problem formulation is proposed for solving the problem whether a given point is within or without the reachable work space of a manipulator. It may further be used to test the feasibility of a planned Cartesian trajectory for a manipulator. In Section 3, the problem formulation in Section 2 is generalized to solve the reachability test problem for a robot to an object. In Section 4, numerical algorithms are developed to calculate the boundaries of manipulator reachable work space recursively and heuristicly. Some further research topics with the same ideas are discussed in Section 5. Section 6 is the conclusion.

Note: the word "robot" and "manipulator" are used interchangeably in this chapter.

2. FORMULATION OF THE BASIC PROBLEMS

When feasibility of a planned trajectory for a manipulator is discussed, the essential problem encountered is to determine whether a given point on the trajectory is within or without the manipulator reachable work space.

According to the manipulator kinematics, the relationship between the manipulator joint coordinates and Cartesian coordinates is given by:

$$P_x = f_x\,(q_1, \ldots, q_n) \tag{1}$$

$$P_y = f_y\,(q_1, \ldots, q_n) \tag{2}$$

$$P_z = f_z\,(q_1, \ldots, q_n) \tag{3}$$

where n is the number of the joint variables of the manipulator; $(q_1, \ldots, q_n)$ and (P_x, P_y, P_z) are the joint coordinates and Cartesian coordinates of a reference point on the manipulator end-effector respectively; and f_x, f_y, f_z are continuous single-value functions which map the point from the m-dimensional joint space to the three-dimensional Cartesian space.

The problem of whether a given point (X, Y, Z) is within or without the manipulator reachable work space may be solved by formulating the following problem.

Problem 1. Solving equations:

$$P_x = X \tag{4}$$

$$P_y = Y \tag{5}$$

$$P_z = Z \tag{6}$$
$$\text{subject to: } q_{i,\min} \leqslant q_i \leqslant q_{i,\max}$$

where P_x, P_y, P_z are given by (1)–(3); and $q_{i,\min}$ and $q_{i,\max}$, $i = 1, \ldots, n$ are the lower and upper bounds of the manipulator joint variables.

To solve Problem 1 is to find a set of $(q_1, q_2, \ldots, q_n)$ which satisfies equations (4)–(6) and the constraint conditions. If there is at least one solution for the above problem, then the given point (X, Y, Z) is within the reachable work space of the manipulator; if no solution exists, (X, Y, Z) is out of the reachable work space of the manipulator. Although in some special cases analytical solution could be obtained [12], it is generally difficult to solve the problem analytically due to the nonlinear nature of f_x, f_y, f_z. One feasible way to overcome the difficulty is to use numerical methods.

2.1. The Main Idea

The main idea of using numerical methods is really simple and clear. Since the motions of the manipulator joint variables are constrained, the work space W composed of all the reachable points is a limited three-dimensional body. If the point is reachable by the manipulator, it must be within the work space and the shortest distance between the point (X, Y, Z) and the work space W must be zero. Define:

$$J = (X - P_x)^2 + (Y - P_y)^2 + (Z - P_z)^2 \tag{7}$$

as the measurement of the distance between (X, Y, Z) and W.
Problem 1 can be transformed into
Problem 2.

$$\text{minimize } J \tag{8}$$

$$\text{s.t. } q_{i,\min} \leqslant q_i \leqslant q_{i,\max} \quad i = 1, \ldots, n \tag{9}$$

Problem 2 is a nonlinear programming problem with linear inequality constraints. If a set of q_i $(i = 1, \ldots, n)$ can be found such that $J = 0$, then the given point $(X, Y, Z,)$ is within the reachable work space of the manipulator and vice versa.

2.2. Basic Algorithm

To solve Problem 2, the complex method [13, 14], which is an extension of simplex method, is used.

Let

$$q_i^k = q_{i,\min} + r_i^k (q_{i,\max} - q_i) \qquad i = 1, \ldots, n, \quad k = 1, \ldots, N \tag{10}$$

Define:

$$Q_k = (q_1^k, \ldots, q_n^k) \tag{11}$$

$$J(Q_H) = \max \{J(Q_1), J(Q_2), \ldots, J(Q_N)\} \tag{12}$$

$$J(Q_L) = \min \{J(Q_1), J(Q_2), \ldots, J(Q_N)\} \tag{13}$$

$$\bar{Q} = \frac{1}{N-1} \sum_{\substack{i=1 \\ i \neq H}}^{N} Q_i \tag{14}$$

where r_i^k are pseudorandom numbers, rectangularly distributed over the open interval (0, 1); $J(Q_i)$ is the value of index function with respect to Q_i; and N is the number of the complex's vertices.

Algorithm 1. Complex method.

Step 1: Let $N > n + 2$, form first vertex of the original complex.
Step 2: Use (10) to generate other $N - 1$ vertices of the original complex.
Step 3: According to (12), find $J(Q_H)$.
Step 4: Reflect Q_H to Q_R by:
$Q_R = \bar{Q} + \alpha(\bar{Q} - Q_H)$,
where α is a positive constant, called the reflection coefficient.
Step 5: (i) If $J(Q_L) > J(Q_R)$ then take an expansion step by computing $Q_E = \bar{Q} + \gamma' (Q_R - \bar{Q})$, where the expansion coefficient $\gamma' > 1$, is a given constant. If $J(Q_R) > J(Q_E)$ then replace Q_H with Q_E to obtain a new complex.
(ii) If $\max \{J(Q_i), Q_i \neq Q_H\} \geq J(Q_R) > J(Q_L)$ then replace Q_H with Q_R to construct a new complex.
(iii) If $J(Q_R) > \max_i \{J(Q_i), Q_i \neq Q_H\}$ then define:
$J(Q'_H) = \min \{J(Q_H), J(Q_R)\}$ and take a contraction step $Q_c = \bar{Q} + \beta (Q'_H - \bar{Q})$, where $0 < \beta < 1$ is the contraction coefficient. If $J(Q_c) \leq J(Q_H^1)$, Q_H is replaced by Q_c in the new complex, otherwise all Q_i are replaced by the new points $\hat{Q}_i$, which are defined as:
$\hat{Q}_i = Q_i + 0.5 (Q_L - Q_i) \qquad i = 1, 2, \ldots, N.$
Step 6: If $\{1/N \sum_{i=1}^{N} [J(Q_i) - J(\bar{Q})]^2\}^{1/2} < \epsilon$ or $|J(Q_i)| < e$ then end the algorithm otherwise go to Step 3, where ϵ is a predetermined small positive number and e is a positive number determined according to the desired accuracy.

Since in the above algorithm only three functions f_x, f_y, f_z and the upper and lower bounds of q_i $(i = 1, \ldots, n)$ are involved, it is easy to implement. Problem 2 may thus be solved.

2.3. Test of the Reachability of a Manipulator to a Trajectory

The concept proposed above can further be used for testing the feasibility of a planned trajectory for a manipulator numerically. Assume that the trajectory of the manipulator end-effector has been planned according to the task and given by the following continuous curve varying with the time t:

$$\Gamma : \{x(t), y(t), z(t), t_o \leq t \leq t_f\} \tag{15}$$

It should be tested whether every point in Γ is reachable by the manipulator.

In general, this problem should be solved numerically. To do this, the continuous trajectory Γ is first discretized into the point set ϕ :$\{x(t_o + iT), y(t_o + iT), z(t_o + iT), 0 \leq i \leq N\}$, where $T = (t_f - t_o)/N$ is the discrete interval selected according to the accuracy requirement. For each point $P = (X, Y, Z) \subset \phi$, where $X = x(t_o + iT), Y = y(t_o + iT), Z = z(t_o + iT), i \in [0, N]$, Problem 2 should be solved for testing whether P is reachable by the manipulator. This testing procedure will be performed repeatedly for all points corresponding to i from 0 to N. In this point-to-point testing fashion, it may be shown whether the planned trajectory Γ is reachable by the manipulator and which parts of Γ should be revised.

In particular, when only the answer that a given trajectory Γ is not reachable by the manipulator is involved, the problem can be formulated much more simpler than the point-to-point testing fashion. Similar to Problem 2, it may be formulated as

$$\begin{aligned} &\text{minimize } J = [x(t) - P_x]^2 + [y(t) - P_y]^2 + [z(t) - P_z]^2 \\ &\text{subject to:} \quad q_{i,\min} \leq q_i \leq q_{i,\max}, \quad i = 1, \ldots, n \\ &\qquad\qquad\quad t_o \leq t \leq t_f \end{aligned} \tag{16}$$

where P_x, P_y, P_z are the same as in Problem 2. If min $J \neq 0$, it can be concluded immediately that the trajectory Γ is not reachable by the manipulator, In (16), the index function J is somewhat different from that in Problem 2 and the number of linear constraints are one more than that in Problem 2. This is not critical for problem solving. What is meaningful is that the optimization problem should only be solved once, which greatly simplifies the test procedure.

3. TEST OF THE REACHABILITY OF A ROBOT TO AN OBJECT

In research on intelligent robot systems, it is often necessary to test whether an object is reachable by a certain robot and whether a robot is going to collide with other robots or obstacles. The main problem here is to test whether an object has intersections with the work space of a robot. In this section, the reachability test problem is called the RT problem.

It is obvious that the most direct way to solve the RT problem is to determine the work space of the robot because it is composed of all the points to which the end effector of the robot is reachable. However, the work space of a robot has generally complex geometric shape because of the nonlinear relationship between the joint coordinates and the end effector's Cartesian coordinates. In most cases, it is impossible to represent the work space in an analytical way. Therefore, numerical methods are often used and the work space will be point-to-point stored in digital computer, which makes the RT problem based on determination of the work space difficult and time-consuming.

In this section, a new approach for solving the RT problem without determining the work space of the robot is proposed. Based on the same idea as in Section 2, the RT problem is normalized as a nonlinear programming problem with linear constraints. The generalized reduced gradient (GRG) algorithm is also revised and used to solve it. Two examples are given to show the effectiveness of the new method. Since no explicit information of the work space is required, this new method provides an easy way to solve the RT problem directly. It is suitable to testing the reachability of a robot to arbitrary specified cuboid with desired accuracy.

3.1. Mathematical Description of Reachability of a Robot to an Object

The relationship between the work space of a robot and an object is a special case of that between two spatial geometric bodies. There are three possible cases: inclusion, intersection, and separation, corresponding to fully overlapping, partially overlapping, and not overlapping respectively. Testing whether an object is reachable by the robot is equivalent to testing whether the object and the work space of robot overlap each other.

In Section 2, an approach to test whether a given point is within the work space of the robot was proposed. It can be used to test whether a vertex of the object is within the work space of the robot. However, for the RT problem, even when the object is a cuboid, the algorithm presented in the last section should be used for at least eight vertices of the cuboid and is not efficient. Furthermore, if the whole work space is within the object, then all the vertices of the object are out of the work space, which will lead to wrong conclusion.

In order to obtain a simpler approach, the characteristics when two objects overlap each other should first be analyzed. Suppose A and B are two objects and P_1 and P_2 are two arbitrary points within A and B respectively, that is:

$$P_1 \in A \tag{17}$$

$$P_2 \in B \tag{18}$$

When A and B fully or partially overlap, at least one pair of (P_1, P_2) can be found such that

$$\| P_1 - P_2 \| = 0 \tag{19}$$

When two objects are separate, for arbitrary $P_1 \in A$ and $P_2 \in B$, we have:

$$\min \| P_1 - P_2 \| > 0 \tag{20}$$

Suppose that the object C is described by the cuboid:

$$C: \quad x_{\min} \leqslant x \leqslant x_{\max} \tag{21}$$

$$y_{\min} \leqslant y \leqslant Y_{\max} \tag{22}$$

$$z_{\min} \leqslant z \leqslant z_{\max} \tag{23}$$

and the reachable work space of a robot is described as:

$$W: \quad P_x = f_x(q_1, q_2, \ldots, q_n) \tag{24}$$

$$P_y = f_y(q_1, q_2, \ldots, q_n) \tag{25}$$

$$P_z = f_z(q_1, q_2, \ldots, q_n) \quad 26)$$

$$q_{i,\min} \leqslant q_i \leqslant q_{i,\max} \qquad i = 1, 2, \ldots, n \tag{27}$$

where P_x, P_y, P_z, and f_x, f_y, f_z, $q_1, q_2, \ldots, q_n$ and $q_{i,\min}$, $q_{i,\max}$ are of the same definitions as in Section 2.1.

According to (19) and (20), with $P_1 = (x, y, z) \in C$, $P_2 = (P_x, P_y, P_z) \in W$, an index function in the same form as (7) can be defined:

$$J = \| P_1 - P_2 \|^2 = (x - P_x)^2 + (y - P_y)^2 + (z - P_z)^2 \tag{28}$$

and the RT problem can be similarly transformed into the following optimization problem:
Problem 3.

$$\text{minimize } J \tag{29}$$

$$\text{s.t.} \quad q_{i,\min} \leqslant q_i \leqslant q_{i,\max} \qquad i = 1, \ldots, n \tag{30}$$

$$x_{\min} \leqslant x \leqslant x_{\max} \tag{31}$$

$$y_{\min} \leqslant y \leqslant y_{\max} \tag{32}$$

$$z_{\min} \leqslant z \leqslant z_{\max} \tag{33}$$

Problem 3 is similar to Problem 2 in form, with a difference only in the number of constraints. The same algorithm as in Section 2 can be used to solve it. If a set of $(q_1, q_2, \ldots, q_n)$ and (x, y, z) could be found such that the index function $J = 0$, then the cuboid overlaps with the work space of the robot because there is at least one point of the cuboid within the reachable work space of the robot.

When the given cuboid is of arbitrary orientation, a new reference coordinate system should be used with its origin at one vertex of the cuboid and the x, y, z, coordinate axes to be the three edges of the cuboid stretching from the vertex. Since the relative position between the cuboid and the robot's base coordinate system is fixed, a constant matrix $\mathbf{T}_o$ transforming the base coordinate system into the reference coordinate system can easily be found. Multiplying $\mathbf{T}_o$ to the left-hand sides of equations (24)–(27) will result in (P_x, P_y, P_z) defined in the new reference coordinate system.

The cuboid defined by (21), (22), and (23) is a general form of cuboid, called generalized cuboid, it may represent rectangles, line segments, and points. For example, when the lower and upper bounds of (21) are set to the same value, that is $x_{\min} = x_{\max} =$ const., a rectangle is obtained. When the lower and upper bounds of two or three of the inequalities in (21), (22), and (23) are made equal respectively, a line segment or a point could be obtained. So Problem 3 is a generalization of Problem 2, and the algorithm used to solve Problem 3 is obviously applicable to Problem 2.

It should also be mentioned that although the solution of the nonlinear programing problem defined by (29)–(33) is not unique due to the multisolution property of trigonometric functions, it is enough to solve the RT problem. $J = 0$ means that there exists at least one point of the cuboid within the reachable work space of the robot while $J \neq 0$ means not. So the problem of whether two objects overlap can be definitely solved by the nonlinear programming defined by (29)–(33). It is particularly valuable in the collision-free planning problem.

3.2. Algorithm for the RT Problem

In Section 2, whether a given point is within the work space of the robot is tested. The problem is directly transformed into a nonlinear programming problem with X, Y, Z in (7) fixed. The complex method introduced in Section 2.2 seems suitable to solving it. However, if a given cuboid instead of a point is tested, the complex method will be no longer appropriate because the increased number of the variables and constraints makes the computation time much more and the convergence quality poorer. In this section, a more effective algorithm—generalized reduced gradient (GRG) algorithm [15] is chosen and revised for solving the RT problem to achieve better results.

The original GRG algorithm is used to solve the nonlinear programming problem of the form:

$$\min J(X) \qquad X \in E^n \tag{34}$$

$$\text{s.t.} \quad h_i[X] = 0 \qquad i = 1, \ldots, m \tag{35}$$

$$l_j \leqslant x_j \leqslant u_j \qquad j = 1, \ldots, n \tag{36}$$

Problem 3 is a special case, that is, without the equality constraint $h_i\,[X] = 0$, of the above problem. In this case, the generalized reduced gradient is just the ordinary gradient.
Define:

$$\begin{aligned} X(k) &= [x_1\,(k), \ldots, x_n\,(k), x_{n+1}\,(k), x_{n+2}\,(k), x_{n+3}\,(k)]^T_{N\times 1} \\ &\triangleq [q_1\,(k), \ldots, q_n\,(k), x, y, z]^T_{N\times 1} \end{aligned} \tag{37}$$

$$R(k) = [r_1\,(k), \ldots, r_N\,(k)]^T_{N\times 1} \tag{38}$$

$$D(k) = -R(k) \tag{39}$$

$$\begin{aligned} L &= [l_1, \ldots, l_n, l_{n+1}, l_{n+2}, l_{n+3}]^T_{N\times 1} \\ &\triangleq [q_{1,\min}, \ldots, q_{n,\min}, x_{\min}, y_{\min}, z_{\min}]^T_{N\times 1} \end{aligned} \tag{40}$$

$$\begin{aligned} U &= [u_1, \ldots, u_n, u_{n+1}, u_{n+2}, u_{n+3}]^T_{N\times 1} \\ &\triangleq [q_{1,\max}, \ldots, q_{n,\max}, x_{\max}, y_{\max}, z_{\max}]^T_{N\times 1} \end{aligned} \tag{41}$$

where $R(k)$ is the gradient vector; $D(k)$ is the minus gradient vector; L and U are the lower and upper bounds of vector $\mathbf{X}(k)$; and k is the iteration counter.

In the search procedure, vector $\mathbf{X}$ is iteratively computed by:

$$\mathbf{X}(k+1) = X(k) + \lambda\, D(k) \tag{42}$$

where λ is a scalar variable.

The original GRG algorithm performs as follows. First suppose

$$L \leqslant X(k) + \lambda\, D(k) \leqslant U \tag{43}$$

to find a range for λ, $\lambda_{\min} \leq \lambda \leq \lambda_{\max}$, then use one-dimension search for λ in (42) and get an optimal λ^* such that $J(X(\lambda^*)) = \min_{\lambda_{\min}\leq\lambda\leq\lambda_{\max}} J(X(\lambda))$. Put λ^* in (42) to constitute $X(k + 1)$. This procedure will be continued until the convergence condition is satisfied. Since λ must satisfy all the inequality constraints, the possible range for λ would be very small, which makes the convergence rate of the search procedure very slow and sometimes even leads to oscillation at one point. In order to overcome the difficulty, the following revision is made:
Rewrite (43) as:

$$L \leqslant X(k) + \Lambda\, D(k) \leqslant U \tag{44}$$

where $\Lambda = \text{diag}\,(\lambda_1, \ldots, \lambda_N)$ with all λ_i scalar variables.

For every Λ_i = diag [0, . . . , 0, λ_i, 0, . . . 0) there exists an optimal Λ_i^* where Λ_i^* = diag (0, . . . , 0, λ_i^*, 0, . . , 0) and the corresponding J_i^* may be obtained. With N Λ_i^* obtained, form Λ_{N+1}^* = diag (λ_1^*, . . . , λ_N^*) and obtain $J_{N+1}^* = J[X(k) + \Lambda_{N+1}^* D(k)]$. Compute all J_i^*, $i = 1, 2, \ldots, N+1$, select m according to $J_m = \min_i J_i$, $X(k + 1)$ is then obtained by replacing λ in (42) with the corresponding Λ_m^*, that is

$$X(k + 1) = X(k) + \Lambda_m^* D(k) \tag{45}$$

In this revised form, although $X(k)$ does not step forward strictly along the minus gradient direction, larger steps may be taken and the iteration procedure will convergent faster.

Algorithm 2. Revised generalized reduced gradient method (RGRG).

Step 1: Set initial $X(0)$ whose elements satisfy (30)–(33). Let $k = 0$.
Step 2: Compute gradient vector $\mathbf{R} = [r_1, r_2, \ldots, r_N]^T$, $r_i = \delta J/\delta x_i \mid x_i(k)$.
Step 3: If $x_j(k) = l_j$, $r_j(k) > 0$ or $x_j(k) = u_j$, $r_j(k) < 0$ then $d_j(k) = 0$ otherwise $d_j(k) = -r_j(k)$, $j = 1, \ldots, N$.
Step 4: Let $\Lambda_i = (0, \ldots, 0, \lambda_i, 0, \ldots, 0)$ and search for Λ_i^* and obtain J_i^*. $i = 1, \ldots, N$. Let $\Lambda_{N+1}^* = (\lambda_1^*, \ldots, \lambda_i^*, \ldots, \lambda_N^*)$ and obtain the corresponding J_{N+1}^*.
Step 5: Determine m such that $J_m = \min_{1\leq i\leq N+1} J_i^*$, substitute the corresponding Λ_m^* for λ in (43) and constitute $x(k + 1)$.
Step 6: Test whether the convergence condition $J < e$ is satisfied, if satisfied then end the algorithm.
Step 7: If ΔJopt $> \epsilon$ then let $k = k + 1$ go to Step 2, otherwise end the algorithm.

Note that the convergence condition $J < e$ characterizes the desired accuracy and may be set according to the problem, while ΔJopt represents the difference between two iteration results.

3.3. Examples

In this section, two robots, RHINO and Planar 2R, are chosen as examples for which the results are perceivable and may easily be examined.

For the convenience of table arrangement, the generalized cuboid (G-cuboids) is expressed as follows:

Cuboids: $(x_{\min}, y_{\min}, z_{\min}) - (x_{\max}, Y_{\max}, z_{\max})$
Rectangles: $(x_{\min}, y_{\min}, z_o) - (x_{\max}, y_{\max}, z_o)$ or $(x_{\min}, y_o, z_{\min}) - (x_{\max}, y_o, z_{\max})$ or $(x_o, y_{\min}, z_{\min}) - (x_o, y_{\max}, z_{\max})$

Line segments: (x_{min} to x_{max}, y_o, z_o) or (x_o, y_{min} to y_{max}, z_o) or (x_o, y_o, z_{min} to z_{max})
Points: (x_o, y_o, z_o)

3.3.1. Example.

The kinematic structure of the Planar 2R robot is shown in Figure 2.1. The relationship between the Cartesian coordinates and joint variables is given by:

$$P_x = 10\cos(q_1) + 4\cos(q_1 + q_2) \tag{46}$$

$$P_y = 10\sin(q_1) + 4\sin(q_1 + q_2) \tag{47}$$

Suppose the lower and upper bounds of the joint variables are: $0° \leqslant q_1 \leqslant 90°$ and $0° \leqslant q_2 \leqslant 90°$. The work space of the robot may easily be drawn as in Figure 2.2 with four points *a:* (−4, 10), *b:* (0, 14), *c:* (10, 4), *d:* (14, 0). The RT problem of the robot to some rectangles, line segments, and points listed in Table 2.1 are solved by the above RGRG algorithm. The results can easily be verified in Figure 2.2.

3.3.2. Example.

The kinematic relationship of the RHINO robot shown in Figure 2.3 is given by:

$$P_x = \cos(q_1)\,\{9\cos(q_2) + 9\cos(q_3) + 4\cos(q_4)\} \tag{48}$$

$$P_y = \sin(q_1)\,\{9\cos(q_2) + 9\cos(q_3) + 4\cos(q_4)\} \tag{49}$$

$$P_z = 9\sin(q_2) + 9\sin(q_3) + 4\sin(q_4) + 10.5 \tag{50}$$

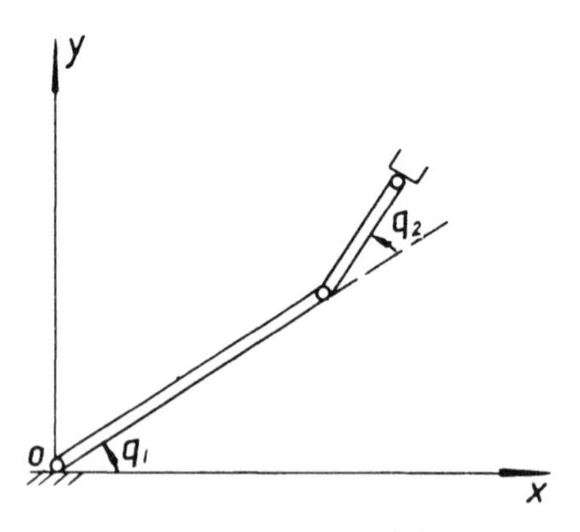

FIGURE 2.1. Kinematic structure of the Planar 2R robot.

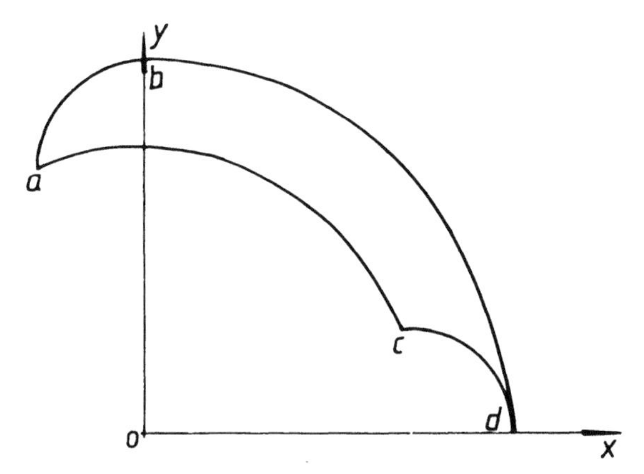

FIGURE 2.2. The reachable work space of the Planar 2R Robot with two joint variables: $0° \leqslant q_1 \leqslant 90°$; $0° \leqslant q_2 \leqslant 90°$.

Suppose the joint variables of the RHINO robot are not restricted, that means they may change from 0° to 360°, then the work space of the RHINO robot is a ball with its origin (0, 0, 10.5) and radius 22. For the generalized cuboids listed in Table 2.2, the RGRG algorithm is used to test whether they overlap with the reachable work space of the robot. The correctness of the RT results can easily be verified with the help of the known work space.

TABLE 2.1. The G-Cuboids and the RT Test Result for Planar 2R Robot ($0° \leq q_1 \leq 90°$; $0° \leq q_2 \leq 90°$)

G-Cuboids[1]	Computation Cycles	J_{min}	RT Result[2]
(−4, 9.4, 0)–(6, 14, 0)	4	9.09E-13	R
(13, 13, 0)–(14, 14, 0)	15	19.23	N
(6, −100, 0)–(8, 100,	1	1.78E-8	R
0)	5	2.09E-6	R
(13, −1, 0)–(15, 0, 0)	4	3.15E-6	R
(−10 to −4, 10, 0)	13	5.91E-6	R
(10 to 12, 4, 0)	6	13.783	N
(5, 0 to 5, 0)	5	1.66E-7	R
(−4, 10, 0)	4	5.81E-6	R
(14, 0, 0)	5	0.329	N
(2, 10, 0)			

[1] G-cuboids stands for generalized cuboids.
[2] R stands for reachable; N stands for unreachable.

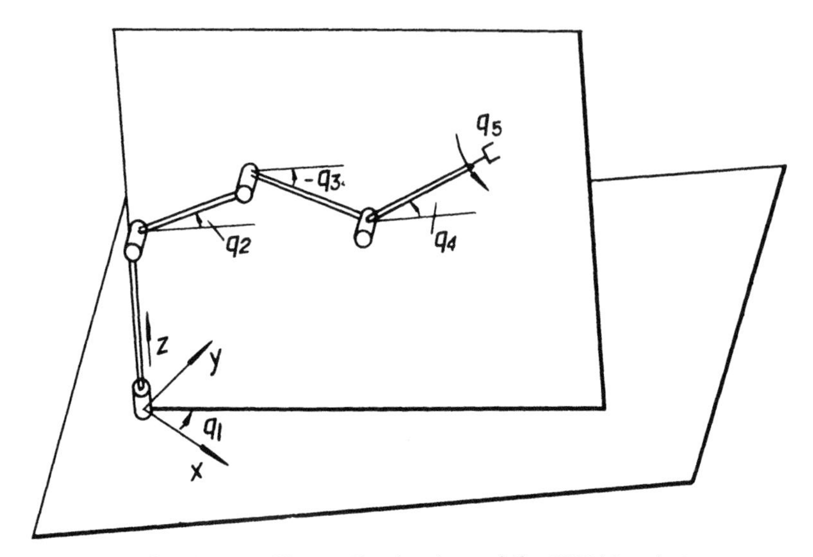

FIGURE 2.3. Kinematic structure of the RHINO robot.

The above two examples have also been worked out by using the complex method and GRG algorithms. It has been shown that for some of the above G-cuboids the GRG algorithm took a longer time to converge and sometimes even stepped back and forth at the same point, while the complex method, in some cases, lead to wrong results. In comparing with them, the RGRG algorithm proposed in this section seems more effective. It has good convergence quality and the number of iterations is small. Therefore it is suitable for solving the RT problem.

TABLE 2.2. The G-Cuboids and the RT Test Results for RHINO Robot (G-Cuboids: Generalized Cuboids, R: Reachable, N: Unreachable)

G-Cuboids[1]	Computation Cycles	J_{min}	RT Result[2]
(7, 7, 2)–(12, 12, 4)	7	3.40E-7	R
(−1, −1, 33)–(1, 1, 35)	21	0.509	N
(9, −22, 9)–(11, −22, 11)	6	3.132	N
(−1, −1, 32)–(1, 1, 32)	28	7.06E-6	R
(0, 0, 20)	6	5.80E-6	R
(0, 30, 0)	14	95.735	N

[1] G-cuboids stands for generalized cuboids.
[2] R stands for reachable; N stands for unreachable.

4. COMPUTATION OF THE BOUNDARIES OF REACHABLE WORK SPACE

In the above sections, it has been pointed out that the reachable work space of a manipulator is very critical in robot kinematics. In order to obtain detailed information about the work space, it is often necessary to determine the boundaries of the reachable work space. In this section, based on the basic problem in Section 2, some efficient numerical algorithms for computing the boundaries of a manipulator are developed. The problem to be solved is decomposed into simpler subproblems and realized in recursive form, which makes the algorithm more structured. Moreover, a heuristic search algorithm instead of the one-by-one search is presented, which reduces the computation burden greatly. To make the new algorithms suitable for more problems, the mathematical description in this section will be taken in a general form with work space problem as its special case.

4.1. Preliminary Concepts and Definitions

Consider an m-dimensional space R^m with coordinates x_i $(i = 1, \ldots, m)$. For a limited m-dimensional object O with holes and voids in it, there always exists a set $\{x_{i,\min}, x_{i,\max}, i = 1, \ldots, m\}$ such that

$$x_{i,\min} = \inf \{x_i \mid x = (x_1, \ldots, x_m) \in O\} \tag{51}$$

$$x_{i,\max} = \sup \{x_i \mid x = (x_1, \ldots, x_m) \in O\} \tag{52}$$

$$i = 1, 2, \ldots, m$$

Define the polyhedron D in m-dimensional space R^m as

$$D = \{X = (x_1, \ldots, x_m) \mid x_{i,\min} \leqslant x_i \leqslant x_{i,\max}, i = 1, \ldots m\} \tag{53}$$

this polyhedron D is said to surround the object O tightly.

For each point in D, a color may be assigned to it according to the following rules:

Assign Red to the point X if $X \in$ O,
Assign Green to the point X if $X \notin$ O.

The polyhedron D is then divided into several red and green blocks. The sets of points where red and green blocks meet are defined as the boundaries of D and O.

In a three-dimensional space R^3, the reachable work space of a manipulator may be recognized as an object W, so that there exists a cuboid C which surrounds the work space W tightly. The relationship between C and W is just like that of D and O in the above general case, and the boundaries of C and W, denoted as BW, are defined as the boundaries of the manipulator work space. Since there exist sometimes holes and/or voids in the work space of the manipulator, the work space may include two or more separate boundaries both internally and externally.

The work space of a manipulator is composed of all the points (x_1, x_2, x_3) which satisfy:

$$x_1 = f_1 (q_1, q_2, \ldots, q_n) \tag{54}$$

$$x_2 = f_2 (q_1, q_2, \ldots, q_n) \tag{55}$$

$$x_3 = f_3 (q_1, q_2, \ldots, q_n) \tag{56}$$

$$q_{i,\min} \leqslant q_i \leqslant q_{i,\max} \qquad j = 1, \ldots, n \tag{57}$$

where q_i, $(i = 1, \ldots, n)$ are the joint variables. $(x_1, x_2, x_3), f_1, f_2, f_3$ correspond to (P_x, P_y, P_z) and f_x, f_y, f_z in Section 2 respectively.

Generally speaking, it is rather difficult to obtain the boundaries of the work space BW analytically from the nonlinear equations (54)–(56). As a practical and feasible way, numerical methods are often resorted. When the cuboid C surrounding the work space W is equally divided into n_1, n_2, n_3 divisions along three orthogonal coordinate axes x_1, x_2, x_3 respectively, altogether $n_1n_2n_3$ minicuboids and thus $(n_1 + 1)(n_2 + 1)(n_3 + 1)$ vertices may be obtained. When the vertices which belong to W are assigned with red color and the others with green color, the work space can be represented by a set of vertices with red color. Our purpose is to find all the red vertices just on the boundaries of W and store them in a proper database so that they may be used to plot the contours of the work space on a specific plane.

4.2. Fundamental Algorithms

In this section, two fundamental algorithms are developed, which results in a one-by-one search algorithm to determine the boundaries of the reachable work space. These fundamental algorithms are given in general forms.

4.2.1. Determination of the Color of a Point in a Hyperbody P.

Let us first consider a general problem of determining whether a point $X^0 = [x_1^0, \ldots, x_m^0] \in R^m$ is within a special m-dimensional hyperbody p:

$$P = \{X = (x_1, \ldots, x_m) \mid x_i = f_i (q_1, \ldots, q_n), i = 1, \ldots, m\} \tag{58}$$

with $q_{j,\min} \leqslant q_j \leqslant q_{j,\max}$, $j = 1, \ldots, n$.

Referring to Section 2, define the following optimization problems:

$$OP_l \qquad \min J = \sum_{i=1}^{m} [x_i - f_i(q_1, \ldots, q_n)]^2 \qquad l = 0, \ldots, m \tag{59}$$

$$\text{s.t.} \qquad q_{j,\min} \leq q_{j,\max} \qquad j = 1, \ldots, n \tag{60}$$

$$x_i = x_i^0 \qquad i = 1, \ldots, 1 \tag{61}$$

It is obvious that OP_m is just the problem of determining whether $X^0 \in P$. If min $J = 0$ then $X^0 \in P$ and vice versa.

Algorithm 3. Testing whether $X^0 \in P$.

Step 1: Use any efficient nonlinear programming method to solve the problem OP_m.
Step 2: Find $J^* = \min J$. If $J^* < \epsilon$ then $X^0 \in P$ and assign red color to X^0, otherwise assign green color to X^0, where ϵ is a small positive number selected according to the accuracy requirement.

4.2.2. *Determination of the Boundaries of a Hyperbody on a Hypersurface.*

Suppose that X is constrained in a specific hypercuboid D

$$D = \{X = (x_1, \ldots, x_m) \mid x_{i,\min} \leq x_i \leq x_{i,\max}, i = 1, \ldots, m\} \tag{62}$$

where the maximum and minimum values of x_i may be determined by the following optimization problem

$$OD_i: \qquad \max x_i = f_i(q_1, \ldots, q_n) \tag{63}$$

$$\min x_i = f_i(q_1, \ldots, q_n) \qquad i = 1, \ldots, m \tag{64}$$

$$\text{s.t. } q_{j,\min} \leq q_j \leq q_{j,\max} \qquad j = 1, \ldots, n \tag{65}$$

which may be solved by usual nonlinear programming method.

In order to determine the boundaries of the hyperbody P on the hypersurface $S_i \subset D$

$$S_i = \{X = (x_1^0, \ldots, x_i^0, x_{i+1}, \ldots, x_m) \mid x_j^0 \text{ fixed for } j = 1, \ldots, i\} \tag{66}$$

S_i is first decomposed into a series of S_{i+1} through discretization:

$$S_i \Leftrightarrow \text{all } S_{i+1} \text{ for } x_{i+1}^0 (k_{i+1}) = x_{i+1,\min} + k_{i+1}\, \Delta x_{i+1}, k_{i+1} = 0, 1, \ldots n_{i+1} \tag{67}$$

with the discrete step length $\Delta x_{i+1} = (x_{i+1,\max} - x_{i+1,\min})/n_{i+1}$.

Let $\Omega(P, S_i)$ denote the problem of determining $BP(S_i)$, the boundaries of P on S_i, then it is clear that $\Omega(P, S_i)$ is equivalent to a series of problems $\Omega(P, S_{i+1})$ for $x^0_{i+1}(k_{i+1})$, $k_{i+1} = 0, \ldots, n_{i+1}$. Define $X^0_i = (x^0_1, \ldots, x^0_i, x_{i+1}, \ldots x_j, \ldots, x_m)$, the following recursive algorithm can be introduced for solving $\Omega(P, S_i)$.

Algorithm 4. Solving $\Omega(P, S_i)$.

Step 1: If $i = m$ use Algorithm 3 to test whether $X^0 \in P$, assign X^0 with appropriate color (red or green). Exit.
Step 2: Set $k_{i+1} = 0$.
Step 3: Let $x^0_{i+1}(k_{i+1}) = x_{i+1,\min} + k_{i+1}\,\Delta x_{i+1}$, solve $\Omega(P, S_{i+1})$.
Step 4: Let $k_{i+1} = k_{i+1} + 1$, if $k_{i+1} \leqslant n_{i+1}$, go to Step 3.
Step 5: If $X^0_i = (x^0_1, \ldots, x^0_i, x_{i+1,\max}, \ldots, x_{m,\max})$, for all $j = i + 1, \ldots, m$, compare the color of x^0_j during the sequence $k_j = 0, \ldots, n_j$. When a color variation takes place, store this changing point as a boundary point.

4.2.3. Determination of the Boundaries of the Reachable Work Space.

Let us now return to the problem of manipulator work space. It is clear that the problem of determining the boundaries of reachable work space is the special problem $\Omega(P, S_0)$ with $m = 3$. Algorithm 4 can of course be used for it. In fact, this algorithm tests all the discrete points in D sequentially and is a one-by-one search method. The basic optimization problem OP_m should be solved up to $(n_1 + 1)(n_2 + 1)(n_3 + 1)$ times. It is time-consuming and inefficient. Therefore, it is worthwhile to develop more efficient methods to determine the boundaries of reachable work space.

4.3. A New Heuristic Method for Determining the Boundaries of the Reachable Work Space

The fact that the boundaries of the work space are continuous, that is, the boundaries $BP(S_i)$ for $x^0_i(k_i + 1)$ are not far from that for $x^0_i(k_i)$, indicates that there is no necessity to determine $BP(S_i \mid x^0_i(k_i + 1))$ through the one-by-one test for all x^0_j, $j = i + 1, \ldots, m$. With the knowledge of $BP(S_i \mid x^0_i(k_i))$ it is possible to search $BP(S_i \mid x^0_i(k_i + 1))$ by only testing the nearby points of $BP(S_i \mid x_i(k_i))$, see Figure 2.4.

In this section, a knowledge-based algorithm is developed which pushes the boundaries forward and simplifies the search procedure.

4.3.1. Knowledge-Based Determination of the Boundaries.

Suppose that the known boundaries $BP(S_i \mid x_i(k_i))$ are composed of N points $X_\ell(x^0_i(k_i)) = (x^0_1 \ldots, x^0_{i-1}, x^0_i, (k_i), x^\ell_{i+1}, \ldots, x^\ell_m)$, $l = 1, \ldots, N$, based on them the following algorithm may be used to obtain $BP(S_i \mid x_i(k_i+1))$.

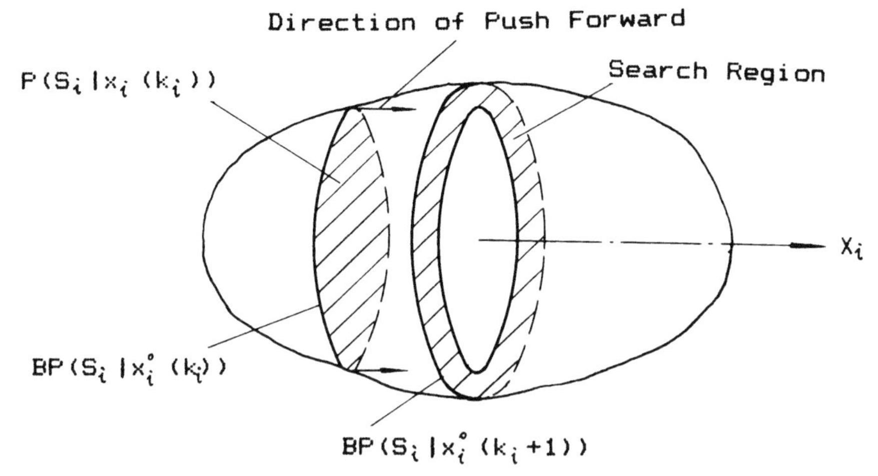

FIGURE 2.4. Knowledge-based push-forward search.

Algorithm 5. Determination of $BP(S_i \mid x_i^0(k_i + 1))$ *based on* $BP(S_{i+1} \mid x_i^0(k_i))$ *(push forward algorithm).*

Step 1: Set $l = 1$.

Step 2: Use algorithm 3 to test whether $x_\ell^0 = (x_1^0, \ldots, x_{i-l}^0, x_i^0(k_i + 1), x_{i+1}^\ell, \ldots, x_m^\ell) \in P$. Assign appropriate color to X_ℓ^0. Let $h = i + 1$.

Step 3: Set $x_h^{\ell 1} = x_h^{\ell 2} = x_h^\ell$.

Step 4: Let $x_h^{\ell 1} = x_h^{\ell 1} + \Delta x_h$, $x_h^{\ell 2} = x_h^{\ell 2} - \Delta x_h$, use algorithm 3 to test whether

$$X_\ell^1 = (x_1^0, \ldots, x_{i-1}^0, x_i^0\,(k_i + 1), \ldots, x_h^{\ell 1} \ldots) \in P$$
$$X_\ell^2 = (x_1^0, \ldots, x_{i-1}^0, x_i^0\,(k_i + 1), \ldots, x_h^{\ell 2} \ldots) \in P.$$

Assign appropriate colors to X_ℓ^1 and X_ℓ^2.

Step 5: If the colors of X_ℓ^1, X_ℓ^2 are the same as that of X_l^0, go to Step 4, otherwise determine the boundary points according to color variation.

Step 6: If $h < m$, let $h = h + 1$, go to Step 3.

Step 7: If $1 < N$, let $l = l + 1$, go to Step 2.

Compare Algorithm 5 with Algorithm 4, it is obvious that the computation burden will be greatly reduced by the former algorithm since the tests are made from the previous boundaries until new boundaries are obtained. The search procedure does not involve all points. However, in using algorithm 5 to determine the boundaries of the work space, some difficulties may appear. When there are holes and/or voids in the work space so that their boundaries sometimes do not appear in the first search result, they will get lost by the push-forward procedure. To avoid the loss of the boundaries for holes and voids, the push-forward results should be examined by performing a limited number of full search procedures. When some new boundaries do appear, a push-backward correction must be performed to determine the entire boundaries of the holes and/or voids. The principle of this heuristic search procedure is shown in Figure 2.5.

4.3.2. *Heuristic Search for the Boundaries.*

Let $n_i = J_i K_i$, where J_i is the number of full search procedures and K_i is the number of push-forward and push-backward procedures between two full searches, as shown in Figure 2.5.

Algorithm 6. Heuristic algorithm for solving $\Omega(P, S_i)$.

Step 1: If $i = m - 1$, use Algorithm 4 directly. Exit.

Step 2: Set $j_{i+1} = 0$, let $x^0_{i+1} = x_{i+1,\min}$, heuristically solve $\Omega(P, S_{i+1})$ to obtain $\overline{BP}(S_{i+1} \mid x^0_{i+1}(j_{i+1})) = \overline{BP}(S_{i+1} \mid x^0_{i+1}(0))$, where $\overline{BP}$ denotes the boundaries obtained by full search.

Step 3: Set $k_{i+1} = 0$, $BP(S_{i+1} \mid X^0_{i+1}(j_{i+1}, 0)) = \overline{BP}(S_{i+1} \mid x^0_{i+1}(j_{i+1}))$.

Step 4: Let $x^0_{i+1} = x_{i+1,\min} + (j_{i+1} K_{i+1} + k_{i+1}) \Delta x_{i+1}$, use Algorithm 5 to obtain $BP(S_{i+1} \mid x^0_{i+1}(j_{i+1}, k_{i+1} + 1))$ from $BP(S_{i+1} \mid X^0_{i+1}(j_{i+1}, k_{i+1}))$.

Step 5: Let $k_{i+1} = k_{i+1} + 1$, if $k_{i+1} < K_{i+1}$, go to Step 4.

Step 6: Let $x^0_{i+1} = x_{i+1,\min} + (j_{i+1} + 1) K_{i+1} \Delta x_{i+1}$, heuristically solve $\Omega(P, S_{i+1})$ to obtain $\overline{BP}(S_{i+1} \mid x^0_{i+1}(j_{i+1} + 1))$.

Step 7: Compare $\overline{BP}(S_{i+1} \mid x^0_{i+1}(j_{i+1} + 1))$ with $BP(S_{i+1} \mid x^0_{i+1}(j_{i+1}, K_{i+1}))$, if they are equal, go to Step 10, otherwise let $BP(S_{i+1} \mid x^0_{i+1}(j_{i+1}, K_{i+1})) = \overline{BP}(S_{i+1} \mid x^0_{i+1}(j_{i+1} + 1))$.

Step 8: Let $x_{i+1} = x_{i+1,\min} + (j_{i+1} K_{i+1} + k_{i+1}) \Delta x_{i+1}$, use Algorithm 5 to obtain $BP(S_{i+1} \mid x^0_{i+1}(j_{i+1}, k_{i+1} - 1))$ from $BP(S_{i+1} \mid x^0_{i+1}(j_{i+1}, k_{i+1}))$.

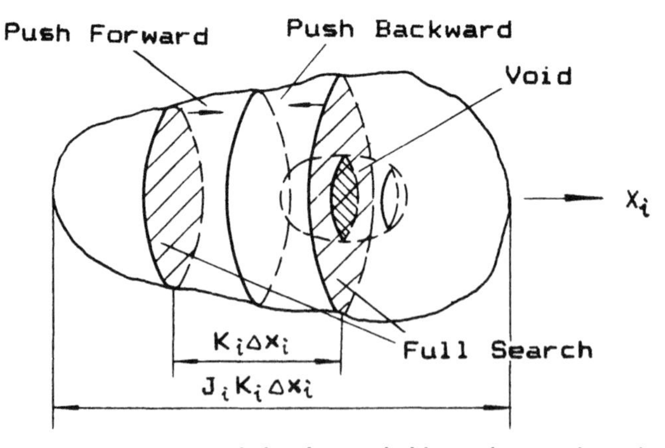

FIGURE 2.5. Heuristic search for the reachable work space boundaries.

Step 9: Let $k_{i+1} = k_{i+1} - 1$, if $k_{i+1} > 0$, go to Step 8.
Step 10: Let $j_{i+1} = j_{i+1} + 1$, if $j_{i+1} < J_{i+1}$, go to Step 3.

Note that the Algorithm 6 is similar to Algorithm 4. Both of them are recursive algorithms.

4.3.3. *Determination of the Reachable Work Space Boundaries.*

When $P = W$, $i = 0$, and $m = 3$, Algorithm 6 may be used to determine the boundaries of the reachable work space. The boundaries for all the holes and voids can also be determined if the parameter K_i is so selected that

$$K_i \Delta x_i < \min (D(H, d_i), D(V, d_i)) \tag{68}$$

where $D(H, d_i)$, $D(V, d_i)$ are the maximal distances of the projections of the holes and voids on direction x_i respectively.

As Algorithm 6 involves the principles of push-forward search, heuristic examination and push-backward correction, it is highly structured and capable of searching for the boundaries of the work space either on a plane or on a line. No prior analysis for the reachable work space of the manipulator is required. It is a general algorithm independent of the number of freedom of the manipulator and the type of the joints. By using the proper database, the boundaries of the reachable work space may be stored to accumulate the necessary data for plotting the internal and external contours of the work space on a specified plane.

4.4. Example

In order to illustrate the efficiency of our new method, the same 2R Planar robot is taken as a computation example, see Figure 2.6.

The kinematic relationship between the robot end effector's Cartesian coordinates and the joint variables is given by (46) and (47).

Assume that the joint variables are constrained by:

$$-180^\circ \leqslant q_1 \leqslant 180^\circ; \qquad -180^\circ \leqslant q_2 \leqslant 180^\circ \tag{69}$$

that is, q_1, q_2 may vary between -180° and 180° respectively. The reachable work space of the manipulator is then constrained in the square region:

$$R: -14 \leqslant x \leqslant 14, \qquad -14 \leqslant y \leqslant 14$$

Assume that the computation accuracy in the numerical solution is taken as $\epsilon = 0.2$, then there will be $(28/0.2) = 19{,}600$ minisquares in R. With the one-by-one search method as described by Algorithm 4, altogether $(28/0.2+1) =$

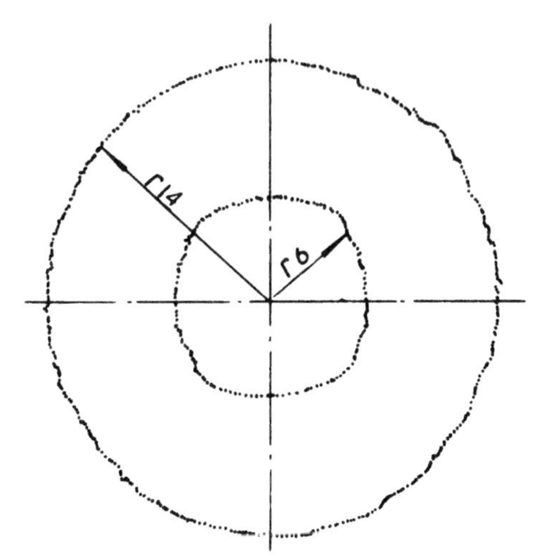

FIGURE 2.6. The boundaries of the reachable work space of the 2R Planar robot by the heuristic search method, with joint variables $-180° \leqslant q_1 \leqslant 180°$, $-180° \leqslant q_2 \leqslant 180°$.

19,881 points need to be tested by Algorithm 3. But with the new heuristic search method, only 3925 points need to be tested and the same result will be achieved (see Figure 6). The computation burden in this case is reduced by about 80%. It is undoubtedly that for higher required accuracy or for more joint variables, the reduction of computation burden would be much more.

4.5. Storing and Displaying the Data of Reachable Work Space

An important goal of computing the reachable work space of a robot is to display and evaluate its properties. Previously, the problem of how to compute the reachable work space efficiently has been successfully solved. The next problem is how to store the data economically and efficiently. It is evident that the point-to-point manner contains complete information of the reachable work space, but the memory storage would be very large. One economical way is to use the following data structure:

$$mX^* + Y^* \rightarrow Z_1^* \text{ to } Z_2^*, Z_3^* \text{ to } Z_4^*, \ldots \tag{70}$$

where m is a predetermined integer; X^*, Y^*, Z^* are the quantization of the coordinates of points X, Y, and Z_i respectively; $mX^* + Y^*$ is used as pointer to indicate the memory address; each pair of Z_i^* to Z_{i+1}^* indicates a line segment that belongs to the robot's reachable work space; and Z_i^* ($i = 1, 2, \ldots$) are the only data to be stored.

By using this data structure, in which data are stored in a line-to-line manner, it is easy to carry out the following tasks:

- To test whether a given point is within the reachable work space of the robot,
- To test whether a given line segment Z_a to Z_b is fully inside the work space of the robot,
- To display the cross sections of the reachable work space such as $X = X_o$, $aY + bZ = c$, or $Y = Y_0$, $dX + eZ = f$,
- To calculate the volume of the entire reachable work space or the volume of voids or holes.

It has been estimated that when the accuracy is selected as one-eighth of the summation of all the robot links, 80% or more of the memory storage could be saved by using such a data structure in comparing with the point-to-point storage manner.

5. DISCUSSIONS

In the previous sections, different problems related to the manipulator reachability or work space are investigated. Generally speaking the basic problem involved here is to test whether one object (point) is overlapped with another, which can be formulated as a uniform nonlinear optimization with linear inequality constraints. In some special cases such as Problem 2, the solution of $J = 0$ is necessary and sufficient for testing whether a given point is within the reachable work space of a manipulator. However, it should be pointed out that in general, $J \neq 0$ is sufficient for the negative answer, such as "two objects do not overlap each other" (collision free) or "a region is not reachable by a manipulator," while $J = 0$ only provides a poor conclusion for the positive answer because the existence of a solution is not enough for obtaining detailed information such as "a trajectory is fully reachable by a manipulator" or "a region is fully within the manipulator work space." In order to answer these problems, the original optimization problem must be further decomposed and solved recursively, as shown in Section 4.

The general problem formulation makes it possible to use the same concept to handle other problems. Either the index function or the constraint conditions in optimization can be revised to fit different problems. For example, if equality constraints are added to Problem 2, then it can be used to solve questions such as:

- Conditional reachability of a robot to an object or a point,
- Computation of dexterous work space [16] of a robot.

If Problem 2 is revised into the form:

$$\min J = (P_{1x} - P_{2x})^2 + (P_{1y} - P_{2y})^2 + (P_{1z} - P_{2z}) \quad (71)$$

$$\text{s.t.} \quad q_{1i,\min} \leq q_{1i} \leq q_{1i,\max} \qquad i = 1, \ldots, n \quad (72)$$

$$q_{2j,\min} \leq q_{2j} \leq q_{2j,\max} \qquad j = 1, \ldots, m \quad (73)$$

where (P_{1x}, P_{1y}, P_{1z}) and (P_{2x}, P_{2y}, P_{2z}) represent the Cartesian coordinates of the end effectors of robot 1 and robot 2 respectively; q_{1i}, q_{2j}, $i = 1, \ldots, n$, $j = 1, \ldots, m$ are the joint variables of robot 1 and robot 2 respectively; and n, m are the numbers of joints that robot 1 and robot 2 have, then it may be used to test whether two robots will collide with each other. If the similar algorithms in Section 4 are used, the common reachable work space of robot 1 and robot 2 can be obtained.

6. CONCLUSION

In this chapter, a general numerical approach for solving the reachable work space related problems is presented. Some typical problems are solved in detail and further potential applications of the approach have been discussed. It has been shown that some simple methods, such as nonlinear programming, when used properly, may lead to many interesting and significant results.

REFERENCES

1. Y.C. Tsai and A.H. Soni, "Accessible region and synthesis of robot arms," *ASME J. Mechanical Design,* vol. 103, pp. 803–811, October, 1981.
2. Y.C. Tsai and A.H. Soni, "An algorithm for the workspace of general N-R robot," *ASME J. Mechanical Trans. Autom. Design,* vol. 105, pp. 52–57, March, 1983.
3. K.C. Gupta and B. Roth, "Design considerations for manipulator workspace," *ASME J. Mechanical Design,* vol. 104, pp. 704–711, October, 1982.
4. K.C. Gupta, "On the nature of robot workspace," *Int. J. Robotics Research,* vol. 5, pp. 112–121, Summer, 1986.
5. T.W. Lee and D.C.H. Yang, "On the evaluation of manipulator workspace, *ASME J. Mechanical Trans. Autom. Design,* vol. 105, pp. 70–77, March, 1983.
6. D.C.H. Yang and T.W. Lee, "On the workspace of mechanical manipulators," *ASME J. Mechanical Trans. Autom. Design,* vol. 105, pp. 62–69, March, 1983.
7. K. Sugimoto and J. Duffy, "Determination of extreme distance of a robot hand–part 1: A general theory," *ASME J. Mechanical Design,* vol. 103, pp. 631–636, July, 1981.
8. A. Kumar and M.S. Patel, "Mapping the manipulator workspace using interactive computer graphics," *Int. J. Robotics Research,* vol. 5, pp. 122–130, May, 1986.

9. J. Rastegar and P. Deravi, "Methods to determine workspace, its subspaces with different numbers of configurations and all possible configurations of a manipulator," *Mechanical Machine Theory,* vol. 22, pp. 343–350, May, 1987.
10. Ming-shu Hsu and Dilip Kohli, "Boundary surfaces and accessibility regions for regional structures of manipulators," *Mechanical Machine Theory,* vol 22, pp. 227–289, May, 1987.
11. Chi-haur Wu and Kun-Young Young, "Robot workspace geometry for trajectory feasibility study," in *Proc. of IEEE Int. Conf. on Systems, Man and Cybernetics,* 1988, pp. 238–241.
12. R.P. Paul, *Robot Manipulators: Mathematics, Programming and Control.* Cambridge, MA: MIT Press, 1981, pp. 41–83.
13. M.J. Box, "A new method of constrained optimization and a comparison with other methods," *Computer J.*, vol. 8, pp. 42–52, April, 1965.
14. Mordecai Avriel, *Nonlinear Programming: Analysis and Methods.* Englewood Cliffs, NJ: Prentice-Hall, Inc., 1976, pp. 130–236.
15. D.M. Himmelblau, *Applied Non-linear Programming.* New York: McGraw-Hill Book Company, 1972, pp. 221–286, 366–390.
16. A. Kumar and K.J. Waldron, "The workspace of a mechanical manipulator," *ASME J. Mechanical Design,* vol. 103, pp. 665–672, July, 1981.

3

Expert Adaptive Control: Method and Medical Application

Howard Kaufman
Rensselaer Polytechnic Institute
Electrical, Computer, and Systems Engineering Department,
Troy, NY

Gregory W. Neat
Jet Propulsion Laboratory
Pasadena, CA

1. INTRODUCTION

This chapter presents a new adaptive control method that adjusts its structure in accordance with its knowledge of the regulated process. An example illustrating the application of the method is presented with the design of a regulator for a drug delivery system.

When designing a controller for a process control problem, a range of algorithms exists. At one end of the spectrum are rule-based approaches that rely on qualitative plant information to determine the control action. At the other end of this range are analytically based adaptive approaches that require the plant transfer function to meet certain requirements to guarantee stable system operation. Figure 3.1 illustrates this range of algorithms. Given different degrees of knowledge of the process, certain control approaches spanning this range will perform the control task better than others. Rule-based controllers are the choice when the process is ill defined. This type of algorithm systematically encodes how a human controller would handle the control challenge and is thus described as a highly intelligent regulation method. This approach proves more robust than analytical counterparts under these conditions. However, it may fail to provide the most precise control as compared with analytically based approaches if some knowledge of plant description is available. Here precision describes the degree to which the controller maintains the process at the set point. Linear controllers such as proportional-integral-derivative (PID), linear quadratic regulator (LQR), etc. provide adequate control when the plant is linear time-invariant with known coefficients. These methods may fail to give the desired closed-loop response, if

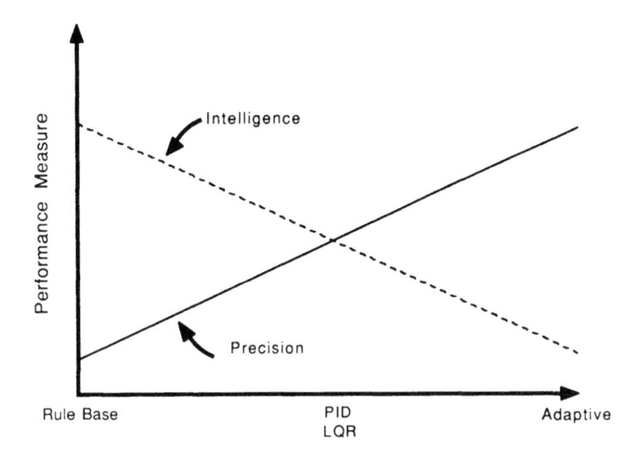

FIGURE 3.1. Plot of range of control algorithms with respect to their precision and intelligence.

the parameter values have a region of uncertainty, or have a wide variation with time. To address plants of this nature, many direct and indirect adaptive control methods exist that may provide the desired precision [1]. In contrast to the highly intelligent, low-precision rule-based methods, adaptive control algorithms provide precise control action while lacking intelligence about the regulated plant. However, these approaches become invalid when the conditions under which the algorithms have been applied are violated. This occurs for example when a process varies substantially from the structure of the model on which the adaptive laws were based.

In the ideal situation, the type of controller chosen from the range of possible controllers should be the most precise given the available knowledge about the process. The control problem becomes difficult for processes that exhibit a variable structure. In this case, the process may behave similarly to the model on which its controller design was based or may vary substantially from the model, resulting in poor regulation. For processes such as these, a controller that combines a range of algorithms is necessary in order to address the variation undergone by the plants. This type of controller, which expands the capabilities of conventional controllers, requires a supervisor or expert to orchestrate the controllers as the conditions of the process change. Figure 3.2 shows a block diagram of this control approach.

The dashed controller box in Figure 3.2 represents the range of algorithms available to determine the control action to the process. The key element distinguishing this closed-loop system from a conventional approach is the expert system. This element makes heuristic decisions about the control structure

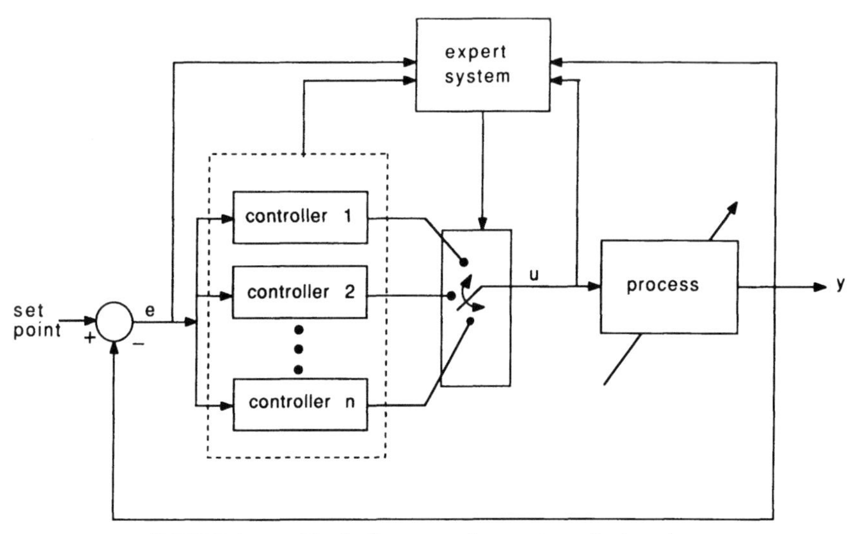

FIGURE 3.2. Block diagram of expert control system.

and/or controller parameters. The fundamental concepts in designing an expert control system include:

- Definition of conditions under which each controller in the bank performs adequately and when it will fail
- Definition of criteria that describe these conditions
- Design of a supervisory function that determines if the conditions for a given controller have been met and then instigates the switching of control procedures.

This chapter discusses control approaches that can be cast into the framework represented in Figure 3.2.

1.1. Background

This section delineates the evolution of control systems containing reconfigurable capabilities. Several reconfiguration schemes have been proposed that assume the plant characteristics are known and linear [2–4]. A generalized digital reconfigurable control approach based on fault-tolerant system techniques was suggested by Ornedo [5, 6]. The 5 MWt MIT Research Reactor served as a testing ground for this control structure. The controller reconfiguration logic in this scheme identified the plant operating condition, selected the appropriate control law, and verified the control law choice through an on-line performance evaluator. Five different control laws were incorporated into the controller bank: a PID controller, a nonlinear digital control algorithm, a fuzzy control algorithm, a relay control algorithm, and a heuristic control law based on rules derived from operator experience. Referring back to Figure 3.1, all these schemes span the range of algorithms from the most intelligent (rule-based) to the center of the curve (PID). Emphasis was placed on ranking alternative control schemes in case of a control law anomaly.

On the adaptive front, Isermann and Lachmann [7] reported merging supervisory functions with adaptive control methods. Subsequently, Astrom and coworkers [8] introduced the concept of expert control. They suggested combining a range of algorithms under the supervision of an expert system to obtain a controller with new capabilities. The possible controllers in this controller bank were: a PID (backup), a minimum variance controller (constant parameter), and a self-tuning regulator (parameter adaptive). The actual control scheme from this bank of three was chosen by an expert system that considered control input excitation level, accuracy of plant parameter estimates, assessment of closed-loop system stability, and a check to verify that the plant is minimum phase. In the context of Figure 3.1, these controllers span the range from the center of the curve (PID), to the most precise control laws (adaptive).

A natural consideration when combining algorithms is to use them in a complementary manner. Bernard [9] suggested this idea in an in-depth comparison of rule-based and analytical control approaches. Recognizing the robust qualities of rule-based control methods, Bernard suggests using them as backups to analytic controllers.

1.2. Objectives

The major contribution of this work is a methodology for designing a controller whose capabilities span the entire range depicted in Figure 3.1. The controllers that compose the controller bank function in a dependent, hierarchical manner. The emphasis is on combining control strategies in such a way that each successive stage of the controller provides a control value to the plant while aiding in satisfying conditions required by the subsequent controller stages. These conditions include obtaining some knowledge about the plant or providing compensation to ensure stability of the closed-loop system. The hierarchy ranges from (highest stage) a rule-based controller, to an adaptive controller (lowest stage), and thus spans the complete spectrum of Figure 3.1. The chapter discusses the individual controllers, cites successful applications of the respective methods, and describes how the elemental control schemes work together in the regulation process.

An expert system orchestrates the controller structure. The goal of this supervisor is to select the best control scheme given the present and past knowledge about the plant. The chapter discusses the basic components of a rule-based expert system with regard to the regulation of the controller structure.

In addition, the chapter discusses the versatility of this control approach by describing how the hybrid control structure can successfully regulate a wide range of process conditions including the ability to:

- Determine precise control actions when the process is perfectly known
- Provide a control action when nothing is known about the process
- Provide a control action during the identification period required by the adaptive techniques.

To illustrate this methodology, the chapter presents details on the design of a controller for a drug delivery system. This process contains the nonlinear, time-varying characteristics that require a control scheme with such capabilities.

1.3. Hierarchical Control Structure

The hybrid control structure presented here combines control approaches that range from a crude control scheme where the actions of the operator are represented in a rule-base format to a model-based adaptive approach where the output of the plant is forced to follow the output of a desired reference model. The controller bank as shown in Figure 3.3 is made up of the following three controllers: (1) Fuzzy controller (FUZZ); (2) multiple model adaptive controller

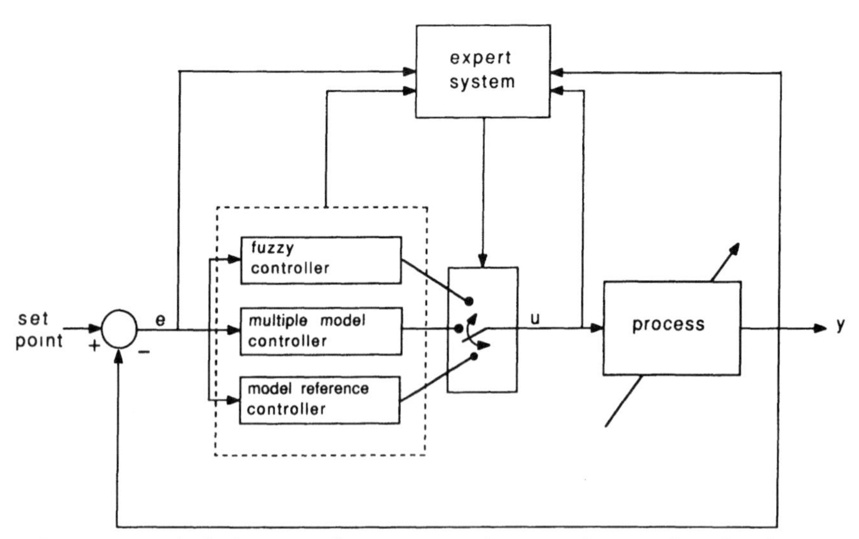

FIGURE 3.3. Block diagram of expert control system discussed in this chapter.

(MMAC); and (3) model reference adaptive controller (MRAC). The controller presented here combines the abovementioned controllers in a dependent fashion in order to gain the benefits of each control approach while overcoming their shortcomings.

The fuzzy control scheme determines a control input for the process based on how a human operator would perform the task. This method is chosen when little is known about the regulated process. However, this approach suffers from the inherent problem of only being able to do as well as the operator since that is the basis of its design.

The multiple model adaptive control method assumes that the process will resemble the response of one of a finite set of predefined models. Once a model is selected, a predefined linear controller designed for the chosen model provides the control input for the process. The approximate identification allows this approach to provide reasonable control actions with incomplete knowledge of the process. However, in the case of regulating a multiple-input multiple-output (MIMO) process, an exorbitant number of models may be required in order to provide satisfactory control for the process.

The model reference adaptive control method determines a control value that forces the process to follow the trajectory of a desired reference model. This algorithm provides the most precise control action to the process. The drawback of this method is the need for the process to satisfy stringent sufficiency conditions in order to ensure closed-loop stability.

The expert system makes heuristic decisions needed to regulate the control structure as it learns about the process. The form of the expert system is a production system and uses a backward chainer for an inference engine. The expert system attempts to select the best control method in accordance with its knowledge about the process. In the case of little process knowledge, the actions of a human operator will be implemented (fuzzy controller—most intelligent control action). In the case of confident parameterization of the process, the controller will force the process to follow a desired predetermined trajectory (model reference adaptive controller—most precise control action).

1.4. Process Control Plants

Examples of complex processes that challenge automatic control methods include: cement kilns, patient physiological status, blast furnaces, batch chemical reactors, and basic oxygen steel making. Many factors contribute to the difficulties in regulating these types of processes. These include: uncertainties in process parameters, nonlinearities, poor quality of available measurements, time-varying nature of process parameters, and time delays in the responses to control actions. The motivation for the development of the hierarchical control structure presented here was the desire to automatically regulate the physiological status of critical care patients. The process in this case is the patient, and the controller determines the infusion rate(s) of drugs and/or fluids required to maintain a desired patient status. A simplified linear time-invariant model representing the transfer function between a physiological variable and a drug infusion shown in Equation (1), gives a basis on which to design analytically based controllers for such a system [10, 11]. In such a design, the parameters k, τ_p, and T_d are considered unknown and time-varying.

$$G(s) = \frac{k \exp(-T_d s)}{(\tau_p s + 1)} \qquad (1)$$

where k represents the sensitivity of the patient to the specific drug; the dynamic delay of the drug action is modeled by τ_p; and the time delay T_d models the time for the drug to reach the site of action and have its initial effect. From the adaptive control perspective (most precise control), the challenging characteristic in Equation (1) is the time delay. In the development of the model-based methods of the hierarchical controller, considerations in maintaining a stable closed-loop system in the presence of process time delays are addressed.

1.5. Chapter Overview

The chapter is broken down into two parts. The first section presents in a general fashion the basic design method for this hierarchical control scheme. This involves discussing each of the individual components shown in Figure 3.3 and how the components collaborate in the regulation process. The chapter presents the elemental controllers by describing how each performs the transformation from e to u as shown in Figure 3.3. Here e represents the difference in the process output (y) from the desired set point and u represents the control value applied to the process.

The second section presents the application of this design method to the regulation of a drug delivery system, the goal of which is to maintain stable physiological status of critical care patients suffering from cardiac failure.

2. HIERARCHICAL CONTROL FRAMEWORK

2.1. Fuzzy Control

The ideas of fuzzy control have grown from the work of Zadeh [12, 13] who introduced the concept of fuzzy sets and linguistic variables. This method for mathematically describing terms such as *small, high,* and *low* allows for a systematic manner to encode the rules of thumb that experts use in controlling poorly modeled processes. Thus, the application of a fuzzy controller to regulating a process is ideal when little is known about the plant and the goal is to model the actions of the expert.

2.1.1. Basic Elements.

The fuzzy controller is actually a low-level expert system. It contains a rule base, a database, and an inference engine. The basic elements of a fuzzy controller are shown in Figure 3.4. These components together define the transformation between the error ($e(k)$) and change in error ($\Delta e(k)$) in the controlled process variable(s) and the change in control value ($\Delta u(k)$) applied to the process. The $e(k)$ and $\Delta e(k)$ determine the state of the process at each sample time (k) as represented in Equations (2) and (3), where $y(k)$ is the measured process output.

$$e(k) = y(k) - \text{set point} \tag{2}$$

$$\Delta e(k) = e(k) - e(k - 1) \tag{3}$$

Note that this discussion considers $\Delta u(k)$ as the output of the controller rather than $u(k)$ for reasons involving the incorporation of this controller into the hierarchical framework. However, the basic fuzzy controller concepts presented hold independent of the selected output variable ($\Delta u(k)$ or $u(k)$).

The first step in the design of the fuzzy controller is to define the process state in terms of fuzzy numbers. This is done by defining fuzzy numbers such as "positive big" (*pb*) or "negative medium" (*nm*) in terms of membership functions ($\mu(\cdot)$) that can take on any value between 0 and 1 inclusive. For example, let X, as in Equation (4), represent the domain or "universe of discourse" of $e(k)$ and $\Delta e(k)$, and let *pb* and *nm* represent fuzzy numbers. Figure 3.5 presents these fuzzy numbers graphically in terms of their corresponding membership functions, $\mu_{pb}(x)$ and $\mu_{nm}(x)$, which are listed in Equations (5) and (6) respectively.

$$X = (-5, -4, -3, -2, -1, 0, 1, 2, 3, 4, 5) \tag{4}$$

$$\mu_{pb}(x) = (0.0, 0.0, 0.0, 0.0, 0.0, 0.0, 0.0, 0.0, 0.5, 1.0, 1.0) \tag{5}$$

$$\mu_{nm}(x) = (0.2, 0.6, 1.0, 1.0, 0.5, 0.0, 0.0, 0.0, 0.0, 0.0, 0.0) \tag{6}$$

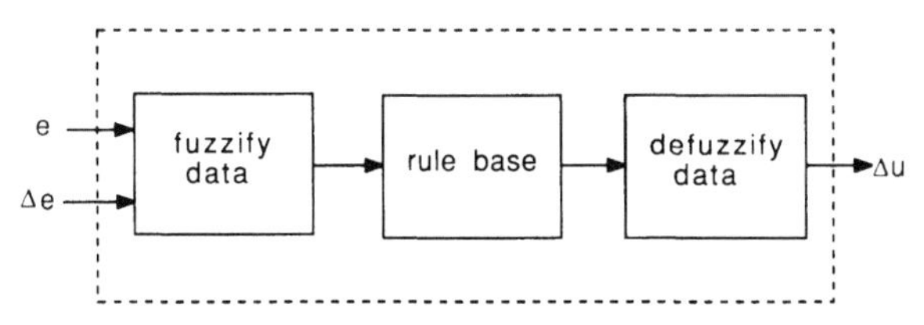

FIGURE 3.4. The transformation from *e* to Δu in terms of the basic components of a fuzzy controller.

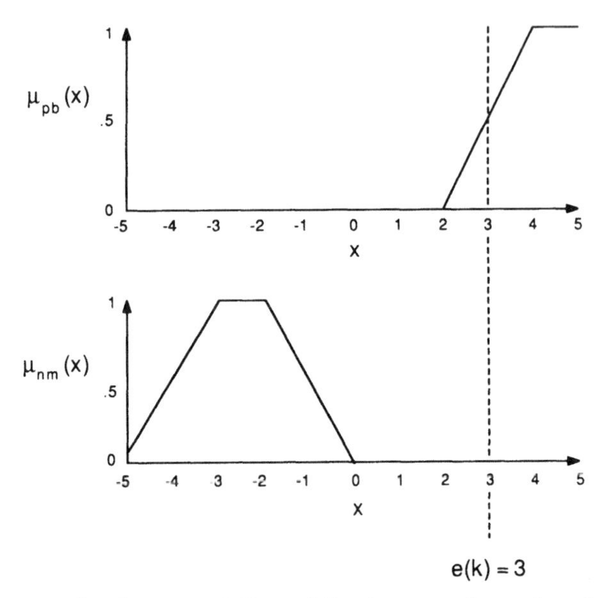

FIGURE 3.5. Graphical representation of the fuzzy numbers *pb* and *nm* in terms of their respective membership functions, $\mu_{pb}(x)$ and $\mu_{nm}(x)$. An example illustrating the degree to which an error value is *pb* and *nm* is shown for $e(k) = 3$.

These membership functions represent the degree to which elements of the domain are *pb*/*nm*. For example, as shown in Figure 3.5, if $e(k) = 3.0$, then the degree to which this element is *pb* is 0.5 and the degree to which this element is *nm* is 0.0. This process of translating a nonfuzzy number (element of the universe of discourse) into a fuzzy number is called fuzzifying the data.

The rule base contains the expert's rules of thumb encoded in terms of the defined fuzzy numbers. These rules have the form: "⟨*premise*⟩⟨*consequent*⟩." The premise may be composed of multiple clauses connected by the conjunction operator. For example:

$$\begin{array}{ll} \text{premise:} & \text{IF } \overbrace{e(k) \text{ is } nm}^{\text{clause 1}} \text{ and } \overbrace{\Delta e(k) \text{ is } pb}^{\text{clause 2}} \\ \text{consequent:} & \text{THEN change the control } nm \end{array}$$

The designer must ask the expert questions about how he/she regulates the process in the above format. It is necessary that the rule base cover the entire universe of discourse of $e(k)$ and $\Delta e(k)$.

The inference engine for the fuzzy controller decides which rules in the rule base contribute and how they contribute to the determination of the control value calculated each sample time. The specific rule(s) that fire depend on the present state of the process and the formulated rule base. Max-min composition is the method of resolution or inference engine used in the fuzzy controller calculation. Figure 3.6 demonstrates graphically this approach for a simple controller containing two rules. For simplicity in this illustration, the change in control value ($\Delta u(k)$) is defined over the same universe of discourse (X) as $e(k)$ and $\Delta e(k)$. Stated in terms of fuzzy numbers, the two rules are:

- Rule 1: IF $e(k)$ is *nb* and $\Delta e(k)$ is *pm* THEN change control *ze*.
- Rule 2: IF $e(k)$ is *nm* and $\Delta e(k)$ is *pb* THEN change control *pm*.

Here *ze*, *nb*, and *pm* are fuzzy numbers representing 0, negative big, and positive medium, respectively. Figure 3.6 shows these two rules graphically in terms of the membership functions representing the fuzzy numbers.

As the name indicates, max-min composition involves performing a maximum operation and a minimum operation on the membership functions that form the rules. The minimum function operates on each rule individually determining the degree to which a specific rule fires based on how its premise matches the

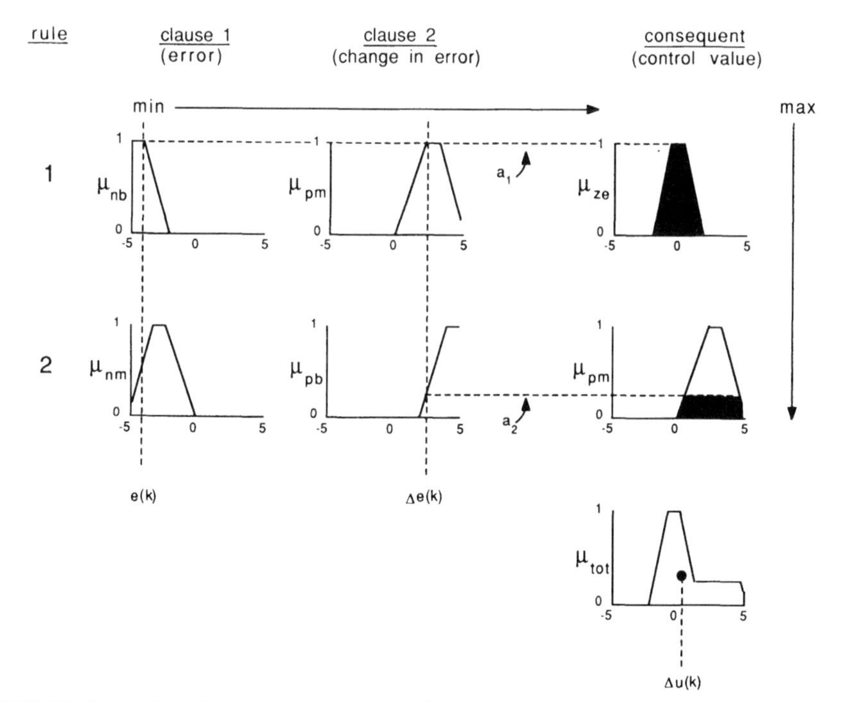

FIGURE 3.6. Graphical representation of the max-min composition inference method.

incoming data ($e(k)$ and $\Delta e(k)$). This measure of rule firing or contribution of the rule to the ultimate control output is represented by a_i for rule i in Figure 3.6. Equation (7) represents the possible values of a_i, where $a_i = 1$ signifies complete satisfaction of the premise of rule i and $a_i = 0$ indicates rule i will not contribute at all to the determination of the control output.

$$0 \leq a_i \leq 1 \tag{7}$$

In the example shown in Figure 3.6, the premise is made up of two clauses involving $e(k)$ and $\Delta e(k)$, respectively. The minimum value between the satisfaction of clause 1 and the satisfaction of clause 2 determines the satisfaction of the premise for that particular rule. For Rule 1, the degree to which $e(k)$ is *nb* is one, and the degree to which $\Delta e(k)$ is *pm* is one. Thus, $a_1 = 1$ as denoted. In the case of Rule 2, neither clause is totally satisfied. The degree to which $\Delta e(k)$ is *pb* represents the least satisfaction and thus defines a_2. Since $a_1 > a_2$, Rule 1 will have more of an effect on the ultimate control value for this example. Figure 3.6 demonstrates this fact. The minimum operation is performed between the scalar value a_i and the membership function in the consequent of rule i. Since $a_1 = 1$, the consequent of Rule 1 is unchanged. However, in the case of Rule 2, since $a_2 < 1$, the membership function in the rule's consequent is altered by the minimum operation as demonstrated in Figure 3.6.

In contrast to the horizontal minimum operation which considers each rule individually, the subsequent maximum operation of this inference method considers all the rules simultaneously and is performed vertically. By performing a point-by-point maximum operation (over the universe of discourse of $\Delta u(k)$), considering all the consequents of the rules, a new membership function ($\mu_{tot}(x)$) results. This fuzzy number (tot) as defined by its membership function ($\mu_{tot}(x)$), represents the contributions from all the membership functions in the consequents of the individual rules. Equation (8) represents the entire procedure of determining this fuzzy number, where "·" and "+" denote the minimum and maximum operations, respectively. *ze* and *pm* denote the fuzzy numbers in the consequent of the two rules.

$$\text{tot} = a_1 \cdot ze + a_2 \cdot pm \tag{8}$$

The final step in the fuzzy controller calculation generates a *crisp* number ($\Delta u(k)$) to apply to the process from the function $\mu_{tot}(x)$. This operation is known as defuzzification. Many defuzzification methods exist. The approach discussed here is the center of area method (COA). Mathematically, this operation involves finding the centroid of $\mu_{tot}(x)$ which results in a nonfuzzy value, $\Delta u(k)$.

Figure 3.6 presents in a graphical manner how the fuzzy controller determines $\Delta u(k)$ based on $e(k)$ and $\Delta e(k)$. It is important to realize that the fuzzy controller in this simple example only contains two rules. In addition, the example only shows fuzzy controller operation for one point in the $e(k)$, $\Delta e(k)$ space. Consider the result of repeating the process presented in Figure 3.6 for all possible values (over the universe of discourse) of $e(k)$ and $\Delta e(k)$. The result would be a multidimensional surface, where for each $e(k)$ and $\Delta e(k)$ pair there exists a corresponding $\Delta u(k)$. Figure 3.7 represents graphically this concept for a fuzzy controller with 36 rules. The x- and y-axes represent all the possible $e(k)$ and $\Delta e(k)$ values. The z-axis represents the corresponding $\Delta u(k)$ value for the given $e(k)$ and $\Delta e(k)$ pair. The example discussed in Figure 3.6 represents a single point in this multidimensional view of the transformation between the inputs to the fuzzy controller ($e(k)$ and $\Delta e(k)$) and the output ($\Delta u(k)$). Two qualities of the fuzzy controller become apparent when viewing this transformation as in Figure 3.7. First, this transformation is free to be nonlinear. This is in contrast to a model-based linear controller. In this case, the inputs ($e(k)$ and $\Delta e(k)$) and output ($\Delta u(k)$) would be the same, but the surface in Figure 3.7 would be a plane. The second important point is this look-up table format is easy to implement.

In summary, the designer of a fuzzy controller must initially define fuzzy numbers to describe the state of the process. Subsequently, the designer must acquire a complete set of rules from the expert, expressed in terms of the defined fuzzy numbers. Given the rule base and the universe of discourse for $e(k)$, $\Delta e(k)$, and $\Delta u(k)$, a multidimensional surface can be generated off-line using the max-

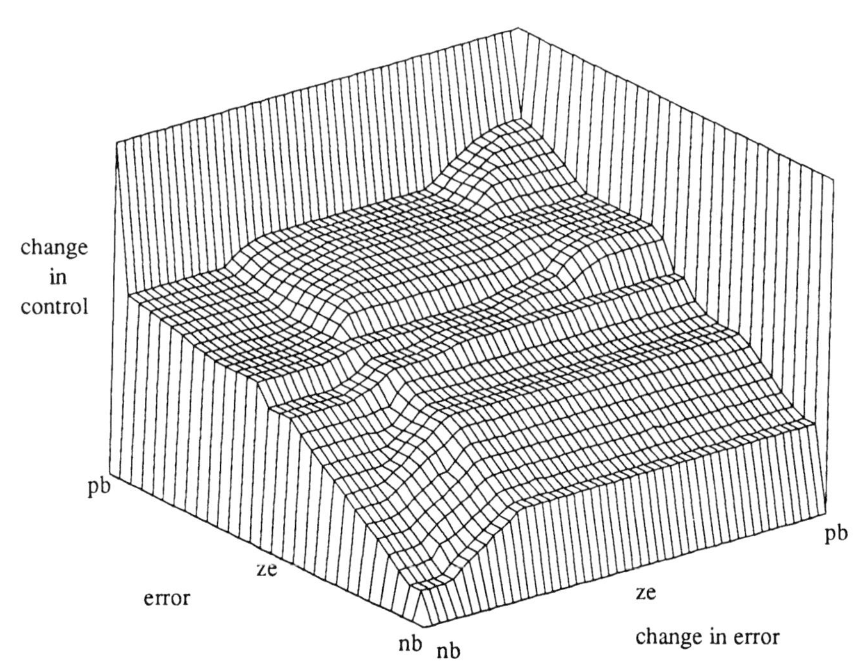

FIGURE 3.7. 3-D representation of the relation between $e(k)$, $\Delta e(k)$, and $\Delta u(k)$.

min composition inference method. Finally, control values for the process can easily be determined from this a priori generated look-up table for real-time operation.

2.1.2. *Literature Survey of Fuzzy Control Applications.*

Mamdani [14] and Mamdani and Assilian [15] were first to report on a successful application of a fuzzy controller to a process. The goal of their study was to regulate the state of a steam engine by adjusting the heat applied to the boiler and engine throttle setting. The boiler steam pressure and engine speed determined the state of the system. Kickert and Lemke successfully regulated the temperature of water leaving a warm-water plant using fuzzy control methods [16, 17]. A notoriously challenging industrial process to regulate is that of an operating cement kiln. A number of investigations applying fuzzy control to this regulation problem have been reported [18–20]. Vishnoi and Gingrich Vishnoi recently presented results on the regulation of anesthesia delivery by using a fuzzy controller [21]. The controller adjusted the inhalation rate of halothane gas based on the patient's heart rate and mean arterial pressure. Most of these successful applications claim improvements or comparable performance in comparison to the regulation provided by a tuned PID controller. Gaines and Kohout [22] present an extensive listing of other applications and developments in the fuzzy control field.

2.1.3. *Use of Fuzzy Control within Hierarchical Control Framework.*

Capturing the heuristics used by the human operator to regulate a process is the function of a fuzzy controller. Therefore, the performance provided by this approach at best will match that of the expert. The fuzzy controller method successfully moves the state of the process toward the set point, but once close, a steady-state oscillation frequently occurs about the set point arising from the nonlinear nature of the algorithm. In the hierarchical control structure, this approach operates during three possible process conditions. This method is responsible for determining the control input(s) when nothing is known about the process. In addition, the hierarchical controller uses the fuzzy controller to make gross control adjustments to drive the plant's output(s) to the region of the set point. And finally, the fuzzy controller is called upon by the expert system to determine the control to the process in the event of an instability in the model-based approaches.

2.2. Multiple Model Adaptive Control

The next level in this hierarchical control structure uses a dramatically different control scheme from that of the coarse fuzzy control approach. In contrast to modeling the operator, this method classifies the plant as one model out of a bank of predefined models and implements a linear controller designed about that nominal model. Lainiotis [23, 24] proposed this adaptive control approach, known as multiple model adaptive control. This coarse identification scheme differs from those that require the identification of all the plant parameters to determine the control value [1]. With this method, the number of possible classifications is finite, and the designer selects this quantization.

2.2.1. Basic Elements.

Figure 3.8 shows a block diagram of the multiple model adaptive control method. A bank of models (M_1, M_2, . . . , M_n) with constant parameters are predefined so as to span the range of parameter variation expected by the plant. Calculated residuals, defined as shown in Equation (9) represent the difference between each model in the bank and the plant in response to the same applied control value.

$$R_j(k) = \frac{[y_{mj}(k) - y(k)]}{y(0) - \mathrm{SP}} \qquad (j = 1, \ldots, n) \tag{9}$$

where $y_{mj}(k)$ is the output of model j; $y(k)$ is the plant output; n is the number of models in the model bank; and SP is the set point.

Each model in the model bank has a corresponding controller (G_{ci}) in the controller bank designed so that each controller-model pair meets specified design criteria. For instance, the design could require the nominal model and corresponding controller to have a specified rise time and amount of overshoot.

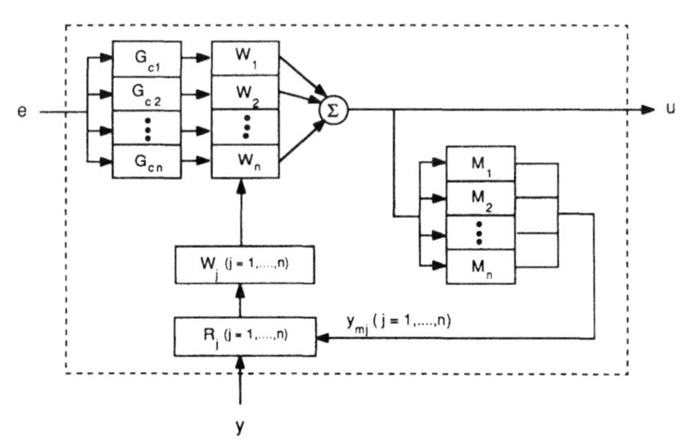

FIGURE 3.8. The transformation from *e* to *u* in terms of the basic components of a multiple model adaptive controller.

Acceptable deviations from these nominal values would then dictate the coarseness of the model bank.

Any linear control design method can be used to formulate the multiple model controller bank (i.e., PID, LQR). A weighted combination of controller bank outputs compose the actual control value applied to the plant. These weights, constructed from the residuals represented in Equation (9), are found by the following procedure [25]. Equation (10) represents the recursive update for each weight

$$W_j'(k) = \frac{\exp\left[\frac{-R_j^2(k)}{2V^2}\right] W_j(k-1)}{\Sigma_{i=1}^{n} \exp\left[\frac{-R_i^2(k)}{2V^2}\right] W_i(k-1)} \tag{10}$$

where V is a chosen constant value. These weights are bounded away from zero by δ as shown in Equation (11).

$$W_j(k) = \begin{cases} W_j'(k), & W_j'(k) \geq \delta \\ \delta, & W_j'(k) \leq \delta \end{cases} \tag{11}$$

Finally, the weights are normalized as shown in Equation (12).

$$W_j(k) = \frac{[W_j^2(k)]}{\Sigma_{i=1}^{n}[W_i^2(k)]} \qquad (j = 1, \ldots, n) \tag{12}$$

The resulting control value, $u(k)$, is given in Equation (13) where $u_j(k)$ and $W_j(k)$ represent the individual controller value and its corresponding weight factor respectively.

$$u(k) = \sum_{j=1}^{n} W_j(k)u_j(k) \tag{13}$$

For the case when the process lies on the border between two model spaces, the residuals represented by Equation (9) must match. Satisfaction of this requirement avoids selection of the incorrect controller for the process. Proper quantization of model parameters assures meeting this constraint.

In the ideal case, the plant would match a particular model in the model bank and the resulting closed-loop system would meet precisely the predetermined design specifications. In actuality, there is a range of responses for a plant within one parameter space. The coarseness of the model bank dictates the ultimate performance of the controller.

2.2.2. *Time Delay Compensation.*

Methods of compensating for time delays through linear control techniques apply here since this control approach includes a bank of linear compensators. In general, processes containing time delays require a lower loop gain in order to maintain stability, and thus have degraded performance. Time delay compensation methods attempt to allow a higher loop gain while maintaining closed-loop stability despite the delays. One method is the Smith predictor scheme [26]. Bahill [27] presents a nice review of this approach in conjunction with adaptive methods. Figure 3.9 shows the basic elements of the Smith predictor approach. G_c represents the linear compensator (as in the multiple model controller bank), G_m represents the model of the process without the process time delay(s), and T_m represents the time delay of the model. The requirement of knowing the process model exactly is the major constraint of this method. If this is unknown at the start of regulation or time-varying, this compensation method in combination with the multiple model adaptive control algorithm could yield promising results.

Another time delay compensation technique involves calculating the control value x steps in the future. This predictive horizon approach [28], avoids the requirement of the Smith predictor to know the process model exactly. Within the multiple model framework, each controller G_c would determine the control x steps in the future, based on its corresponding model for the process.

2.2.3. *Literature Survey on Multiple Model Applications.*

The first reported application of this method to regulation of a medical process was by He and coworkers [25]. The goal was to maintain mean arterial pressure by regulating the infusion rate of sodium nitroprusside. Successful animal experiments and simulation results verified the approach. Martin and coworkers [29] extended this method addressing the same process. They augmented the controller structure in this study with a Smith predictor to compensate for the time delays in the system. In another medical application, Yu and coworkers [30] applied the multiple model controller to maintaining a desired respiratory status in mechanically ventilated animals. This controller adjusted the inspired oxygen fraction to attain desired oxygen saturation. This study demonstrated the improvement using this method over a well-tuned PI controller.

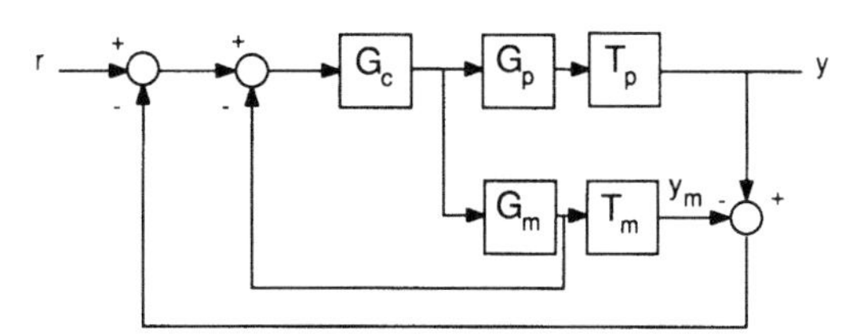

FIGURE 3.9. Block diagram of a Smith predictor.

Athans and coworkers [31] used this approach to regulate the longitudinal and lateral dynamics of the F-8C aircraft during an equilibrium flight condition. The compensator determined elevator rate for the longitudinal dynamics, and aileron and rudder rate for the lateral dynamics. They designed the controller bank entries based on the linear-quadratic-Gaussian (LQG) method to cover each flight condition. Each LQG controller in the controller bank utilized a Kalman filter to reconstruct state variables. The difference between the actual measured states and the predicted measured states calculated by each Kalman filter formed the residual vector in Equation (9) in this design.

Leahy and Tellman [32] applied the multiple model approach to a robotic manipulator control problem. Their goal was to maintain trajectory tracking accuracy in uncertain payload environments. Simulation studies verified the potential of this approach.

2.2.4. *Use of Multiple Model Adaptive Control within the Hierarchical Control Framework.*

The multiple model adaptive controller selects a controller of fixed structure based on partial knowledge of the process. Once the controller attains this knowledge (i.e., convergence of Equation (12)), and the state of the process is in the region of the set point, the hierarchical controller selects this algorithm to apply the control to the plant. One drawback of the method is the potential requirement of a large number of model-controller pairs in order to meet performance specifications. Maybeck and Hentz address this problem by incorporating a moving model bank [33]. The hierarchical structure deals with this issue by using the multiple model controller and the model reference adaptive controller (described in the next section) in tandem. The coarsely quantized multiple model controller acting alone provides a coarse model-based control action to the process. The model reference adaptive controller increases the precision of the control action applied to the process. In exchange for improving the precision, the model reference adaptive controller requires initialization and satisfaction of sufficiency conditions to assure closed-loop stability. The multiple model adaptive controller provides these functions.

2.3. Model Reference Adaptive Control

A direct model reference adaptive control algorithm performs the most precise control calculation for the hybrid controller. Many model reference adaptive control approaches exist [1]. This method, originally proposed by Sobel and coworkers [34], is based on command generator tracker theory [35]. The approach directly calculates the control value without the use of adaptive observers or the need for full state feedback. Though one of the attributes of this method is the applicability to multiple-input multiple-output processes, this discussion will focus on the single-input single-output case in order to establish the basic foundations of the method. In addition, the following presents the continuous time version of the control law leaving the designer to make the necessary adjustments for implementation [36].

2.3.1. Basic Elements.

This model reference adaptive controller evolved from the foundations of command generator tracker theory. The goal of both these methods is to determine the control input for a process given by Equations (14) and (15).

$$\dot{x}_p(t) = A_p x_p(t) + B_p u(t) \tag{14}$$

$$y(t) = C_p x_p(t) \tag{15}$$

so that the output of the process asymptotically follows the output of the chosen reference model given by Equations (16) and (17).

$$\dot{x}_m(t) = A_m x_m(t) + B_m u_m(t) \tag{16}$$

$$y_m(t) = C_m x_m(t) \tag{17}$$

Under the assumptions of command generator tracker theory, which includes knowledge of A_p, B_p, and C_p, an ideal plant trajectory $x_p^*(t)$ exists such that

$$C_p x_p^*(t) = C_m x_m(t) \tag{18}$$

$$\dot{x}_p^*(t) = A_p x_p^*(t) + B_p u^*(t). \tag{19}$$

The control law to achieve perfect model following for this nonadaptive case ($u^*(t)$), is a linear combination of reference model states and inputs as shown in Equation (20).

$$u^*(t) = \mathbf{K}_x^* x_m(t) + \mathbf{K}_u^* u_m(t) \tag{20}$$

The gain matrices $\mathbf{K}_x^*$ and $\mathbf{K}_u^*$ are functions of the known matrices $\mathbf{A}_m$, $\mathbf{B}_m$, $\mathbf{C}_m$, $\mathbf{A}_p$, $\mathbf{B}_p$, and $\mathbf{C}_p$. The model reference adaptive controller discussed here extends this framework to the case where $\mathbf{A}_p$, $\mathbf{B}_p$, and $\mathbf{C}_p$ are not explicitly known [37].

Figure 3.10 shows a block diagram of the basic components of this model reference adaptive controller. Note this figure presents the model reference adaptive controller in the same format from e to u as the other controllers in the hybrid structure. However, in this case e in Figure 3.10 represents the difference between the output of the reference model (y_m) and the plant output (y). The plant control input $u(t)$ is a combination of the output error $y_m(t) - y(t)$, reference model states $x_m(t)$, and the reference model input command $u_m(t)$,

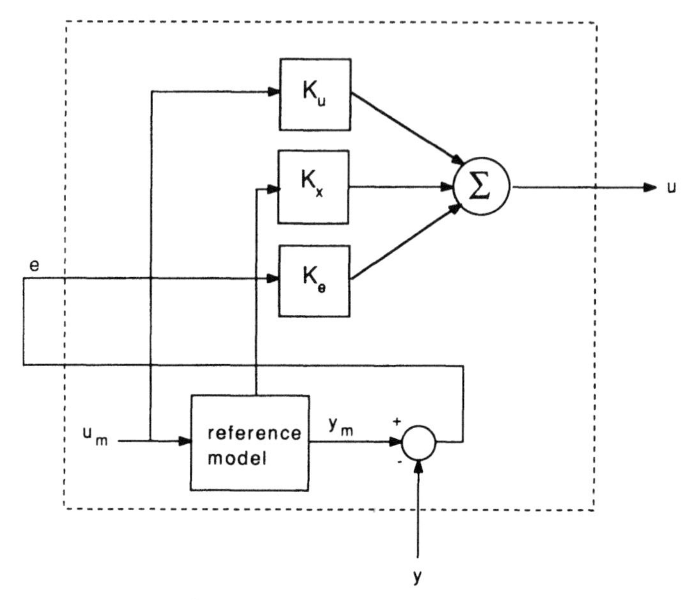

FIGURE 3.10. The transformation from *e* to *u* in terms of the basic components of the model reference controller.

$$u(t) = K_e(t)[y_m(t) - y(t)] + K_x(t)x_m(t) + K_u(t)u_m(t). \tag{21}$$

Simplifying, let $\mathbf{K}_r(t)$ represent the matrix of gains,

$$\mathbf{K}_r(t) = [K_e(t), K_x(t), K_u(t)], \tag{22}$$

and let $\mathbf{r}(t)$ represent the residual vector,

$$\mathbf{r}(t) = [y_m(t) - y(t), x_m(t), u_m(t)]^T, \tag{23}$$

resulting in $u(t)$ as shown in Equation (24).

$$u(t) = \mathbf{K}_r(t)r(t) \tag{24}$$

The adaptive gain matrix $\mathbf{K}_r(t)$ is a combination of proportional gain $K_p(t)$, and integral gain $K_i(t)$,

$$\mathbf{K}_r(t) = K_p(t) + K_i(t). \tag{25}$$

The proportional and integral gains that form $\mathbf{K}_r(t)$ are proportional to the output error or quadratic output error as shown in Equations (26) and (27).

$$K_p(t) = [y_m(t) - y(t)]r^T(t)\bar{\mathbf{T}} \tag{26}$$

$$\dot{K}_i(t) = [y_m(t) - y(t)]r^T(t)\mathbf{T} \tag{27}$$

$\mathbf{T}$ and $\bar{\mathbf{T}}$ are gain-weighting matrices, chosen by the designer, to yield desired transient behavior for the closed-loop system.

As with many adaptive algorithms, the characterization of the process as strictly positive real[1] (SPR) is the issue regulating the bounds on allowable plants. For the algorithm presented above, the conditions for stability require $\mathbf{T}$ to be positive definite and $\bar{\mathbf{T}}$ positive semidefinite with the plant required to be feedback positive real. This means for the plant represented by Equations (14) and (15) there exists a matrix $\mathbf{K}_e^*$ (not necessary for implementation) such that the new stabilized plant described by Equations (28) and (29) is SPR.

$$\dot{x}_p(t) = (\mathbf{A}_p - \mathbf{B}_p\mathbf{K}_e^*\mathbf{C}_p)x_p(t) \tag{28}$$

$$y(t) = \mathbf{C}_p x_p(t) \tag{29}$$

Under these conditions, all states and gains are bounded and the $y_m(t) - y(t)$ decreases asymptotically to zero.

Bar-Kana [38] extended the class to include those plants that can be forced to become feedback positive real by augmenting the original plant ($G_p(s)$) with a feedforward filter ($D(s)$) as represented in Equation (30).

$$G_a(s) = G_p(s) + D(s) \tag{30}$$

The output of this augmented system, $y_a(t)$, is given in Equation (31).

$$y_a(t) = y(t) + f(u(t)) \tag{31}$$

In this case, the feedforward element is considered part of the controller. Unfortunately, this approach forces $[y_m(t) - y_a(t)] \rightarrow 0$ rather than the true error defined by $y_m(t) - y(t)$ resulting in a steady-state error. However, if the $D(s)$ filter is chosen sufficiently small, then the $y_a(t) \cong y(t)$ and the augmented plant error approximates the original plant error. Bar-Kana has addressed the problem of developing an adaptive method of choosing the feedforward structure to force the system to remain SPR in the face of a changing plant, but this is still an area of research due to a lack of stability proof [39].

In summary, the designer must select $\mathbf{T}$ and $\bar{\mathbf{T}}$ to achieve the desired performance of the regulated process. With regard to stability of the closed-loop system, the feedback positive real constraints must be satisfied.

---

[1] A process is said to be strictly positive real if all its poles are in the open left half-plane and its Nyquist plot only has positive real values.

2.3.2. *Time Delay Compensation.*

This section presents an adjustment to the control law given in Equation (27) in order to ensure closed-loop stability for processes containing time delays. In general, process time delays trouble adaptive controllers due to their inherent nonminimum phase nature [1, 40]. Figure 3.11 presents an illustration of the effect on the stability, dimensionality, and feedback positive realness characterization of a simple process ($G_1(s)$) by adding a time delay ($G_2(s)$). The addition of the time delay does not change the open-loop stability of the plant. However, the T_d does cause the new plant ($G_2(s)$) to become infinite-dimensional. The original plant $G_1(s)$, is strictly positive real as seen from its Nyquist plot shown in Figure 3.11. A plant that is SPR is also feedback positive real since K_e^* can be selected as zero, therefore satisfying the condition described in Equations (28) and (29). In contrast, $G_2(s)$ is not SPR, as demonstrated by its Nyquist plot in Figure 11, and no value of K_e^* exists such that $G_2(s)$ is feedback positive real. Thus, the guarantee of global asymptotic stability with the adaptive gains as in Equation (27) no longer exists for $G_2(s)$. To ensure closed-loop stability for plants of this form, the control law must be adjusted.

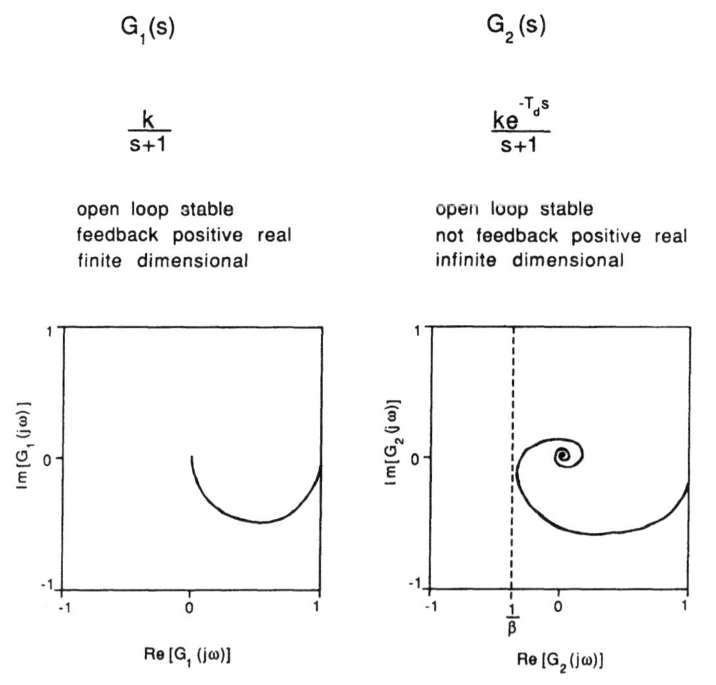

FIGURE 3.11. Comparison between two transfer functions ($G_1(s)$, $G_2(s)$) in terms of their corresponding Nyquist plots, demonstrating the effect of a time delay.

Using the hyperstability theorem extended to infinite dimensional space [41], global asymptotic stability was proven for processes of the form $G_2(s)$ resulting in the modification of the control law given in Equation (27) to the one given in Equation (32). Details on the proof are given in [42].

$$\dot{K}_i(t) = (y_m(t) - y(t) + \gamma(u^*(t) - u(t)))r^T(t)\mathbf{T} \tag{32}$$

Compare this control law with the one described by Equation (27). Note that this adaptive control law requires selecting γ whose value is bounded by $1/\beta$, which is defined in Equation (33) and graphically represented in the Nyquist plot for $G_2(s)$.

$$\frac{1}{\beta} = \min[\mathrm{Re}[G(j\omega)]] \tag{33}$$

The restriction on the value of γ is given in Equation (34).

$$\gamma \geq \mathrm{abs}\left(\frac{1}{\beta}\right) \tag{34}$$

Also note that the control law defined in Equation (32) reverts to the one described by Equation (27) in the case when $1/\beta \rightarrow 0$. This is analogous to letting $T_d \rightarrow 0$ in $G_2(s)$ resulting in the feedback positive real process $G_1(s)$. Unfortunately, the adjusted control law presented in Equation (32) requires knowledge of the ideal control ($u^*(t)$) given in Equation (20). The hierarchical control structure alleviates this problem by combining the model-based control structures.

2.3.3. Literature Survey on Model Reference Applications.

Kaufman and coworkers [43] applied this algorithm to the problem of regulating mean arterial blood pressure. The controller determined the proper infusion rate of sodium nitroprusside in order to maintain a desired set point. To compensate for the inherent process time delays, the reference model included a time delay equaling the nominal value of the process time delay. Though this augmentation lacked a formal stability proof, it provided promising experimental results.

A research group at the Jet Propulsion Laboratory has extensively applied this method to the regulation of large flexible space structures. Ih and coworkers [44] used this control method to regulate a space station under severe disturbance conditions such as shuttle docking. They augmented the controller with a fixed low-gain inner loop to compensate for unstable rigid-body dynamics. Simulation results using a six-degree-of-freedom finite element model verified the performance of the controller. A further advancement by Bayard and coworkers [45] augmented the basic algorithm with branch filters used to suppress measurement noise. The addition of these filters did not affect the global stability of the algorithm. Subsequently, Ih and coworkers [46] reported promising experimental results demonstrating the success of the model reference adaptive controller with the noted augmentations when applied to the regulation of a large space antenna-like structure.

2.3.4. *Use of Model Reference Adaptive Control within the Hierarchical Control Framework.*

Given knowledge of the adaptive gains $K_i(t)$, proper selection of $\mathbf{T}$, $\bar{\mathbf{T}}$, and either knowledge of γ (Equation (34)) or a guarantee that the process satisfies strictly positive real considerations, the model reference adaptive controller will provide a stable, precise control action forcing $[y_m(t) - y(t)] \rightarrow 0$. As with any adaptive algorithm, this method requires a period of time to identify the control law gains ($K_i(t)$). In the hierarchical control structure, the multiple model adaptive controller performs this identification. For each model in its model bank, a model reference initialization vector is stored. This vector includes the initial model reference adaptive controller gains $K_i(0)$, the performance matrices $\mathbf{T}$ and $\bar{\mathbf{T}}$, and the γ value. Once the multiple model adaptive controller identifies the process as one of the models in its bank, and the process state is close to the set point, the model reference adaptive controller will begin performing the precise regulation.

The model reference control corrects for any variation that may exist between the actual process response and the selected nominal model from the multiple model bank. Figure 3.12 presents a block diagram of the controller structure when the model reference controller determines the control for the process. The chosen multiple model compensator in combination with the process form an inner loop which becomes the "augmented plant" for the model reference controller. Its precise control action adjusts the former reference signal (set point command) of the multiple model compensating scheme. It is this inner loop configuration that allows for the complete implementation of the model reference control algorithm as given in Equation (32). The ideal control for this augmented plant is the set point value as given in Equation (35).

$$u^* = \text{set point} \tag{35}$$

It is important to note that the multiple model controller must provide the necessary compensation so that the inner loop configuration appears as a type 1 system[2] to the model reference controller. Under these conditions the model reference controller will operate successfully.

In addition to providing knowledge of u^*, this control architecture presented

[2] A type 1 system is a feedback system that can track a step input with zero error in steady state.

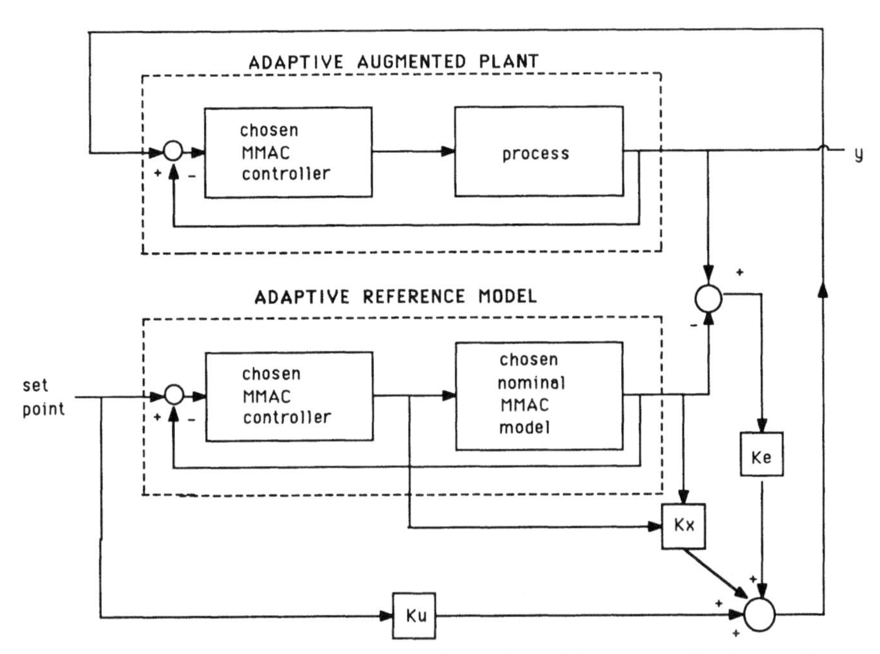

FIGURE 3.12. The most precise configuration of the hierarchical controller.

in Figure 12 accommodates an adaptive reference model selection process. The dynamics of the reference model are a function of the multiple model classification of the process. The system is designed so that a process with slow dynamics will not be forced to follow an extremely fast model, and a process with fast dynamics will not be forced to follow an excessively slow model. Once the multiple model controller converges on a model-controller pair, this nominal closed-loop system becomes the reference model for the augmented plant to follow. Once the expert system warrants the advancement of the system to the model reference stage, this controller will stay in command until the model reference gains inflate excessively or the plant output strays too far from the set point value. In this case, the system will return to the multiple model controller—or perhaps the fuzzy controller—for control actions and then repeat the procedure by determining a new augmented inner loop configuration more suitable for the plant state. Once the model reference begins its fine-tuning adjustment of the augmented plant, the inner loop no longer changes in accordance with changes in the process. However, the multiple model controller continues to classify the process by calculating the weights given in Equation (12).

This hybrid control structure enables these two model-based control schemes to improve upon the abilities of each method used individually. The multiple model procedure provides two functions for the model reference control algorithm. If the process alone does not satisfy the feedback positive real requirements represented in Equation (28), the multiple model controller can be used to compensate the process so that these conditions are satisfied. In addition, the multiple model controller eliminates the need for the model reference controller to identify controller parameters. The precise adjustment of the model reference controller allows a decrease in the multiple model quantization, thus assuring a realizable control structure in order to meet performance specifications.

2.4. Expert System

The expert system orchestrates the operation of the hybrid controller in accordance with the dynamic process. This element allows for a systematic approach in dealing with heuristic decisions involved in determination of the proper controller. Many approaches to representing knowledge in the expert system format exist [47–49]. The approach used here is that of a production system or rule-based expert system.

2.4.1. Basic Elements.

Figure 3.13 shows the basic components of a rule-based expert system. These include a database, a rule base, and an inference engine. The goal of the expert system is to perform the transformation between a real-time measurement vector ($\mathbf{Z}(k)$) quantifying system status at $t = k$, and the controller best suited to regulate the process. The following discusses this transformation in terms of the constituent parts of the expert system.

The database contains both static and dynamic elements describing the state of the system. The measurement vector ($\mathbf{Z}(k)$) acquired from the process and controllers, represents the dynamic elements updated at each sample time. Table 3.1 lists the vector elements. Dynamic elements also include history of the process behavior as described by previous values of $\mathbf{Z}(k)$ stored in the database.

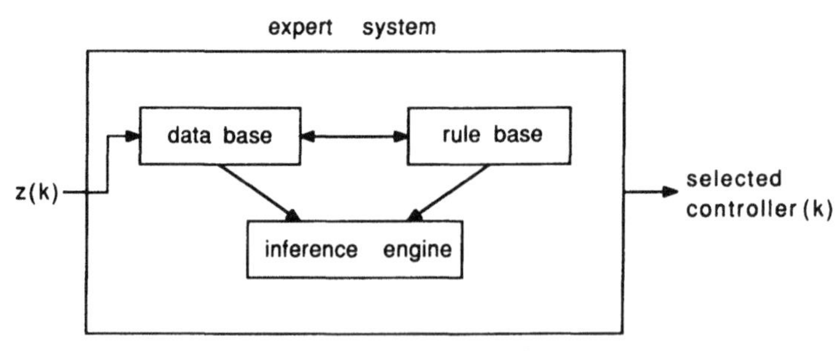

FIGURE 3.13. Basic components of an expert system.

TABLE 3.1. Elements of $Z(k)$.

Parameter	Description
$u(k)$	general
$e(k)$	general
$\Delta e(k)$	general
$W_i(k)$, $i = 1, \ldots, n$	MMAC
$K_r(k)$	MRAC
$K_p(k)$	MRAC
$K_i(k)$	MRAC
γ	MRAC
$y_m(k)$	MRAC
present controller	general
set point	general
$y(k)$	general
max rule fired	FUZZ

Static data include thresholds to determine convergence of control algorithm parameters or different levels of system performance. For example, the multiple model adaptive controller is defined as converged when one of the weights represented in Equation (12) surpasses some defined static threshold value α. Similarly, static conditions on $e(k)$ and $\Delta e(k)$ yield a means of defining operating regions for the different controllers in terms of a common performance measure. Ornedo [5] used this method in the design of a reconfigurable controller for a nuclear reactor. Figure 3.14 demonstrates this concept in terms of functional restrictions on $e(k)$ and $\Delta e(k)$.

The rule base contains rules in the ⟨*premise*⟩⟨*consequent*⟩ format, which deal with conditions involving changing control schemes. A premise may contain multiple clauses joined by the conjunction operator. For example,

premise: IF: $e(k)$ is less than β_1 and $\Delta e(k)$ is less than β_2 and present controller is FUZZ and $W_{max}(k)$ is greater than α
consequent: THEN: switch from FUZZ to MMAC

The rule base structure allows for including rules specific to the regulated process. For instance, the rule base could contain diagnostic rules which would require the process to exhibit a detectable behavior before the controller would initiate control law calculations.

A backward chainer performs the inference function for this production system. Winston [49] presents a detailed description on the fundamentals of this resolution method. This goal-driven approach assumes a hypothesis and attempts to prove it using the database and the rule base. In this context, the three hypotheses for this application are:

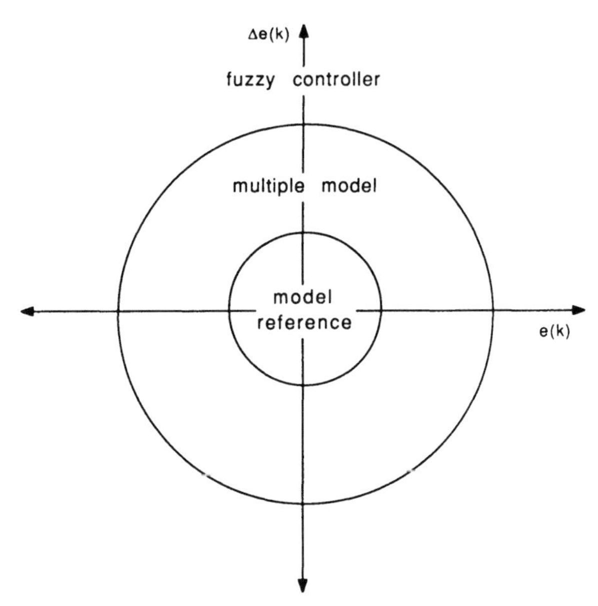

FIGURE 3.14. Graphical representation of the different operating regions of the elemental controllers based on $e(k)$ and $\Delta e(k)$.

- The controller is the fuzzy controller.
- The controller is the multiple model adaptive controller.
- The controller is the model reference adaptive controller.

The backward-chaining procedure is analogous to generating an and/or tree as shown in Figure 3.15. The operation originates at the root of the tree (H), and recursively unfolds until one of the three hypotheses is proven true. These hypotheses are assumed to be mutually exclusive and exhaustive. The function of the backward chainer is to check the validity of the hypotheses and subgoals (individual clauses from each rule) in the recursive process. Connected to the root (H) through an OR node is the subset of rules ($R(H)$) from the rule base, each of which contains the hypothesis under search as its consequent. Once any rule in $R(H)$ is found to be true, the procedure is completed. In order to determine the firing of a rule, the clauses in the rule's premise must be checked. Figure 3.15 illustrates this concept for r_1 in $R(H)$, which is connected to the clauses in its premise through an AND node. All of these clauses ($c_{11}, c_{12}, \ldots, c_{1m}$) must be true in order for r_1 to fire, and thus prove the hypothesis true. At this level in the and/or tree, the procedure repeats. Each clause or subgoal of r_1 becomes a new root. For instance, the backward-chaining procedure searches the rule base to find the subset of rules ($R(c_{11})$), that have c_{11} as a consequent. If any of the rules

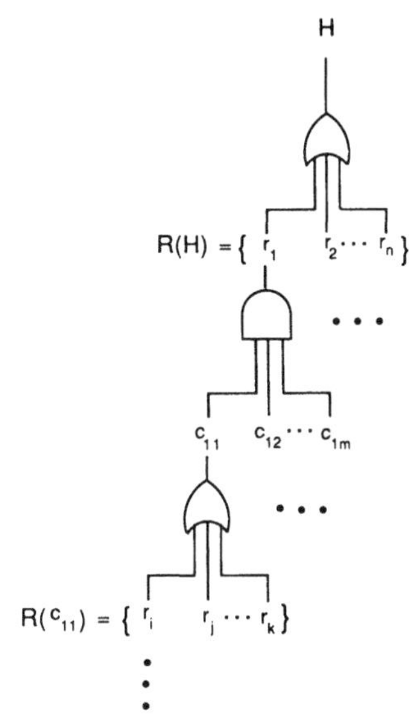

FIGURE 3.15. An and/or tree representing the backward-chaining inference method.

in $R(c_{11})$ fire, indicating c_{11} is true, the search is stopped along this branch of the tree. Figure 3.15 represents this concept with an OR node connecting $R(c_{11})$ to c_{11}. To determine whether any of the rules in $R(c_{11})$ fire, the clauses in the premise of each rule must be checked and the procedure again repeats.

Once a hypothesis is proven true, the procedure halts as the expert system has completed its task. The brute force alternative of evaluating all the rules in the rule base to make an inference about the hypotheses convincingly demonstrates the benefits of the backward-chaining approach.

2.5. Operation of the Hierarchical Controller

The complete hybrid control structure is shown in Figure 3.16. The system builds on itself in a learning type manner. At the outset of controller operation, when the system knows nothing about the process, the fuzzy controller provides the initial crude control by imitating the actions of the operator (switch in position 1). This initial control algorithm allows the multiple model adaptive controller to determine which model in its model bank most closely matches the processes' dynamic response. Control switches to the multiple model controller when the expert system determines that conditions warrant the change. At this point, the compensator designed a priori for the chosen model dictates the control

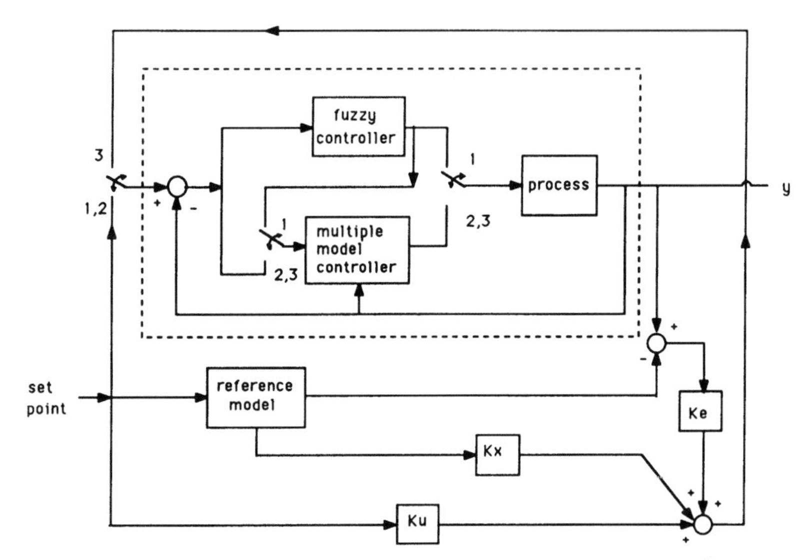

FIGURE 3.16. Block diagram of the overall expert controller.

value (switch in position 2). The multiple model controller and plant at this stage become the inner loop (outlined) for the model reference adaptive controller in order to satisfy sufficiency conditions for its application. The final switch, as determined by the expert system, allows the model reference controller to adjust in a fine-tuning manner the reference signal to the inner loop in order that the process follows the desired trajectory given by the reference model (switch in position 3).

The effect on the process of switching between different control schemes is an important operational consideration. Most likely, the control signal determined by the rule-based method will differ from that of the model-based methods. Therefore, consideration must be given in making the transition from one control scheme to another a "bumpless transfer." Figure 3.17 shows how the hierarchical control scheme addresses this issue. Both the fuzzy controller and the multiple model controller calculate change in control value at each sample time. For the fuzzy controller, this requires obtaining rules from the operator in terms of change in control to the process. For the multiple model adaptive controller, an incremental or velocity form of the control law is computed from the position form as in Equation (36).

$$\Delta u(k) = u(k) - u(k - 1) \tag{36}$$

In contrast to the position form, where the algorithm calculates the total control value applied at each sample time, the velocity form computes the change in

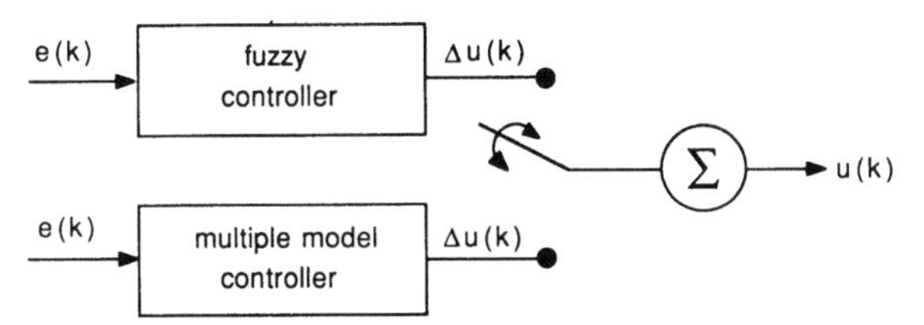

FIGURE 3.17. Block diagram illustrating how the hierarchical controller ensures bumpless transfer.

control signal. A common integrator transforms these incremental control calculations from either the model-based or rule-based controllers into the total control to apply to the process. The origin of the $\Delta u(k)$ in this strategy is transparent to this integrator and thus ensures a bumpless transfer.

Since the model reference adaptive controller adjusts the set point of the compensated inner loop, its adjustments will not cause abrupt changes to the control value applied to the process. Thus, bumpless transfer is not an issue at this stage of the hybrid controller.

3. EXPERT ADAPTIVE CONTROLLER APPLICATION—DRUG DELIVERY SYSTEM

3.1. Overview and Problem Statement

The goal of this application of the hierarchical control structure is the design of an automated drug delivery system for the treatment of critically ill patients suffering from cardiac failure. The primary motivation for this work is to reduce the workload of the attending personnel. Many factors contribute to the fact that the regulatory challenge faced by such a controller is a difficult one. These include unmodeled dynamics of the process, time-varying nature of plant parameters, nonlinearities and cross-interaction of drug responses, narrow bounds of stable existence, and the inherent time delays of the patient's response to the drug infusion. The evolution of control schemes for tackling such problems is demonstrated by a nonadaptive blood pressure controller [50], an adaptive blood pressure controller [25], an adaptive blood pressure and cardiac output controller [51], and an adaptive arterial pressure and central venous pressure controller [52]. However, these approaches have been plagued by problems, which include the question of stability of adaptive controllers applied to processes with unmodeled dynamics [53], deterioration of controller performance in the presence of poor parameter estimates, the need for an initial parameter estimation period, and failure to apply the controller to the targeted sick heart condition, where such a scheme would serve its purpose.

Application of the hierarchical controller to this regulation problem represents the next stage in the evolution of automated drug delivery systems. Stated simply, the controller initially imitates the coarse actions of the physician (fuzzy controller) when little is known about the patient's pharmacokinetics. Subsequent switching to the more finely tuned model-based approaches occurs when the expert system warrants the change. As a means of illustration, the performance of the expert adaptive controller was tested in simulations by comparing its regulatory abilities to the individual controllers that make up its structure in the regulation of blood pressure. The example presented here is an elaboration of that discussed in [54]. The simplified model for the process used in this design is given in Equation (37) and represents the transfer function between the change in mean arterial pressure in response to sodium nitroprusside infusion [11].

$$G(s) = \frac{Y(s)}{U(s)} = \frac{k_p \exp(-T_d)}{\tau_p s + 1} \tag{37}$$

$$P(s) = Y(s) + P_o \tag{38}$$

where $Y(s)$ is the change in blood pressure from its initial value P_o; $U(s)$ is the drug infusion rate; and $P(s)$ is the actual blood pressure.

The parameters in Equation (37) were found to be [11]: $k_p = -.25$ to -9.0 (nominal $= -1$), $\tau_p = 30$ to 60 (nominal $= 40$), $T_d = 20$ to 60 (nominal $= 40$).

3.2. Hierarchical Controller Design

The following delineates the design of a hierarchical control scheme to regulate blood pressure. The method follows the concepts presented in the first part of the chapter.

3.2.1. Fuzzy Controller.

Initially, eight fuzzy numbers were defined as shown in Figure 3.18. The universe of discourse for the three controller variables $e(k)$, $\Delta e(k)$, and $\Delta u(k)$ are shown in Equations (39), (40), and (41), respectively.

$$U_e = [-30, 30] \tag{39}$$

$$U_{\Delta e} = [-50, 50] \tag{40}$$

$$U_{\Delta u} = [-30, 30] \tag{41}$$

To simplify implementation, all three variables were normalized to $[-1, 1]$. The controller rules were generated by asking physicians what infusion rates of sodium nitroprusside they would use in treating the spectrum of conditions

nb = negative big
nm = negative medium
ns = negative small
nz = negative zero
pz = positive zero
ps = positive small
pm = positive medium
pb = positive big

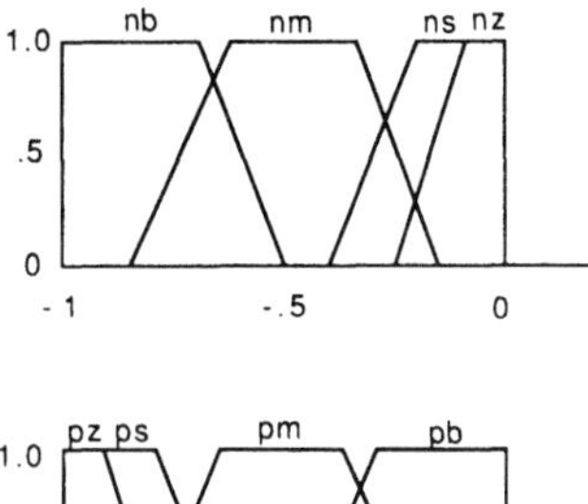

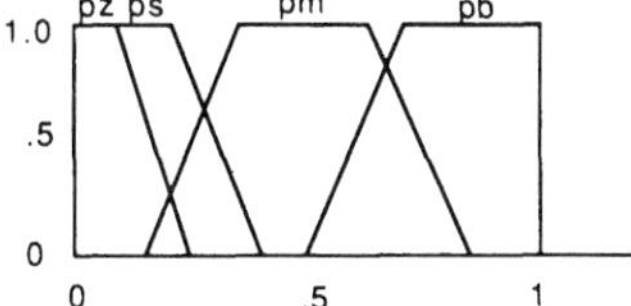

FIGURE 3.18. Defined fuzzy numbers for the blood pressure regulation problem.

spanning the entire universe of discourse of $e(k)$ and $\Delta e(k)$ in terms of the defined fuzzy numbers. Figure 3.7 represents graphically the resulting controller transformation from its inputs ($e(k)$ and $\Delta e(k)$) to the output ($\Delta u(k)$) in the look-up table format as described in the text. With the intent of modeling accurately the physician's actions, the sample rate of the controller was set at 1.5 minutes, the approximate rate given by those interviewed.

3.2.2. Multiple Model Adaptive Controller.

Previous simulation studies [25] revealed that a successful multiple model controller design resulted by quantizing the model bank according to patient gain while selecting nominal and maximum values for the process $\tau_p = 45.0$ and $T_d = 60.0$, respectively. Proportional integral (PI) controllers ($G_c(s)$) of the form given in Equation (42) composed the controller bank.

$$G_c(s) = k_c\left(1 + \frac{1}{\tau_c S}\right) \tag{42}$$

Four controller-model pairs were designed to span the possible process gain range. Controller gains (k_c) were selected so that each controller with its corresponding nominal plant within its subspace met specific rise time and settling time specifications as in [25].

Models in the model bank ($G_m(s)$) took the form of Equation (43).

$$G_m(s) = \frac{k_m \exp(-T_{dm}s)}{1 + \tau_m(s)} \tag{43}$$

The model gains (k_m) were determined by guaranteeing that the residuals represented in Equation (9) met at the boundaries between two parameter spaces as discussed in the text. Table 3.2 lists the resulting controller and model parameters used in the simulation. For both model-based controllers, the sample rate of the controller was set at 10 s.

3.2.3. Model Reference Adaptive Controller.

The closed-loop combination of each controller with its corresponding nominal model was designed to meet the same performance specifications. Therefore, independent of the selected model space, each augmented plant (formed by the closed-loop combination of the process and the selected multiple model controller) followed the same reference model given by Equation (44).

$$\frac{y_m(s)}{\text{pchange}} = \frac{1}{\tau_{\text{ref}}s + 1} \tag{44}$$

where pchange represents the desired change in pressure from the initial value; and $\tau_{\text{ref}} = 200.0$. This reference model represents the desired dynamics defining the change in blood pressure for the augmented process, which includes the original process and the multiple model compensator. When the model reference controller is chosen by the expert system as the control method, the reference model is initialized to the value of the plant at that time. The control law took the form of Equation (45).

$$u(t) = K_e(t)\,(y_m(t) - y(t)) + K_x(t)y_m(t) + K_u(t)\text{pchange} \tag{45}$$

TABLE 3.2. MMAC Controller and Model Parameters.

Number	k_c	τ_c	k_m	τ_m	T_{dm}	Plant Gain
1	0.696	45.0	−0.43	50.0	60.0	−0.25 to −0.61
2	0.284	45.0	−0.795	50.0	60.0	−0.61 to −1.49
3	0.116	45.0	−2.2	50.0	60.0	−1.49 to −3.63
4	0.047	45.0	−5.06	50.0	60.0	−3.63 to −9.0

The initial values for the gains were selected so that the input to the inner loop was the set point or, $u(t)$ = pchange, i.e., $K_e(0) = 0$, $K_x(0) = 0$, and $K_u = 1.0$. The $1/\beta$ values (as defined in Equation (33) and Figure 3.11), for each of the model spaces are given in Table 3.3. To assure closed-loop stability over all model spaces, γ was selected as the maximum value of 0.4. The weighting matrices were selected as $\mathbf{T} = .0001$ and $\bar{\mathbf{T}} = 0.0$ yielding the desired performance from the controller. Finally, the ideal control law (u^*) needed for calculation of control law gains from Equation (32) is the commanded change, pchange. This is possible since the multiple model adaptive controller provides the necessary compensation (PI control) to force the new augmented plant to be a type 1 system.

3.2.4. Expert System.

The expert system took the form of that outlined in the text. Rules involved conditions on the elements in the measurement vector ($\mathbf{Z}(k)$). The backward chainer used the rule base and database to resolve on each sample time which of the three control schemes was best suited to perform the regulation. The three hypotheses were mutually exclusive and exhaustive. Figure 3.15 represents the recursive backward-chaining algorithm used in the simulation.

3.3. Simulation Results

In this simulation, a comparison is made between the regulatory ability of the fuzzy controller, the fuzzy controller plus the multiple model controller, and the entire hierarchical control structure (fuzzy controller plus multiple model controller plus model reference controller). Equation (37) represents the model used in this simulation with $\tau_p = 60$ s, $T_d = 50$ s, and k_p a function of time, as shown in Figure 3.19. The variable gain, which models the patients's decrease in sensitivity to the drug, and the time delay represent two characteristics that make this a challenging control problem.

The initial blood pressure was $P_o = 170.0$ mmHg and the set point was 140 mm Hg; thus pchange = 30.0 mmHg. Figures 3.20 and 3.21 show the resulting

TABLE 3.3 $\frac{1}{\beta}$ Values for the Different Model Spaces.

Number	$\frac{1}{\beta}$
1	−0.40
2	−0.28
3	−0.33
4	−0.30

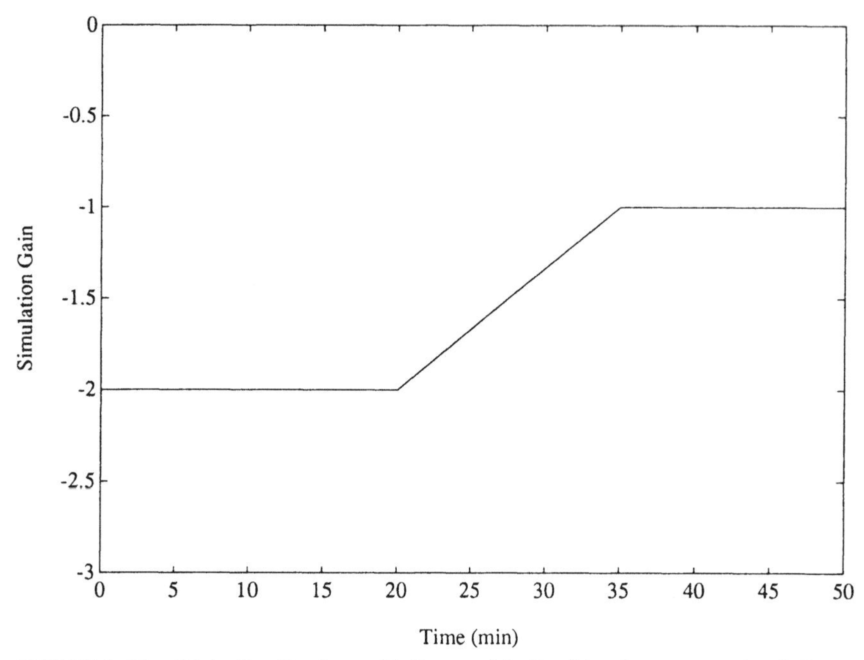

FIGURE 3.19. Plot of patient sensitivity used in the blood pressure simulation.

pressure responses and nipride infusion rates respectively for the three controller structures. The dot-dash line at the base of each figure signifies when the expert system chose to switch between different control schemes. The simulation demonstrates that the complete hybrid structure yields superior regulation ability. The fuzzy controller alone displays the difficulty that humans have in regulating systems with time delays and parameter variations. In addition, this simulation shows the coarseness of this control action since no control adjustment occurs if the pressure is in the region of the set point. However, in the hybrid control structure, the fuzzy controller action provides gross adjustments to attain the state of the system near the set point, while simultaneously giving a signal to the multiple model adaptive controller to search its model bank in order to find the closest linear compensator for the given plant.

With the multiple model controller in charge, the system attains the following benefits: the control is more precise than the fuzzy controller, it provides improved compensation in dealing with the plant time delay, and helps to satisfy the sufficiency conditions required by the model reference controller. In contrast to the coarse fuzzy controller, the more precise multiple model controller makes adjustments in the control value whenever the process is not at the set point. The

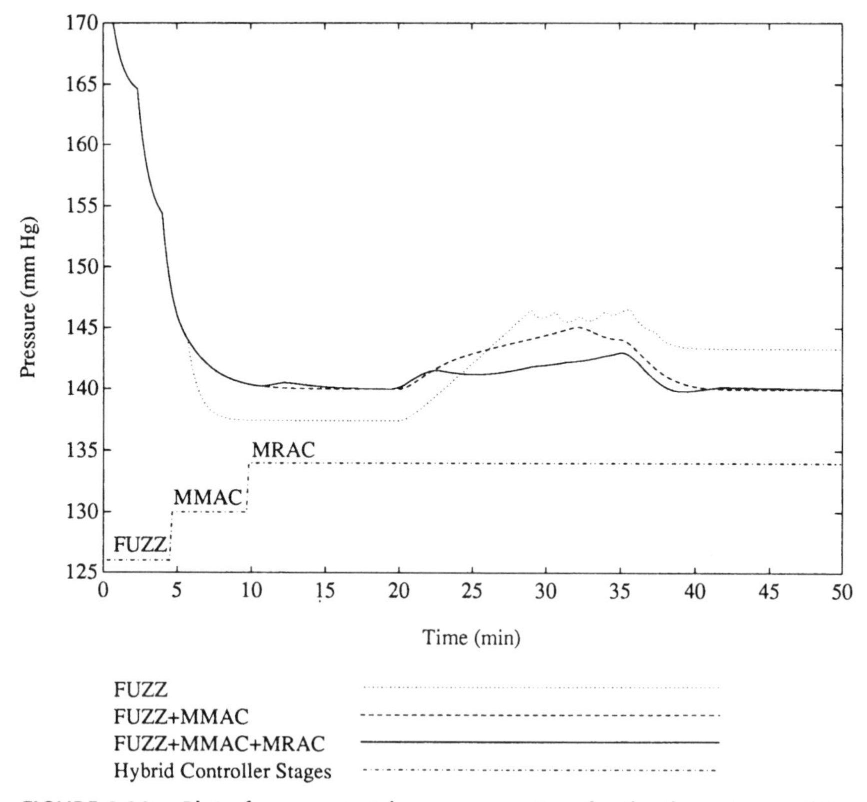

FIGURE 3.20. Plot of mean arterial pressure vs. time for the three stages of the hierarchical controller. The dash-dot curve at the base of the figure indicates when the hierarchical controller switched control schemes.

multiple model controller compensates for the gain variation by switching its linear controllers when the plant moves from one model space to another. Figure 3.22 shows the model weights (as determined by Equation (12)) indicating this transition. However, the multiple model approach has difficulty dealing with the patient gain variation as compared with the model reference controller.

The model reference control method compensates for the gain variation by adjusting the set point to the inner loop formed by the closed-loop combination of the patient and the selected multiple model compensator. Figure 3.23 shows the signal (i.e., $u(t)$ in Equation (45)) applied to this inner loop for the case when the controller advanced to the model reference stage. Figure 3.20 demonstrates that this controller structure yields the most precise control action as compared with the individual controller components acting alone.

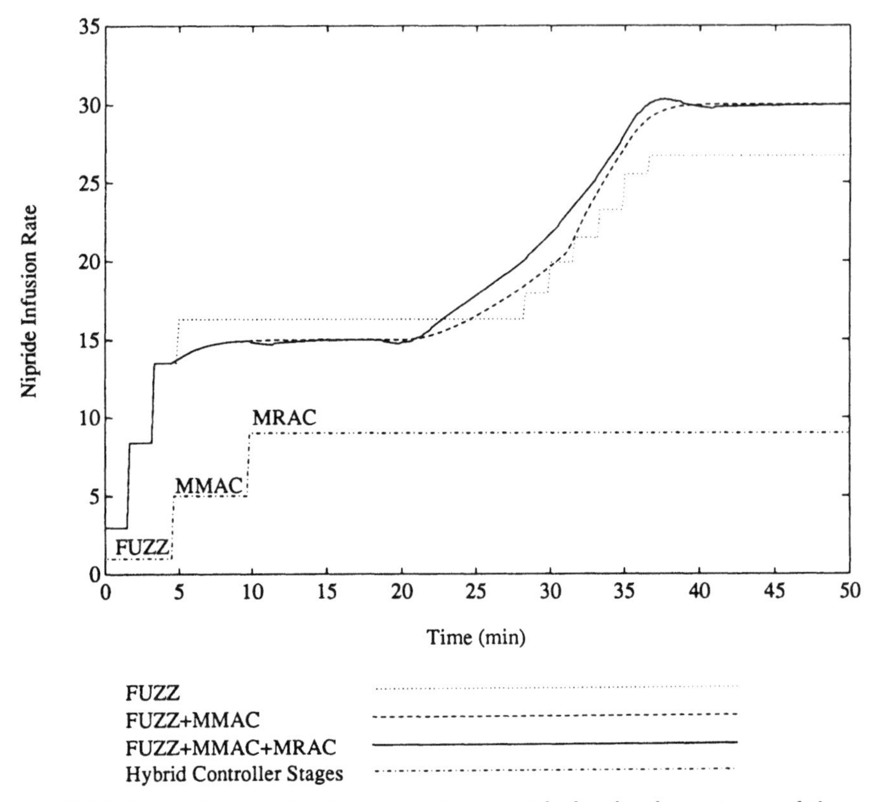

FIGURE 3.21. Infusion rate of sodium nitroprusside for the three stages of the hierarchical controller. The dash-dot curve at the base of the figure indicates when the hierarchical controller switched control schemes.

4. CONCLUSION

This chapter presents a methodology for designing an expert hybrid control scheme. The approach utilizes a range of control algorithms in order to provide the best control for a broad range of plant conditions. Three control procedures, ranging from a coarse, intelligent control action (fuzzy control), to a precise control (model reference control), are uniquely combined to improve the regulation capabilities of each controller individually. In the worst case, this method has the potential to perform as well as a human operator; at best the state of the process will be driven to the desired set point in a manner prescribed by the chosen reference model.

The chapter presents details on the design methods for the elemental control-

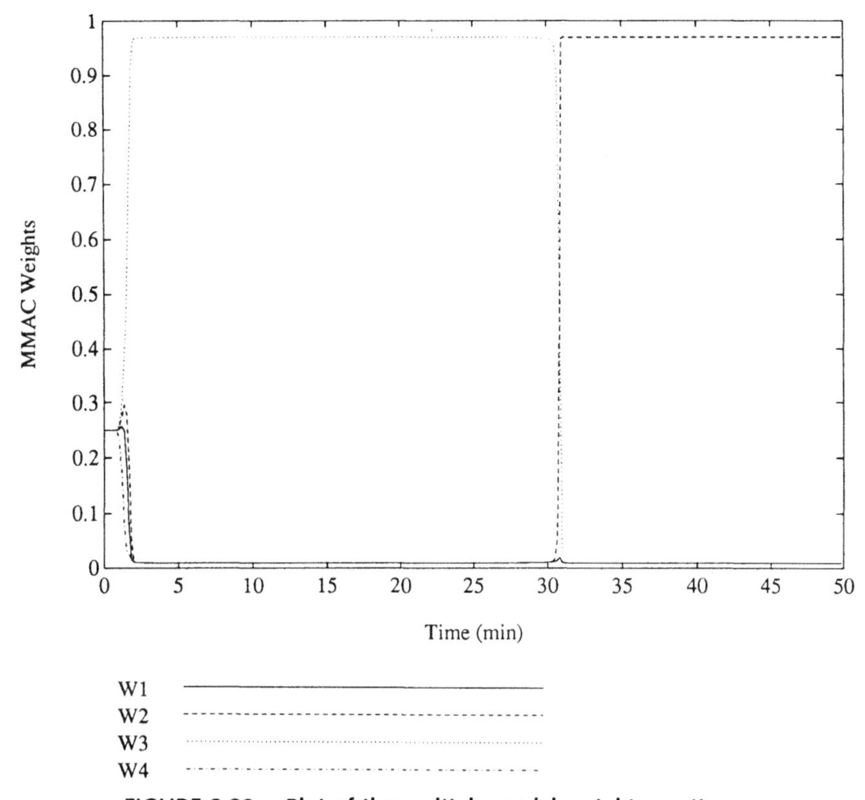

FIGURE 3.22. Plot of the multiple model weights vs. time.

lers describing how each method determines the transformation from e to u as depicted in Figure 3.3. Depending on the system's knowledge of the process and the present operating condition, the control to the process could be regulated by a nonlinear control law lacking a guarantee of closed-loop stability (fuzzy controller), a linear control law assuming the algorithm has converged on a specific model (multiple model adaptive control), or a nonlinear control law that guarantees closed-loop stability (model reference controller).

In order to attain the full benefit of this method, the process must possess certain qualities. Traditional regulation of the process should rely on a human operator for control adjustments. In addition, an approximate model of the process with ranges on parameter variations must be available. The example of regulating a drug delivery system presented in this chapter meets these conditions. Another study assessing the promise of this regulation scheme involves the control of a robot arm under different payload conditions [55].

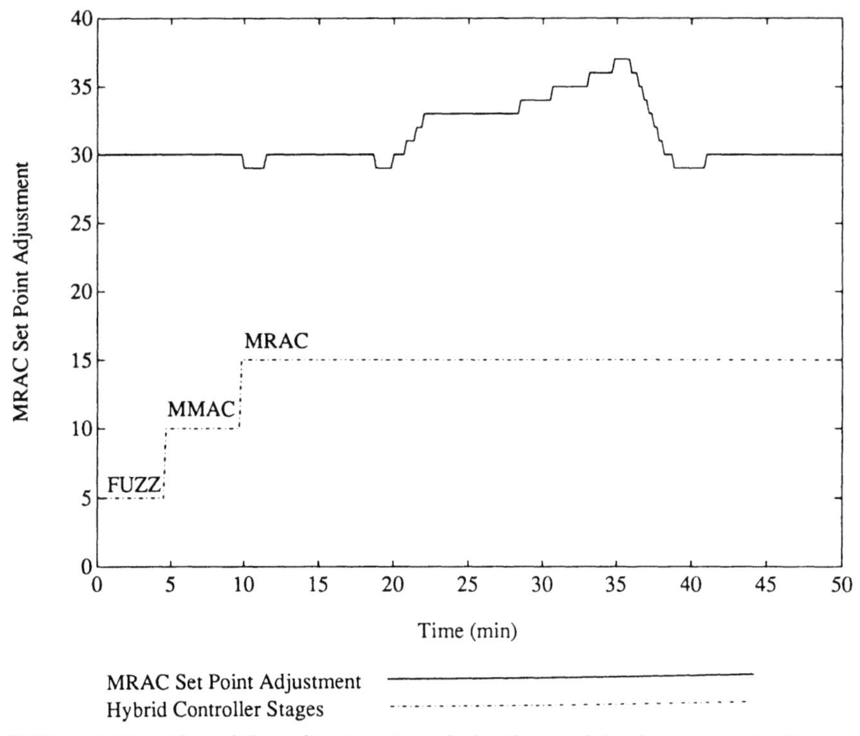

FIGURE 3.23. Plot of the adjustment made by the model reference controller to the set point of the inner loop. The dash-dot curve at the base of the figure indicates when the hierarchical controller switched control schemes.

This chapter addresses the regulation and supervision at the control level. The inclusion of the expert system structure allows for the extension to multiple levels of decision making. In the medical example for instance, the expert system knowledge base could be augmented to include diagnostic decisions. This would include for instance, assessment of the physiological state of the patient in the example given. Depending on the patient's status, the expert system may activate a hierarchical controller (operating at the control level) to move the patient's state into a desired physiological region. This capability of making multilevel decisions in conjunction with the analytical control methods results in a control scheme capable of regulating extremely complex processes.

REFERENCES

1. K. Astrom and B. Wittenmark, *Adaptive Control*. Reading, MA: Addison-Wesley, 1988.

2. D. Looze, "Automatic redesign approach for restructurable control systems," *IEEE Control Systems Magazine*, Vol. 5, No. 2, pp. 16–22, May 1985.
3. A. Ostroff and R. Hueschen, "Investigation of control law reconfiguration to accommodate a control element failure on a commercial aircraft," in *Proceedings of the American Control Conference*, pp. 1746–1754, 1984.
4. W. Vander Velde, "A control system reconfiguration," in *Proceedings of the American Control Conference*, pp. 1741–1745, 1984.
5. R. Ornedo, "Concept and Demonstration: A reconfigurable control system for nuclear power plants," Massachusetts Institute of Technology, Cambridge, MA, PhD thesis, 1986.
6. R. Ornedo, J. Bernard, D. Lanning, and J. Hopps, "Design and experimental evaluation of an automatically reconfigurable controller for process plants," in *Proceedings of the American Control Conference*, pp. 1662–1668, 1987.
7. R. Isermann and K. Lachmann, "Parameter-adaptive control with configuration aids and supervision functions," *Automatica*, Vol. 21, No. 6, pp. 625–638, November 1985.
8. K. Astrom, J. Anton, and K. Arzen, "Expert control," *Automatica*, Vol. 22, No. 3, pp. 277–286, May 1986.
9. J. Bernard, "Use of a rule-based system for process control," *IEEE Control Systems Magazine*, Vol. 8, pp. 3–13, Oct. 1988.
10. K. Gingrich, R. Vishnoi, R. Roy, C. Yu, and G. Neat, "Modeling the hemodynamic response to dopamine in acute heart failure," in *IEEE Trans. on BioMedical Eng*, Vol. 38, No. 3, pp. 267–272, March 1991.
11. J. Slate, "Model-based design of a controller for infusing sodium nitroprusside during postsurgical hypertension," University of Wisconsin-Madison, Madison, WI, PhD thesis, 1980.
12. L. Zadeh, "Fuzzy sets," *Information and Control*, Vol. 8, No. 3, pp. 338–353, June 1965.
13. L. Zadeh, "Outline of a new approach to analysis of complex systems and decision processes," *IEEE Trans. Systems, Man and Cybernetics*, Vol. SMC-3, No. 1, pp. 28–44, January 1973.
14. E. Mamdani, "Application of fuzzy algorithms for control of a simple dynamic plant," *Proc. IEEE*, Vol. 121, pp. 1585–1588, 1974.
15. E. Mamdani and S. Assilian, "A fuzzy logic controller for a dynamic plant," *Int. J. Man-Machine Studies*, Vol. 7, No. 1, pp. 1–13, January 1975.
16. W. Kickert and H. van Nauta Lemke, "Application of a fuzzy controller in a warm water plant," *Automatica*, Vol. 12, pp. 301–308, July 1976.
17. W.J.M. Kickert, "Application of Fuzzy Set Theory to Warm Water Control," thesis, Delft Univ of Technology, Dept of Electrical Eng, Lab of atuotmatic control, Nov. 1974.
18. L. Homblad and J. Ostergaard, "Control of a cement kiln by fuzzy logic," *Fuzzy Information and Decision Processes*, M. Gupta and R. Sanchez, Eds. Amsterdam: North Holland Publishing Company, 1982, pp. 389–399.
19. F. Smidth, "Kiln control . . . using fuzzy logic," *FLS-Newsfront*, pp. 10–15, 1979.
20. S. Sheridan and P. Skjoth, "Automatic kiln control at portland oregon cement company's durkee plant utilizing fuzzy logic," *IEEE Trans. Industrial Applications*, Vol. 1A–20, pp. 562–568, 1984.

21. R. Vishnoi and K. Gingrich, "Fuzzy controller for gaseous anesthesia delivery using vital signs," in *Proc. 27th Conference on Decision and Control,* pp. 346–347, 1987.
22. B. Gaines and L. Kohout, "The fuzzy decade: A bibliography of fuzzy systems and closely related topics," *Int. J. Man-Machine Studies,* Vol. 9, pp. 1–68, January 1977.
23. D. Lainiotis, T. Upadhyay, and J. Deshpande, "A non-linear separation theorem," in *Proc. Symposium on Nonlinear Estimation Theory,* pp. 184–187, 1971.
24. D.L. et al., "Partioning: A unifying framework for adaptive systems, ii: Control," *Proc. IEEE,* Vol. 64, No. 8, pp. 1182–1198, August 1976.
25. W. He, H. Kaufman, and R. Roy, "Multiple model adaptive control procedure for blood pressure control," *IEEE Trans. Biomedical Engineering,* Vol. BME-33, pp. 10–19, January 1986.
26. O. Smith, "Closer control of loops with dead time," *Chemical Engineering Progress,* Vol. 53, pp. 217–219, 1957.
27. A. Bahill, "A simple adaptive smith-predictor for controlling time-delay systems," *IEEE Control Systems Magazine,* Vol. 3, pp. 16–22, May 1983.
28. D. Clarke, "Application of generized predictive control to industrial processes," *IEEE Control Systems Magazine,* Vol. 8, pp. 49–55, April 1988.
29. J. Martin, A. Schneider, and T. Smith, "Multiple-model adaptive control of blood pressure using sodium nitroprusside," *IEEE Trans. Biomedical Engineering,* Vol. BME-34, No. 8, pp. 603–611, August 1987.
30. C. Yu, W. He, J. So, R. Roy, H. Kaufman, and J. Newell, "Improvement in arterial oxygen control using multiple-model adaptive control procedures," *IEEE Trans. Biomedical Engineering,* Vol. BME-34, pp. 567–574, August 1987.
31. M. Athans, D. Castanon, K. Dunn, C. Greene, W. Lee, N. Sandell, and A. Willsky, "The stochastic control of the f-8c aircraft using a multiple model adaptive control (mmac) method—Part 1: Equilibrium flight," *IEEE Trans. Automatic Control,* Vol. AC-22, pp. 768–780, October 1977.
32. M. Leahy and L. Tellman, "Multiple model-based control of robotic manipulators: Theory and simulation," in *Proceedings of the IEEE International Symposium on Intelligent Control 1989,* 1989, pp. 53–58.
33. P. Maybeck and K. Hentz, "Investigation of moving-bank multiple model adaptive algorithms," *J. Guidance, Control and Dynamics,* Vol. 10, pp. 90–96, Jan./Feb. 1987.
34. K. Sobel, H. Kaufman, and L. Mabius, "Implicit adaptive control for a class of mimo systems," *IEEE Trans. Aerospace and Electronic Systems,* Vol. 18, No. 5, pp. 576–590, Sept. 1982.
35. J. Broussard and M. O'Brien, "Feed-forward control to track the output of a forced model," in *Proc. 17th Conference on Decision and Control,* 1979, pp. 1149–1155.
36. K. Astrom and B. Wittenmark, *Computer Controlled Systems.* Englewood Cliffs, NJ: Prentice Hall, 1984.
37. K. Sobel and H. Kaufman, "Direct model reference adaptive control for a class of mimo systems," *Control and Dynamic Systems : Advances in Theory and Applications,* C. Leondes, Ed. New York: Academic Press, 1987, pp. 245–314.
38. I. Bar-Kana and H. Kaufman, "Global stability and performance of a simplified adaptive control algorithm," *Int. J. Control,* Vol. 42, pp. 1491–1505, June 1985.
39. I. Bar-Kana, "Adaptive control—A simplified approach," in *Control and Dynamic*

Systems : Advances in Theory and Applications, C. Leondes, Ed. New York: Academic Press, 1987, pp. 182–235.

40. G. Franklin, J. Powell, and A. Naeini, *Feedback Control of Dynamic Systems*. Reading, MA: Addison Wesley, 1986.
41. J. Wen, "Robustness analysis for evolution systems in hilbert space: A passivity approach," Rensselaer Polytechnic Institute Troy, NY, Tech. Rep. 13, CIRSSE, 1988. ELJE Dept.
42. G. Neat, J. Wen, and H. Kaufman, "Expert hierarchical adaptive control," in *Proc. 1989 American Control Conference*, 1989, pp. 13–18.
43. H. Kaufman, R. Roy, and X. Xu, "Model reference adaptive control of drug infusion rate," *Automatica*, Vol. 20, pp. 205–210, March 1984.
44. C. Ih, S. Wang, and C. Leondes, "Adaptive control for the space station," *IEEE Control Systems Magazine*, Vol. 7, pp. 29–34, Feb. 1987.
45. D. Bayard, C. Ih, and S. Wang, "Adaptive control for flexible space structures with measurement noise," in *Proc. 1987 American Control Conference*, pp. 368–379, 1987.
46. C. Ih, D. Bayard, and S.W.D. Eldred, "Adaptive control experiment with a large flexible structure," in *Proc. American Institute of Aeronautics and Astronautics Guidance, Navigation and Control Conference*, pp. 832–851, 1988.
47. L. Brownston, R. Farrell, E. Kant, and N. Martin, *Programming Expert Systems in OPS5 An Introduction to Rule-Based Programming*. Reading, MA: Addison Wesley, 1985.
48. F. Hayes-Roth, D. Waterman, and D. Lenat, *Building Expert Systems*. Reading, MA: Addison Wesley, 1983.
49. P. Winston, *Artificial Intelligence*. Reading, MA: Addison Wesley, 1984.
50. A. Koivo, "Automatic continuous-time bp control in dogs by means of hypotensive drug injection," *IEEE Trans. Biomedical Engineering*, vol. BME-27, pp. 574–581, Oct. 1980.
51. G. Voss, P. Katona, and H. Chizeck, "Adaptive multivariable drug delivery: Control of arterial pressure and cardiac output in anesthesized dogs," *IEEE Trans. Biomedical Engineering*, vol. BME-34, pp. 617–623, August 1987.
52. B. McInnis and L. Deng, "Automatic control of blood pressures with multiple drug inputs," *Annals of Biomedical Engineering*, Vol. 13, No. 3/4, pp. 217–225, 1985.
53. C. Rohrs, L. Valavani, M. Athans, and G. Stein, "Robustness of continuous-time adaptive control algorithms in the presence of unmodeled dynamics," *IEEE Trans. on Automatic Control*, Vol. AC-30, No. 9, pp. 881–888, Sept. 1985.
54. G. Neat, H. Kaufman, and R. Roy, "Expert adaptive control for drug delivery systems," *Control Systems Magazine*, Vol. 9, No. 4, pp. 20–24, June 1989.
55. D. Minnick, H. Kaufman, and G. Neat, "Expert hierarchical adaptive controller for robotic systems," in *Proc. IEEE Int. Symp. Intelligent Control 1989*, pp. 8–11, 1989.

ACKNOWLEDGMENT

This paper is based on research performed under NSF grant EET-8620246 and NASA grant NAGW-1333.

4

Integrated Piezoelectric Sensor/Actuator Design for Distributed Identification and Control of Smart Machines and Flexible Robots Part 1: Theory and Experiments

H.S. Tzou
Department of Mechanical Engineering
Center for Robotics and Manufacturing Systems
University of Kentucky
Lexington, KY 40506-0046

1. INTRODUCTION

Smart robotic manipulators and machine structures/components with integrated self-monitoring and control capabilities are becoming more important in the development of new-generation intelligent manufacturing systems. In this report, a generic integrated piezoelectric sensor/actuator design and its applications to smart machines/robots are presented. The report is divided into two parts (chapters): (1) Theory and experiments; and (2) finite element formulation and analysis. In Part 1, a generic multilayered piezoelectric sensor/actuator theory for a generic distributed parameter system (a continuum) is derived using Love's theory, Hamilton's principle, and piezoelectric theory. The derived equation can be simplified to account for many other common geometries (e.g., plates, cylinders, beams, etc.), which are the basic components of robotic and machine structures/components. A simple reduction procedure is proposed and examples are demonstrated in case studies. Physical systems which include a flexible manipulator and a smart high-precision microactuator are studied analytically and experimentally.

1.1. Background

In recent years, a strong demand for high-performance manufacturing systems, e.g., robots and machines, has driven innovations of technologies and materials to design new-generation *smart* machines and robots [1]. (The *smartness* is defined in a hardware way, i.e., the machines and robot structures are integrated with sensors and actuators.) This report presents a new integrated piezoelectric sensor/actuator design for structural identification and active vibration control of smart mechanical systems.

In general, machine and robot structures are distributed in nature, i.e., structural dynamics are functions of *spatial* and *time* variables, which are classified as *distributed-parameter systems* (DPSs). In this report, a generic shell structure is regarded as a general DPS from which many other structures, e.g., plates, cylinders, spheres, etc., can be derived [2, 3]. Moreover, conventional sensors and actuators usually measure and control spatially *discrete* locations. Serious problems can occur when these devices are installed at modal nodes or lines. One solution to this problem is to use *distributed* sensors and actuators for structural identification and control of DPSs [4]. Thus, new development on active distributed vibration control and structural identification of smart DPSs using distributed piezoelectric sensors/actuators are proposed and evaluated in this report.

New development in theoretical and practical aspects of piezoelectricity has drawn much interest in recent years. Tzou and Gadre designed a dual-purpose piezoelectric exciter and vibration isolator [5, 6]. Test data showed a close comparison with the theoretical solutions. This technique was also applied to rotordynamic vibration controls [7]. Use of a distributed piezoelectric actuator in control of flexible beam oscillation was studied [8–12]. Tzou and Gadre derived a generic layered shell actuator theory for distributed vibration control of flexible shells [13]. Tzou derived an integrated distributed sensing and control theory for thin shells [1, 4]. Tzou and Tseng also formulated a new piezoelectric finite element for distributed sensing and control of shells and plates [14]. A generic theory on structural identification and vibration control of DPSs using electroded piezoelectric layers still needs to be further developed. Applications of the theory to smart machines and robots also need to be explored and implemented.

1.2. Objectives

In this study, based on a generic DPS—a generic shell element—theories on the distributed vibration control and structural identification using distributed piezoelectric sensors and actuators are derived. Simplification of the theories to other geometries will also be discussed. The modal expansion method is incorporated with the theories to express the general identification and control equations in modal coordinates. Active distributed vibration control and identification of a simple DPS—a flexible arm—is demonstrated by laboratory experiments. A prototype microdisplacement piezoelectric actuator is also designed and tested to evaluate its performance.

2. PIEZOELECTRICITY THEORY

Piezoelectric phenomena were first observed by Curies, et al. in 1880 [15]. In general, when a mechanical force is applied to a piezoelectric material, an electrical voltage or charge is generated; this electromechanical phenomenon is referred to as the *direct piezoelectric effect.* Conversely, when an electric field is applied to the material, a mechanical stress or strain is induced; this is called the *converse piezoelectric effect.* In this study, the *direct* effect is used for distributed structural identification and the *converse* effect for the active distributed vibration control of DPSs. There are two constitutive equations representing the direct and converse piezoelectric effects respectively [16],

$$\begin{cases} \{S\} = [s^E] \cdot \{T\} + [d]^t \cdot \{E\}, & (1) \\ \{D\} = [\epsilon^T] \cdot \{E\} + [d] \cdot \{T\}; & (2) \end{cases}$$

and

$$\begin{cases} \{T\} = \{c^D\} \cdot \{S\} - [h]^t \cdot \{D\}, & (3) \\ \{E\} = [\beta^S] \cdot \{D\} - [h] \cdot \{S\}; & (4) \end{cases}$$

where {S} is the strain vector (i.e., $\{S\} = \{\, S_{11}\ S_{22}\ S_{33}\ S_{23}\ S_{31}\ S_{12}\,\}^t$); {T} is the stress vector (i.e., $\{T\} = \{\, T_{11}\ T_{22}\ T_{33}\ T_{23}\ T_{31}\ T_{12}\,\}^t$); $[s^E]$ is the elastic compliance matrix measured at constant electric field; $[\epsilon^T]$ is the dielectric matrix evaluated at constant strain; [d] is the piezoelectric constant matrix; $[c^D]$ is the elasticity matrix evaluated at constant dielectric displacement; [h] is the piezoelectric constant matrix; {D} is the electric displacement vector; $[.]^t$ indicates the matrix transpose; {E} is the electric field vector; and $[\beta^S]$ is the dielectric impermeability matrix evaluated at constant strain. The constitutive equations formulated in Equations (1)–(4) are assumed to be instantaneously mechanically and electrically balanced and the two effects can be decoupled.

In this research, the piezoelectric layer is assumed to be polarized in the thickness direction (α_3), i.e., isotropic in the transverse direction (α_3) and anisotropic in the in-plane directions (α_1 and α_2). Polymeric polyvinylidene fluoride (PVDF) and some piezoceramics are in this category. The piezoelectric matrix [d] of this category can be expressed as [16, 17]

$$[d_{ij}] = \begin{bmatrix} 0 & 0 & 0 & 0 & d_{15} & 0 \\ 0 & 0 & 0 & d_{24} & 0 & 0 \\ d_{31} & d_{32} & d_{33} & 0 & 0 & 0 \end{bmatrix}. \quad (5)$$

If the piezoelectric material is electrically polarized not mechanically stretched, the piezoelectric coefficient d_{24} is equal to d_{15}.

3. DISTRIBUTED STRUCTURAL IDENTIFICATION THEORY

In this section, a distributed piezoelectric sensing theory for a generic DPS—shell—is developed. The output signal equation is extended to incorporate with the modal expansion method from which a *modal voltage* concept is introduced.

3.1. System Definition and Assumptions

A generic shell structure, defined in a triorthogonal curvilinear coordinate system, is regarded as a generic DPS in the derivation of distributed vibration control and structural identification theories. (This shell structure has been proved so that its system equations can be directly simplified to account for many other commonly occurring geometries, such as plates, spheres, cylinders, beams, arches, cylindrical panels, etc. [2, 3].) The generic shell structure is made of a passive elastic material sandwiched between two thin piezoelectric layers (Figure 4.1). Note that the top layer serves as a distributed actuator for distributed vibration control and the bottom a distributed sensor for structural identification.

It is assumed that the piezoelectric layers are relatively thin and flexible compared with the elastic shell. Thus, the structural dynamics of the shell can be approximated by Kirchhoff-Love assumptions [3]. In this section, a distributed

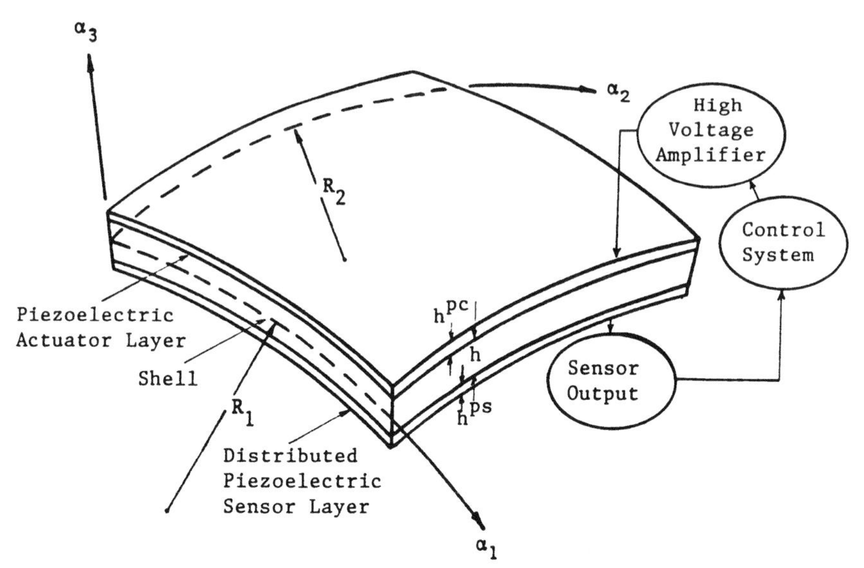

FIGURE 4.1. A generic DPS (a shell) with distributed piezoelectric sensor/actuator.

structural identification theory for thin shells is developed using Maxwell theory [1, 4]. Note that in the distributed sensing application, only the direct piezoelectric effect is considered. Also note that only the transverse electric field E_3 is considered in the shell configuration. Thus, the strains and dielectric displacement D_3 are independent of thickness, α_3.

3.2. Distributed Sensing Formulation

Using Maxwell's equation, one can derive the electric field E_i related to the electric potential ϕ^* as

$$\{E_i\} = -\nabla \phi^* \tag{6}$$

where $\nabla = \{ \partial/\partial\alpha_1, \partial/\partial\alpha_2, \partial/\partial\alpha_3 \}^t$. The voltage across the electrodes can be obtained by integrating the electric field over the thickness of the piezoelectric sensor layer, i.e.,

$$\phi = \wedge \int^{h^{ps}} E_3 d\alpha_3, \tag{7}$$

where the superscript ps denotes the distributed piezoelectric sensor layer; and h^{ps} is the thickness of the piezoelectric sensor layer. (A list of nomenclature is provided in Appendix 1.) Since the piezoelectric layer is relatively thin and flexible, it is assumed that the strains in the layer are constant and equal to the outer surface strains of the shell.

Considering the induced in-plane strains (S_1^{ps} and S_2^{ps}) and the piezoelectric coefficients (h_{31} and h_{32}), one can derive an electrical voltage equation as

$$\phi = h^{ps}(h_{31}S_1^{ps} + h_{32}S_2^{ps} - \beta_{33}D_3), \tag{8}$$

where S_1^{ps} and S_2^{ps} are the in-plane strains in the α_1- and α_2-directions, respectively. The in-plane strains in the piezoelectric sensor layer can be calculated by [3, 4]

$$S_1^{ps}(\alpha_1, \alpha_2, t) = d_1^{ps} \left\{ \frac{1}{A_1} \frac{\partial}{\partial\alpha_1} \left(\frac{u_1}{R_1} - \frac{1}{A_1} \frac{\partial u_3}{\partial\alpha_1} \right) + \frac{1}{A_1 A_2} \frac{\partial A_1}{\partial\alpha_2} \left(\frac{u_2}{R_2} - \frac{1}{A_2} \frac{\partial u_3}{\partial\alpha_2} \right) \right\}, \tag{9}$$

$$S_2^{ps}(\alpha_1, \alpha_2, t) = d_2^{ps} \left\{ \frac{1}{A_2} \frac{\partial}{\partial\alpha_2} \left(\frac{u_2}{R_2} - \frac{1}{A_2} \frac{\partial u_3}{\partial\alpha_2} \right) \right.$$

$$+ \frac{1}{A_1 A_2} \frac{\partial A_2}{\partial \alpha_1} \left(\frac{u_1}{\mathbb{R}_1} - \frac{1}{A_1} \frac{\partial u_3}{\partial \alpha_1} \right) \bigg\} , \tag{10}$$

where d_1^{ps} and d_2^{ps} are the distances measured from the shell neutral surface; A_1 and A_2 are Lame's parameters; u_1 and u_2 are the in-plane displacements; $\mathbb{R}_1$ and $\mathbb{R}_2$ are the radii of curvature; and u_3 is the transverse displacement.

The electric displacement D_3^{ps} can be expressed in terms of the voltage ϕ

$$D_3^{ps} = \frac{1}{\beta_{33}} \left(\frac{-\phi}{h^{ds}} + h_{31} S_1^{ps} + h_{32} S_2^{ps} \right) . \tag{11}$$

The distributed piezoelectric sensor outputs can be calculated in two ways: (1) A discrete voltage ϕ_d^{ps}; and (2) an averaged ϕ_a^{ps}. For discrete voltage calculation, it is assumed that there are infinite electrodes on the surface of the piezoelectric layer. Thus, knowing the discrete voltage amplitude of every discrete location, one can construct a distributed voltage contour corresponding to any time instant or state. In the second method, there are a limited number of electrodes made on the surface of the layer so that an averaged voltage for a specific electrode segment can be calculated.

Integrating Equation (11) over the electrode surface A^{ps} yields a charge. Setting the resulting charge expression equal to zero gives an averaged open-circuit voltage ϕ_a^{ps} of the sensor,

$$\phi_a^{ps} = \frac{h^{ps}}{A^{ps}} \int_{A^{ps}} (h_{31} S_1^{ps} + h_{32} S_2^{ps}) A_1 A_2 d\alpha_1 d\alpha_2 \tag{12a}$$

$$\begin{aligned} &= \frac{h^{ps}}{A^{ps}} \int_{A^{ps}} \bigg\{ h_{31} d_1^{ps} \bigg[\frac{1}{A_1} \frac{\partial}{\partial \alpha_1} \left(\frac{u_1}{\mathbb{R}_1} - \frac{1}{A_1} \frac{\partial u_3}{\partial \alpha_1} \right) \\ &+ \frac{1}{A_1 A_2} \frac{\partial A_1}{\partial \alpha_2} \left(\frac{u_2}{\mathbb{R}_2} - \frac{1}{A_2} \frac{\partial u_3}{\partial \alpha_2} \right) \bigg] \\ &+ h_{32} d_2^{ps} \bigg[\frac{1}{A_2} \frac{\partial}{\partial \alpha_2} \left(\frac{u_2}{\mathbb{R}_2} - \frac{1}{A_2} \frac{\partial u_3}{\partial \alpha_2} \right) \\ &+ \frac{1}{A_1 A_2} \frac{\partial A_2}{\partial \alpha_1} \left(\frac{u_1}{\mathbb{R}_1} - \frac{1}{A_1} \frac{\partial u_3}{\partial \alpha_1} \right) \bigg] \bigg\} A_1 A_2 d\alpha_1 d\alpha_2 . \end{aligned} \tag{12b}$$

Note that α_1 and α_2 constitute a two-dimensional curvilinear plane. For discrete point voltage ϕ_d^{ps}, the above equation can be modified as

$$\phi_d^{ps} = h^{ps} (h_{31} S_1^{ps} + h_{32} S_2^{ps}) \tag{13a}$$

$$\phi_d^{ps} = h^{ps} \bigg\{ h_{31} d_1^{ps} \bigg[\frac{1}{A_1} \frac{\partial}{\partial \alpha_1} \left(\frac{u_1}{\mathbb{R}_1} - \frac{1}{A_1} \frac{\partial u_3}{\partial \alpha_1} \right)$$

$$+ \frac{1}{A_1A_2}\frac{\partial A_1}{\partial \alpha_2}\left(\frac{u_2}{\mathbb{R}_2} - \frac{1}{A_2}\frac{\partial u_3}{\partial \alpha_2}\right)\Bigg]$$

$$+ h_{32}d_2^{ps}\left[\frac{1}{A_2}\frac{\partial}{\partial \alpha_2}\left(\frac{u_2}{\mathbb{R}_2} - \frac{1}{A_2}\frac{\partial u_3}{\partial \alpha_2}\right)\right.$$

$$\left.+ \frac{1}{A_1A_2}\frac{\partial A_2}{\partial \alpha_1}\left(\frac{u_1}{\mathbb{R}_1} - \frac{1}{A_1}\frac{\partial u_3}{\partial \alpha_1}\right)\right]\Bigg\}. \tag{13b}$$

3.3. Modal Expansion Method

Dynamic response of the shell can be expressed in modal coordinates λ_k using the *modal expansion method* [3]. That is, the dynamic response $u_i(\alpha_1, \alpha_2, t)$ is a summation of all participating modes U_{ik} with individual modal participation factor λ_k,

$$u_i(\alpha_1, \alpha_2, t) = \sum_{k=1}^{\infty} \lambda_k(t)U_{ik}(\alpha_1, \alpha_2), \tag{14}$$

where $i = 1, 2, 3$ indicating three axes. Since for a distributed system, the number of modes are infinite, $k = 1 - \infty$. Substituting the modal expansion equation into the (averaged) distributed sensing equation yields

$$\phi_a^{ps} = \frac{h^{ps}}{A^{ps}}\int_{A^{ps}}\left\{h_{31}d_1^{ps}\left[\frac{1}{A_1}\frac{\partial}{\partial \alpha_1}\left(\frac{1}{\mathbb{R}_1}\sum_{k=1}^{\infty}\lambda_k U_{1k} - \frac{1}{A_1}\frac{\partial}{\partial \alpha_1}\sum_{k=1}^{\infty}\lambda_k U_{3k}\right)\right.\right.$$

$$\left.+ \frac{1}{A_1A_2}\frac{\partial A_1}{\partial \alpha_2}\left(\frac{1}{\mathbb{R}_2}\sum_{k=1}^{\infty}\lambda_k U_{2k} - \frac{1}{A_2}\frac{\partial}{\partial \alpha_2}\sum_{k=1}^{\infty}\lambda_k U_{3k}\right)\right]$$

$$+ h_{32}d_2^{ps}\left[\frac{1}{A_2}\frac{\partial}{\partial \alpha_2}\left(\frac{1}{\mathbb{R}_2}\sum_{k=1}^{\infty}\lambda_k U_{2k} - \frac{1}{A_2}\frac{\partial}{\partial \alpha_2}\sum_{k=1}^{\infty}\lambda_k U_{3k}\right)\right.$$

$$\left.\left.+ \frac{1}{A_1A_2}\frac{\partial A_2}{\partial \alpha_1}\left(\frac{1}{\mathbb{R}_1}\sum_{k=1}^{\infty}\lambda_k U_{1k} - \frac{1}{A_1}\frac{\partial}{\partial \alpha_1}\sum_{k=1}^{\infty}\lambda_k U_{3k}\right)\right]\right\}A_1A_2 d\alpha_1 d\alpha_2. \tag{15}$$

And the (discrete) voltage output can be written in a similar way,

$$\phi_d^{ps} = h^{ps}\left\{h_{31}d_1^{ps}\left[\frac{1}{A_1}\frac{\partial}{\partial \alpha_1}\left(\frac{1}{\mathbb{R}_1}\sum_{k=1}^{\infty}\lambda_k U_{1k} - \frac{1}{A_1}\frac{\partial}{\partial \alpha_1}\sum_{k=1}^{\infty}\lambda_k U_{3k}\right)\right.\right.$$

$$+ \frac{1}{A_1 A_2} \frac{\partial A_1}{\partial \alpha_2} \left(\frac{1}{\mathbb{R}_2} \sum_{k=1}^{\infty} \lambda_k U_{2k} - \frac{1}{A_2} \frac{\partial}{\partial \alpha_2} \sum_{k=1}^{\infty} \lambda_k U_{3k} \right) \Bigg]$$

$$+ h_{32} d_2^{ps} \left[\frac{1}{A_2} \frac{\partial}{\partial \alpha_2} \left(\frac{1}{\mathbb{R}_2} \sum_{k=1}^{\infty} \lambda_k U_{2k} - \frac{1}{A_2} \frac{\partial}{\partial \alpha_2} \sum_{k=1}^{\infty} \lambda_k U_{3k} \right) \right.$$

$$\left. + \frac{1}{A_1 A_2} \frac{\partial A_2}{\partial \alpha_1} \left(\frac{1}{\mathbb{R}_1} \sum_{k=1}^{\infty} \lambda_k U_{1k} - \frac{1}{A_1} \frac{\partial}{\partial \alpha_1} \sum_{k=1}^{\infty} \lambda_k U_{3k} \right) \right] \Bigg\} . \qquad (16)$$

It is observed that the distributed sensor output signal (either averaged or discrete formulation) is contributed by all shell vibration modes. That is, the distributed sensor can measure or identify all vibration modes of the shell structure. Also note that the modal participation factor λ_k could be zero for any kth mode not participating in the shell oscillation.

However, for a fully electroded shell, antisymmetrical modes can induce zero-averaged voltage because the positive output is canceled out with the negative output in the DPS. In this case, segmenting or shaping surface electrodes is required to produce nonzero outputs.

3.4. Distributed Modal Voltage

As discussed previously, a discrete point voltage output $\phi_d^{ps}(\alpha_1^*, \alpha_2^*)$ can also be calculated by neglecting the surface average. (α_1^*, α_2^*) denotes a specific location on the DPS. Connecting all discrete local voltages forms a voltage distribution contour representing the current vibration state of the DPS. A kth *distributed modal voltage* $\phi_{dk}^{ps}(\alpha_1, \alpha_2, t)$ function, which describes modal voltage distribution, can be calculated in this way,

$$\phi_{dk}^{ps}(\alpha_1, \alpha_2, t) = h_{31} d_1^{ps} \left[\frac{1}{A_1} \frac{\partial}{\partial \alpha_1} \left(\frac{1}{\mathbb{R}_1} (\lambda_k U_{1k}) - \frac{1}{A_1} \frac{\partial}{\partial \alpha_1} (\lambda_k U_{3k}) \right) \right.$$

$$\left. + \frac{1}{A_1 A_2} \frac{\partial A_1}{\partial \alpha_2} \left(\frac{1}{\mathbb{R}_2} (\lambda_k U_{2k}) - \frac{1}{A_2} \frac{\partial}{\partial \alpha_2} (\lambda_k U_{3k}) \right) \right]$$

$$+ h_{32} d_2^{ps} \left[\frac{1}{A_2} \frac{\partial}{\partial \alpha_2} \left(\frac{1}{\mathbb{R}_2} (\lambda_k U_{2k}) - \frac{1}{A_2} \frac{\partial}{\partial \alpha_2} (\lambda_k U_{3k}) \right] \right.$$

$$\left. + \frac{1}{A_1 A_2} \frac{\partial A_2}{\partial \alpha_1} \left(\frac{1}{\mathbb{R}_1} (\lambda_k U_{1k}) - \frac{1}{A_1} \frac{\partial}{\partial \alpha_1} (\lambda_k U_{3k}) \right) \right] , \qquad (17)$$

in which the modal participation factor λ_k is unity. (Note that any U_{ik} could be zero if it does not contribute the mode.) Note that the modal voltage distribution contour is constructed by connecting all discrete point voltage amplitudes.

4. DISTRIBUTED VIBRATION CONTROL THEORY

In the trilayer shell structure, the top piezoelectric layer serves as a distributed actuator for active vibration suppression and control. For a biaxially polarized piezoelectric actuator, a voltage applied to the distributed actuator layer results in two in-plane strains (S_1^{pc} and S_2^{pc}) due to the converse piezoelectric effect, where the superscript *pc* denotes the distributed piezoelectric controller/actuator layer. Since these strains are located a distance, a *moment arm,* away from the shell neutral surface, two distributed counteracting control moments are then introduced. Figure 4.2 illustrates the microscopic and macroscopic actions of the distributed piezoelectric actuator.

It is assumed that the piezoelectric actuator layer is not constrained and free from external in-plane forces, the stress effects are neglected in the analysis. Besides, the applied control voltage ϕ^{pc} is much larger than the induced voltage ϕ due to the direct effect in the distributed actuator. Thus, this induced ϕ is neglected in the derivation.

In this section, control moments of the open-loop and closed-loop feedback controls of the DPS are derived. In the closed-loop control systems, two control algorithms (1) Negative-velocity proportional feedback control; and (2) constant-amplitude negative-velocity feedback control, are proposed [4, 14]. System dynamic equations and the state equation of the open- and closed-loop controls are also formulated.

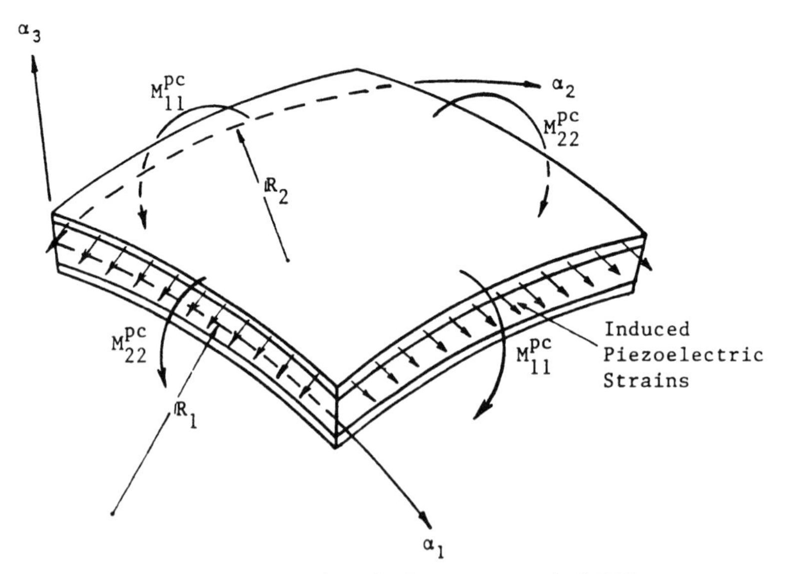

FIGURE 4.2. Distributed vibration control of DPS.

4.1. Open-Loop Control

In the open-loop control, a reference signal ϕ^{pc} can be injected into the distributed piezoelectric actuator. (Note this ϕ^{pc} is not taken from the sensor layer.) Due to the converse piezoelectric effect, two induced strains, S_1^{pc} and S_2^{pc}, in the distributed piezoelectric actuator can be calculated,

$$\begin{cases} S_1^{pc} = d_{31}\phi^{pc}/h^{pc}, & (18a) \\ S_2^{pc} = d_{32}\phi^{pc}/h^{pc}, & (18b) \end{cases}$$

where h^{pc} is the piezoelectric actuator thickness. In order to keep force equilibrium in the in-plane directions (defined by α_1- and α_2-axes), the induced strains in the piezoelectric layer cause effective in-plane strains. These strains result in resultant forces and moments. The moments are produced through the moment arm from the midplane of each layer to the neutral surface of the shell. Since a biaxially oriented piezoelectric actuator is used for the distributed sensor/actuator and its thickness is much thinner than that of the shell, the produced line moment M_{11}^{pc} in α_1-direction can be calculated by

$$M_{11}^{pc} = d_1^{pc} d_{31}\, Y_p\, \phi^{pc}, \tag{19}$$

where d_i^{pc} is the moment arm (distance measured from the neutral surface to the midplane of the piezoelectric actuator); d_{31} is the piezoelectric coefficient; and Y_p is the modulus of the piezoelectric actuator. Similarly, the line moment in the α_2-direction can be calculated by

$$M_{22}^{pc} = d_2^{pc}\, d_{32}\, Y_p\, \phi^{pc}. \tag{20}$$

Note that the injected voltage needs to be controlled in a way that the produced moments can counteract and suppress the shell oscillation. In a feedback control system, this reference signal is directly taken from the distributed sensor measurements. Thus, a closed-loop feedback control system is formed.

4.2. Closed-Loop Feedback Control

In this case, it is assumed that the distributed sensor output ϕ^{ps} is directly used in a feedback loop. In the later derivation, two feedback algorithms are considered: (1) negative-velocity proportional feedback control; and (2) constant-amplitude negative-velocity feedback control [4, 14]. Note that the piezoelectric sensor output (ϕ^{ps} used in the later derivation could be either ϕ_a^{ps} or ϕ_d^{ps}.

4.2.1. Negative-Velocity Proportional Feedback Control.

In this case, the sensor output voltage is directly differentiated, amplified, and fed back into the distributed actuator, which results in two counteracting moments in the α_1- and α_2-directions as discussed previously. The feedback control voltage ϕ^{pc} and the sensor output voltage ϕ^{ps} can be related by

$$\phi^{pc} = -\mathbb{C}\,\phi^{ps}, \tag{21}$$

where $\mathbb{C}$ is the control gain (voltage amplified ratio). Substituting the distributed (averaged) sensing expression into the above equation, one can write a feedback voltage in terms of displacements.

$$\begin{aligned}\phi^{pc} = -\mathbb{C}\frac{d}{dt}\Bigg(\frac{h^{ps}}{A^{ps}}\int_{A^{ps}}\Bigg\{h_{31}d_1^{ps}\Bigg[\frac{1}{A_1}\frac{\partial}{\partial\alpha_1}\left(\frac{u_1}{\mathbb{R}_1}-\frac{1}{A_1}\frac{\partial u_3}{\partial\alpha_1}\right)\\ +\frac{1}{A_1A_2}\frac{\partial A_1}{\partial\alpha_2}\left(\frac{u_2}{\mathbb{R}_2}-\frac{1}{A_2}\frac{\partial u_3}{\partial\alpha_2}\right)\Bigg]\\ +h_{32}d_2^{ps}\Bigg[\frac{1}{A_2}\frac{\partial}{\partial\alpha_2}\left(\frac{u_2}{\mathbb{R}_2}-\frac{1}{A_2}\frac{\partial u_3}{\partial\alpha_2}\right)\\ +\frac{1}{A_1A_2}\frac{\partial A_2}{\partial\alpha_1}\left(\frac{u_1}{\mathbb{R}_1}-\frac{1}{A_1}\frac{\partial u_3}{\partial\alpha_1}\right)\Bigg]\Bigg\}A_1A_2\,d\alpha_1\,d\alpha_2\Bigg).\end{aligned} \tag{22}$$

Substituting this into the counteracting control moments equations yields

$$\begin{aligned}M_{11}^{pc} &= -\mathbb{C}d_1^{pc}d_{31}Y_p\\ &\cdot\frac{d}{dt}\Bigg(\frac{h^{ps}}{A^{ps}}\int_{A^{ps}}\Bigg\{h_{31}d_1^{ps}\Bigg[\frac{1}{A_1}\frac{\partial}{\partial\alpha_1}\left(\frac{u_1}{\mathbb{R}_1}-\frac{1}{A_1}\frac{\partial u_3}{\partial\alpha_1}\right)\\ &+\frac{1}{A_1A_2}\frac{\partial A_1}{\partial\alpha_2}\left(\frac{u_2}{\mathbb{R}_2}-\frac{1}{A_2}\frac{\partial u_3}{\partial\alpha_2}\right)\Bigg]\\ &+h_{32}d_2^{ps}\Bigg[\frac{1}{A_2}\frac{\partial}{\partial\alpha_2}\left(\frac{u_2}{\mathbb{R}_2}-\frac{1}{A_2}\frac{\partial u_3}{\partial\alpha_2}\right)\\ &+\frac{1}{A_1A_2}\frac{\partial A_2}{\partial\alpha_1}\left(\frac{u_1}{\mathbb{R}_1}-\frac{1}{A_1}\frac{\partial u_3}{\partial\alpha_1}\right)\Bigg]\Bigg\}A_1A_2\,d\alpha_1\,d\alpha_2\Bigg),\end{aligned} \tag{23}$$

and

$$M_{22}^{pc} = -\mathbb{C}d_2^{pc}d_{32}Y_p$$

$$\cdot \frac{d}{dt}\left(\left(\frac{h^{ps}}{A^{ps}}\int_{A^{ps}}\left\{h_{31}d_1^{ps}\left[\frac{1}{A_1}\frac{\partial}{\partial\alpha_1}\left(\frac{u_1}{\mathbb{R}_1}-\frac{1}{A_1}\frac{\partial u_3}{\partial\alpha_1}\right)\right.\right.\right.\right.$$

$$\left.+\frac{1}{A_1A_2}\frac{\partial A_1}{\partial\alpha_2}\left(\frac{u_2}{\mathbb{R}_2}-\frac{1}{A_2}\frac{\partial u_3}{\partial\alpha_2}\right)\right]$$

$$+\ h_{32}d_2^{ps}\left[\frac{1}{A_2}\frac{\partial}{\partial\alpha_2}\left(\frac{u_2}{\mathbb{R}_2}-\frac{1}{A_2}\frac{\partial u_3}{\partial\alpha_2}\right)\right.$$

$$\left.\left.\left.+\frac{1}{A_1A_2}\frac{\partial A_2}{\partial\alpha_1}\left(\frac{u_1}{\mathbb{R}_1}-\frac{1}{A_1}\frac{\partial u_3}{\partial\alpha_1}\right)\right]\right\}A_1A_2d\alpha_1d\alpha_2\right). \tag{24}$$

Note that the feedback gain $\mathbb{C}$ is constant for all vibration modes. This concept can be extended to a more general case in which a modal control gain $\mathbb{C}_k$ is considered so that the control effort for each mode can be different. This concept will be demonstrated in the next control algorithm. (However, determination of modal control gains is not the topic in this report; therefore, it is not discussed.) The counteracting control moments can also be rewritten in modal coordinates as discussed previously.

Note that if only a single point velocity is considered in the feedback control, the counteracting control moment equations can be simplified by neglecting surface average as discussed in the previous section.

4.2.2. Constant-Amplitude Negative-Velocity Feedback Control.

In this case, the amplitude of feedback voltage is constant and the sign is opposite the velocity. Thus, the feedback voltage can be expressed as

$$\phi^{pc} = -\mathbb{C}\ \mathrm{SGN}\ \frac{d}{dt}(\phi^{ps}). \tag{25}$$

where $\mathbb{C}$ is the feedback gain and SGN is a signum function, i.e.

$$\mathrm{SGN}\ [\ u\ ] = \mathbf{-1} \text{ if } u < 0;\ \mathbf{0} \text{ if } u = 0;\ \mathbf{+1} \text{ if } u > 0. \tag{26}$$

More explicitly, the control moments are written as

$$M_{11}^{pc} = -\mathbb{C}_1^*\ \mathrm{SGN}\ \frac{d}{dt}(\phi^{ps}); \tag{27a}$$

$$M_{22}^{pc} = -\mathbb{C}_2^*\ \mathrm{SGN}\ \frac{d}{dt}(\phi^{ps}), \tag{27b}$$

where $\mathbb{C}_i^* = \mathbb{C}\, d_i^{pc}\, d_{3i}\, Y_p$. Note this velocity could be a single-point transverse velocity $\dot{u}_3(t, \alpha_1^*, \alpha_2^*)$ in the feedback control, where (α_1^*, α_2^*) denotes a specific location on the DPS. Thus,

$$\phi^{pc} = -\mathbb{C}\ \mathrm{SGN}\ [\ \dot{u}_3(t, \alpha_1^*, \alpha_2^*)\ ]. \tag{28}$$

And

$$M_{11}^{pc} = -\mathbb{C}^*\ \mathrm{SGN}\ [\ \dot{u}_3(t, \alpha_1^*, \alpha_2^*)\ ]; \tag{29a}$$

$$M_{22}^{pc} = -\mathbb{C}^*\ \mathrm{SGN}\ [\ \dot{u}_3(t, \alpha_1^*, \alpha_2^*)\ ]. \tag{29b}$$

Note that the feedback gains are constant for all vibration modes in the above equations. To extend the theory, an individual modal control gain $\mathbb{C}_k$ can be introduced into the equations to represent different feedback effects for various vibration modes. Thus, substituting the modal expression into the above equations yields

$$M_{11}^{pc} = -d_1^{pc} d_{31} Y_P \left(\sum_{k=1}^{\infty} \mathbb{C}_k\ \mathrm{SGN}[\dot{\lambda}_k(t)]\ U_{3k}(\alpha_1^*, \alpha_2^*) \right), \tag{30a}$$

$$M_{22}^{pc} = -d_2^{pc} d_{32} Y_P \left(\sum_{k-1}^{\infty} \mathbb{C}_k\ \mathrm{SGN}[\dot{\lambda}_k(t)]\ U_{3k}(\alpha_1^*, \alpha_2^*) \right). \tag{30b}$$

4.3. System Dynamic Equations

Substituting these induced piezoelectric normal forces and counteracting moments into the equation of motions yields

$$-\frac{\partial(N_{11}^* A_2)}{\partial\alpha_1} - \frac{\partial(N_{21}^* A_1)}{\partial\alpha_2} - N_{12}^* \frac{\partial A_1}{\partial\alpha_2} + N_{22}^* \frac{\partial A_2}{\partial\alpha_1} - A_1 A_2 \frac{Q_{13}^*}{\mathbb{R}_1} + A_1 A_2 \rho h \ddot{u}_1 = A_1 A_2 F_1, \tag{31a}$$

$$-\frac{\partial(N_{12}^* A_2)}{\partial\alpha_1} - \frac{\partial(N_{22}^* A_1)}{\partial\alpha_2} - N_{12}^* \frac{\partial A_2}{\partial\alpha_1} + N_{11}^* \frac{\partial A_1}{\partial\alpha_2} - A_1 A_2 \frac{Q_{23}^*}{\mathbb{R}_2} + A_1 A_2 \rho h \ddot{u}_2 = A_1 A_2 F_2, \tag{31b}$$

$$-\frac{\partial(Q_{13}^* A_2)}{\partial\alpha_1} - \frac{\partial(N_{23}^* A_1)}{\partial\alpha_2} + A_1 A_2 \left[\frac{N_{11}^*}{\mathbb{R}_1} + \frac{N_{22}^*}{\mathbb{R}_2} \right] + A_1 A_2 \rho h \ddot{u}_3 = A_1 A_2 F_3; \tag{31c}$$

where Q_{13}^* and Q_{23}^* are defined by

$$\left[\begin{array}{l} Q_{13}^* = \dfrac{1}{A_1A_2}\left[\dfrac{\partial(M_{11}^*A_2)}{\partial\alpha_1} + \dfrac{\partial(M_{12}^*A_1)}{\partial\alpha_2} + M_{12}^*\dfrac{\partial A_1}{\partial\alpha_2} - M_{22}^*\dfrac{\partial A_2}{\partial\alpha_1}\right], \quad (32a) \\ Q_{23}^* = \dfrac{1}{A_1A_2}\left[\dfrac{\partial(M_{12}^*A_2)}{\partial\alpha_1} + \dfrac{\partial(M_{22}^*A_1)}{\partial\alpha_2} + M_{12}^*\dfrac{\partial A_2}{\partial\alpha_1} - M_{11}^*\dfrac{\partial A_1}{\partial\alpha_2}\right]; \quad (32b) \end{array}\right.$$

where $N_{ij}^* = N_{ij} - N_{ij}^{pc}$; $M_{ij}^* = M_{ij} + M_{ij}^{pc}$; and $N_{ij}^{pc} = d_{ij}h^{ps}Y_p\phi^{ps}$. Note these resultant forces and moments are modified to include the induced normal forces and counteracting moments as discussed previously. In a special case in which the in-plane twisting effect is neglected, i.e., $M_{12}^* = M_{12}$ and $N_{12}^* = N_{12}$, the transverse vibration equation becomes

$$\begin{aligned} &-\frac{\partial}{\partial\alpha_1}\left(\frac{1}{A_1}\left[\frac{\partial(M_{11}^*A_2)}{\partial\alpha_1} + \frac{\partial(M_{12}A_1)}{\partial\alpha_2} + M_{12}\frac{\partial A_1}{\partial\alpha_2} - M_{22}^*\frac{\partial A_2}{\partial\alpha_1}\right]\right) \\ &-\frac{\partial}{\partial\alpha_2}\left(\frac{1}{A_2}\left[\frac{\partial(M_{12}A_2)}{\partial\alpha_1} + \frac{\partial(M_{22}^*A_1)}{\partial\alpha_2} + M_{12}\frac{\partial A_2}{\partial\alpha_1} - M_{11}^*\frac{\partial A_1}{\partial\alpha_2}\right]\right) \\ &+ A_1A_2\left[\frac{N_{11}^*}{\mathbb{R}_1} + \frac{N_{22}^*}{\mathbb{R}_2}\right] + A_1A_2\rho h\ddot{u}_3 = A_1A_2F_3. \end{aligned} \quad (33)$$

Note that the superscript * terms include the feedback control effects induced by the converse piezoelectric effect. Detailed definition of N_{ij} and M_{ij} can be found in [2, 3, 13].

Note that in all control algorithms, the moment arm, the actuator stiffness, and piezoelectric coefficients are all important in determining the magnitude of counteracting moments. In addition, the feedback gains $\mathbb{C}$ can also be extended to be modal-dependent modal gains, which manipulate differential modal control effects.

4.4. State Equation

In this section, the system dynamic equation is rewritten in a state equation form. Separating the control moments and induced piezoelectric resultant forces, moving them to the right side, and defining them as $H(\alpha_1, \alpha_2, t)$ gives

$$\begin{aligned} &H(\alpha_1, \alpha_2, t) \\ &= \frac{1}{A_1A_2\rho h}\left\{\frac{\partial}{\partial\alpha_1}\left[\frac{1}{A_1}\frac{\partial(M_{11}^{pc}A_2)}{\partial\alpha_1} - M_{22}^{pc}\frac{1}{A_1}\frac{\partial A_2}{\partial\alpha_1}\right]\right. \\ &\left.+ \frac{\partial}{\partial\alpha_2}\left[\frac{1}{A_2}\frac{\partial(M_{22}^{pc}A_1)}{\partial\alpha_2} - M_{11}^{pc}\frac{1}{A_2}\frac{\partial A_1}{\partial\alpha_2}\right]\right\} + \frac{1}{\rho h}\left[\frac{N_{11}^{pc}}{\mathbb{R}_1} + \frac{N_{22}^{pc}}{\mathbb{R}_2}\right]. \end{aligned} \quad (34)$$

The original elastic terms associated with the DPS are defined as

$$
\begin{aligned}
Lu_3 \\
= \frac{1}{A_1 A_2 \rho h} \Bigg\{ & - \frac{\partial}{\partial \alpha_1} \left(\frac{1}{A_1} \left[\frac{\partial (M_{11} A_2)}{\partial \alpha_1} + \frac{\partial (M_{12} A_1)}{\partial \alpha_2} + M_{12} \frac{\partial A_1}{\partial \alpha_2} - M_{22} \frac{\partial A_2}{\partial \alpha_1} \right] \right) \\
& - \frac{\partial}{\partial \alpha_2} \left(\frac{1}{A_2} \left[\frac{\partial (M_{12} A_2)}{\partial \alpha_1} + \frac{\partial (M_{22} A_1)}{\partial \alpha_2} + M_{12} \frac{\partial A_2}{\partial \alpha_1} - M_{11} \frac{\partial A_1}{\partial \alpha_2} \right] \right) \\
& + A_1 A_2 \left[\frac{N_{11}}{\mathbb{R}_1} + \frac{N_{22}}{\mathbb{R}_2} \right] \Bigg\}, \qquad (35)
\end{aligned}
$$

where L is a differential operator. Thus, the original equation can be simplified to

$$
\ddot{u}_3 + Lu_3 = \frac{F_3}{\rho h} + H(\alpha_1, \alpha_2, t). \qquad (36)
$$

Using state variable transformation, one can rewrite the above equation in a state equation form,

$$
\frac{\partial}{\partial t} \begin{bmatrix} u_3 \\ \dot{u}_3 \end{bmatrix} = \begin{bmatrix} 0 & 1 \\ -L & 0 \end{bmatrix} \cdot \begin{bmatrix} u_3 \\ \dot{u}_3 \end{bmatrix} + \begin{bmatrix} 0 \\ 1 \end{bmatrix} \cdot \{(F_3/\rho h) + H\}, \qquad (37a)
$$

or

$$
\frac{\partial \mathbf{X}}{\partial t} = \mathbf{A}\,\mathbf{X} + \mathbf{B}\,\mathbf{m}, \qquad (37b)
$$

where

$$
\begin{aligned}
& \mathbf{X} = \begin{bmatrix} u_3 \\ \dot{u}_3 \end{bmatrix}; \mathbf{A} = \begin{bmatrix} 0 & 1 \\ -L & 0 \end{bmatrix}; \mathbf{B} = [0 \quad 1]^t; \text{ and} \\
& \mathbf{m} = \{F_3/\rho h) + H\}. \qquad (38)
\end{aligned}
$$

4.5. Reduction Procedure

The system dynamic equations and the state equation can be further simplified to account for many other geometries if four geometric parameters (A_1, A_2, $\mathbb{R}_1$, and $\mathbb{R}_2$) are given, e.g., (1) $A_1 = 1$ (dx), $A_2 = 1$ (dy), $\mathbb{R}_1 = \infty$, and $\mathbb{R}_2 = \infty$ for a rectangular plate; (2) $A_1 = 1$ (radial direction dr), $A_2 = r$ (angular direction $d\theta$), $\mathbb{R}_1 = \infty$, and $\mathbb{R}_2 = \infty$ for a circular plate; (3) $A_1 = 1$ (longitudinal direction dx),

A_2 = a (circumferential direction $d\theta$), $\mathbb{R}_1 = \infty$, and $\mathbb{R}_2 = a$ (radius) for a cylinder (or a cylindrical panel); etc [2, 3].

5. CASE STUDIES

There are three cases presented in this section. The first case demonstrates a simple reduction procedure illustrating how to apply the generic distributed vibration sensing and control theories to a flexible manipulator case [1]. In the second case, a prototype flexible beam with distributed piezoelectric sensor/actuator is design and tested. The third case presents a piezoelectric micro-displacement actuator, in which its performance is evaluated analytically and experimentally [5, 6].

5.1. Application to Flexible Manipulators

In this case, a reduction procedure is demonstrated. By defining four geometrical parameters, one can easily reduce the original DPS equations to a variety of mechanical systems. In the flexible robot application, it is assumed that the effective piezoelectric direction α_1 is aligned with the longitudinal direction x of the manipulator as illustrated in Figure 4.3. (Note that the transverse direction is α_3 or z). Since this is a uniaxial application of the distributed sensing theory, h_{32} is neglected from the equation. Besides, the Lame's parameters for a beam structure are $A_1 = 1$ and $A_2 = 1$. The radii of curvature are $\mathbb{R}_1 = \infty$ and $\mathbb{R}_2 = \infty$.

Thus, the distributed (averaged) sensor output from the distributed piezoelectric sensor can be simplified as

$$\phi_a^{ps} = -\frac{h^{ps}}{A^{ps}} \int_{A^{ps}} \left(h_{31} d_1^{ps} \frac{\partial^2 u_3}{\partial x^2} \right) dA^{ps} \tag{39}$$

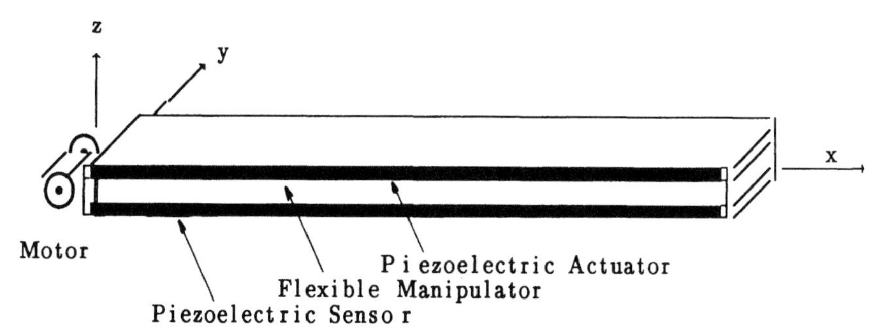

FIGURE 4.3. A flexible manipulator with distributed piezoelectric sensor and actuator.

Whenever a structural oscillation occurs, the distributed piezoelectric sensor can sense the oscillation and generate output signals. In the active distributed vibration control, this output voltage can be further processed and amplified. Then, the amplified high voltage is injected into the distributed piezoelectric actuator layer, generating a feedback force to control the oscillation of distributed systems. The counteracting control moments can be simplified as

1. Negative-velocity proportional feedback control:

$$M_{11}^{pc} = -\mathbb{C} d_1^{pc} d_{31} Y_p \frac{h^{ps}}{A^{ps}} \int_{A^{ps}} \left(h_{31} d_1^{ps} \frac{\partial^2 u_3}{\partial x^2} \right) dA^{ps}; \tag{40}$$

2. Constant-amplitude negative-velocity feedback control:

$$M_{11}^{pc} = -d_1^{pc} d_{31} Y_p \left[\sum_{k=1}^{\infty} \mathbb{C}_k \, \mathrm{SGN}[\dot{\lambda}_k(t)] U_{3k}(\alpha_1^*, \alpha_2^*) \right]. \tag{41}$$

For other standard smart mechanical components or structures, e.g., plates, cylinders, rings, etc., one can follow the same procedure to derive the sensing/control equations.

5.2. Distributed Structural Identification and Control

To demonstrate the proposed distributed vibration control and identification theories, a physical model was designed and tested in Dynamics and Systems Laboratory. One of the simplest DPSs, a cantilever beam, was used as an example. A Plexiglas cantilever beam (0.15m × 0.01m × 0.0018m) was sandwiched between two polymeric piezoelectric layers (40 μm). The top piezoelectric layer serves as a distributed actuator and the bottom a distributed sensor (Figure 4.3).

5.2.1. Laboratory Setup.

The physical model was clamped at one end on a fixture which was directly mounted on a shaker. A random noise signal was input into the shaker which provided a direct excitation to the beam model. The distributed piezoelectric sensor responded to the excitation and generated an output signal (which is a function of all distributed strains), and this signal was phase-shifted, amplified, and then fed back into the distributed piezoelectric actuator. The level of excitation was kept constant. Thus, the counteracting control moments were solely controlled by a gain control on the high-voltage amplifier. Figure 4.4 illustrates the laboratory setup.

5.2.2. Distributed Vibration Control and Identification.

Original frequency response (no feedback) of the cantilever beam subjected to random excitations was recorded first. Then, various feedback gains were ap-

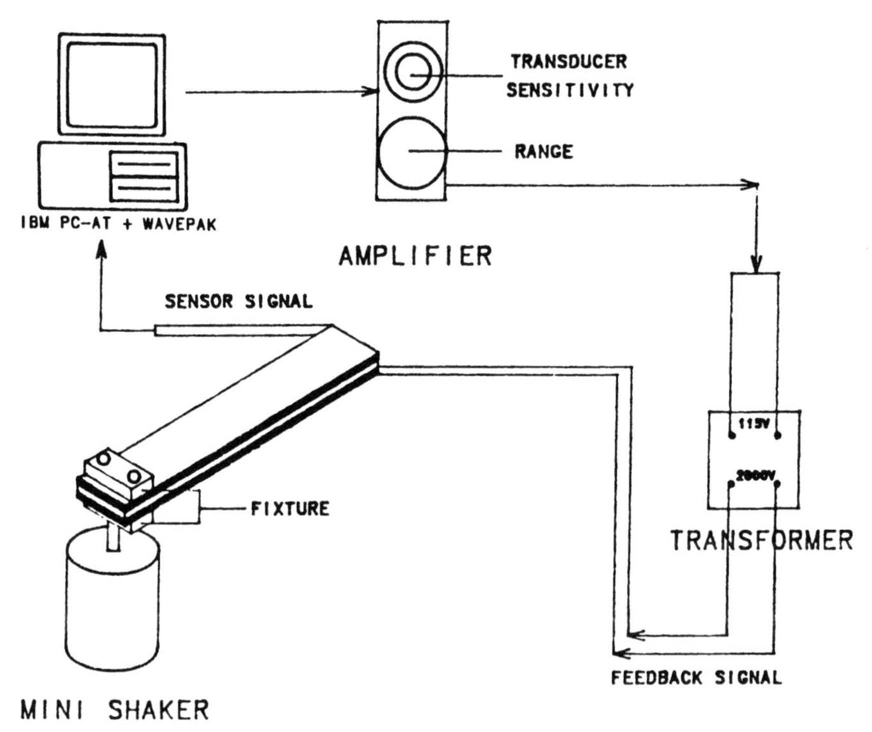

FIGURE 4.4. Laboratory setup for the feedback control system.

plied to the feedback control system and the attenuations after feedback were also recorded and compared with the original (uncontrolled) response. Figure 4.5 shows a test sample of two frequency responses of the first three modes, in which the higher represents the uncontrolled response and the lower represents the controlled response with a feedback gain of 1700. Note that the frequencies were also slightly reduced due to the enhanced system damping in feedback controls.

Figure 4.5 shows 60-Hz line noise and its harmonics which were induced by the feedback control electronics. This noise could be filtered out if the appropriate filter was used. Note that the distributed actuator controlled all three modes simultaneously. Table 4.1 summarized the distributed control effectiveness of the first three modes, five tested cases, with control gains from 100 to 1700.

It is observed that at lower feedback gains the control effectiveness was not as significant as those at higher gains. This is because there was a combined electromechanical effect of the direct and converse piezoelectric effects at lower feedback voltages. The control voltages were neutralized and the feedback control effects were reduced. At high feedback voltages, however, the feedback was

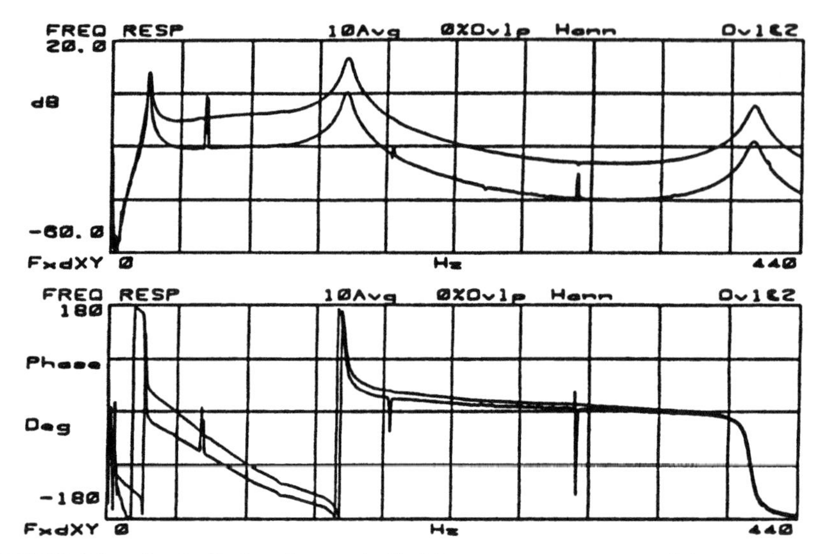

FIGURE 4.5. Controlled and uncontrolled frequency responses of the cantilever beam.

much larger than the self-generated voltage; and the direct effect was suppressed. Thus, the control effectiveness was much more significant.

Moreover, the control effectiveness of each mode is also different. It is observed that the controlled attenuation at higher modes is more effective than the lower modes. This is because the distributed actuator can effectively constrain the mode shape variation at higher modes. It should also be noted that in the above measurements no charge amplifier was used to amplify the sensor output signals.

TABLE 4.1. Vibration Control for the First Three Beam Vibration Modes (db).

Feedback Gain	22 Hz	149 Hz	411 Hz
	1st Mode	2nd Mode	3rd Mode
0	—	—	—
100	−0.265	−0.165	−0.361
650	−3.444	−11.164	−11.689
870	−3.766	−11.311	−12.326
1300	−4.915	−11.874	−12.647
1700	−6.694	−12.027	−13.410

5.3. A Piezoelectric Microdisplacement Actuator

High-precision and high-speed machine operation is very much in demand in the modern manufacturing industry. This section presents an active microposition control technique using a piezoelectric actuator [5, 6]. A general theory for the piezoelectric actuator subjected to mechanical excitations and feedback voltages is first developed. Effectiveness of the piezoelectric microposition attenuation is evaluated analytically and experimentally.

5.3.1. Model Definition.

The general idea is to utilize the converse piezoelectric effect generated in a piezoelectric slab to counteract the induced base excitation and to improve operation accuracy. A prototype model was designed and tested in laboratory to validate the theory. Figure 4.6 shows the prototype model with a layer of piezoelectric PVDF actuator. The model has a 0.25-in. thick steel base with a standard 10-32 stud which can be mounted on a shaker. A l-mm-thick PVDF polymer with an effective surface area of $4 \times 10^{-4} \mathrm{m}^2$ is sandwiched between two 0.25-in. Plexiglas layers which provide the same boundary conditions to the piezoelectric actuator. The bottom Plexiglas is epoxied to the steel base, and an interchangeable metal plate is screwed onto the top Plexiglas layer. A mini-accelerometer is attached above this metal plate. Thus, the seismic mass consists of all the items above the piezoelectric actuator—the Plexiglas and metal plates and the miniaccelerometer. The vibration of this seismic mass was monitored by the miniaccelerometer. The acceleration signal was phase-shifted, amplified, stepped up using a transformer and then injected into the piezoelectric polymer to achieve active feedback position control.

5.3.2. Theoretical Formulation.

Using general energy and force equilibrium concept, one can derive a general equation of motion for the piezoelectric actuator,

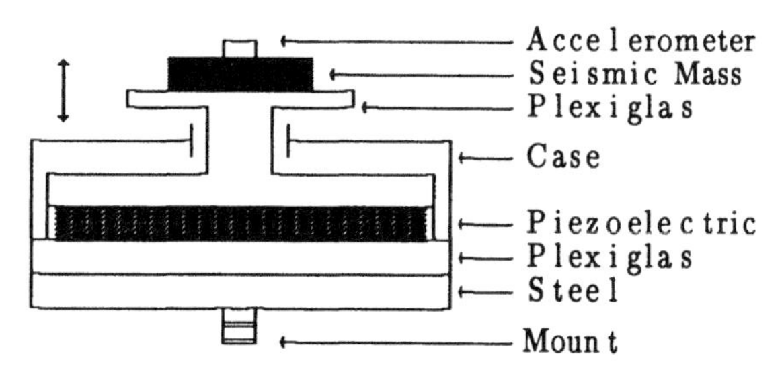

FIGURE 4.6. A prototype piezoelectric actuator.

$$\frac{\partial^2 u_3}{\partial t^2} = \frac{1}{\rho\kappa}\left(\frac{\partial S_{33}}{\partial \alpha_3} - d_{33}\frac{\partial E_3}{\partial \alpha_3}\right), \tag{42}$$

where ρ is the mass density; and κ is an elasticity constant. It is assumed that the electric field is constant over the thickness, i.e., $\partial E_3/\partial\alpha_3 = 0$; and the strain can be expressed as $S_{33} = \partial u_3/\partial\alpha_3$. Thus, the one-dimensional system equation for the piezoelectric actuator becomes

$$\ddot{u}_3 = \lambda^2 \frac{\partial^2 u_3}{\partial \alpha_3^2}, \tag{43}$$

where λ is the velocity of wave propagation in the piezoelectric actuator, $\lambda = \sqrt{1/\rho\kappa}$. Consider the general mechanical and electrical boundary conditions at $\alpha_3 = 0$ and h (h is the thickness of the piezoelectric actuator)

1. $\alpha_3 = 0$

$$\begin{cases} T_{33} = T_{33_0}\sin\omega t & \text{(44a)} \\ E = E_0 \sin\omega t & \text{(44b)} \end{cases},$$

2. $\alpha_3 = h$

$$\begin{cases} T_{33} = T_{33_0}\sin\omega t & \text{(45a)} \\ E = E_0 \sin\omega t & \text{(45b)} \end{cases},$$

where ω is the excitation frequency; and $E_3 = V_{fb_0}/h$ and V_{fb_0} is the feedback voltage. The steady-state solution of u_3 can be derived as

$$\begin{aligned} u_3(t) &= \int_0^h S_{33}(\alpha_3, t)\,d\alpha_3 \\ &= \int_0^h (d_{33}E_0 + \kappa T_{33_0})\left[\cos\left(\frac{\omega}{\lambda}\alpha_3\right) + \frac{1-\cos(\phi)}{\sin(\phi)}\sin\left(\frac{\omega}{\lambda}\alpha_3\right)\right]\cdot \sin\omega t \cdot d\alpha_3 \\ &= h(d_{33}E_0 + \kappa T_{33_0})\frac{\tan(\phi/2)}{(\phi/2)}\sin\omega t, \end{aligned} \tag{46}$$

where $\phi = (\omega h/\lambda)$. The feedback acceleration $G_{fb}(t)$ is produced by the piezoelectric polymer resulting from the converse piezoelectric effect and the g-level can be expressed as

$$\begin{aligned} G_{fb}(t) &= \frac{1}{g}\ddot{u}_3(t) \\ &= -\frac{\omega^2 h}{g}[d_{33}E_0 + \kappa T_{33_0}]\cdot \sin\omega t\,\frac{\tan(\phi/2)}{(\phi/2)}, \end{aligned} \tag{47}$$

where g is the gravity. It is assumed that the piezoelectric actuator is used to isolate the motion of a seismic mass m_s. After substituting $E_0 = V_{fb_0}/h$ into Equation (47), the converse piezoelectricity induced force $F_{fb}(t)$ resulting from the feedback voltage V_{fb_0} (controlled by the feedback gain) becomes

$$\begin{aligned} F_{fb}(t) &= m_s \cdot g \cdot G_{fb}(t) \\ &= -m_s \cdot \omega^2 \cdot [d_{33} V_{fb_0} + \kappa h T_{33_0}] \cdot \sin\omega t \, \frac{\tan(\phi/2)}{(\phi/2)}. \end{aligned} \tag{48}$$

Similarly, the equivalent force $F_b(t)$ introduced by the base excitation $G_{b_0}(t)$ is given by

$$F_b(t) = m_T \cdot g \cdot G_{b_0}(t), \tag{49}$$

where m_T is the total mass (including the piezoelectric actuator mass). The resultant acceleration G_r due to the combining effects of excitations G_{b_0} and feedback V_{fb_0} can be obtained by balancing the forces,

$$G_r = \frac{1}{g} \frac{F_b + F_{fb}}{m_T}. \tag{50}$$

Substituting Equations (48) and (49) into Equation (50) yields a general equation for the piezoelectric actuator subjected to the base excitations and feedback,

$$G_r(t) = \left[G_{b_0} - \omega^2 \cdot \left(\frac{d_{33}}{g} \cdot V_{fb_0} + \kappa h \, G_{b_0} \frac{m_s}{A} \right) \cdot \frac{\tan(\phi/2)}{(\phi/2)} \frac{m_s}{m_T} \right] \cdot \sin\omega t. \tag{51}$$

This is a general equation for a piezoelectric actuator with a seismic mass and it can also be used for other piezoelectric actuators. Integrating the acceleration twice yields a general stroke equation. In our case, we use a polymeric piezoelectric PVDF as the actuator.

Substituting the physical properties of the model, we found $\tan(\phi/2)/(\phi/2) \cong 1$. Thus,

$$G_r(t) = \sin\omega t \left[G_{bo} - \omega^2 \cdot \left(\frac{d_{33}}{g} V_{fbo} + \kappa G_{bo} \frac{m_s}{A} \right) \frac{m_s}{m_T} \right]. \tag{52}$$

If an excitation can be measured by an accelerometer (or built-in accelerometer), the accelerometer output can be processed and fed back into the piezoelectric actuator counteracting the oscillation and eliminating the disturbance. The active vibration isolation due to the feedback-induced converse effect can then be defined as the difference between the resultant acceleration $G_r(t)$ and the base excitation $G_b(t)$. Note that it is assumed the residual stress in the piezoelectric actuator is negligible in the active vibration isolation application. The isolation percentage $\mathbb{R}(\%)$ can be defined as

$$\mathbb{R}(\%) = \frac{G_b(t) - G_r(t)}{G_b(t)} \times 100 \tag{53a}$$

$$= \frac{\omega^2\, m_s}{G_{b_0}\, m_\tau} \left(\frac{d_{33}}{g} \cdot V_{fb_0} + \kappa h G_{b0} \frac{m_s}{A} \right) \times 100. \tag{53b}$$

The gradient of the isolation surface with respect to the excitation frequency and the feedback voltage is evaluated when the base excitation is constant ($G_{b_0} = G$) and feedback gain $\mathbb{C}$ varies ($V_{fb0} = \mathbb{C}V_O$ where V_O is the transducer output). When the base excitation is a constant $G_{b_0} = G$,

$$\mathbb{R}(\%) = \frac{\omega^2 m_s}{G_{b_0} m_\tau} \left(\frac{d_{33}}{g} \cdot V_{fb_0} + G \cdot \kappa h \frac{m_s}{A} \right) \times 100$$

$$= \frac{\omega^2 m_s}{G_{b_0} m_\tau} \left(\frac{d_{33}}{g} \mathbb{C}\, V_O + G \cdot \kappa h \frac{m_s}{A} \right) \times 100. \tag{54}$$

It is found that the second term is small compared with the first term after substituting all material properties into the equation. Thus,

$$\frac{\partial \mathbb{R}(\%)}{\partial V_{fb_0}} \cong \left(\frac{100}{G} \frac{d_{33}}{g} \frac{m_s}{m_\tau} \right) \cdot \omega^2 \tag{55a}$$

$$= (\text{constant}) \cdot \omega^2 \tag{55b}$$

$$\propto \omega^2 \tag{55c}$$

$$\frac{\partial \mathbb{R}(\%)}{\partial \omega} \cong \left[\frac{200}{G} \frac{d_{33}}{g} \frac{m_s}{m_\tau} \right] \cdot V_{fb_0} \cdot \omega$$

$$= (\text{constant}) \cdot V_{fb_0} \cdot \omega \tag{56a}$$

$$\propto V_{fb_0} \cdot \omega \tag{56b}$$

$$\propto \mathbb{C}\, V_O \cdot \omega \cdot \tag{56c}$$

Equation (55) shows that isolation is a quadratic function of the frequency. As the frequency increases, the isolation will increase as frequency squared. Equation (56) shows that the isolation varies linearly with feedback voltage at constant frequency.

5.3.3. Experimental Setup.

The prototype model (Figure 4.6) was mounted on a shaker that could be excited at various frequencies using a function generator. The seismic mass acceleration was sensed by the miniaccelerometer. The acceleration signal was phase-shifted, amplified, and applied across the piezoelectric layer in such a way that the piezoelectric vibration was 180° out of phase with that of the base. The shaker was excited at various frequencies and amplitudes from 250 Hz to 2.5k Hz. The experimental setup is shown in Figure 4.7.

5.3.4. Results and Discussion.

A sample spectra of the resultant accelerations before and after feedback at 500 Hz are given in Figures 4.8 and 4.9. (A detailed performance of the actuator will be discussed later.) Figure 4.8 shows the spectrum of the undamped signal while Figure 4.9a shows the spectrum of the attenuated signal and Figure 4.9b for the feedback (before a step-up transformer at 100 gain) at 500 Hz. The magnitude is 0.09241 V for uncontrolled response (Figure 4.9), and 0.09105 V for controlled response (Figure 4.9a). It should be noted that the feedback voltage was too high for the signal processing equipment to handle. The spectrum had to be observed before the final feedback voltage.

The base excitation was kept constant in this set of experiments. Four excitation amplitudes to the shaker were chosen for each frequency. Since the base excitation is a constant, the feedback gain is varied so that the feedback voltage injected into the piezoelectric polymer can be controlled. The spectrum of the undamped seismic mass acceleration signal was observed and the frequency and amplitude recorded. Then the feedback was applied and the spectra of the feed-

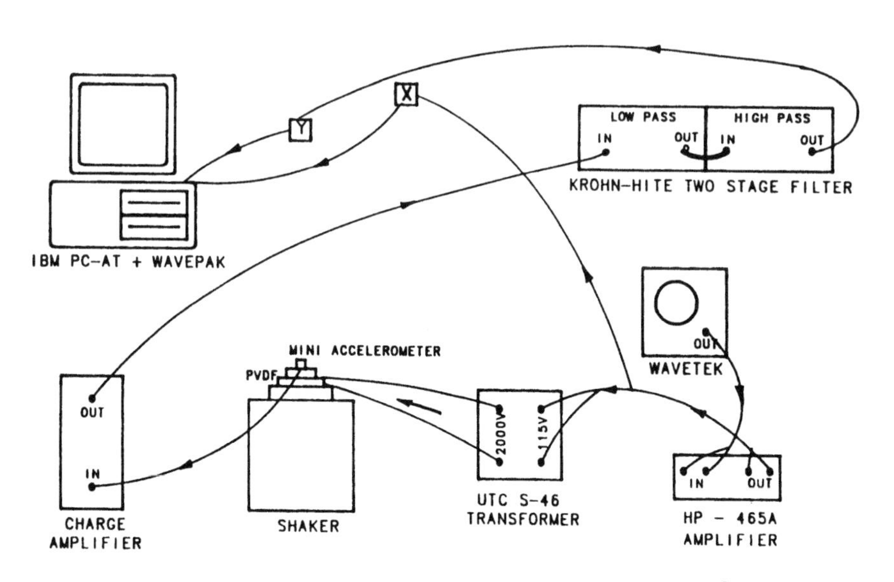

FIGURE 4.7. Apparatus for the active feedback position control.

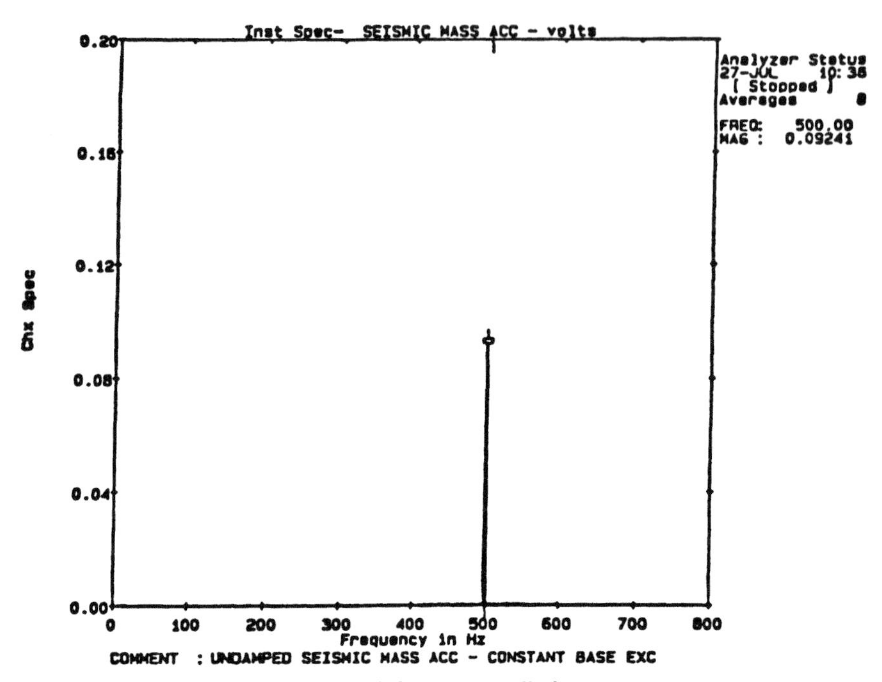

FIGURE 4.8. Spectrum of the uncontrolled response at 500 Hz.

back and attenuated seismic mass acceleration signals were observed and the peaks recorded. In this way data for a three-dimensional surface plot was collected.

Base excitations G_{b0} (g), excitation frequencies f(Hz), feedback voltages V_{fb_0} (volts), and resultant accelerations G_r (experimental) can be acquired from the experiments and isolation percentage $\mathbb{R}(\%)$(exp) and $\mathbb{R}(\%)$(thy) can be calculated and then plotted (Figure 4.10). Note that "exp" denotes experimental and "thy" denotes theoretical data.

Experimentally, the active vibration isolation is found to vary between 0.3% at 500 Hz and a feedback voltage of 12 V_{rms} to a maximum of 48% at 2500 Hz and a feedback voltage of 85 V_{rms}. Analytically, the isolation ranges from 0.07% at 250 Hz and a feedback voltage of 11.4 V_{rms} to 47.5% at 2500 Hz and feedback of 53.4 V_{rms}. The theory predicts that the isolation gradient should be a linear function of excitation voltage for a given frequency and a quadratic function of frequency for a given excitation voltage. Both the experimental and analytical data plots show this tendency. Due to an equipment limitation and the system stability, it was only tested up to 25k Hz. The performance could be even better if higher feedback voltages were available.

The absolute percentage differences (errors) $\epsilon(\%)$ between the theoretical

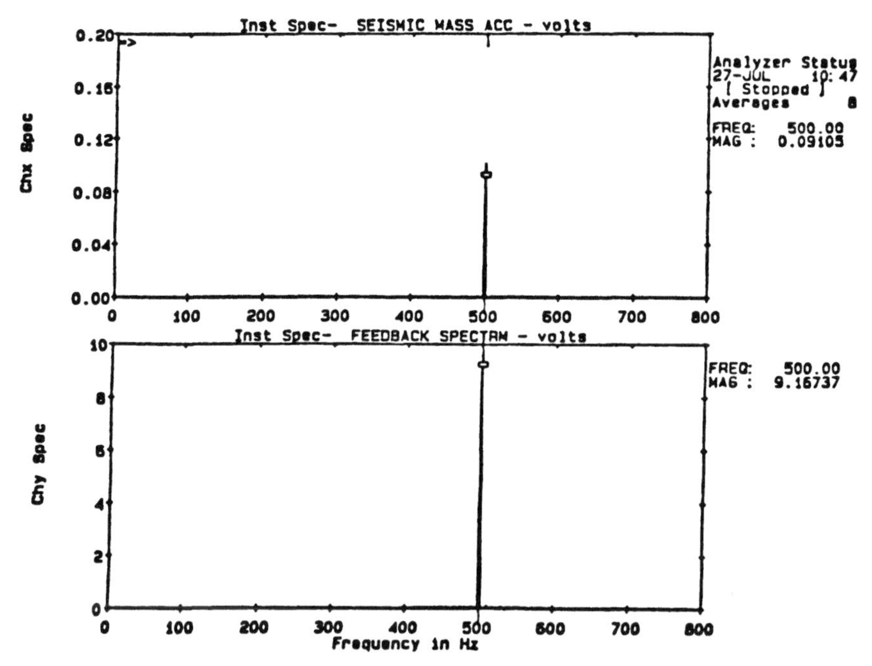

FIGURE 4.9. Spectra of (a) A controlled response; and (b) the feedback signal.

predictions and experimental results at a given frequency and a feedback voltage is given by

$$\epsilon(\%) = \frac{|G_r(\text{theoretical}) - G_r(\text{experimental})|}{G_r(\text{theoretical})} \times 100. \tag{57}$$

The absolute errors between the analytical and experimental data of the resultant seismic mass acceleration can be plotted in Figure 4.11. The maximum was about 7%.

It is observed that the theoretical data are higher than those obtained experimentally at high frequency due to the energy dissipation at high feedback gains. The other possible sources of the errors could be (1) An interaction between the direct effect and the converse effect in the piezoelectric PVDF actuator; (2) the feedback signal not being exactly 180° out of phase with the seismic mass acceleration; and (3) the nonlinearity associated with the experimental model (e.g., epoxy). Note that the theory developed and verified is for transverse direction only. Besides, the operation should be kept within the linear range of the mechanical system with the piezoelectric actuator. Otherwise, nonlinear control techniques should be used.

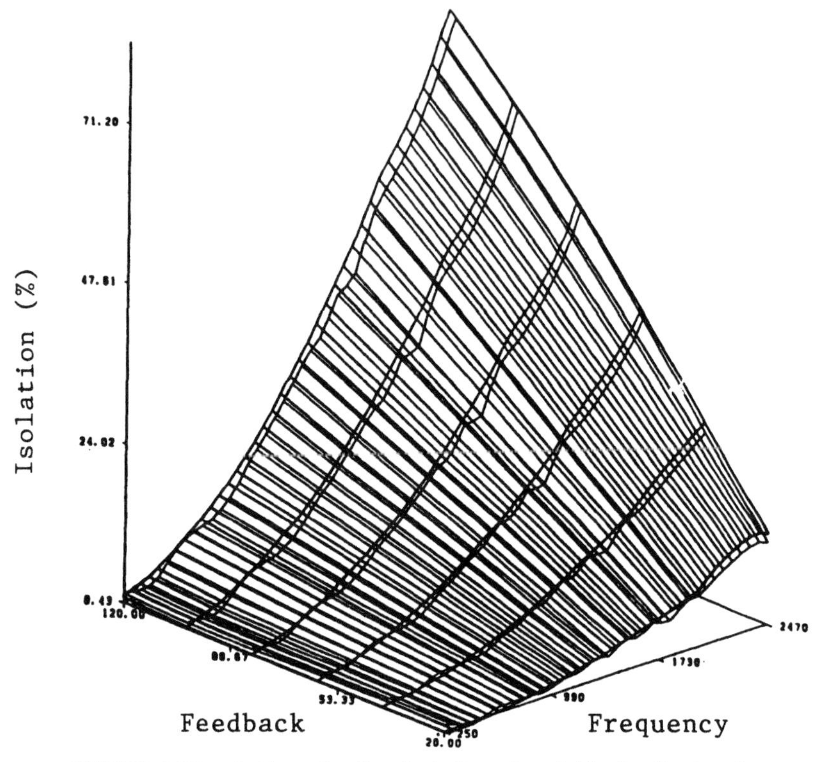

FIGURE 4.10. Active vibration isolation at variable feedback gains.

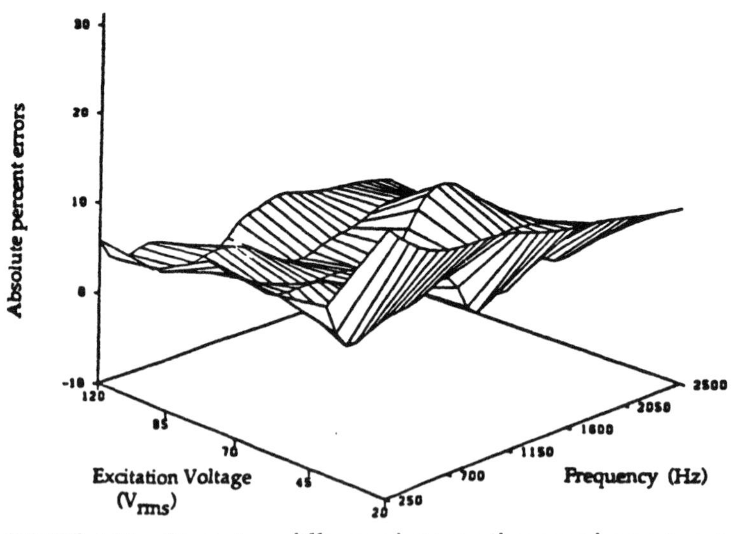

FIGURE 4.11. Percentage difference between theory and experiment.

6. SUMMARY AND CONCLUSIONS

A strong demand for smart high-performance manufacturing systems, e.g., robots and machines, has driven innovations of technologies and materials to design new-generation smart machines and robots. This report presents a new integrated piezoelectric sensor/actuator design for structural identification and active vibration control of smart mechanical systems. Since the majority of mechanical systems are distributed in nature, a new distributed vibration control theory and a distributed structural identification theory based on a generic DPS—a shell element coupled with distributed piezoelectric sensors and actuators—were proposed. If four essential geometrical parameters, two Lame's parameters A_is and two radii of curvature $\mathbb{R}_i$s are derived, the developed theories can be directly simplified to many commonly occurring mechanical components or structures, e.g., spheres, cylinders, plates, beams, etc.

Based on Maxwell's principle, piezoelectricity theory, and the Kirchhoff-Love theory, a distributed structural identification theory was first derived for a generic DPS—a shell. The derived theory shows that the DPS responses can be measured by the distributed piezoelectric sensor whose output is contributed by all vibration modes. A modal voltage equation was also derived, which shows that the modal voltage distribution (contour) is a function of space and time. The detailed modal voltage distribution pattern can be constructed by connecting all calculated point voltage amplitudes. An equation of an (area) averaged output was also derived. Note that for distributed structural identification, segmenting, and/or shaping sensor electrodes would be necessary.

A theory on active distributed vibration control of the DPSs was also derived using the same shell. The control effect was introduced and contributed by the voltage-induced piezoelectric strains (due to the converse piezoelectric effect) which result in counteracting control moments to suppress the DPS vibrations. The control moments for the DPS were formulated and a new set of system dynamic equations as well as state equations were also derived. Two feedback controls, namely, (1) Negative-velocity proportional feedback; and (2) constant-amplitude negative-velocity feedback, were derived.

A simple reduction procedure to simplify the generic theories to a flexible robot manipulator was proposed and demonstrated. Following the same procedure, one can apply the theories to other mechanical structures and components, e.g., plates, beams, cylinders, etc [2].

Distributed structural identification and control of a smart cantilever beam was demonstrated in laboratory experiments. Test results showed that the distributed piezoelectric sensor measured multimodes and the distributed piezoelectric actuator controlled multimodes. However, the modal control effectiveness was different for each mode, i.e., more effective for higher modes in this case. At low feedback voltages, it showed that the vibration control was not significant due to a combined electromechanical effect of the direct and converse piezoelectricity.

When the feedback voltage overcame the self-generated voltage, the distributed vibration controls were very effective.

An active microposition feedback control technique using piezoelectric actuators was also studied analytically and experimentally. A general mathematical model of the piezoelectric actuator was first formulated and the theory associated with the model was also proposed. Theoretical solutions for the actuation and isolation control at variable feedback gains were also derived. A prototype piezoelectric actuator made of a β-phase piezoelectric polyvinylidene fluoride (PVDF) polymer was designed and evaluated. A miniaccelerometer was used to monitor the system responses; and the output signals were processed, phase-shifted, fed back, and then injected into the piezoelectric actuator to counteract the base excitations. Effectiveness of the actuation and isolation was proved significant. Note that this technique is primarily for microposition feedback controls of high-precision operations. For relatively large stroke, it would require much thicker piezoelectric actuator. Some other factors, such as breakdown voltage, temperature effects, noise, phase shift, time delay, etc. could also influence the control effectiveness. The potential applications of the technique include positioning, grinding, laser mounts, polishing, machining, etc.

In distributed structural identification, using averaged distributed voltage formulation can introduce zero voltage when involving antisymmetrical modes of DPSs. Should this problem occur, a point reference voltage or a local averaged voltage of an interested bounded area can be used in feedback control systems. Note that if a bounded sensor area is considered, segmentation of surface electrodes would be necessary. Determination of various modal control gains to achieve optimal controls of DPSs is also an interesting and challenging problem which needs to be further studied and explored.

APPENDIX 1. NOMENCLATURE

$\{\cdot\}$ = vector.
$[\cdot]$ = matrix.
$[\cdot]^t$ = matrix transpose.
A^{ps} = area.
A_1, A_2 = Lame's parameter.
$\mathbb{C}, \mathbb{C}^*$ = feedback gains.
$[c^D]$ = elasticity matrix evaluated at constant dielectric displacement.
$[d_{ij}]$ = piezoelectric constant matrix.
$\{D_i\}$ = dielectric displacement vector.
d_i^{pc} = distance measurement measured from neutral surface.
$\{E_i\}$ = electric field vector.
F_i = external mechanical force in α_i-direction.

h = shell thickness.
h^{pc} = piezoelectric controller/actuator thickness.
h^{ps} = piezoelectric sensor thickness.
$[h_{ij}]$ = piezoelectric constant matrix.
M_{ij} = resultant moments.
N_{ij} = resultant forces.
$\mathbb{R}_1$, $\mathbb{R}_2$ = radii of curvature of α_1- and α_2-axes, respectively.
SGN = signum function.
$[s^E]$ = elastic compliance matrix measured at constant electric field.
$\{S_i\}$ = strain vector.
t = time.
$\{T_i\}$ = stress vector.
u_i = displacement in α_i-direction.
$\dot{u}_i$ = velocity in α_i-direction.
$\ddot{u}_i$ = acceleration in α_i-direction.
U_{ik} = kth modal function in ith direction.
Y_p = Young's modulus of piezoelectric material.
α_1, α_2, α_3 = three principal axes in a curvilinear coordinate system.
$[\beta^S_{ij}]$ = dielectric impermeability matrix evaluated at constant strain.
∇ = differential operator.
ϕ = electric potential.
ϕ^{ps} = sensor output.
ϕ^{pc} = feedback to actuator.
$[\epsilon^T]$ = dielectric matrix evaluated at constant strain.
λ_k = kth modal participation factor.
ρ = density.

ACKNOWLEDGMENT

This research was supported by a grant from the National Science Foundation (No. RII-8610671) and the Kentucky Commonwealth, and a grant on flexible assembly and intelligent machines (1988–1989) from the Center for Robotics and Manufacturing Systems (CRMS) at the University of Kentucky. All my past graduate students (C.I. Tseng, S. Pandita, K.J. Liu, and M. Gadre) who contributed to this research are also gratefully acknowledged.

REFERENCES

1. H.S. Tzou, "Integrated distributed sensing and active vibration suppression of flexible manipulators using distributed piezoelectrics," *J. Robotic Systems*, vol. 6.6, pp. 745–767, Dec. 1989.

2. H.S. Tzou, *Piezoelectric Shells (Distributed Sensing and Control of Continua)*, Chapters 3–5, Kluwer Academic Publishers, Dordrecht/Boston/London, February 1993.
3. W. Soedel, *Vibrations of Shells and Plates*. Chap. 3, New York: Marcel Dekker, 1981.
4. H.S. Tzou, "Distributed modal identification and vibration control of continua: theory and applications," *ASME J. of Dynamic Systems, Measurement, and Control*, vol. 113, pp. 494–499, Sept. 1991.
5. H.S. Tzou and M. Gadre, "Active vibration isolation by piezoelectric polymer with variable feedback gain," *AIAA J.*, vol. 26, pp. 1014–1017, Aug. 1988.
6. H.S. Tzou and M. Gadre, "Active vibration isolation and excitation by a piezoelectric slab with constant feedback gains," *J. Sound and Vibration*, vol. 136, pp. 477–490, Feb. 1990.
7. A.B. Palazzolo, R.R. Lin, R.R. Kascak, and R.M. Alexander, "Active control of transient rotordynamic vibration by optimal control methods," *ASME Journal of Engineering for Gas Turbines and Power*, vol. 111, p. 265, Sept. 1989.
8. H.S. Tzou, "Active vibration control of flexible structures via converse piezoelectricity," *Developments in Mechanics*, vol. 14-C, pp. 1201–1206, Aug. 1987.
9. E.F. Crawley and J. de Luis, "Use of piezoelectric actuators as elements of intelligent structures," *AIAA J.*, vol. 25, pp. 1373–1385, Oct. 1987.
10. S. Hanagud and M.W. Obal, "Identification of dynamic coupling coefficients in a structure with piezoelectric sensors and actuators," AIAA paper No. 88-2418, SDM Conference, April 1988.
11. A. Baz and S. Poh, "Performance of an active control system with piezoelectric actuators," *J. Sound and Vibration*, vol. 126, pp. 327–343, April 1988.
12. J.M. Plump, J.E. Hubbard, and T. Baily, "Nonlinear control of a distributed system: Simulation and experimental results," *ASME J. Dynamic Systems, Measurement, and Control*, vol. 109, pp. 133–139, June 1987.
13. H.S. Tzou and M. Gadre, "Theoretical analysis of a multi-layered thin shell coupled with piezoelectric shell actuators for distributed vibration controls," *J. Sound and Vibration*, vol. 132, pp. 433–450, Aug. 1989.
14. H.S. Tzou and C.I. Tseng, "Distributed piezoelectric sensor/actuator design for dynamic measurement/control of distributed parameter systems: A finite element approach," *J. Sound and Vibration*, vol. 138, pp. 17–34, April 1990.
15. J. Curie and P. Curie, Acad. Science (Paris), 91, 1880, 294 and 383.
16. H.F. Tiesten, *Linear Piezoelectric Plate Vibrations*. Chap. 2, New York: Plenum Press, 1969.
17. H.S. Tzou and S. Pandita, "A multipurpose dynamic and tactile sensor for robot manipulators," *J. Robotic Systems*, vol. 4.6, pp. 719–741, Dec. 1987.

5

Integrated Piezoelectric Sensor/Actuator Design for Distributed Identification and Control of Smart Machines and Flexible Robots Part 2: Finite Element Formulation and Analysis

H. S. Tzou
Department of Mechanical Engineering, Center for Robotics and Manufacturing Systems, University of Kentucky, Lexington, KY 40506-0046

C. I. Tseng
Department of Mechanical Engineering
Feng-Chia University, Taichung,
Taiwan, R.O.C.

1. INTRODUCTION

This report presents a new piezoelectric sensor/actuator design for new smart robots and machine structures/components. The integrated piezoelectric sensor/actuator provides self-monitoring and control capabilities for smart mechanical systems. This is the second part of the report, concentrating on numerical aspects—finite element formulation, analysis, and applications. In this report, a new piezoelectric finite element is derived for modeling piezoelectric sensor/actuator systems. Distributed structural identification and vibration control of distributed parameter systems (DPSs) are also derived and formulated. Applications of the theory to distributed modal identification and control, flexible robots, high-precision machine operations, etc. are demonstrated in case studies.

1.1. Background

The rapid development of high-speed computers has facilitated the use of computation techniques in a variety of engineering applications. Today, the finite element method is one of the most popular techniques in modern engineering design and analysis of complicated structures and multifield problems. In recent years, however, research on piezoelectric smart machines and structures has primarily focused on experimental and theoretical studies. General piezoelectric finite element development was relatively limited [1–4]. In general, experimental models are limited by size, cost, noise, and many other laboratory unknowns. Theoretical models can be more general, however, analytical solutions are restricted to relatively simple geometries and boundary conditions. When the geometry and/or boundary conditions become relatively complicated, difficulties occur both on theoretical and experimental models. Thus, the finite element development becomes very important in modeling and analysis of advanced distributed parameter systems (DPSs) with integrated distributed piezoelectric sensors and/or actuators.

Piezoelectric solid finite elements were formulated and applied to piezoceramic transducer and oscillator designs [3, 4]. However, the derived isoparametric hexahedron and tetrahedral elements are too thick and very inefficient for thin DPS applications. Thus, Tzou and Tseng formulated a new thin piezoelectric solid finite element and applied to the modeling and analysis of large flexible DPSs—shells and plates with distributed piezoelectric sensors/actuators [1, 2, 5].

1.2. Objectives

In this chapter, a new improved piezoelectric finite element with three internal degrees of freedom (DOFs) is formulated. Then structural identification and control of DPSs using the piezoelectric finite element is derived. State variable transformation of the dynamic equation is also presented. A micropositioning device and a zero-curvature shell (plate) are investigated in case studies. Distributed structural identification and control of DPSs—elastic structures—are studied and evaluated.

2. NEW PIEZOELECTRIC FINITE ELEMENT FORMULATION

In this section, a new thin piezoelectric finite solid element with internal DOFs is formulated using a variational method and Hamilton's principle. The system matrix equation and control strategies are also derived.

2.1. Variational Equation

The Lagrangian $\mathcal{L}$ of a bounded piezoelectric body is defined by a summation of all kinetic energy $\mathcal{T}$ and potential energy $\mathcal{U}$ (including strain and electrical energies) [5],

$$\mathcal{L} = \int_{\mathcal{V}} (\mathcal{T} - \mathcal{U}) d\mathcal{V}$$

$$= \int_{\mathcal{V}} \left(\frac{1}{2} \rho \cdot \{\dot{u}\}^t \{\dot{u}\} - \frac{1}{2} [\{S\}^t \{T\} - \{E\}^t \{D\}] \right) d\mathcal{V}, \tag{1}$$

where $\dot{u}$ is the velocity (time derivative of the displacement u); $\mathcal{L}$ is the Lagrangian; and $\mathcal{V}$ is the piezoelectric volume. The virtual work $\delta\mathcal{W}$ done by the external forces and the applied surface charge σ is

$$\delta\mathcal{W} = \int_{\mathcal{V}} \{\delta u\}^t \{P_b\} d\mathcal{V} + \int_{S_1} \{\delta u\}^t \{P_S\} dS_1 + \{\delta u\}^t \{P_C\} - \int_{S_2} \delta\phi \, \sigma \, dS_2, \tag{2}$$

where $\{P_b\}$ is the body force, S_i is the surface area; $\{P_s\}$ is the surface force; $\{P_c\}$ is the concentrated load; and σ is the surface charge. Based on the Lagrangian and the virtual work defined above, the dynamic equations of a piezoelectric structure can be derived using Hamilton's principle

$$\int_{t_1}^{t_2} \delta(\mathcal{L} + \mathcal{W}) dt = 0, \tag{3}$$

where t_1 and t_2 define the time interval, and all variations must vanish at $t = t_1$ and $t = t_2$. Thus, substituting individual terms in Hamilton's equation and taking the variational yield

$$\int_{\mathcal{V}} [\rho\{\delta\dot{u}\}^t\{\dot{u}\} - \{\delta S\}^t[c]\{S\} + \{\delta S\}^t[e]\{E\}$$
$$- \{\delta E\}^t[e]\{S\} - \{\delta E\}^t[\epsilon]\{E\} + \{\delta u\}^t[P_b]] d\mathcal{V}$$
$$+ \int_{S_1} \{\delta u\}^t \{P_S\} dS_1 - \int_{S_2} \delta\phi \, \sigma \, dS_2 + \{\delta u\}^t \{P_C\} = 0. \tag{4}$$

To derive the electroelastic matrix relationship for a piezoelectric finite element, the displacement $\{u\}$ and electric potential ϕ are defined in terms of nodal variables via the shape function matrices $[N_u]$ and $[N_\phi]$,

$$\begin{cases} \{u\} = [N_u]\{u_i\}, & (5) \\ \{\phi\} = [N_\phi]\{\phi_i\}. & (6) \end{cases}$$

Note that the subscript i denotes the nodal variables. The strains {S} are defined by the first derivative of nodal displacement vector {u} using a differential operator matrix $[L_u]$ [6]. The electric field vector {E} is defined by the electrical potential energy ϕ using a gradient operator ∇. Writing the strain {S} and {E} in nodal variables yields

$$\begin{cases} \{S\} = [B_u]\{u_i\}, & (7) \\ \{E\} = -[B_\phi]\{\phi_i\}, & (8) \end{cases}$$

where $[B_u] = [L_u][N_u]$ and $[B_\phi] = \nabla\, [N_\phi]$.

2.2. New Piezoelectric Element with Internal DOFs

Conventional isoparametric solid elements have two significant deficiencies in thin structural analysis. First, if the element thickness is thin compared with element span, an excessive shear strain energy is stored in the thickness direction. Second, the stiffness coefficients in the thickness direction become much higher than those in planar directions. This results in relatively poor estimations and inaccurate results [6, 7]. An important technique to improve the behavior of isoparametric elements is to introduce internal degrees of freedom (DOFs) to correct the above problems.

The original piezoelectric solid element was an eight-node hexahedron element [1]. The added three internal nodal DOFs are numbered from 9 to 11. By adding these internal DOFs to the dependent variables, one can derive the displacement {d} vector as

$$\{d\} = [N_u]\{u_i\} + [Z]\{a_j\}, \tag{9}$$

where $[N_u]$ is the displacement shape function matrix for nodal displacements $\{u_i\}$; and [Z] is the extra mode shape function matrix for the added internal DOFs $\{a_j\}$. [Z] and $\{a_j\}$ are represented as

$$[Z] = \sum_{j=9}^{11} \begin{bmatrix} N_j & 0 & 0 \\ 0 & N_j & 0 \\ 0 & 0 & N_j \end{bmatrix}, \tag{10}$$

$$\{a_j\} = \sum_{j=9}^{11} [u_j \quad v_j \quad w_j]^t. \tag{11}$$

Note that the added DOFs $\{a_j\}$ are not physical displacements, which can be regarded as generalized coordinates, or as displacements relative to the nodal displacements. These displacements vanish at all element edges, so that these DOFs are internal and have no effect on interelement compatibility. The strain-displacement equation {S} can be written as

$$\{S\} = [B_u]\{u_i\} + [G]\{a_j\}, \tag{12}$$

where

$$[B_u] = [L_u][N_u], \tag{13}$$

$$[G] = \sum_{j=9}^{11} \begin{bmatrix} \partial N_j/\partial x & 0 & 0 \\ 0 & \partial N_j/\partial y & 0 \\ 0 & 0 & \partial N_j/\partial z \\ 0 & \partial N_j/\partial z & \partial N_j/\partial y \\ \partial N_j/\partial z & 0 & \partial N_j/\partial x \\ \partial N_j/\partial y & \partial N_j/\partial x & 0 \end{bmatrix}. \tag{14}$$

Note $[L_u]$ is a differential operator matrix. Using the variational principle, one can derive the electric enthalpy $\mathcal{U}$ of the piezoelectric finite element with the internal DOFs as

$$\begin{aligned} \delta\mathcal{U} = & \int_{\mathcal{V}} \begin{Bmatrix} [B_u]^t \\ [G]^t \end{Bmatrix} [c]\{[B_u][G]\} \begin{bmatrix} \{u_i\} \\ \{a_j\} \end{bmatrix} d\mathcal{V} \delta \begin{bmatrix} \{u_i\} \\ \{a_j\} \end{bmatrix} \\ & - \int_{\mathcal{V}} \begin{Bmatrix} [B_u]^t \\ [G]^t \end{Bmatrix} [e]^t[B_\phi] d\mathcal{V}\{\phi_i\} \cdot \delta \begin{bmatrix} \{u_i\} \\ \{a_j\} \end{bmatrix} \\ & - \int_{\mathcal{V}} [B_\phi]^t[e]\{[B_u][G]\} d\mathcal{V} \begin{bmatrix} \{u_i\} \\ \{a_j\} \end{bmatrix} \cdot \delta\{\phi_i\} \\ & - \int_{\mathcal{V}} [B_\phi]^t[\epsilon][B_\phi] d\mathcal{V}\{\phi_i\} \cdot \delta\{\phi_i\}, \end{aligned} \tag{15}$$

where [c] is the elasticity matrix; $\mathcal{V}$ denotes the integration volume; $[e]^t$ is a transpose of the dielectric permittivity matrix; $[B_\phi]$ is the shape function for electric potential $\{\phi_i\}$; and $[\epsilon]$ is the dielectric matrix. The static homogeneous system equations of the piezoelectric element are derived as

$$\begin{bmatrix} [k_{uu}] & [k_{ua}] \\ [k_{au}] & [k_{aa}] \end{bmatrix} \begin{bmatrix} \{u_i\} \\ \{a_j\} \end{bmatrix} + \begin{bmatrix} [k_{u\phi}] \\ [k_{a\phi}] \end{bmatrix} \{\phi_i\} = 0, \tag{16}$$

$$[[k_{\phi u}][k_{\phi a}]] \begin{bmatrix} \{u_i\} \\ \{a_j\} \end{bmatrix} + [k_{\phi\phi}]\{\phi_i\} = 0, \tag{17}$$

where the elemental matrices are defined as

$$[k_{ua}] = \int_{\mathcal{V}} [B_u]^t[c][G]d\mathcal{V}, \tag{18a}$$

$$[k_{au}] = [k_{ua}]^t, \tag{18b}$$

$$[k_{aa}] = \int_{\mathcal{V}} [G]^t[c][G]d\mathcal{V}, \tag{18c}$$

$$[k_{\phi a}] = \int_{\mathcal{V}} [B_\phi]^t[e][G]d\mathcal{V}, \tag{18d}$$

$$[k_{a\phi}] = [k_{\phi a}]^t. \tag{18e}$$

Note the lowercase letters represent the element properties (e.g., stiffness, damping, charge, force, etc.) and the uppercase letters denote the corresponding assembled global properties. The uppercase letters will be used later.

2.3. System Equations

Since the internal DOFs do not contribute any physical significance, they can be condensed from the system equations to improve computation efficiency. Employing Guyan's reduction method [8], one can obtain the modified element static matrix equations as

$$[k^*_{uu}]\{u_i\} + [k^*_{u\phi}]\{\phi_i\} = \{0\}, \tag{19}$$

$$[k^*_{\phi u}]\{u_i\} + [k_{\phi\phi}]\{\phi_i\} = \{0\}; \tag{20}$$

where

$$[k^*_{uu}] = [k_{uu}] - [k_{ua}][k_{aa}]^{-1}[k_{au}], \tag{21}$$

$$[k^*_{u\phi}] = [k_{u\phi}] - [k_{ua}][k_{aa}]^{-1}[k_{a\phi}]. \tag{22}$$

Note that $[k^*_{uu}]$ is the modified elastic stiffness matrix and $[k^*_{u\phi}]$ is the piezoelectric stiffness matrix. The modified dynamic equations are derived accordingly,

$$[m_{uu}]\{\ddot{u}\} + [c]\{\dot{u}\} + [k^*_{uu}]\{u\} + [k^*_{u\phi}]\{\phi\} = \{f_i\}, \tag{23}$$

$$[k^*_{\phi u}]\{u\} + [k_{\phi\phi}]\{\phi\} = \{q_i\}, \tag{24}$$

where $[m_{uu}]$ is the element mass matrix; [c] is the element proportional damping matrix. Note that $\{f_i\}$ is the mechanical excitation and $\{q_i\}$ is the electric excitation of the piezoelectric element. If the piezoelectric properties are not considered in the equation, the element is reduced to a conventional thin elastic element.

3. EIGENVALUE ANALYSIS

The mode shapes and natural frequencies can be obtained by reducing assembled global matrices into a standard eigenvalue form. For an eigenvalue analysis, undamped homogeneous system matrices are used, i.e.,

$$[M]\{\ddot{u}\} + [K_{uu}]\{u\} + [K_{u\phi}]\{\phi\} = 0, \tag{25}$$

$$[K_{\phi u}]\{u\} + [K_{\phi\phi}]\{\phi\} = 0. \tag{26}$$

The unspecified potentials are condensed from the system matrices. Equations (25) and (26) can be reduced to a standard eigenvalue equation,

$$([K^*] - \omega^2[M])\{u\} = 0, \tag{27}$$

where

$$[K^*] = [K_{uu}] - [K_{u\phi}][K_{\phi\phi}]^{-1}[K_{\phi u}]. \tag{28}$$

4. STRUCTURAL IDENTIFICATION AND CONTROLS

Piezoelectric elements can be used as sensing elements for structural identification via the direct piezoelectric effect and active control elements—actuators—via the converse piezoelectric effect. Figure 5.1 illustrates a generic elastic DPS coupled with a distributed piezoelectric sensing layer and a piezoelectric actuator layer. The corresponding governing identification and control equations are formulated in this section.

Assembling all element matrices and applying Guyan's reduction technique, one can decouple the displacement and electric potential dynamic equations as:

$$[M]\{\ddot{u}\} + [C]\{\dot{u}\} + [K^*]\{u\} = \{F\} + \{F_e\}, \tag{29}$$

$$\{\phi\} = [K_{\phi\phi}]^{-1}(\{Q\} - [K_{\phi u}]\{u\}) \tag{30}$$

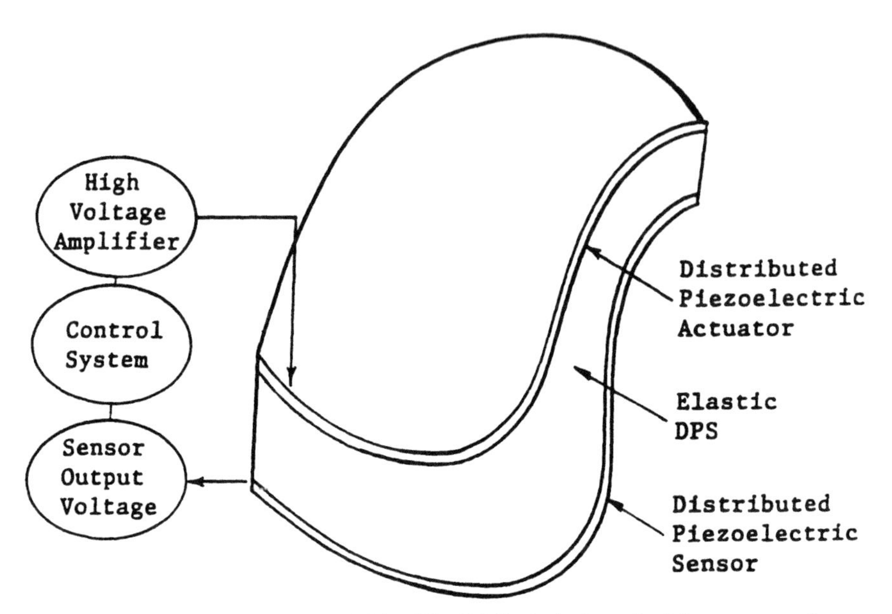

FIGURE 5.1. A generic smart DPS with distributed piezoelectric sensor and actuator.

with

$$[K^*] = [K_{uu}] - [K_{u\phi}][K_{\phi\phi}]^{-1}[K_{\phi u}], \tag{31}$$

$$\{F_e\} = -[K_{u\phi}][K_{\phi\phi}]^{-1}\{Q\}, \tag{32}$$

where [M], [C], $[K_{uu}]$, and [K*] are $(n \times n)$ matrices; $[K_{u\phi}]$, $[K_{\phi\phi}]$, and $[K_{\phi u}]$ are, respectively, $(n \times j)$, $(j \times j)$, and $(j \times n)$ matrices; {F}, $\{F_e\}$, {Q}, and {u} are, respectively, $(n \times 1)$, $(n \times 1)$, $(j \times 1)$, and $(n \times 1)$ vectors. Note that there are two force components, {F} and $\{F_e\}$, in Equation (29); the former is the mechanical excitation force and the latter is the feedback control force. Moreover, the sensor output is also contributed by two components: (1) feedback {Q}; and (2) displacement {u} in the general expression. However, usually there is no feedback input to the sensor layer, i.e., $\{Q\} = 0$, in practical applications.

The electric potentials calculated in Equation (29) are regarded as the output signals of piezoelectric sensors for structural identification; and they can be further processed to provide feedback to peizoelectric actuators for active vibration controls. Note that the sensor output is only contributed by the mechanical displacement {u} because no external charge, $\{Q\} = 0$, is applied to the distributed piezoelectric sensor layers.

It is also assumed that the direct effect in the piezoelectric actuator is negligible because the feedback voltage is much higher than the self-generated voltage. Thus, separating the actuator DOFs from the sensor DOFs, one can further partition the matrix as

$$\begin{pmatrix} \{F_e\}_a \\ \{0\} \end{pmatrix} = \begin{bmatrix} [K_{u\phi}]_a[K_{\phi\phi}]_a^{-1} & [0] \\ [0] & [K_{u\phi}]_s[K_{\phi\phi}]_s^{-1} \end{bmatrix} \begin{pmatrix} \{Q\}_a \\ \{0\} \end{pmatrix}, \tag{33}$$

where $[K_{u\phi}]_a$, $[K_{\phi\phi}]_a$, $[K_{u\phi}]_s$, and $[K_{\phi\phi}]_s$ are, respectively, $(n - k) \times (j - p)$, $(j - p) \times (j - p)$, $(k \times p)$, and $(p \times p)$ matrices. $\{F\}_a$ and$\{Q\}_a$ are, respectively, $(n - k) \times 1$, and $(j - p) \times 1$ vectors. Note the subscript a denotes the DOFs associated with the piezoelectric acuator layer, and s denotes sensor DOFs.

The sign and magnitude of the feedback is of importance to velocity feedback controls. Thus, the external charge $\{Q\}_a$ applied to the actuator needs to be properly controlled in order to produce adequate and desirable feedback control forces. Using Equation (29), one can derive the induced electric potential due to mechanical displacement,

$$\begin{pmatrix} \{\phi\}_a \\ \{\phi\}_s \end{pmatrix} = \begin{bmatrix} [K_{\phi\phi}]_a^{-1} & [0] \\ [0] & [K_{\phi\phi}]_s^{-1} \end{bmatrix} \left\{ \begin{pmatrix} \{Q\}_a \\ \{Q\}_s \end{pmatrix} - \begin{bmatrix} [K_{\phi u}]_a & [0] \\ [0] & [K_{\phi u}]_s \end{bmatrix} \begin{pmatrix} \{u\}_a \\ \{u\}_s \end{pmatrix} \right\}, \tag{34}$$

where $\{\phi\}_a$ and $\{\phi\}_s$ are, respectively, $(j - p) \times 1$, and $(p \times 1)$ vectors. Since the charge applied to the sensor layer is zero, the voltage from the sensor layer can be written as

$$\{\phi\}_s = -[K_{\phi\phi}]_s^{-1}[K_{\phi u}]_s\{u\}_s. \tag{35}$$

In order to provide actuators proper velocity information, voltage induced from the sensor layer is differentiated. Moreover, the magnitude of feedback voltage is chosen as large as possible, but below the breakdown voltage. Hence a feedback control gain is used to enhance the feedback voltage and also to change the sign before injecting it into the piezoelectric actuators. Note the feedback gain and capacitance for each actuator are not necessarily identical. Assume G_1, G_2, . . . G_g represent the feedback control gains and C_1, C_2, . . . C_g the capacitances. Multiplying the feedback voltage by a capacitance, one can derive the feedback charge injected into the ith actuator. Substituting the feedback charge into Equation (33) and multiplying a feedback gain, one obtains the equivalent feedback force generated by the ith actuator. Two control algorithms, namely (1) *Constant-amplitude negative-velocity feedback control;* and (2) *constant-gain negative-velocity feedback control* are formulated [5, 9]. Figure 5.2 shows the feedback voltages of the two control algorithms. Note that the velocity is opposite the control voltage.

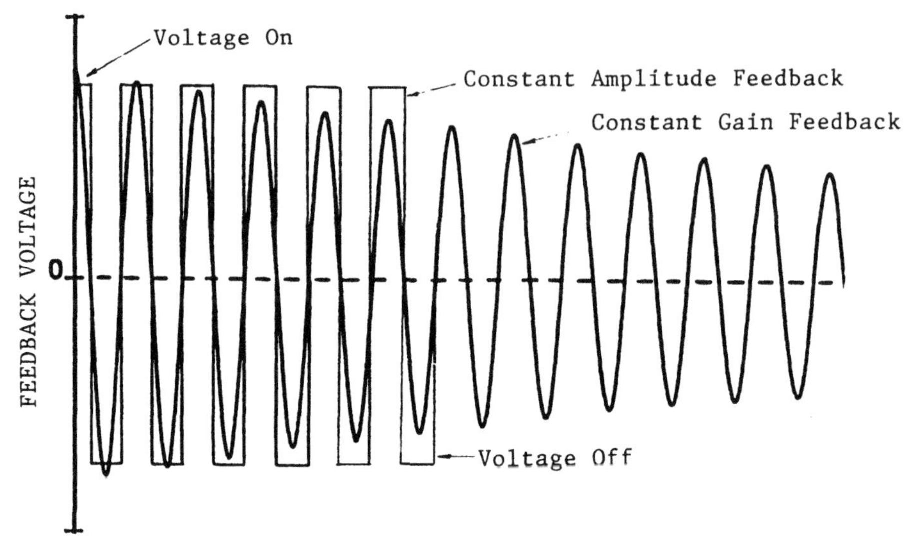

FIGURE 5.2. Feedback of the constant-gain and constant-amplitude controls.

4.1. Constant-Amplitude Negative-Velocity Feedback Control

In the constant-amplitude negative-velocity feedback control, the feedback amplitude is constant and the sign is opposite the velocity,

$$\{F_e\}_{a,i} = -G_i \cdot C_i \cdot [K_{u\phi}]_{a,i}[K_{\phi\phi}]_{a,i}^{-1}\,[K_{\phi u}]_{s,i}^{-1}\,[K_{\phi u}]_{s,i}\text{SGN}\,(\{\dot{u}\}_{s,i}), \tag{36}$$

where SGN is a sign function,

$$\text{SGN}[u] = \begin{bmatrix} -1 & u < 0 \\ 0, \text{ if } & u = 0. \\ +1 & u > 0 \end{bmatrix} \tag{37}$$

Note that the subscript i indicates the ith sensor or actuator if multiple sensors and actuators are used in the DPSs.

4.2. Constant-Gain Negative-Velocity Feedback Control

In the constant-gain negative-velocity feedback control, the feedback control force is calculated by

$$\{F\}_{a,i} = -G_i \cdot C_i \cdot [K_{u\phi}]_{a,i}[K_{\phi\phi}]_{a,i}^{-1}\,[K_{\phi\phi}]_{s,i}^{-1}\,[K_{\phi u}]_{s,i}\{\dot{u}\}_{s,i}. \tag{38}$$

In the finite element analysis, the time history responses of the piezoelectric system are calculated using a time-domain direct integration algorithm, the modified Wilson-θ method and a pseudoforce method [10] to accommodate the control force derived from the applied surface charges.

Note that the constant-amplitude negative-velocity feedback control is nonlinear and discontinuous. The constant-gain negative-velocity feedback control is linear and continuous; however, as the vibration amplitude decays, so does the feedback control voltage.

5. STATE VARIABLE TRANSFORMATION

State variable representation of system equations is highly desirable in design and analysis of control systems. In this section, the derived piezoelectric/elastic system equations are transferred into state space. Define state variables as

$$x_1 = \{u\}, \tag{39}$$

$$x_2 = \{\dot{u}\}. \tag{40}$$

For an *open-loop* system, the system is represented by a set of two first-order differential equations in terms of state variables x_1 and x_2,

$$\dot{x}_1 = x_2, \tag{41}$$

$$\dot{x}_2 = -[M]^{-1}[K^*]x_1 - [M]^{-1}[C]x_2 + [M]^{-1}\{F\}, \tag{42}$$

where the superscript -1 denotes the matrix inversion. Writing in a matrix form gives

$$\begin{bmatrix} \dot{x}_1 \\ \dot{x}_2 \end{bmatrix} = \begin{bmatrix} [0] & [I] \\ -[M]^{-1}[K^*] & -[M]^{-1}[C] \end{bmatrix} \begin{bmatrix} x_1 \\ x_2 \end{bmatrix} + \begin{pmatrix} [0] \\ [M]^{-1} \end{pmatrix} \{F\}, \tag{43}$$

$$\{\phi\}_s = [([0], -[K_{\phi\phi}]_s^{-1}[K_{\phi u}]_s)\ [0] \begin{bmatrix} x_1 \\ x_2 \end{bmatrix}; \tag{44}$$

or,

$$\dot{\mathbf{x}} = \mathbf{Ax} + \mathbf{Br}, \tag{45}$$

$$\mathbf{y} = \mathbf{Cx}, \tag{46}$$

where $\mathbf{x}$ is a $(2n \times 1)$ state vector; $\mathbf{A}$ is a $(2n \times 2n)$ system matrix; $\mathbf{B}$ is a $(2n \times n)$ control matrix; $\mathbf{r}$ is a $(n \times 1)$ input vector; $\mathbf{y}$ is a $(p \times 1)$ output vector; and $\mathbf{C}$ is a $(p \times 2n)$ matrix. They are respectively defined as

$$\mathbf{x} = \begin{bmatrix} x_1 \\ x_2 \end{bmatrix}, \tag{47a}$$

$$\mathbf{A} = \begin{bmatrix} [0] & [I] \\ -[M]^{-1}[K^*] & -[M]^{-1}[C] \end{bmatrix}, \tag{47b}$$

$$\mathbf{B} = \begin{pmatrix} [0] \\ [M]^{-1} \end{pmatrix}, \tag{47c}$$

$$\mathbf{r} = \{F\}, \tag{47d}$$

$$\mathbf{y} = \{\phi\}_s, \tag{47e}$$

$$\mathbf{C} = [([0], -[K_{\phi\phi}]_s^{-1}[K_{\phi u}]_s) \quad [0]]. \tag{47f}$$

Note that the electromechanical force $\{F_e\}$ does not appear in the above equations; this is because $\{F_e\}$ is used as a feedback control force in a closed-loop control system.

Considering a closed-loop system with **u** representing the control force with a dimension of $(n \times 1)$ one can derive the state equation as

$$\dot{\mathbf{x}} = \mathbf{Ax} + \mathbf{Bu}, \tag{48}$$

$$\mathbf{y} = \mathbf{Cx}. \tag{49}$$

With the conventional feedback control law, the feedback force is

$$\mathbf{u} = \mathbf{r} - \mathbf{K}^t\mathbf{x}, \tag{50}$$

where $\mathbf{K}^t$ is the feedback gain matrix and it is defined as follows.

1. Constant-amplitude negative-velocity feedback control:

$$\mathbf{K}^t = [[0] \quad ([0], \Sigma(G_i \cdot C_i \cdot [K_{u\phi}]_{a,i}[K_{\phi\phi}]_{a,i}^{-1}[K_{\phi\phi}]_{s,i}^{-1}[K_{\phi u}]_{s,i}\mathrm{SGN}(\cdot))] \begin{bmatrix} x_1 \\ x_2 \end{bmatrix}. \tag{51}$$

2. Constant-gain negative-velocity feedback control:

$$\mathbf{K}^t = [[0] \quad ([0], \Sigma(G_i \cdot C_i \cdot [K_{u\phi}]_{a,i}[K_{\phi\phi}]_{a,i}^{-1}[K_{\phi\phi}]_{s,i}^{-1}[K_{\phi u}]_{s,i}))] \begin{bmatrix} x_1 \\ x_2 \end{bmatrix}. \tag{52}$$

6. MATRIX CONDENSATION AND REDUCTION

After assembling the element properties, a global system matrix equation is derived. Conceptually, it is not necessary to employ a reduction scheme in the analysis. However, in order to improve computation efficiency, Guyan's reduction scheme is applied. The static system equation is written as

$$\begin{bmatrix} [K_{uu}] & [K_{u\phi}] \\ [K_{\phi u}] & [K_{\phi\phi}] \end{bmatrix} \begin{bmatrix} \{u\} \\ \{\phi\} \end{bmatrix} = \begin{bmatrix} \{F\} \\ \{Q\} \end{bmatrix}. \tag{53}$$

Note the capital letters denote the global properties. Partitioning the static equation in terms of retained DOFs $\{u_1\}$ and reduced DOFs $\{u_2\}$ and $\{\phi\}$ gives

$$\begin{bmatrix} [K^{11}_{uu}] & [K^{12}_{uu}] & [K^{1}_{u\phi}] \\ [K^{21}_{uu}] & [K^{22}_{uu}] & [K^{2}_{u\phi}] \\ [K^{1}_{\phi u}] & [K^{2}_{\phi u}] & [K_{\phi\phi}] \end{bmatrix} \begin{bmatrix} \{u\}_1 \\ \{u\}_2 \\ \{\phi\} \end{bmatrix} = \begin{bmatrix} \{F\}_1 \\ \{F\}_2 \\ \{Q\} \end{bmatrix}. \tag{54}$$

Note that $\{\phi\}$ is also condensed from the time-domain integration in the analysis. Solving the condensed displacement and potential DOFs in terms of $\{u\}_1$ gives

$$[\bar{K}_{uu}]\{u\}_1 = \{\bar{F}\}_1, \tag{55}$$

where the modified stiffness matrix $[\bar{K}_{uu}]$ and force vector $\{\bar{F}\}$ are defined as

$$[\bar{K}_{uu}] = [K^{11}_{uu}] - \{[K^{12}_{uu}][K^{1}_{u\phi}]\} \begin{bmatrix} [K^{22}_{uu}] & [K^{2}_{u\phi}] \\ [K^{2}_{\phi u}] & [K_{\phi\phi}] \end{bmatrix}^{-1} \begin{bmatrix} [K^{2}_{uu}] \\ [K^{1}_{\phi u}] \end{bmatrix}, \tag{56}$$

$$\{\bar{F}\}_1 = \{F\}_1 - \{[K^{12}_{uu}]\,[K^{1}_{u\phi}]\} \begin{bmatrix} [K^{22}_{uu}] & [K^{2}_{u\phi}] \\ [K^{2}_{\phi u}] & [K_{\phi\phi}] \end{bmatrix}^{-1} \begin{bmatrix} \{F\}_2 \\ \{Q\} \end{bmatrix}. \tag{57}$$

And the condensed system mass matrix becomes

$$[\bar{M}] = [T_g]^t \begin{bmatrix} [M^{11}] & [M^{12}] & [0] \\ [M^{21}] & [M^{22}] & [0] \\ [0] & [0] & [0] \end{bmatrix} [T_g], \tag{58}$$

where $[T_g]$ is a congruent transformation matrix defined as

$$[T_g] = \begin{bmatrix} [I] \\ -\begin{bmatrix} [K^{22}_{uu}] & [K^{2}_{u\phi}] \\ [K^{2}_{\phi u}] & [K_{\phi\phi}] \end{bmatrix}^{-1} \begin{bmatrix} [K^{21}_{uu}] \\ [K^{1}_{\phi u}] \end{bmatrix} \end{bmatrix}, \tag{59}$$

where [I] is a unit matrix with the dimension equal to the retained DOFs. Note that a consistent mass formulation is used in this study. Following the same procedure, one can derive the condensed damping matrix $[\bar{C}]$ as

$$[\bar{C}] = [T_g]^t \begin{bmatrix} [C^{11}] & [C^{12}] & [0] \\ [C^{21} & [C^{22}] & [0] \\ [0] & [0] & [0] \end{bmatrix} [T_g]. \tag{60}$$

Hence, the condensed dynamic system matrix equation (open-loop) can be written as

$$[\bar{M}]\{\ddot{\bar{u}}\} + [\bar{C}]\{\dot{\bar{u}}\} + [\bar{K}]\{\bar{u}\} = \{\bar{F}\}, \tag{61}$$

where $\{\bar{u}\} = \{u\}_1$, which denotes the displacement of the retained master DOFs; and $\{\bar{F}\} = \{\bar{F}\}_1$, as derived previously.

For efficient computation, a condensation technique is used here. The theoretical derivation of this theory has been discussed previously. Premultiplied by $[M]^{-1}$, we can find the transformation matrix $[T_g]$ similar to that discussed earlier is defined. Considering the closed-loop case, one can derive the system equation

$$[\bar{M}]\{\ddot{\bar{u}}\} + [\bar{C}]\{\dot{\bar{u}}\} + [\bar{K}^*]\{\bar{u}\} = [\bar{F}^*] + [\bar{F}_e^*], \tag{62}$$

$$\{\phi\} = [K_{\phi\phi}]^{-1}(\{Q\} - [K_{\phi u}][T_g]\{\bar{u}\}), \tag{63}$$

with

$$\{u\} = [T_g]\{\bar{u}\}, \tag{64a}$$

$$[\bar{M}] = [T_g]^t[T_g], \tag{64b}$$

$$[\bar{C}] = [T_g]^t[M]^{-1}[C][T_g], \tag{64c}$$

$$[\bar{K}^*] = [T_g]^t[M]^{-1}[K^*][T_g], \tag{64d}$$

$$[\bar{F}^*] = [T_g]^t[F], \tag{64e}$$

$$[\bar{F}_e^*] = [T_g]^t[F_e], \tag{64f}$$

where the overbar represents the condensed matrices or vectors.

In state space form, $\mathbf{z}_1 = \{\bar{u}\}$, $\mathbf{z}_2 = \{\dot{\bar{u}}\}$, Equations (62) and (63) are written as

$$\dot{\mathbf{z}} = \mathbf{F}\mathbf{z} + \mathbf{G}\mathbf{u}, \tag{65}$$

$$\mathbf{y}^r = \mathbf{H}\mathbf{z}, \tag{66}$$

where the vector and matrices are defined as

$$\mathbf{z} = \begin{bmatrix} z_1 \\ z_2 \end{bmatrix}, \tag{67a}$$

$$\mathbf{F} = \begin{bmatrix} [0] & [I] \\ -[\bar{M}]^{-1}[\bar{K}^*] & -[\bar{M}]^{-1}[\bar{C}] \end{bmatrix}, \tag{67b}$$

$$\mathbf{G} = \begin{pmatrix} [0] \\ [\bar{M}]^{-1}[T_g]^t[M]^{-1} \end{pmatrix}, \tag{67c}$$

$$\mathbf{H} = [([0], -[K_{\phi\phi}]_s^{-1}[K_{\phi u}]_s) \quad [0]] \begin{bmatrix} [T_g] & [0] \\ [0] & [T_g] \end{bmatrix}, \tag{67d}$$

with the same feedback control law as

$$\mathbf{u} = \mathbf{r} - \mathbf{K}^t\mathbf{x}, \tag{68}$$

where $\mathbf{r}$ and $\mathbf{K}^t$ are defined previously.

7. SYSTEM AGGREGATION FOR CONTROL OF LARGE-SCALE SYSTEMS

The principle of the aggregation technique is to preserve the dominant modes and use lower-dimension models in a control analysis. The concept is similar to the condensation method, which is usually used in finite element analysis to reduce computation effort and to obtain approximate solutions. In this section, combination of the condensation and aggregation techniques to derive a reduced-order system is presented.

Aggregation techniques involve replacing the initial system model S^i by a reduced-order model S^r, which preserves the dominant system characteristics (Figure 5.3). In general, S^i can be represented by a state vector $\mathbf{x}$ of dimension $2n$ and S^r is represented by a state vector $\mathbf{z}$ of dimensions $2m$ ($m << n$). In this section, it is assumed that (1) S^i and S^r are both linear dynamic models; and (2) the aggregation is linear, namely, $\mathbf{z}$ is a linear function in $\mathbf{x}$.

Thus, the aggregation technique transforms state $\mathbf{x}$ to $\mathbf{z}$ by

$$\mathbf{z} = \mathbf{L}\mathbf{x}, \tag{69}$$

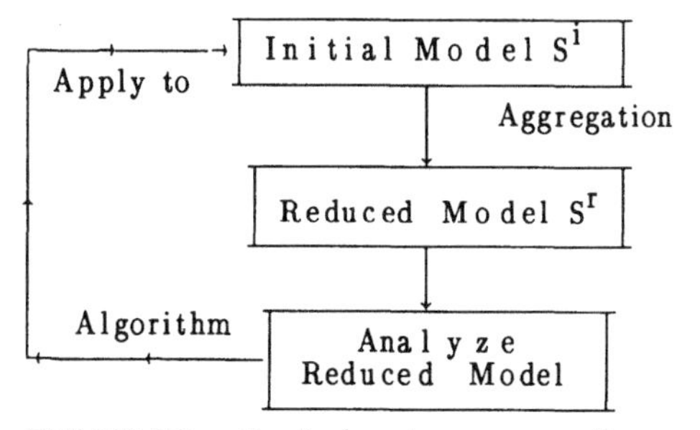

FIGURE 5.3. Control system aggregation.

where $\mathbf{L}$ is the so-called aggregation matrix and is of dimension $2m \times 2n$. The state variable representation of the initial model S^i can be written as

$$\dot{\mathbf{x}} = \mathbf{Ax} + \mathbf{Bu}, \tag{70}$$

$$\mathbf{y}^i = \mathbf{Cx}. \tag{71}$$

A reduced model ($\mathbf{y}^r$) takes the form

$$\dot{\mathbf{z}} = \mathbf{Fz} + \mathbf{Gu}, \tag{72}$$

$$\mathbf{y}^r = \mathbf{Hz}. \tag{73}$$

Note that Equations (72) and (73) represent the aggregate model for the initial model S^i. There are still two problems that need to be answered: (1) The determination of aggregation matrix $\mathbf{L}$; and (2) the determination of control strategies for reduced model S^r suitable for the initial model S^i. It can be derived and proved that

$$\mathbf{z} = \begin{bmatrix} ([T_g]^t[T_g])^{-1}[T_g]^t & [0] \\ [0] & ([T_g]^t[T_g])^{-1}[T_g]^t \end{bmatrix} \mathbf{x} \tag{74}$$

where $[T_g]$ is the congruent transformation matrix and

$$\mathbf{F} = \mathbf{LA}, \tag{75}$$

$$\mathbf{G} = \mathbf{LB}; \tag{76}$$

where

$$\mathbf{L} = \begin{bmatrix} ([T_g]^t[T_g])^{-1}[T_g]^t & [0] \\ [0] & ([T_g]^t[T_g])^{-1}[T_g]^t \end{bmatrix}. \tag{77}$$

Thus, the reduced-order model can sufficiently represent the initial system model in feedback controls. The control algorithms derived previously are still applicable in the reduced system. The detailed discussion and proofs are presented in [11]. Control spillover problems and the system stability for the reduced-order model are also discussed in [11].

8. CASE STUDIES

Two extensive case studies are presented in this section (1) A smart piezoelectric bimorph beam; and (2) a smart plate. The first case, a bimorph structure, can be applied to a number of engineering applications, e.g., a micropositioning device, a piezoelectric gripper, a high-precision actuator, or a flexible robot manipulator. Thus, two aspects of the smart bimorph structure were investigated (1) Microdisplacement actuation; and (2) distributed structural identification.

Plate structures or components are very common in many mechanical and aerospace systems. The second case presents a study on distributed structural identification and vibration control of a smart plate with integrated piezoelectric sensor/actuator layers. Distributed modal voltages of the first three modes were illustrated and control effectiveness of the two control algorithms evaluated.

8.1. A Piezoelectric Bimorph Beam

The piezoelectric bimorph beam was made of two layers of piezoelectric polymeric polyvinylidene fluoride (PVDF) with opposite polarity. Material properties are summarized in Table 5.1.

When an external voltage is applied, the induced internal stresses result in a resultant bending moment which forces the bimorph beam to bend. The first study was a static deflection test in which a strong voltage was applied across the thickness and the beam deflection was studied using theoretical, experimental, and finite element techniques. The second study was a distributed structural identification in which the distributed voltage along the beam was calculated and illustrated using the developed finite element code. The bimorph beam model was discretized into 10 piezoelectric finite elements, five elements on each layer. One end of the bimorph beam was assumed fixed. The physical dimension and polarity are illustrated in Figure 5.4, and the experimental setup is shown in Figure 5.5.

8.1.1. Microdisplacement Actuation.

A unit voltage (1 volt) was applied across the thickness and the static deflections of five nodes were calculated analytically and by the finite element method. A physical model with the same dimension was also tested in the Dynamics and Systems Laboratory. The results are summarized in Table 5.2 and plotted in

TABLE 5.1. Material Properties of the Piezoelectric PVDF.[1]

Dielectric Permittivity		Dielectricity	
e_{31}	0.0460 C/m^2	ϵ_{11}	0.1062×10^{-9} F/m
e_{32}	0.0460 C/m^2	ϵ_{22}	0.1062×10^{-9} F/m
e_{33}	0.0000 C/m^2	ϵ_{33}	0.1062×10^{-9} F/m

[1] Poisson ratio = 0.2900; mass density = 0.1800×10^4 Kg/m^3; and modulus = 0.2000×10^{10} N/m^2.

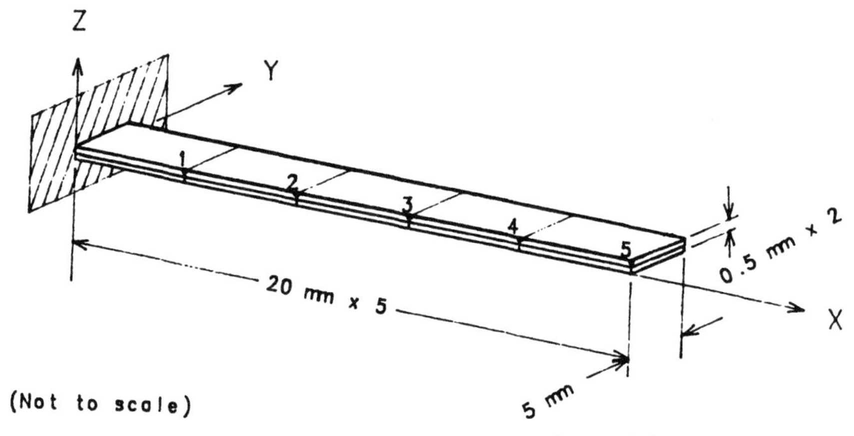

FIGURE 5.4. A piezoelectric polymeric bimorph beam.

Figure 5.6. Note that only the tip deflection was measured in laboratory experiments (node 5 in Figure 5.6).

The error at the fixed end is more significant than that at the free end. This is because the discretized finite element is relatively rigid at the fixed end, which can be improved if more elements were used in the modeling. It is also observed

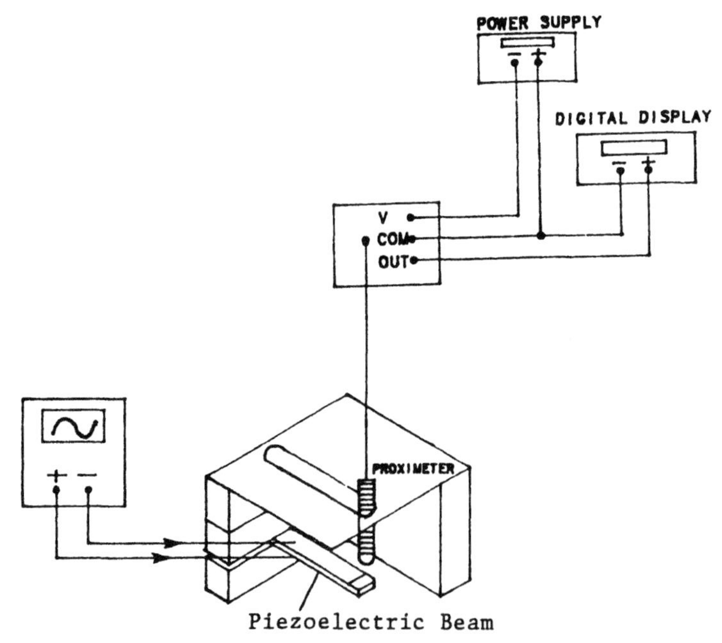

FIGURE 5.5. Experimental apparatus.

TABLE 5.2. Static Deflection of the Piezoelectric Bimorph Beam (10^{-7}m).

Node	Theory	FEM	Error (%)
1	0.138	0.124	10
2	0.552	0.508	8.0
3	1.24	1.16	6.2
4	2.21	2.10	5.1
5	3.45	3.30	4.4

that the experimental results are lower than the other two, which could be introduced by (1) Nonperfect bonding; (2) voltage leakage; (3) energy dissipation, etc.

8.1.2. Distributed Structural Identification.

The piezoelectric bimorph beam was also studied for its voltage response. A snapback with an initial tip displacement of 1 cm was performed and the voltage response was calculated using the developed finite element code. Figure 5.7 shows the displacement response and Figure 5.8 shows the voltage distribution along the beam. Note that only the span from 2 to 10 cm was plotted in Figure 5.8 because relatively fine mesh occurred near the fixed end.

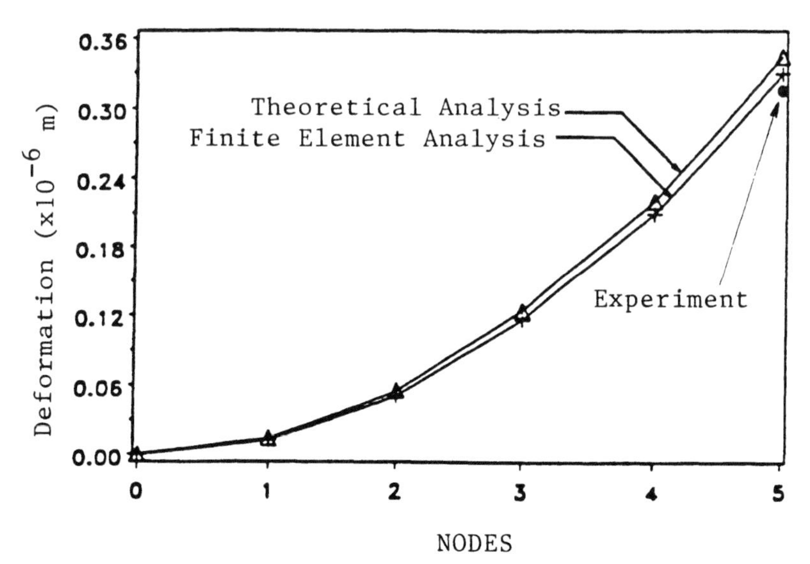

FIGURE 5.6. Nodal deflection of the piezoelectric bimorph beam.

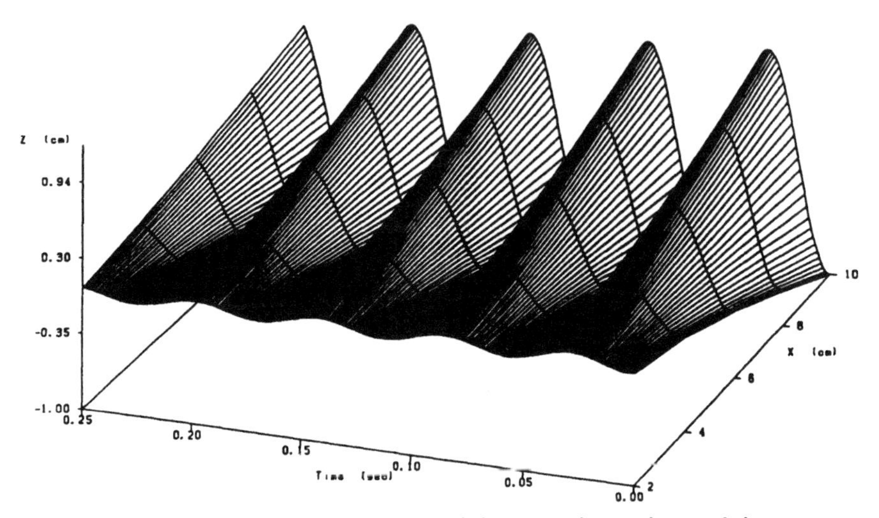

FIGURE 5.7. Snapback response of the piezoelectric bimorph beam.

It is observed that the displacement profile exhibited a reversed first mode shape of the bimorph beam. Thus, the voltage distribution should also represent the *modal voltage* [5] of the first mode. However, transient voltage response was observed in the beginning of the snapback response. Note that finite electrode separations are required to obtain distributed voltages in practice. Measurement data should show a stepped voltage distribution on the beam with segmented electrodes.

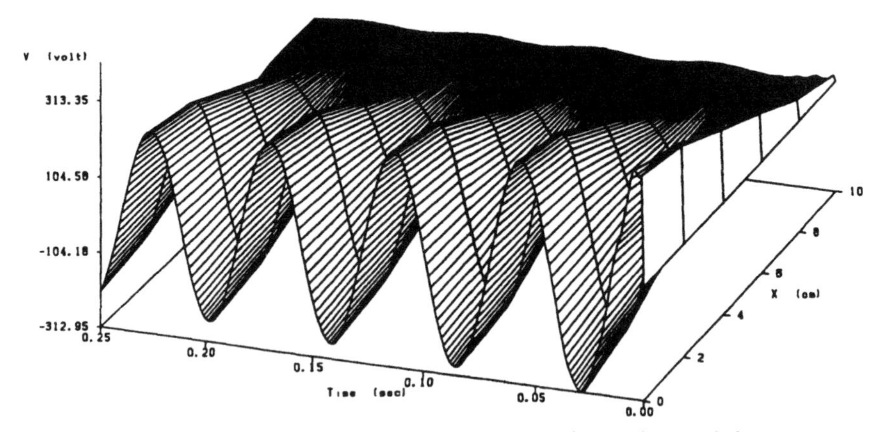

FIGURE 5.8. Distributed voltage of the piezoelectric bimorph beam.

8.2. Distributed Vibration Measurement and Control of a Smart Plate

In this section, distributed structural identification and vibration control of a zero-curvature shell, a plate, are studied. The physical model is a Plexiglas cantilever plate (10 cm × 10 cm × 0.31 cm) with a top distributed piezoelectric PVDF actuator layer (40 μm) and a bottom distributed sensor layer. The plate with the distributed sensor/actuator was divided into 75 elements, 25 for each layer, modeled by the piezoelectric finite elements developed earlier. Figure 5.9 shows the cantilever plate finite element model.

Distributed modal voltage measured by the distributed sensor is demonstrated in the first three modes. The distributed piezoelectric actuator can be made either monoaxially oriented, aligned with the *x*-axis, or biaxially oriented, effective in both *x*- and *y*-axes. This orientation is evaluated in distributed vibration controls of the plate. It should be noted that some assumptions were made in the finite element analysis: (1) The resistance of deposited metal electrode on each side of the PVDF was assumed negligible such that the applied voltage can quickly reach a steady state; (2) the expanded or contracted strains settle very fast without time delay; and (3) the velocity information can be obtained and instantaneously used as a feedback signal.

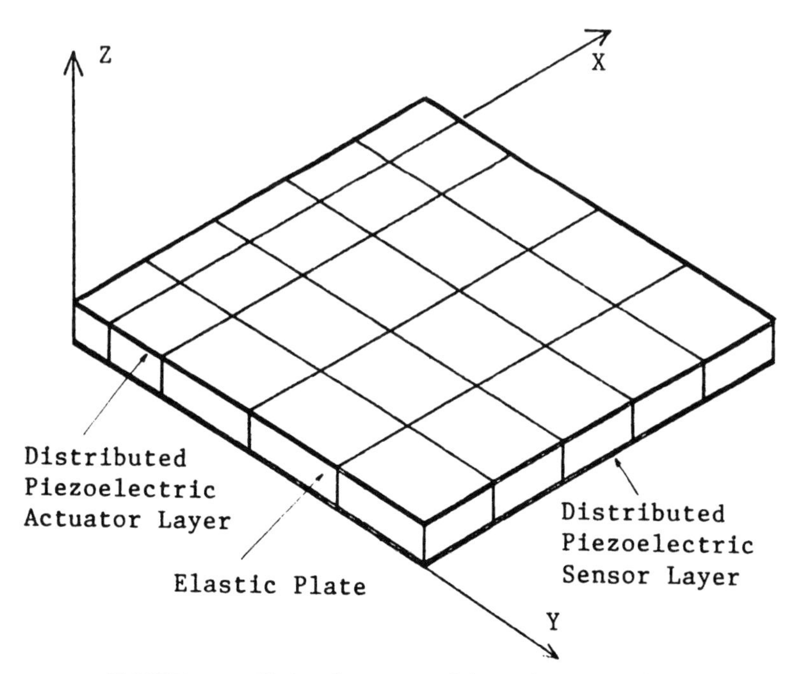

FIGURE 5.9. Finite element modeling of a smart plate.

As discussed previously, the distributed piezoelectric sensing layer should respond to the plate oscillation and generate an electric voltage representing the distributed dynamic strain response of the plate. This distributed sensing phenomena are demonstrated in an eigenvalue analysis. The active distributed vibration suppression and control of the plate is studied and evaluated in a snapback analysis in which an initial displacement (first mode) was imposed.

8.2.1. Distributed Structural Identification: Modal Voltage Distribution.

The output signals of each node on the distributed piezoelectric sensor layer can be calculated as a function of displacements. (Note that the {G} vector is zero in an eigenvalue analysis.) After the nodal voltage is calculated, the overall voltage distribution of the plate can be plotted by connecting all nodal voltage amplitudes. Thus, for a given mode, the modal voltage distribution (distributed sensing phenomena) can be observed. The first three plate mode shapes and modal voltage distributions are illustrated in Figures 5.10–5.12.

It is observed that the first mode is a bending mode, the second mode is a torsion mode, and the third mode is a warping mode. The sensitivity of each mode could also change because the tension and compression vary in different modes. Voltage drops at two corners, $V(0, 0)$ and $V(0, 100)$ are observed in all three figures. This is because the strains are (numerically) smaller at these two boundary nodes in the finite element calculation. According to the theory discussed previously, the output amplitude is inversely proportional to the displacement, i.e., the maximum positive voltage occurs at the maximum strain. The modal voltage distribution could also be refined if more elements and finer mesh were used in the finite element modeling.

8.2.2. Distributed Vibration Controls.

The distributed piezoelectric actuator on the top surface of the plate contracts or expands depending on negative or positive feedback voltages (the converse

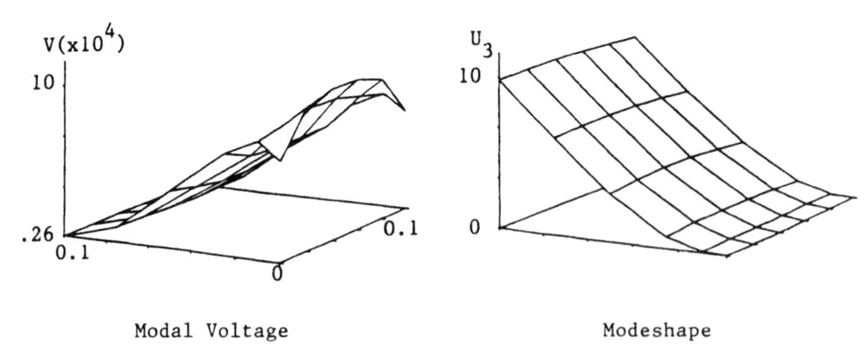

FIGURE 5.10. First mode shape and distributed modal voltage.

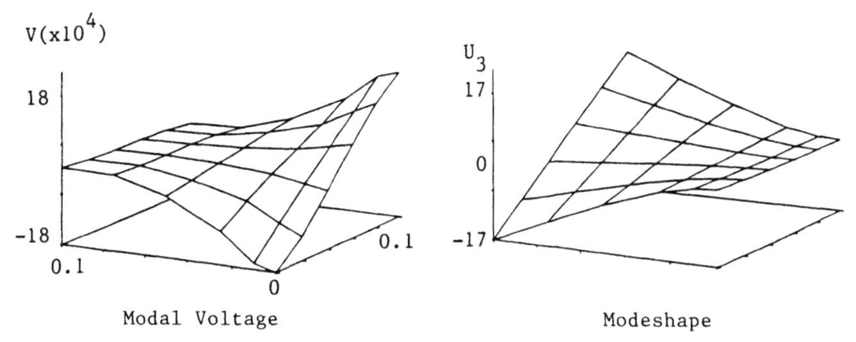

FIGURE 5.11. Second mode shape and distributed modal voltage.

piezoelectric effect). In general, a positive feedback voltage is needed for an upward (positive) displacement. There were two distributed piezoelectric actuators evaluated in the finite element analysis: (1) a monoaxial actuator whose piezoelectric orientation is aligned with the x-axis; and (2) a biaxial actuator whose orientation is aligned with both x- and y-axes.

An initial displacement (first mode) is imposed and then the plate is set free. The vibration amplitude decays depending on the modal damping and the feedback. An initial modal damping was assumed to be 0.9% (based on a laboratory experiment); and the damping ratio change was evaluated when comparing the control effectiveness of the distributed piezoelectric actuator with different control algorithms. It should be noted that the tip velocity (node P) was used in the feedback controls (since in practical applications, this is the simplest way to implement).

8.2.2.1. Negative-Velocity Constant-Gain Feedback Control. In the first case, the feedback voltage amplitude varies with respect to the negative velocity. Since node P's velocity was used in the feedback, the original equation is mod-

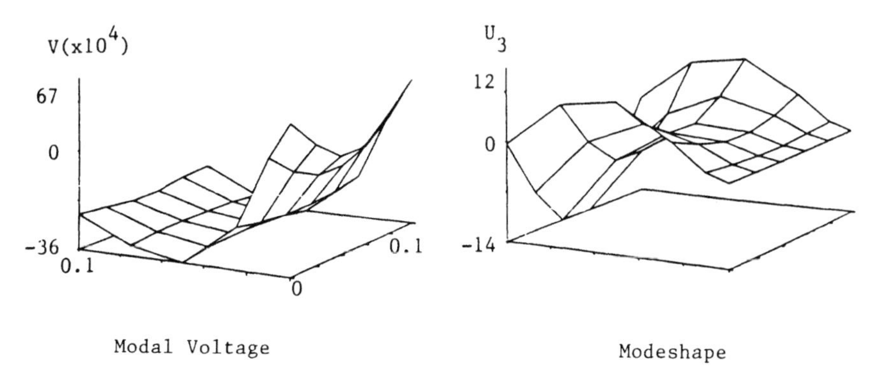

FIGURE 5.12. Third mode shape and distributed modal voltage.

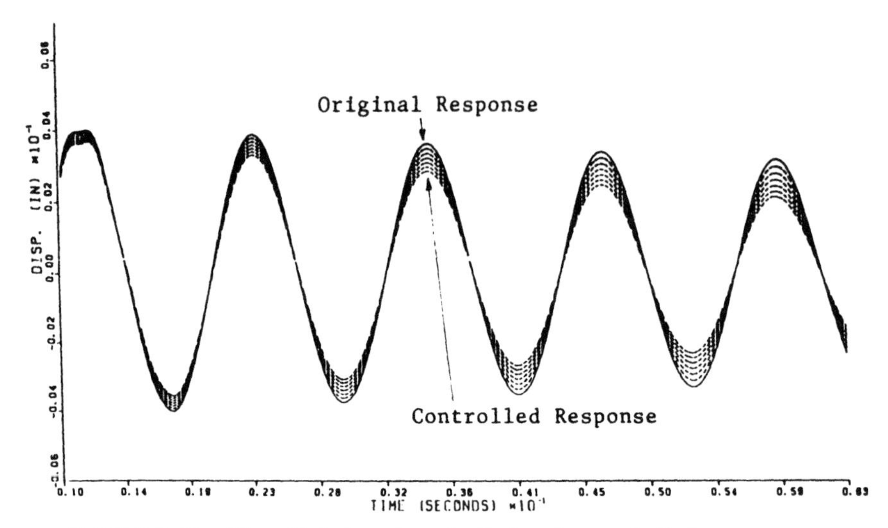

FIGURE 5.13. Time histories of controlled and uncontrolled responses.

ified as: $\{G\} = -[\mathbb{C}]^* \cdot \{\dot{u}_p\}$. The gain matrix $[\mathbb{C}]^*$ is also changed in order to evaluate its control effectiveness. The time histories with and without feedback control are plotted in Figure 5.13.

The vibration amplitudes were suppressed by the distributed piezoelectric actuator and the damping ratio change was calculated and plotted versus feed-

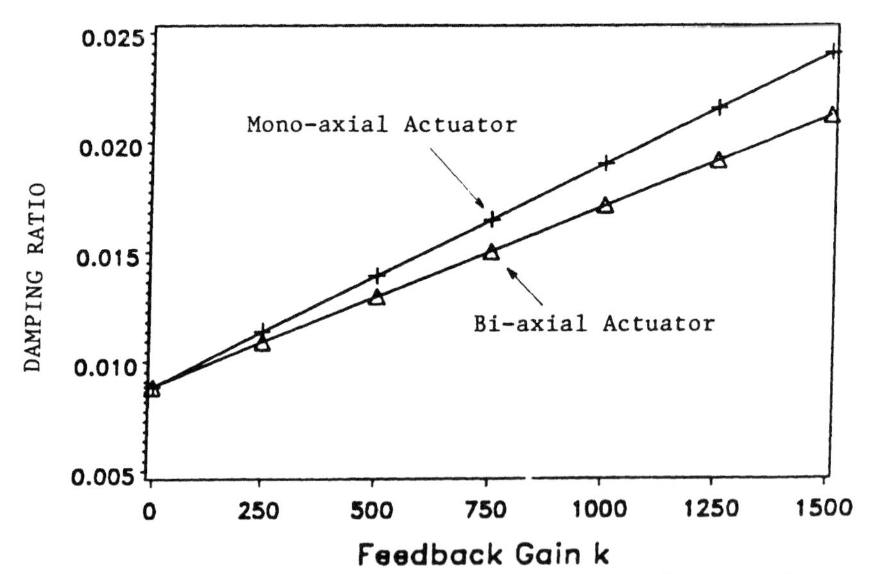

FIGURE 5.14. Damping calculation at constant-gain feedback controls.

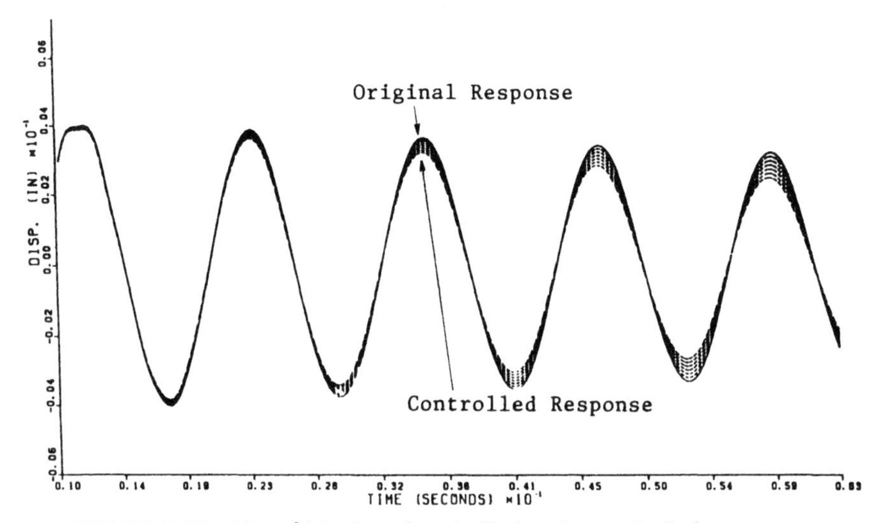

FIGURE 5.15. Time histories of controlled and uncontrolled responses.

back gains as shown in Figure 5.14. It shows that the damping ratio increases when the feedback gain increases. The monoaxial actuator is more effective on first-mode vibration control.

8.2.2.2. Negative-Velocity Constant-Amplitude Feedback Control. In the second case, although the feedback amplitude is constant, the feedback volt-

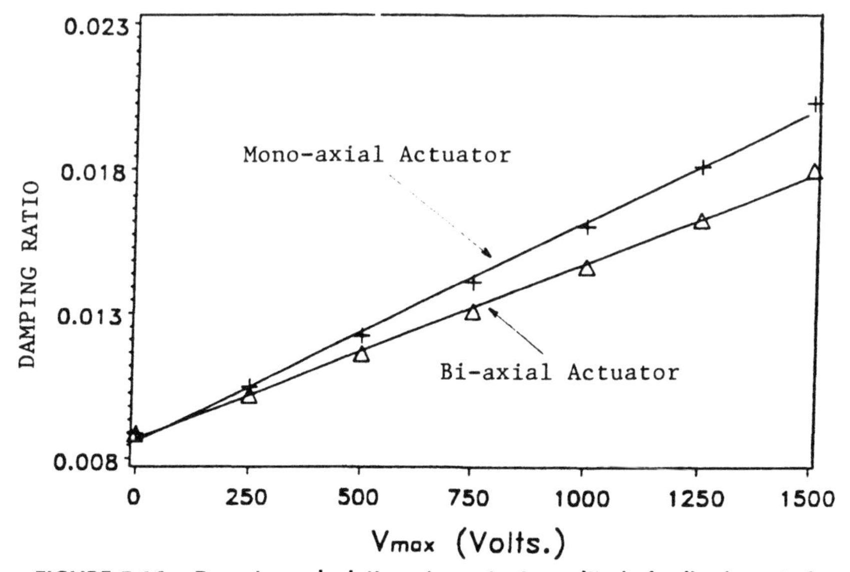

FIGURE 5.16. Damping calculation at constant-amplitude feedback controls.

age changes sign when the nodal velocity $\dot{u}_p$ changes its direction, i.e., {G} = − [ℂ]* SGN [{$\dot{u}_p$}] (nonlinear bang-bang control). The controlled and uncontrolled response histories are plotted in Figure 5.15, and the damping ratio was studied at different feedback voltages as plotted in Figure 5.16.

Note that a single node velocity was used in the feedback controls, even the distributed voltage was discussed in the previous section. This single signal was amplified and fed back to all nodes of the actuator. It is observed that higher [ℂ] results in a higher damping matrix in the system equation. Thus, the plate oscillation should be damped much faster at higher feedback gain as shown in Figures 5.13 and 5.14. It is also noted that the control effectiveness of the monoaxially oriented piezoelectric actuator is better than the biaxially oriented actuator for the first mode control. (It is anticipated that the biaxial actuator could be more effective if mode shapes involve motions in the *y*-direction. However, this is not demonstrated in this study.)

9. SUMMARY AND CONCLUSIONS

Conventional machines and robotic manipulator structures usually are passive in nature, i.e., they do not process any inherent self-sensation and action/reaction capability. Development of new-generation smart structures with integrated sensors, actuators, and control electronics has received increasing attention and interest in recent years [12, 13]. This report presents an integrated distributed piezoelectric sensor/actuator design for smart distributed parameter systems (DPSs) applied to smart machines and robotic manipulators. This is the second part of the report, which is concentrated on a new piezoelectric finite element formulation and applications to smart structures.

Conventional isoparametric hexahedron piezoelectric elements are thick. Thus, it is very inefficient and undesirable for modeling and analysis of thin piezoelectric smart continua. Besides, conventional thin plate/shell elements are deficient to model surface electric charges. Thus, a new thin piezoelectric hexahedron element with internal degrees of freedom (DOFs) was formulated using a variational principle, Hamilton's principle, and piezoelectricity theory. System electromechanical equations of the piezoelectric/elastic coupled DPSs were transformed into state equations. Two control algorithms, namely, constant-amplitude and constant-gain (negative-velocity) feedback controls were proposed and integrated into finite element formulations. Control of original large-scale DPSs was reduced to a control of aggregated DPSs using Guyan's reduction scheme. Two smart piezoelectric structures (1) A micropositioning device (a piezoelectric polyvinylidene fluoride (PVDF) bimorph beam); and (2) a smart machine component (a zero-curvature shell, a plate, sandwiched between two piezoelectric sensor/actuator layers) were investigated in case studies.

In Case 1, the finite element solutions of the beam static deflection were compared closely with the theoretical and experimental results. This proved the dependability of the developed finite element code. Voltage distribution along the beam was also calculated, which showed that the output amplitude is proportional to local strains. Transient output was also observed at the beginning of the snapback response. Applications of this bimorph configuration include (1) Micropositioning; (2) high-precision actuation; (3) robot fingers; (4) flexible robot manipulators, etc.

Distributed modal identification and vibration control of a smart plate were studied and evaluated in Case 2. The distributed piezoelectric sensor responded to plate oscillation and generated distributed modal voltage. The first three modal voltages were plotted with the corresponding mode shapes. Voltage outputs from the distributed sensor were processed and fed back to the distributed actuator via two control algorithms (1) Constant-amplitdue negative-velocity feedback; and (2) constant-gain negative-velocity feedback. It showed that the constant-amplitude feedback control is superior to the constant-gain feedback control in this case.

Use of distributed piezoelectric sensors and actuators for distributed vibration control and identification of DPSs was demonstrated to be an effective technique. The developed piezoelectric finite element and control techniques were also proved to be suitable for modeling and analysis of piezoelectric/elastic coupled electromechanical smart DPSs. In practical applications, however, some potential problems, such as (1) Bonding techniques; (2) resistance of electrodes; (3) voltage leakages; (4) time delay associated with the control electronics; (5) segmenting electrodes and signal transmission, etc. still need further investigation.

APPENDIX A. NOMENCLATURE

[ ] = matrix.
{ } = vector.
$\ldots^t$ = transpose of vector or matrix.
[c] = the elasticity matrix evaluated at constant electric field.
$[c_{uu}]$ = element damping matrix.
$\mathbb{C}$ = feedback gain.
$[\mathbb{C}]$ = feedback gain matrix.
[C] = system damping matrix.
$\{C_n\}$ = equivalent nonlinear damping matrix.
{D} = the electric displacement vector.
{E} = the electric field vector.
[e] = dielectric permittivity matrix.
$\{F_e\}$ = equivalent electric force.
{g} = element charge vector.

$\{G\}$ = external applied charge vector.
$[k_{uu}]$ = element stiffness matrix.
$[k_{u\phi}]$ = element piezoelectric stiffness matrix.
$[k_{\phi\phi}]$ = element dielectric stiffness matrix.
$[k^*]$ = condensed system stiffness matrix.
$\mathcal{L}$ = Lagrangian.
$[L_u]$ = elasticity operator.
$[m_{uu}]$ = element consistent mass matrix.
$[M]$ = system mass matrix.
$[N_u]$ = shape function matrix related to nodal displacement.
$[N_\phi]$ = shape function matrix related to nodal potential energy.
$\{P_b\}$ = body force vector.
$\{P_s\}$ = surface force vector.
$\{P_c\}$ = concentrated load vector.
$\{u_i\}$ = nodal displacement vector.
$\{u\}$ = displacement vector.
$\{\dot{u}\}$ = velocity vector.
$\{\ddot{u}\}$ = acceleration vector.
$\{S\}$ = strain tensor.
S_1 = surface where surface forces applied.
S_2 = surface where surface charges applied.
$\{T\}$ = stress tensor.
$[T_c]$ = transformation matrix.
$\mathcal{U}$ = internal energy.
$\mathcal{V}$ = piezoelectret volume.
$\delta\mathcal{W}$ = virtual work.
α, β = Rayleigh's coefficients.
$[\epsilon^s]$ = dielectric matrix evalutaed at constant strain.
$\{\phi_i\}$ = nodal electrical potential vector.
$\{\phi\}$ = electrical potential vector.
σ = surface charge.
∇ = gradient operator.

ACKNOWLEDGMENT

This research was supported, in part, by a grant from the National Science Foundation (No. RII-8610671) and the Kentucky Commonwealth; and a seed grant (on intelligent machines and robots, 1988–1989) from the Center for Robotics and Manufacturing Systems (CRMS) at the University of Kentucky.

REFERENCES

1. H.S. Tzou and C.I. Tseng, "Active vibration control of distributed parameter systems by finite element method," *Computer in Engineering 1988*, vol. 3, pp. 599–604, Aug. 1988.
2. H.S. Tzou and C.I. Tseng, "Distributed modal identification and vibration control of continua: piezoelectric finite element formulation and analysis," *ASME J. of Dynamic Systems, Measurement, and Control*, Vol. 113, pp. 500–505, September 1991.
3. H. Allik and T.J. Hughes, "Finite element method for piezoelectric vibration," *Int. J. Numerical Methods Engineering*, vol. 2, pp. 151–168, 1979.
4. M. Nailon, R.H. Coursant, and F. Besnier, "Analysis of piezoelectric structures by a finite element method," *ACTA Electronica*, vol. 25, pp. 341–362, 1983.
5. H.S. Tzou and C.I. Tseng, "Distributed piezoelectric sensor/actuator design for dynamic measurement/control of distributed parameter systems: A piezoelectric finite element approach," *J. Sound and Vibration*, vol. 138, pp. 17–34, April 1990.
6. K.J. Bathe and E.L. Wilson, *Numerical Method in Finite Element Analysis*. London: Prentice Hall, 1976.
7. R.D. Cook, *Concepts and Applications of Finite Element Analysis*. New York: John Wiley & Sons, Inc., 1974.
8. R.J. Guyan, "Reduction of stiffness and mass matrices," *AIAA J.*, vol. 3, p. 380, 1965.
9. H.S. Tzou, *Piezoelectric Shells (Distributed Sensing and Control of Continua)*, Chapter 10, Kluwer Academic Publishers, Dordrecht/Boston/London, February 1993.
10. H.S. Tzou and A.J. Schiff, "Development and evaluation of a pseudo-force approximation applied to nonlinear contacts and viscoelastic damping," *Computers & Structures*, vol. 26, pp. 481–493, 1987.
11. C.I. Tseng, "Electromechanical dynamics of a coupled piezoelectric/mechanical system applied to vibration control and distributed sensing," Department of Mechanical Engineering, University of Kentucky, Ph.D. thesis, Lexington, KY, Aug. 1989.
12. H.S. Tzou and T. Fukada, (Ed.), *Precision sensors, actuators, and systems*. Boston: Klower Academic Publishers, 1992.
13. H.S. Tzou and G.L. Anderson, (Ed.) *Intelligent structural systems*. Boston: Klower Academic Publishers, 1992.

6

Automated Repair of Aircraft Transparencies

Dennis J. Wenzel
6434 Amber Oak
San Antonio, TX 78249
dennis.wenzel@worldnet.att.net

James A. Luckemeyer
9591 Cloverdale
San Antonio, TX 78250
210-647-1550

David S. McFalls
603 Williamsport
League City, TX 77573
sculptor@ghg.net

1. INTRODUCTION

1.1. Background and Motivation

Canopy transparencies become scratched, pitted, or otherwise damaged during their service life on aircraft through a number of mechanisms. While it would seem that contact by objects during flight might be a major source, in reality such strikes from objects of substantial size are rare and usually cause irreparable damage. Instead, the types of damage requiring rework fall under the category of normal deterioration while in service: wind-blown sand while parked on the ground gradually reduces clarity, fine abrasives in the air while in flight cause hazing, contact on external surfaces due to maintenance crew carelessness causes scratches and pits, contact on interior surfaces by pilot helmets and clipboards produces predictable damage to the transparency.

Similar types of handling and maintenance damage occur with transparencies during canopy manufacture and assembly of the aircraft. The transparency surface is relatively delicate, and any type of contact with a hard object will usually leave a mark which must be repaired before the canopy transparency or aircraft can be sold. In the case of new manufacture of the transparency itself, some rework may be necessary to minimize distortion caused by the molding process.

There are two basic types of plastics used in the manufacture of transparencies: stretched acrylic and polycarbonate. The term *stretched acrylic* implies that the polymer has been subjected to mechanical stretching while being held at an elevated temperature to soften the material. This additional processing step increases the toughness of the acrylic, making it much more resistant to impact damage. Acrylic has been the dominant choice for canopy transparencies in the past, but polycarbonate is increasingly being used because of its superior toughness and hence improved bird-strike resistance. Polycarbonate canopy transparencies may be either monolithic or laminated. Laminated styles usually include both acrylic and polycarbonate layers interfaced with bonding layers of urethane. Although the polycarbonate is stronger and tougher than acrylic, it is a comparatively softer material, so the surface must be protected from damage with a hard coat. A *hard coat* is a thin layer of material which is deposited on the polycarbonate surface and often possesses reflective and conductive properties as well as contributing to surface abrasion resistance.

In the case of transparency rework for aircraft in service, repair of only acrylic materials has been routinely successful as witnessed by various process specifications, which are correspondingly lacking for transparencies of polycarbonate construction. The acrylic material lends itself to removal using abrasives and can be brought back to acceptable clarity by polishing. Polycarbonate is typically reworked only as necessary during the manufacturing stages before the hard coat is applied. In addition to providing a scratch-resistant surface, the hard coat contributes to the overall clarity of the polycarbonate transparency which cannot be achieved by polishing alone. Rework of hard-coated polycarbonate requires removal of the hard coat from the entire transparency prior to repair of the polycarbonate by sanding and polishing, followed by reapplication of the hard coat. Local removal has not been successful because partial hard-coating would leave a visibly unacceptable coating overlap around the perimeter.

Manual rework of canopy transparencies for flaw removal is a tedious procedure which involves several steps. The transparency must first be inspected to determine the location and severity of the flaws. Certain types of flaws will cause a transparency immediately to be rejected for rework in favor of replacement with a new transparency. Once a transparency has been determined to be a candidate for repair, it is temporarily marked with a grease pencil to indicate the location of the flaws to be reworked. The flaws are removed by polishing with a series of sanding abrasives, beginning with a coarse abrasive to remove the flaw and continuing with finer abrasives to remove scratch patterns and restore clarity. Specific polishing motions are used, with widening strokes to blend the material removal profile. Correct execution of this procedure is critical, since improper application of the abrasives can induce optical distortion to the degree that the transparency would be rejected for service. Once the flaws have been removed, the entire transparency surface is polished with a fine abrasive to remove very small scratches and restore a high gloss to the surface. The transparency is then cleaned to remove abrasive residue and inspected to certify that its optical properties are acceptable for service. The rework procedure is required for both inner and outer transparency surfaces.

Rework of canopy transparencies and aircraft windows occurs to some extent on all aircraft. The private pilot will purchase a kit for rework and polishing of minor scratches. Commercial airlines will polish the passenger windows during major maintenance of the aircraft. As mentioned, canopy transparency manufacturers and new aircraft manufacturers will repair handling damage as required. The military necessarily performs similar repairs to aircraft in service. By far, the most demanding rework occurs on transparencies for fighter aircraft where flaws can interfere with the pilot's vision, thus endangering the success of a mission. Also, polishing of flaws can help to relieve stress concentrations in areas prone to cracking, thus extending the service life of the transparency.

Canopy transparency polishing is an art in which considerable experience is required to achieve proficiency. Many transparencies are destroyed or only marginally repaired during the learning process. As manual labor, the work is tedious and time-consuming, and it requires upper-body strength and endurance. Careers of experienced canopy polishing personnel are often reduced due to back problems developed on the job. Because of the nature of the work, it is increasingly difficult to find people who will learn the trade. Thus, there is a desire to transfer, to the extent possible, the skills of the canopy transparency polishing expert to an automated machine.

In addition to the potential labor savings, there are other reasons for automating the polishing process. The actual polishing is a tedious and repetitive procedure, lending itself to automation. Furthermore, an automatic system should produce much more repeatable results than is possible with manual polishing, thus reducing the number of canopy transparencies which are rejected due to improper rework. Finally, an automatic system should increase the rework throughput over manual techniques because of the unlimited endurance of a machine.

1.2. Technical Approach

There have been previous attempts to automate polishing of acrylic canopy transparencies. The most notable was a robot system developed for Sierracin/Sylmar (Sylmar, Calif.) which used a lead-through-teach (painting) robot to repeat an operator-trained motion for a programmable number of times [1]. The end effectors employed bonded abrasives for heavy material removal and slurries

FIGURE 6.1. The Robotic Canopy Polishing System.

for final polishing. This approach has evolved into a more complex system for Texstar, Inc. (Grand Prairie, Tex.). Related technology has existed for optical finishing of both plastics and glass lenses (which require greater polishing accuracies) and flat panels made possible by definable processes for new manufacture [2, 3]. Some research has also been performed to study the basic polishing process for acrylics and polycarbonates [4, 5].

The Robotic Canopy Polishing system[1] (RCPS) was developed in response to a request by Hill Air Force Base (Ogden, Utah) to build an automated transparency polishing system. Figure 6.1 shows the RCPS in a laboratory during development. This development effort was performed in two phases which spanned from February 1986 through October 1988. During the first phase, research and testing were performed to demonstrate the feasibility of automatically inspecting and repairing the surface of aircraft transparencies, while during the second phase, a three-robot production work cell for performing the automated transparency

[1] U.S. patent no. 5,067,085.

repair procedure was implemented and installed. The RCPS was initially targeted to repair the acrylic transparency surfaces of the F-16A canopy and the forward and aft F-4 canopies. These target canopies are shown in Figure 6.2.

It was assumed that some combination of robotics and machine vision would provide the basic tools necessary for a prototype canopy polishing system. However, the specific techniques to be applied had to be created while routine canopy transparency inspection and manual polishing techniques themselves were being learned. Potential avenues of research required careful selection so that the results could be applied to a realizable industrial system.

The areas of research involved were generally categorized as (1) surface repair; (2) surface inspection; (3) surface modeling; and (4) strategy for optimal repair.

Surface repair entailed the discovery of mechanical processing methods that could be implemented under robotic control to remove flawed areas of the transparancy surface and restore clarity. The first requirement was to understand the basic methods of sanding and polishing for repair of acrylic materials. The techniques learned were then applied to the testing of abrasives to define a process specification for transparency restoration. A special backing pad for application of the abrasives and a unique sanding tool for robotic use was developed and machine-applicable strategies for sanding with minimal induced distortion and polishing to restore clarity were devised and tested.

Initial study of the transparency repair problem revealed a number of key areas for which automated inspection of the transparency surface was needed to provide repair process input and support. An automated inspection system had to be capable of discovering the location of the flaws, should be able to make some kind of determination regarding the severity (depth) of the located flaws, and must determine whether a flaw is on the inside or the outside of the transparency surface. Together, the information obtained from these key areas of investigation could be used to design the best form of repair processing. It was also important to the efficiency and functionality of the rework operation that some form of feedback be provided to gauge the progress of the rework (known as *in-process inspection*).

Surface modeling is concerned with creating a mathematical description of the transparency contours applicable to guiding a robotic manipulator along each surface. Pretaught robot path programs are not practical due to the unlimited locations and arrangements of flaws which might be encountered and so surface modeling allows the freedom of off-line computation of a solution set of coordinate points for execution by the manipulator. A geometric surface modeling technique to accomodate a variety of fighter aircraft transparencies was developed. Fundamental operations on the model are used to project planar points onto the surface and to partition the surface into patches of nearly equal area.

An optimal repair strategy is presented that computes the optimum placement of a minimum number of sanding patterns for removing the identified flaws

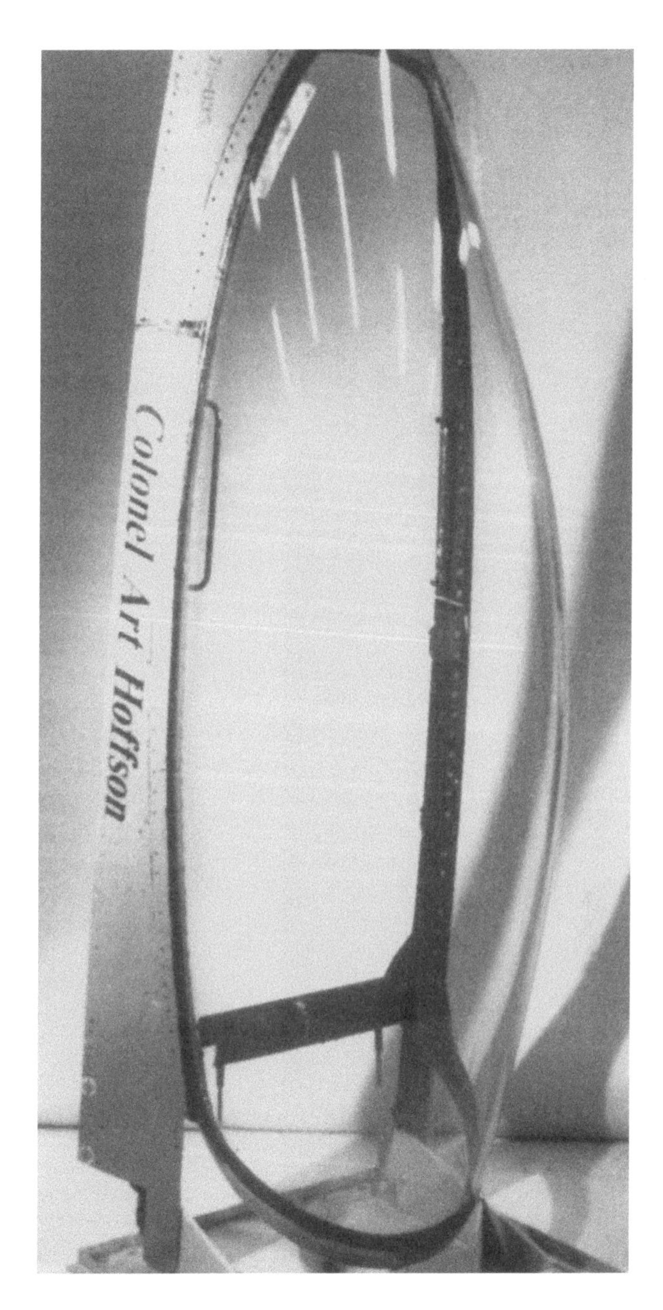

FIGURE 6.2. Target canopies. (a) F-16A canopy. (b) F-4 forward canopy. (c) F-4 aft canopy.

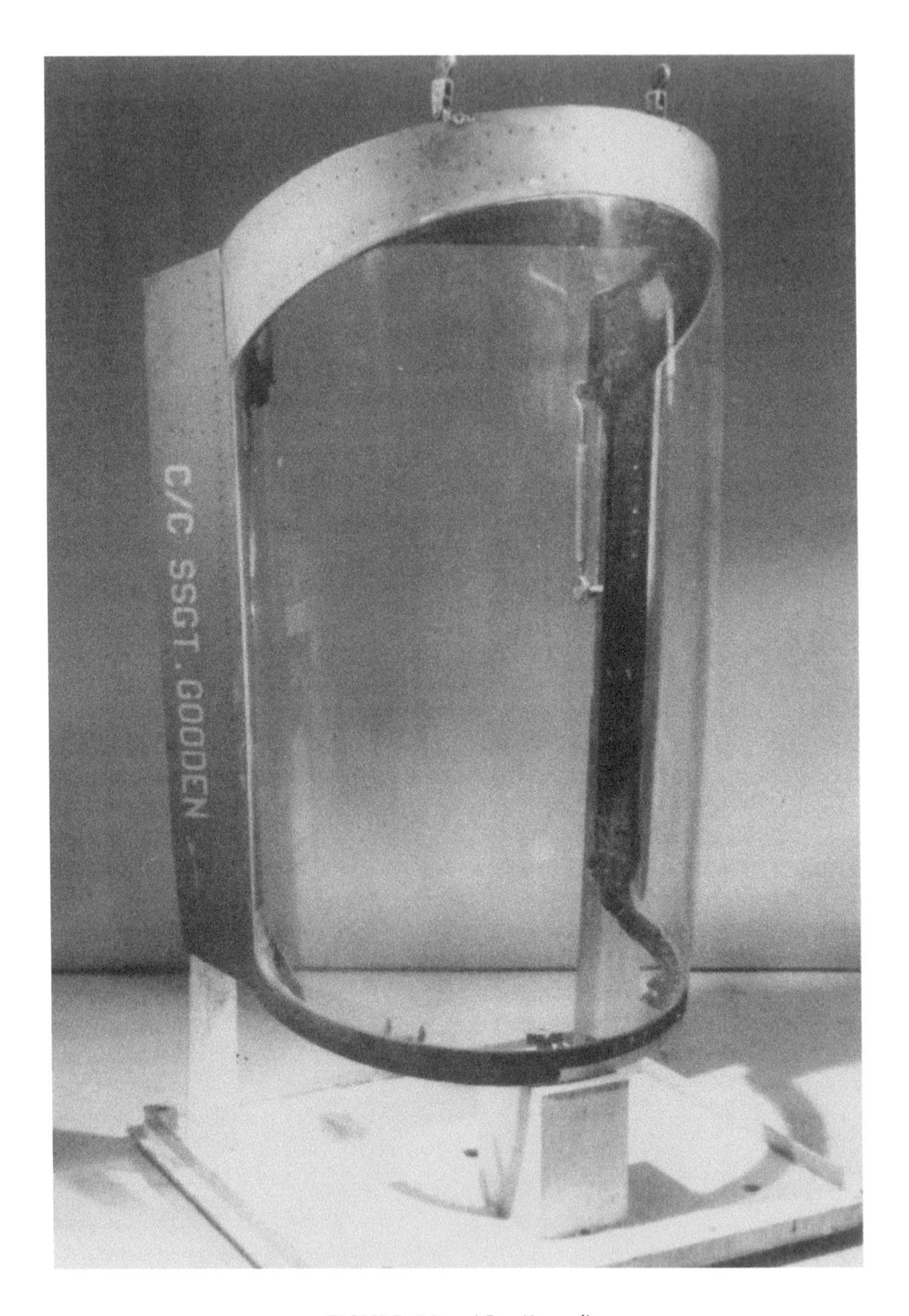

FIGURE 6.2. *(Continued)*

FIGURE 6.2. *(Continued)*

while minimizing optical distortion. The position, size, shape, and severity of flaws proximally located to each other are taken into consideration, as a group of flaws may be removed by a sufficiently large sanding pattern.

2. SURFACE REPAIR

2.1. Terminology

For the purpose of this discussion, a number of terms must be defined which have special meaning in reference to the characterization and rework of damaged aircraft transparency surfaces. The following definitions are derived from a working experience with U.S. Air Force canopies, although some are taken from the industry and military standards literature.

A *scratch* is any damaging mark on the transparency surface which is very narrow but may be of any length [6]. Scratches may be deep or shallow and can result from almost any contact with the transparency by a hard object. A *scrape* is a wider, shorter scratch. The inside of a fighter canopy transparency will normally have scrapes caused by the aircraft pilot's helmet. A *pit* is an indentation in the transparency surface caused by impact [6]. Pits are deceptively difficult to repair because they tend to grow as the surrounding material is sanded and the stress in the damaged material is relieved. A *dig* is a pit or small crater on the transparency surface defined by the diameter [6]. The diameter of an irregularly shaped dig is 0.5 times the sum of the length and width. The notation "50 dig" corresponds to 0.5-mm diameter. *Crazing* consists of minute cracks in the plastic (usually acrylic) material [7]. Crazing may not be visible under certain lighting conditions, but if the light is reflected at the proper angle off the crazing to the eye, the craze will appear to sparkle and will become quite obvious. A small amount of craze near the surface of a transparency can be removed by sanding, but crazing which extends deeper into the plastic will cause the transparency to be rejected. Crazing is commonly associated with ultraviolet or chemical exposure.

Binocular disparity is the difference between the two images on the human retina resulting from the lateral separation between the two eyes when viewing an object at a fixation point or due to the fact that an object is either nearer or farther than the fixation point [9]. A certain amount of disparity is beneficial and natural, leading to the perception of depth. However, when the disparity exceeds the limits for binocular fusion, a pilot may experience doubling of vision, eye fatigue, and headaches occur as the eyes strain to merge the disparate images.

There are a number of terms used to describe the general deterioration of visual quality when looking through a transparency which results when inside and outside surfaces of the transparency are not perfectly parallel in the field of view. *Deviation* is the displacement of a line or object when viewed through the transparent part and is expressed as the angular measurement of the displaced line, [8]. *Distortion* is the rate of change of deviation resulting from an irregularity in a transparent part and may be expressed as the slope of the angle of localized grid line bending, [8]. Distortion may also be expressed as the angular bending of the light ray per unit of length of the part. A *grid board* is an arrangement of orthogonal lines which is used as a reference when viewing through a canopy transparency to locate and judge the severity of distortion. Grid board lines will appear distorted when viewed through a transparency which has been improperly reworked.

Many special terms have been created to describe special types of distortion. *Haze* is the spatial attribute of smokiness or dustiness that interferes with clear vision [10]. Haze is the ratio of diffuse to total transmittance of a beam of light. *Halation* is a spreading of light beyond its proper boundaries. Halation is often a synonym for haze [10]. *Orange peel* is a mottled surface or high-frequency pattern which results from reworking of the transparency using abrasives. Orange peel cannot be eliminted but its objectionable nature can be reduced by the proper choice and application of the abrasives.

With regard to transparency rework, the term *sanding* will be reserved for reference to material removal using bonded abrasives (e.g., sandpaper), both heavy sanding for initial flaw removal and contouring as well as light sanding to remove scratch patterns from previous sanding. Sanding occurs in localized areas, where the goal is to remove the flaw while limiting the sanded area to the minimum required to control distortion. *Grit* will be used to refer to the coarseness of the bonded abrasive, which is usually graded in terms of average micron size of the abrasive particles. The term *polishing* will be reserved for reference to light material removal using an abrasive slurry. Polishing occurs over the entire transparency surface for general improvement of clarity.

2.2. Repairing Transparent Plastic Materials

A difficult aspect in learning about canopy transparency repair is the understanding of specific optical distortion characteristics that can be found in the plastic surface. Terms such as distortion, orange peel, haze, and crazing have intuitive meanings, but proficiency in recognizing these properties in the canopy transparencies is a learned skill requiring many years of experience. Techniques to control and correct such effects are even more difficult to learn (in fact, the techniques employed can only minimize but not eliminate the effects) since consulted experts were usually not in agreement as to the cause or the control thereof.

If a transparency has an overall dull, hazy appearance, it may be possible to save the transparency with polishing alone. This is the simplest repair that can be made because the amount of material removed in any one location should not introduce distortion. An overall final polish with a liquid abrasive slurry may be sufficient to restore clarity. If the haze extends deeper into the surface, a very light sanding of the overall surface area using fine abrasives may be in order before the final polish (a typical final polish will require a minimum of one hour). Often, paste wax is used to fill and hide any remaining minute scratches.

Flaws which are less than 0.25 mm or so in depth can be fully repaired. Deeper flaws may be reworked, but sometimes the deepest part of the flaw may be left in the plastic in order to reduce the time of rework. Most repair procedures allow some flaws to be left in the transparency, as long as all flaws are less than a specified diameter (usually 6 mm) and as long as there are not several in close proximity. The repair procedure involves sanding the plastic in the area until the flaw is removed. The initial sanded area may be on the order of 200 mm in diameter for a 0.25-mm-deep flaw in order to adequately contour the surface to avoid objectionable distortion. A typical distortion specification calls for a 10:1 slope-out of the sanded area, which refers to a maximum 10:1 slope of a grid board line when judging distortion from the pilot's point of view. The subjective nature of this distortion specification can be judged in the example shown in Figure 6.3. An automated procedure for measuring transparency optical distortion has been implemented and is discussed in [11].

The documented procedure used by the U.S. military which specifies how a canopy transparency is to be polished is written as a Technical Order (TO). A TO for canopy transparency rework applies only to one aircraft type, such as the F-4; however, each branch of the military will have its own version of the TO because each aircraft type, although similar in appearance, will incorporate slight differences in detail [12, 13]. Military aircraft manufacturers may also have repair procedures specific to a given weapon system [14] or general in nature but targeted for a specific canopy material [15, 16]. The repair procedures in these references cover manual rework techniques only.

The Air Force TO for rework of the F-4 transparency [12] is virtually identical to the Navy's version [13]. These TOs describe two methods of rework for small and large scratches. Small scratches are polished using a 6-mm felt tip chucked in a drill in conjunction with a fine abrasive polishing compound. This process is followed by lightly blending the area using a 150-mm diameter flannel buffing wheel driven by a low-speed power tool. Large scratches are to be removed by hand sanding with 400-grit sandpaper on a hard rubber block and using a circular motion. This is to be followed with 600-grit sandpaper and then buffing with the flannel wheel and polishing compound.

A McDonnell Douglas process specification for restoration of acrylic transparencies [16] describes the use of Micro-Mesh[2] products. Interestingly enough, it does not allow the use of power tools even for final buffing, and all hand sanding motion is to be along straight lines only, alternating between the fore and aft direction and the hoop direction of the transparency each time the sanding grit

[2] Micro-Mesh is a registered trademark of Micro-Surfaces Finishing Products, Inc.

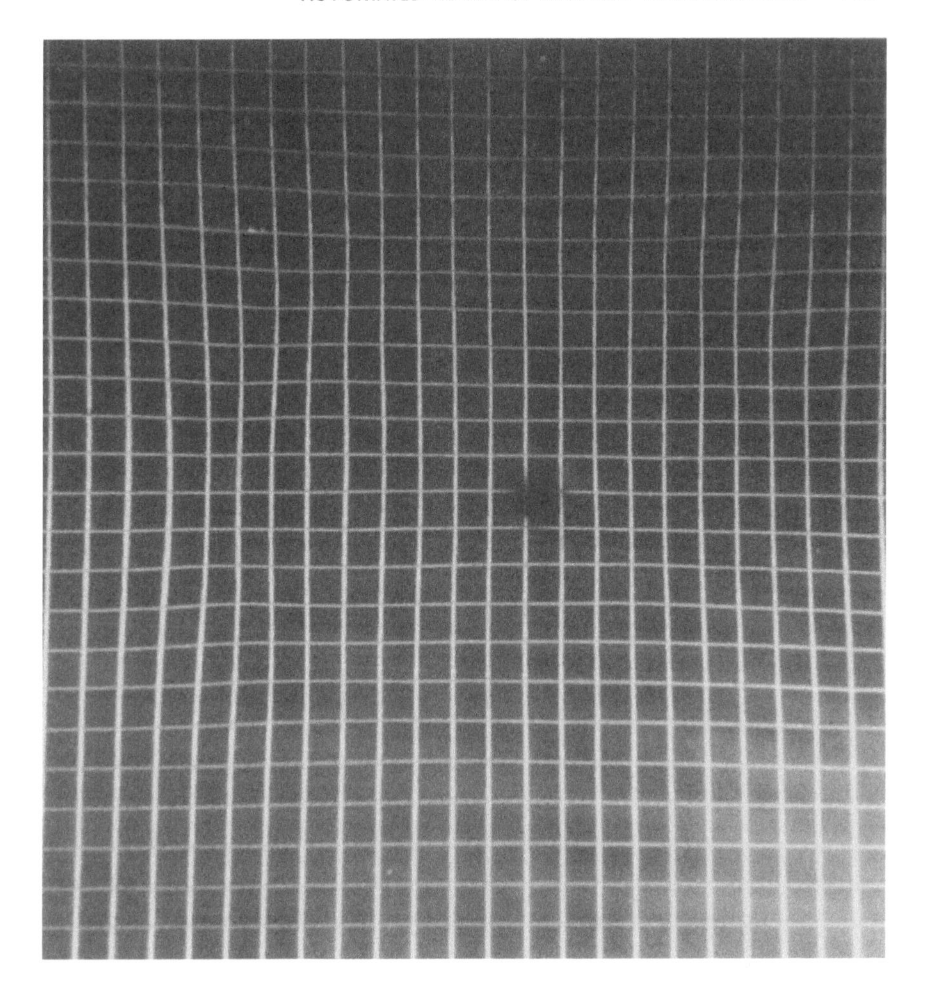

FIGURE 6.3. Example of transparency distortion.

is changed. This is quite contrary to the procedures recommended by the Air Force and Navy TOs for the same transparency. The process described uses 12 abrasive steps to remove the flaw and restore clarity. This is followed by application of a liquid cleaner and a liquid antistatic cream. The criteria for acceptable levels of distortion are somewhat descriptive but still completely objective:

- Abrupt slope reversal of straight-line elements of the target is unacceptable.
- Jumping or abrupt movement of the target during examination by moving the observer's head is unacceptable.

- Vision impairment, defined as a condition which causes the observer to focus on the reworked area instead of the target, is unacceptable.
- Slight, gradual bending of straight-line elements of the target is acceptable.
- Localized distortion (bull's-eyes) located entirely within one inch (25 mm) of any edge attachment strip is acceptable.

Optical distortion specifications did not become a significant measured issue until the F-16 canopy was put into service. The General Dynamics critical item development specification [14] lists performance criteria for angular deviation, optical distortion using grid board measurement techniques, original luminous transmittance, and haze. This document is a procurement specification for the transparency only, so it does not include process information for polishing and rework.

The referenced written procedures for repair of acrylic prohibit the use of power tools except for fine polishing. The danger lies in the control of the power tool to prevent digging into the surface and inducing additional unwanted distortion. There is also an unsubstantiated concern that the use of power tools may cause an orange peel distortion effect in the final polish quality.

2.3. Selecting Abrasive Materials

The choice of abrasive materials and the process with which they were to be applied were a basic undertaking in the early stages of the project. The manual tools and techniques employed at Hill Air Force Base were discussed with the experts there. A review of the literature was made, and suppliers of abrasives were contacted for samples and literature. With information and materials in hand, the manual techniques were applied to scrap transparencies in order to learn to duplicate the results seen at Hill AFB. Various abrasive materials were tested manually to characterize their cutting and polishing action, and eventually, a final selection of abrasive schedule was made based on the ability to quickly remove a flaw while producing a quality final polish [17].

Wet sanding using water as the liquid was selected as a preferred method over dry sanding as a means to control loading of the abrasive paper with residual plastic. Further, wet sanding assisted in minimizing the temperature rise of the plastic which was suspected of contribution to crazing [18]. By forcing water through the center of the backing pad to the surface of the abrasive, a reliable flooding of the abrasion area could be achieved, and the liquid tended to flush away abraded plastic particles.

Micro Mesh cushioned abrasives have been considered the industry standard for repair of plastic transparencies, and thus were tested first. Although they were quite suitable for manual sanding, it was determined that the number of recommended sanding steps involved and the lack of aggressiveness of the cushioned abrasives made their use with an automated system cumbersome. Several similar products were also tested with the same results. The time required to complete a removal process using these materials, plus the number of abrasive grits which would have to be employed, made these materials less suitable for automation.

Imperial[3] Microfinishing Film and Imperial Lapping Films furnished by the Industrial Abrasives Division of 3M were also tested. These are aluminum oxide abrasives bonded to a synthetic backing. The difference between the two products has to do with the way that the abrasive particles are attached to the backing material. Particles on the Microfinishing Films are electrostatically oriented during attachment to orient the sharp points away from the backing surface to give them a more aggressive cutting action. Lapping Films have the abrasive particles laid down flat to lessen their effect. Testing of the materials showed that the Microfinishing Films removed acrylic scratches quickly, but the finer Lapping Films made it difficult to produce a suitable final polish without leaving small scratches. It was found that 60-μm and 40-μm Microfinishing Films left heavy scratches which required much effort to remove. However, 30-μm abrasives used for a longer period of time with higher application pressures did an adequate job of material removal while eliminating two sanding steps, and so 30-μm Microfinishing Film was selected as the abrasive for initial flaw removal. The 15-μm Microfinishing Film was chosen as the second grit in the refinishing sequence.

A less aggressive bonded abrasive with a softer backing was needed for the final sanding step. 3M Wetordry[4] Production Polishing Paper in a 9-μm grade was found to be suitable. This paper has a synthetic paper backing, and the abrasive is bonded to the backing material in stripes for improved flexibility. Testing showed that the abrasive resulted in a less severe scratch pattern than the equivalent grade of Microfinishing Films, and left the plastic surface ready for a final polish.

A liquid slurry was necessary for the final polish. Union Carbide UCAR[5] APC-1 alumina polishing powders utilizing 0.3-μm aluminum oxide abrasives were tested with favorable results. Although constant stirring was necessary, this mixture was able to keep the abrasive in suspension longer than other mixtures tested. The slurry was applied using a 3M Imperial Polishing Pad, which is a flock-coated film pad that serves as a carrier to hold the slurry abrasives at the pressure surface. It was further believed that this pad also contributed an abrasive property of its own which, in this application, proved to be beneficial.

The final abrasive schedule for transparency flaw removal and polishing was

---

[3] Imperial is a registered trademark of 3M.

[4] Wetordry is a registered trademark of 3M.

[5] UCAR is a registered trademark of Union Carbide Corp.

thus chosen. Initial flaw removal would begin with the 3M 30-μm Microfinishing Film followed by 15-μm Microfinishing Film. 3M 9-μm Production Polishing Paper would be used next, followed by final slurry polishing with the Union Carbide 0.3-μm aluminum oxide suspension. This combination of abrasives minimized the number of grit changes required, provided for the heavy material removal required to grind away flaws of 0.20-mm and 0.25-mm depth, and produced a good final polish.

In order to avoid water spotting, deionized water was used during the wet sanding process. Deionization and particulate filtering of the water supply proved valuable for achieving repeatable polishing results. The deionized water was also used as a surfactant to make possible in-process inspection imaging, which will be discussed in a later section.

2.4. Design of Pad for Application of Abrasives

The design of the backing pad for both bonded and loose abrasives was of considerable concern. A soft pad would conform well to the arbitrary contours of the transparency, but extremely soft pads would not adequately transfer the driving action from the pad attachment to the abrasive so that cutting would occur. A rigid pad would not conform to the surface such that all cutting action would occur along the pad edge on the inside surface of the transparency, or cutting would occur only at the center of the pad on the outside surface of the transparency. The possibility of special pads for the inside and outside surfaces was reviewed, where a convex pad would be used on the inside and a flat or slightly concave pad would be used on the outside. The possibility of an adaptive pad, which would conform to the canopy contour, was also studied.

A series of tests were conducted to determine the operational attributes of two pad designs, a flat pad with a 12-mm liquid feed hole in the center and a domed (slightly convex) pad with no holes [19]. By making repeated linear strokes with a robot between two position points, a material removal pattern could be established for the rut thus sanded into the acrylic as shown in Figure 6.4. Actual canopy transparency acrylic material was used for these tests to avoid discrepancies which might result from the use of nonaged, unstretched commercial acrylic sheet. The amount of material removed at five prescribed locations across the sanded area was then measured using before and after optical micrometer readings in order to plot a rough profile of material removal which would correspond to the pressure profile across the pad surface. Data was collected as parameters of pad type, pad force, and abrasive type were varied.

The results of these tests, although tainted with experimental and measurement error, indicated that the flat pad would be preferred since it displayed a better distribution of material removal over the pad area. The domed pad removed more material volume, but the removal pattern was highly biased toward

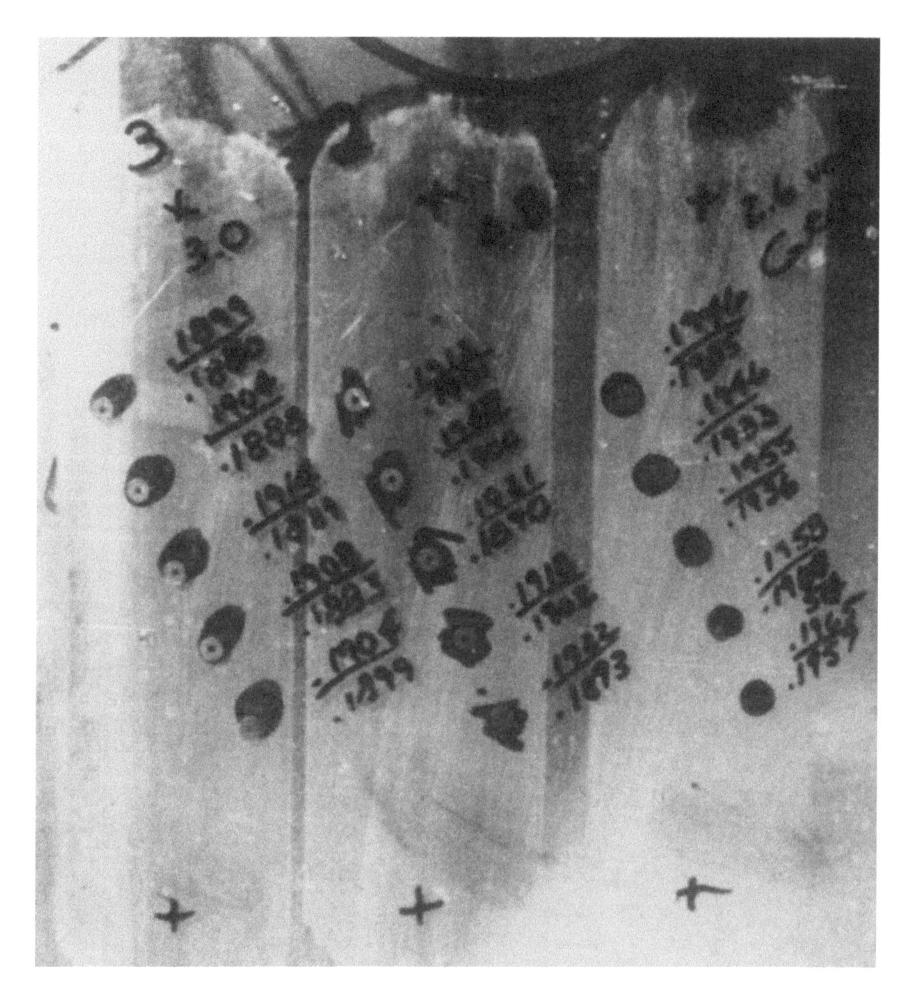

FIGURE 6.4. Material removal tests.

the center of the pad such that there was very little cutting action at the pad edges. The most significant fact learned from these experiments was that the large center hole in the pad for water feed was significantly reducing the material removal rate along the path of the pad center.

The backing pad design which ultimately resulted from testing such as described above incorporated a multilayer construction as shown in Figure 6.5. A rigid plastic material was used for attachment to the sanding tool. Bonded to the attachment plate was a 12-mm layer of closed-cell, 10-lb (density) foam neoprene which provided the spring force behind the sanding abrasive. Next came a

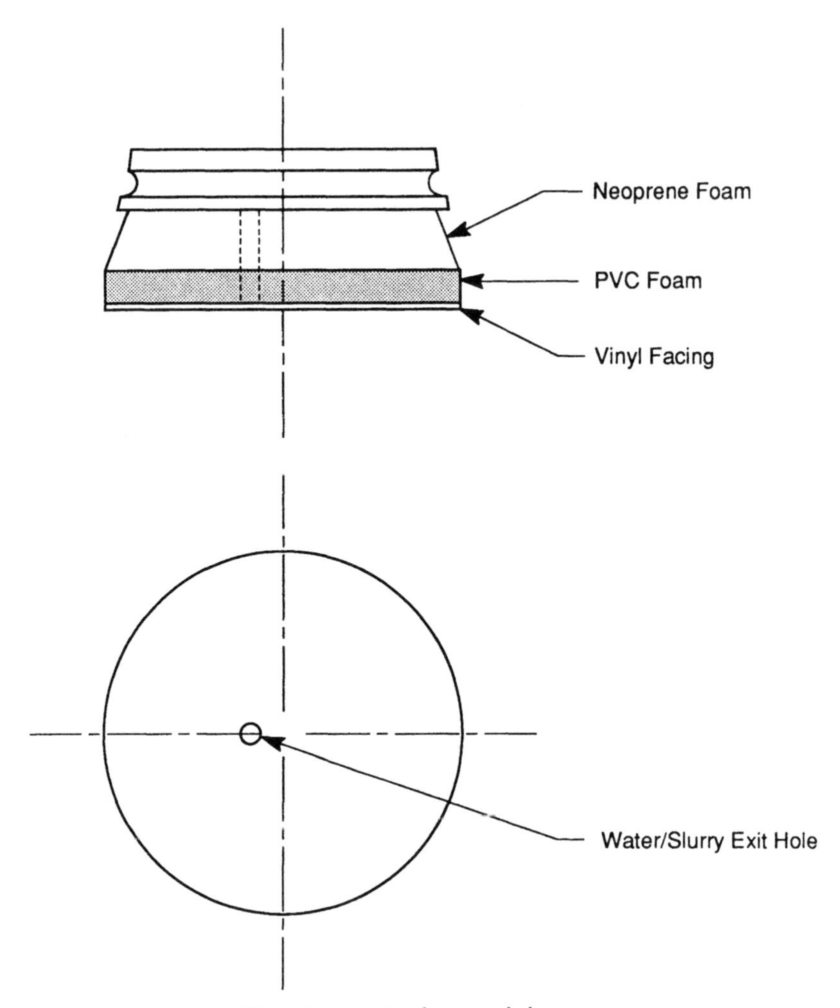

FIGURE 6.5. Backing pad design.

6-mm layer of closed-cell polyvinyl chloride (PVC) foam of a type which recovers slowly from deformation and provided a cushion which conformed to the general surface contours. (This layer of PVC foam was believed to reduce the amount of orange peel displayed after the final polish, but this could not be proven to be fact.) The final layer was a 0.25-mm-thick piece of vinyl used as a tough surface for attachment of abrasive discs using pressure-sensitive adhesive.

A 75-mm-diameter pad was somewhat arbitrarily chosen as a standard size for the production of sanding discs and as a compromise for the contact characteristics with the transparency surface contours of interest. To further improve contact with the surface and reduce the amount of cutting along the perimeter of the round pad, the edge of the foam pad was trimmed at a 60° angle. This causes the pad spring force to approach zero at the perimeter of the pad and to increase approximately linearly along a radial path toward the center of the pad. The center portion of this pad design approximates a uniform pad pressure.

Of course, this piecewise linear idealization of the pad pressure across the surface is not achievable in practice. When the sanding pad is in contact with the surface, the amount of pressure applied varies as a more complex function across the pad. The *pad pressure profile* is a function which yields the amount of pressure applied to the canopy surface at any point on the sanding pad. For the purpose of simulating material removal across a surface, the pad pressure profile is assumed to be radially symmetric and is modeled by a second-order polynomial function of the radius [20], such as shown in Figure 6.6. Force compliance within the sanding tool mechanism was used to maintain a constant, adjustable magnitude pad pressure. The assumption is made that the characteristic pressure profile remains unchanged, but the magnitude of the polynomial coefficients increases as more sanding pressure is applied to the pad.

2.5. Sanding and Polishing Mechanics

Even though discouraged by the process specifications, a powered sanding tool was necessary for an automated system in order to achieve production rates for transparency rework throughput. High material removal rates require a high occurrence of relative motion between the abrasive and the material to be removed. A pneumatic power sander was seen as a means to realize such relative

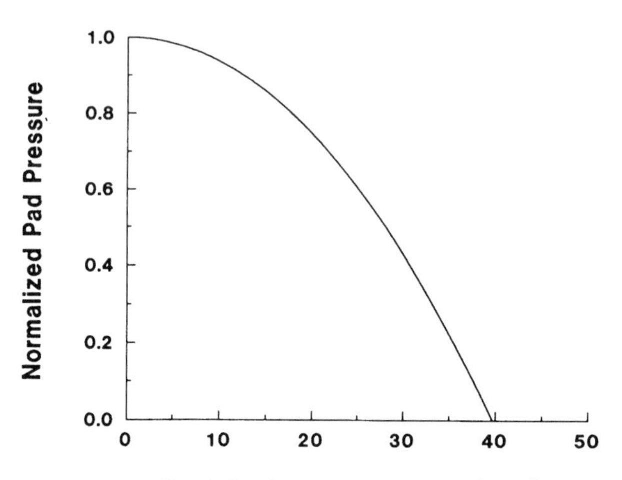

FIGURE 6.6. Pad pressure profile curve.

motion, where the velocity of the sander pad and the application force could be adjusted to control the rate of material removal.

One issue in the use of power tools was the selection of the pad motion type. The basic types readily available include straight-line, rotary, and orbital. Straight-line means that the pad reciprocates with a fixed mechanical stroke. Since the pad must stop at the end of each stroke, the average material removal rate is less than can be achieved using other actions with comparable peak velocities. The rotary motion gives high surface speeds at the perimeter of a round pad, but control of the cutting action can be much more difficult due to the variation in the material removal rate across the surface of the abrasive pad. An orbital motion, in which the pad is oscillated by driving it at a radial offset from a rotating shaft via a rotary bearing gave a uniform pad surface speed, which was important for uniform material removal, while displaying superior cutting action to that of the straight-line motion. (A random orbital motion is a variation which allows the pad itself to freely rotate while being driven in an orbital fashion. This additional randomness was not implemented in the final system simply because of mechanical considerations in delivering water and slurry to the pad.) Thus, the orbital sanding motion was chosen as the best design compromise of rate of material removal, uniformity of material removal across the pad surface, controllability of sander reaction forces, and ease of implementation.

The sanding pad must be held approximately normal to the surface of the canopy in order to implement a uniform, or at least repeatable, cutting action across the surface of the sanding pad. This was particularly important for the initial material removal and contouring using the heavy abrasive grits, becoming less of an issue as the rework proceeded to light polishing. (This concept was understood by the writers of the Air Force and Navy TOs for rework of the F-4 canopy transparency since they allowed the use of power tools only for final polishing.) The gross normality of the sanding tool to the surface of the canopy would have to be accomplished by the manipulation of the sander. Angular compliance of the pad itself over a range of a few degrees could be accomplished by mechanical means in the sanding tool itself.

Since manipulation of a tool along a surface with control of manipulator application force could not be achieved to the required resolution for transparency polishing, the sanding tool required force compliance in the application direction. This would be implemented with a pneumatic pressure loading against a linear degree of freedom. The low-speed response of a pneumatic system would be adequate for the job at hand since motion of the tool assembly across the transparency surface involves relatively low velocities, i.e., the material cutting action would be the result of the high-velocity pad motions where the superimposed gross movements of the sanding tool itself would contribute little to the sanding action.

The linearly compliant axis in the sanding tool would be necessary to accommodate inaccuracies in the mechanical and control systems. A mechanical tolerance buildup occurs due to manufacturing variations in the canopy transparency and frame as well as variations in the fixturing for the canopy. Control inaccuracies occur in the manipulation of machines relative to the approximate mathematical modeling of the transparency surface. A scheme was devised wherein the position of the compliant axis in the sanding tool would be used to correct for such tolerance errors.

A position feedback signal would be used to detect initial contact with the transparency surface and to monitor for sustained surface contact while manipulating the tool across the transparency. In this manner, real-time tool offset corrections could be made to correct for inaccuracy trends whereby the canopy surface would appear to the manipulator controller as moving closer to the tool or farther away. A large range of travel for the compliant motion could be allowed so that these corrections could be applied at a leisurely pace.

The end result of the aforementioned tests and design requirements was a decision for a baseline sanding and polishing tool configuration [21]. This tool is referred to as the *polishing end effector* and is shown in Figure 6.7. The polishing end effector is basically an orbital sander which is mounted on a linear slide. The sanding portion is pushed outward with a bellows actuator so that a controlled pressure in the bellows determines the ultimate force with which the abrasive

FIGURE 6.7. Polishing end effector.

backing pad presses against the canopy transparency surface. The 35 mm of travel in the end effector slide accommodates dimensional errors in fixturing, transparency surface modeling, robot motion execution, etc. to keep the pad against the surface as the robot performs the sanding and polishing motions. Figure 6.8 shows the polishing end effector in contact (normal and compliant) with the transparency surface.

A position feedback device is used to monitor the position of the sander portion on the slide mechanism. This is a Hall effector sensor with integrated amplifier that is held in close proximity to the side of a bar magnet as it travels along the length of the magnet from one pole to another. The output from the device closely resembles a voltage divider output from a linear potentiometer (the original design incorporated a linear potentiometer, but reliability was not acceptable).

The end effector includes a pad change mechanism for autonomous exchange of sanding and polishing pads. This is an inflatable gasket design wherein the gasket (tubing) is evacuated for dropping a pad and inflated for gripping a pad

FIGURE 6.8. Polishing and effector in normal and compliant contact with surface.

with a special attachment adapter. The support for the latex tubing is a molded urethane structure which may be quickly replaced should the tubing develop a leak.

Fluids required by the end effector for the rework process are supplied by hoses which are routed along the robot arm from a remote supply. Compressed air drives the sander motor, inflates the linear slide bellows actuator, and holds the sanding pad. Water is fed through the center of the abrasive pad during wet sanding. Abrasive slurry is also fed through the center of the pad for the final polish. Additional end effector operations are provided by a spray nozzle for washing and air blast nozzles for drying. These process operations are necessary to clean debris from the transparency to prevent it from drying on the surface and to clean the area that is to be viewed by the camera during in-process inspection.

2.6. Sanding Patterns for Localized Flaw Removal

Once the basic sanding technique was determined, it was necessary to develop methods for manipulating the polishing end effector about the transparency surface in order to achieve flaw removal. Initial sanding tests were first attempted on flat panels of acrylic using the polishing end effector with a robot [19]. The goal of these tests was simply to verify the procedural sequence of abrasives used to abrade the surface with a coarse grit for removing flaws and then to bring the surface back to optical clarity using finer abrasives. The tests involved sanding radially symmetric circular patterns and increasing the final radius for each successively finer abrasive grit. This blending of patterns into an enlarging area spreads the removal pattern to minimize distortion and to make sure that each preceding grit is completely removed during the next sanding step with a finer abrasive. The square piece of acrylic sanded with progressively decreasing finer abrasives toward the center of the square is shown in Figure 6.9. The maximum material removed at the center of the pattern was on the order of only 0.25 mm, but it was readily apparent from a visual inspection that material had been removed.

As a result of these sanding tests, it was determined that the flaws on the surface would be locally removed by the application of a sequence of circular patterns with increasing radius, decreasing abrasive grit, and decreasing depth of material removal. Holding the abrasive grit constant, the circular sanding pattern then is a function of the pressure exerted at each point on the sanding pad (as defined by the pad pressure profile) and the movement of the pad along a specified path across the surface. Let $P(x, y)$ be the pad pressure profile function, i.e., the pressure at point (x, y) on the pad surface where $(0, 0)$ represents the pad center. Let $F(t)$ be a function of time which gives the location of the center of the pad along a specified planar path. The amount of material removed at any point on the plane over some continuous time interval is given by

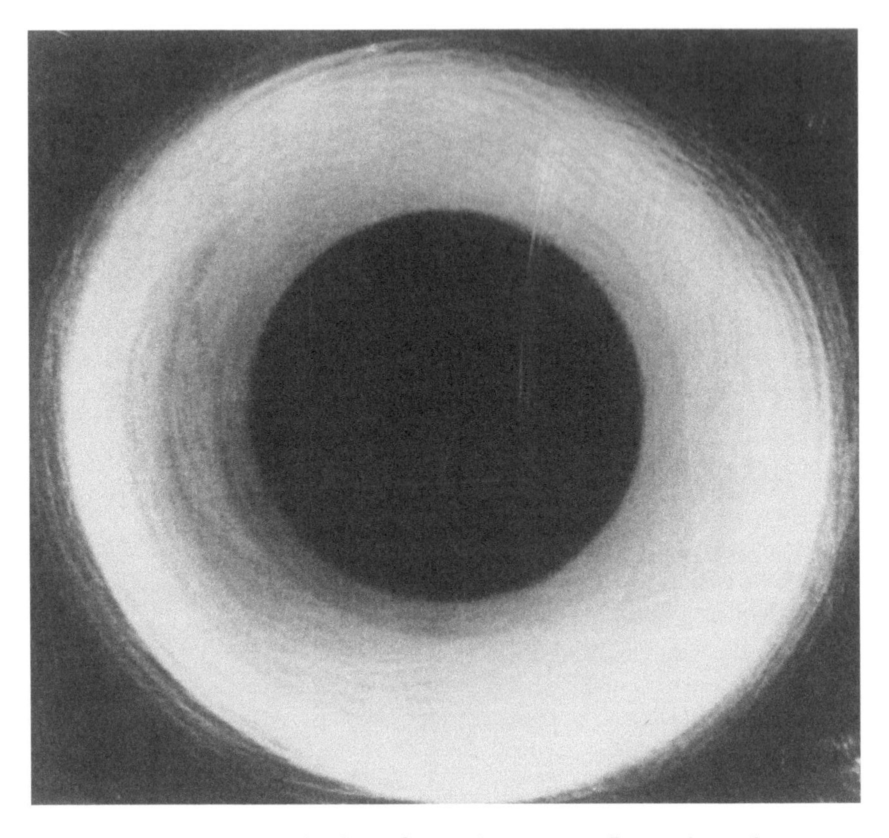

FIGURE 6.9. Result of circular sanding test on flat acrylic surface.

$$m(x, y) = \int P(F_x(t) - x, F_y(t) - y)\, dt$$

The area traversed by the sanding pad and the total amount of surface material removed along the path is called an *area removal pattern*.

Once the pad design was determined and the pad pressure profile was known, the *ideal removal pattern* was defined to be a circular pattern with the following constraints:

- The amount of material removed is radially symmetric; i.e., strictly a function of the radius of the circular pattern.
- The maximum amount of material is removed at the center of the pattern.
- The slope of the material removal as a function of the pattern radius must be zero at the center of the pattern and at the edge of the pattern. This is to avoid any discontinuities in the radially symmetric circular pattern.

- Material removal as a function of the pattern radius must be a smooth continuous function in order to minimize distortion.

Under these conditions, the (assumed) ideal removal pattern can be represented by a cubic polynomial function of pattern radius. Let $r_{max} > 0$ be the maximum pattern radius, and for a given radius r, define the normalized radius as $\hat{r} = r/r_{max}$ and let h be the maximum depth of the pattern. Then the *ideal removal function* of the normalized radius is defined to be the parametric cubic curve given by

$$m(\hat{r}) = h(a_3\hat{r}^3 + a_2\hat{r}^2 + a_1\hat{r} + a_0) \ , \ \hat{r} \in [0, 1]$$

and the function which represents the slope of the curve is given by

$$\dot{m}(\hat{r}) = h(3a_3\hat{r}^2 + 2a_2\hat{r} + a_1)$$

The constraints on the circular pattern as an ideal material removal pattern can be expressed in terms of the material removal function and its derivative as given by

$$\begin{aligned} m(0) &= h = ha_0 \\ \dot{m}(0) &= 0 = ha_1 \\ m(1) &= 0 = h(a_3 + a_2 + a_1 + a_0) \\ \dot{m}(1) &= 0 = h(3a_3 + 2a_2 + a_1) \end{aligned}$$

Solving this set of four simultaneous equations for the four unknown coefficients yields

$$m(\hat{r}) = h(2\hat{r}^3 - 3\hat{r}^2 + 1)$$

with a plot of this function shown in Figure 6.10.

Several path generation functions were investigated which yielded the circular material removal pattern with polynomial function of radius given above. Such functions included the N-leaved rose, the N-leaved double rose, and concentric circles. The results of sanding tests using these path generation functions showed that the primary problem with the rose functions was the inability to control the removal rate along the pattern radius, which, however, could be more readily controlled when using the concentric circles [19, 20].

Computer simulation and optimization were applied in order to determine the concentric circular paths that would result in the ideal area material removal pattern [20]. Each concentric circular path is controlled by two parameters which can vary in order to facilitate the match: circle radius and sanding coefficient. The circle radius positions the sanding pad with respect to the pattern center and the sanding coefficient establishes the amount of time to be spent by the pad in following this path, i.e., the motion velocity.

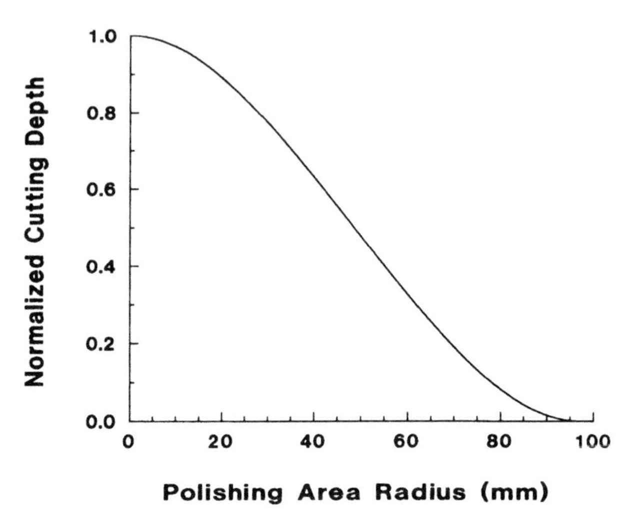

FIGURE 6.10. Ideal material removal curve.

The optimization begins by locating the point at which the maximum deviation in the ideal removal function is allowed. This is at the edge of the ideal removal pattern, so the radius of the outside concentric circle will be one sanding pad radius inside this point. The sanding coefficient for this radius is a multiple of the pad pressure profile function that just causes the multiplied function to exceed the ideal removal function at that radius. Successive concentric circular radii and sanding coefficients are found iteratively stepping from the innermost circle radii established thus far toward the center pattern radius of zero. This technique establishes the circle radii and sanding coefficients in an outside to inside order. It assumes that once the parameters have been established, new circular paths do not cause interaction with previous circular paths. Although this is not theoretically true, tests performed indicated that the effects are minimal. A plot of the removal function that resulted from the optimization for a certain sanding pad radius, circular pattern radius, and maximum allowable error is shown in Figure 6.11.

Circular sanding pattern tests again were performed on flat acrylic sheets using the concentric circular paths determined by the optimization. Although the material removal at each radius from the pattern center was controlled and thus more smooth, the effect of the overlapping material removal patterns during subsequent sanding with finer abrasives and increased pattern diameter introduced orange peel distortion into the surface. In the final implementation, this effect was minimized by constructing less than perfect concentric circular paths by adding a small random error value to the radius of each point on a circle.

The sanding patterns discussed in this section are in reference to being applied

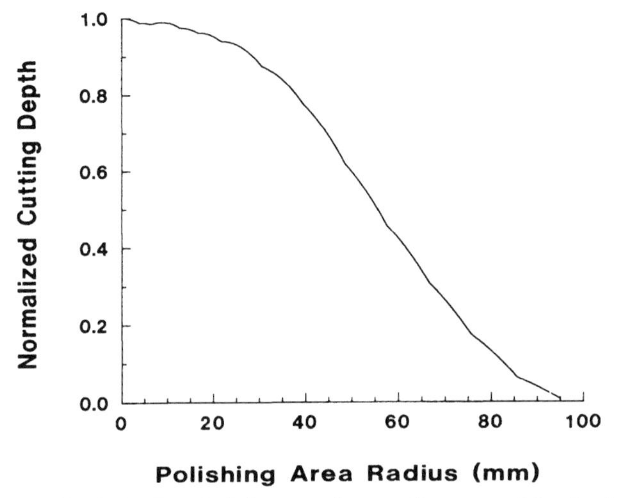

FIGURE 6.11. Optimized material removal curve for concentric circular pattern.

to planar surfaces. The projection of these patterns onto the actual curved transparency surface is an operation of the surface modeling to be discussed in a later section. Different sanding patterns were developed to handle the staged procedure for grinding and refinishing and to handle the range of severity and type of flaws. An optimal strategy for placing combinations of these patterns over a flawed area of the transparency surface will also be discussed in a later section.

2.7. Overall Surface Polishing

Sanding of localized areas to remove flaws was a necessary, but not sufficient requirement. A strategy for polishing the entire transparency surface with a fine abrasive slurry while using a raster motion was also required. Every transparency to be reworked requires an overall polish to restore the general clarity of the acrylic which has gradually degraded while in service. The polishing motion would need to accurately follow the transparency contours while maintaining surface normality and velocity consistency. The motion guidance techniques developed would also be applied for washing and drying the transparency.

A vertical raster motion was selected for this application. The motivation for this preference had largely to do with the fact that the robot stroke lengths would be maximized but mainly because one of the canopy rework references specified that final polishing be performed in the fore-to-aft direction [16], along the pilot's general line of sight. Some benefit was also realized in the application of polishing slurry to reduce the amount consumed due to runoff.

The pad design previously discussed was modified slightly for the polishing application. Here, because of the low cutting rate of the fine abrasive slurry, the pressure profile of the pad was less critical in the control of distortion. Further, the low cutting rate and the general lubricating properties of the slurry made the use of a softer pad possible. Thus, the opportunity was available to use a slightly larger pad size for the final polish in order to cover the surface area more quickly. A soft pad with diameter of 100 mm was selected based on a compromise of diameter versus full pad contact with the transparency surface contours. In practice, the increased pad size (compared to the 75-mm pads used for sanding) was used to increase the overlap of the polishing strokes. This polishing overlap was believed necessary to avoid streaking in the final polish.

The traversal of the upright transparency surface with vertical strokes is achieved by following a sequence of surface points up on one stroke and an adjacent sequence of points down on the next stroke. The method of generating the surface points involves the partitioning of the surface model into equal area patches as described in a later section. The partition points define the locations for placement of the polishing pad center. The traversal of the surface in up and down vertical strokes is accomplished by ordering the partition points accordingly (these partition points for the polishing operation are called *polishing points*). The actual traversal motion involves movement over the surface while linearly interpolating between consecutive polishing points. The horizontal spacing between columns of points is used to control the amount of pad overlap on successive vertical polishing strokes. This parameter was chosen to be 25 mm. The vertical spacing between points along a column of polishing points is used to control the speed at which the polishing pad traverses the surface during a vertical polishing stroke. This parameter was chosen to be 50 mm.

The polishing motion is accomplished via a robotic manipulator, and ideally, it is desirable that the entire transparency surface be polished in one continuous motion. However, at any orientation of the transparency, every point on the transparency surface cannot be reached by a robot due to joint angle limits, singularity, or collision of the arm or end effector with the work piece. As a result, the transparency surface area is divided into a number of polishing sectors [22]. A *polishing sector* is a grouping of polishing points on either side of a transparency surface that can be reached safely by a robot with a single orientation of the transparency.

The strategy and constraints for organizing the polishing sectors for a transparency surface is summarized as follows. Each sector should contain a large number of polishing points so that the amount of time involved in polishing a surface with a few large sectors is shorter than when using many small sectors. Each sector should contain many polishing points in the vertical direction since that is the primary direction of the polishing motion. The amount of horizontal overlap between neighboring polishing sectors should be reduced, i.e., vertically long and horizontally narrow sectors should be avoided. The rectangular box that bounds the polishing points associated with a sector should not contain polishing points associated with another sector. The number of polishing sectors within critical regions of a transparency surface, e.g., the F-16 HUD (heads-up-display) area, should be minimized and the organization of polishing sectors (e.g., shape, size, and location) should reflect the symmetrical nature of the transparency surface. All polishing points in a sector should be reached when the end effector is in contact with the surface for polishing and for washing and drying, these same points must be traversed without robot limitations when the end effector is at an offset of 50 mm from the surface.

Polishing sectors were organized for inside and outside surfaces of all three transparency shapes using a robot and positioner configuration to be described in a later section. The entire inside surface of the F-4 forward and F-4 aft canopy transparencies could be polished using one polishing sector, while the outside surface of the F-4 forward canopy transparency required four polishing sectors, and the outside surface of the F-4 aft canopy transparency required three polishing sectors. Due to the severe constraints on polishing sector organization for the F-16 canopy transparency, the inside surface required five polishing sectors while the outside surface required 10 polishing sectors. The large number of polishing sectors needed on the outside surface of the F-16 transparency surface is somewhat due to reach limitations of the robot, but mostly due to the complexity of the shape. In spite of great difficulty in dealing with the F-16 canopy transparency, this method of polishing via sectors has proven to be effective.

3. SURFACE INSPECTION

3.1. Inspecting Transparent Materials

The lens-making trade includes much experience in the examination and characterization of transparent materials. Glass, and more recently plastic, lenses have standard test procedures for quantifying clarity, pits, bubbles, scratch, and dig [6, 7, 10, 18, 23, 24]. Extremely accurate methods for testing the curvature and regularity of the surfaces have also been developed [8, 9, 25–27]. These techniques are used in the factories where the optics are made, and in laboratories where experiments require knowledge of every detail of the lens imperfection to succeed. Many of the tests require an optical bench for stability, and in most cases, the lens bodies are of a convenient, manageable size. Almost every sensor technology available has been applied to this difficult task. Light-based techniques are almost invariably used due to their extreme accuracy, and due to the fact that ultimately, because the object is transparent, it will be used to transmit light in a predictable manner.

Large transparent surfaces on fighter aircraft present a difficult challenge for inspection, even with the extensive background of technology already developed by the industry for lens making. Transparencies on some of the aircraft are very large: they are long (3000 mm or more), thick (16 mm or more), and can have wraparound shapes (e.g., the F-16 canopy transparency). Primary optical materials include acrylics and polycarbonates, and laminates thereof.

Large, unwieldy transparencies leave the factory with reasonably good optical character, but succumb rapidly to damage and aging artifacts. Inspection at the factory makes use of large jigs with what might be considered in the lens industry relaxed standards. Important variables to be measured include optical deviation and binocular disparity. These qualities ultimately determine how accurately the pilot can target and how likely he is to suffer flight disorientation caused by transparency optical distortion. As the canopy ages and undergoes repair cycles, it is important to gauge distortion, hazing, halation, crazing, cracking, and disbonding. Inspection must reveal these aging artifacts.

The optical quality of transparency material (its clearness) presents both an advantage and a disadvantage. The interior volume of the material can be examined, but it is difficult, for instance, to gauge where in the depth of the material a region of interest lies. Multiple paths through the material, reflections, widely varying characteristics of the differing damage types, and a host of other well known lens-type undesirable characteristics help to make inspection difficult to take from the laboratory into an industrial environment of dirt, damage, and improper repairs.

A wide variety of nondestructive inspection (NDI) techniques have been developed for, or tried for use on, acrylic and polycarbonate materials [28, 29]. Methods for inspecting the transparent plastics include X-ray, ultrasonic, photoelastic, microwave, electrified particle, penetrant, and optical/visual. Radiographic inspection techniques suffer due to the low density difference between the plastic and voids or damage areas within. Ultrasonic inspection of cracks and crazing is possible through the combined use of compression and shear wave techniques. The detection of slight surface flaws on the order of 0.025 mm is difficult at best, and would require specialized focused beam devices creating a very small inspection area on the surface. This small area translates to long inspection time, and also implies direct contact with the transparency surface. Photoelastic methods lend themselves to generalized stress concentration visualization. Due to the low optical sensitivity of polymethylmethacrylates, thickness variations, and geometrical considerations, it is difficult to obtain more than a general indication of the stress fields acting in the transparency surface material. Microwaves can be useful for finding cracks in the material, but are not useful for finding minor surface flaws.

Particle and penetrant schemes are useful in the detection of transparency damage having at least some exposure to the material surface. Experiments with a variety of penetrants showed enhancement for human visual inspection of cracks and other surface flaws, but poor enhancement for machine-type visual inspection using charge-coupled device (CCD) cameras [30]. This result was confirmed at several color frequencies. Successful devices using this technique have been built for turbine blade inspection, but they use high-power lasers as illumination, and use banks of photomultiplier tubes to image the laser-induced fluorescence [31].

One optical inspection method uses the scattering effect that defects in the plastic material have on penetrating laser beams. (This technique was examined in this research and further proposed in [32].) The laser beam is scattered both forward and backward with intensities and patterns which depend entirely upon the nature of the defect. The patterns, in general, resemble conical distributions, but are of a very discrete form, i.e., there is a great deal of speckle and spotting of the beam patterns in each of the conical scatter patterns induced.

There are a variety of ways that the scatter pattern could be detected from either side of the transparency surface. From the back side, a roughly coordinated receiver would integrate photon energy outside of the central (Gaussian distribution) beam pattern. From the front side, a half-silvered mirror could perform a similar task. Actual detection could be as simple as a phototransistor for simple magnitude information, or as complex as a circular CCD sensor or a linear 2D planar detector (both offering magnitude and positional information on beam spotting). While this method would easily detect flaws in the transparency surface, it would be very difficult to determine whether the flaw was on the inside, outside, or somewhere in the middle of the surface. Unfortunately, relatively complex analysis would be necessary to determine other than flaw information, and as might be expected, this method is also very sensitive to dust and dirt on the canopy.

Available visual techniques include both single- and two-surface activity versions of dark-field imaging, laser scattering, and high-resolution field distortion measurements. Visual techniques are both noncontact and nondestructive, and can meet the resolution needs of the flaw detection problem. Direct machine vision imaging uses only the properties of the illumination to highlight flawed areas on or in the surface. Consequently, illumination techniques are the critical issue for this method of inspection.

3.2. Illumination and Flaw Detection

A family of illumination techniques makes use of the fact that a transparent plastic sheet acts very much like a wave guide to tunnel light (as well as ultrasound and microwaves). Once the light has been introduced into the transparency surface, it bounces back and forth between the walls in a regular manner, and due to various forms of absorption and scattering, the intensity slowly decreases with range (as in fiber-optic cable). When, however, the light strikes a flaw in the wave guide, a large amount of the light can escape which can be detected. Some light naturally will be coupled into the transparency surface as it passes through;

however, for optimal flooding, a coupling method is used. As indicated in Figure 6.12, a prism and optical coupling oil can be used to transfer a very high percentage of the incoming light into the transparency surface (e.g., from a helium neon laser).

Both the use of a strobe and a laser with this method were tested and each have their disadvantages [30]. First, the sensing devices must be in physical contact with the transparency surface. Second, due to the varying intensity gradient that forms with range from the coupling site, reliable flaw intensity data is difficult to obtain. This gradient would also mandate the use of several light sources simultaneously for large area coverage, or movement of the sources for each new image acquisition. Third, the method requires a large amount of power to give an acceptable average power level over a large area of the transparency surface. Thus, while CCD cameras are very sensitive to helium neon laser light, to couple the beam and scan it through an F-16 canopy transparency in an effective lighting manner would require delivered light on the order of 10 watts. This last disadvantage for lasers can be overcome by strobes due to the short illumination period, yet the unusual distribution of light in the transparency surfaces during experimentation revealed that the second disadvantage is a major one. Despite its

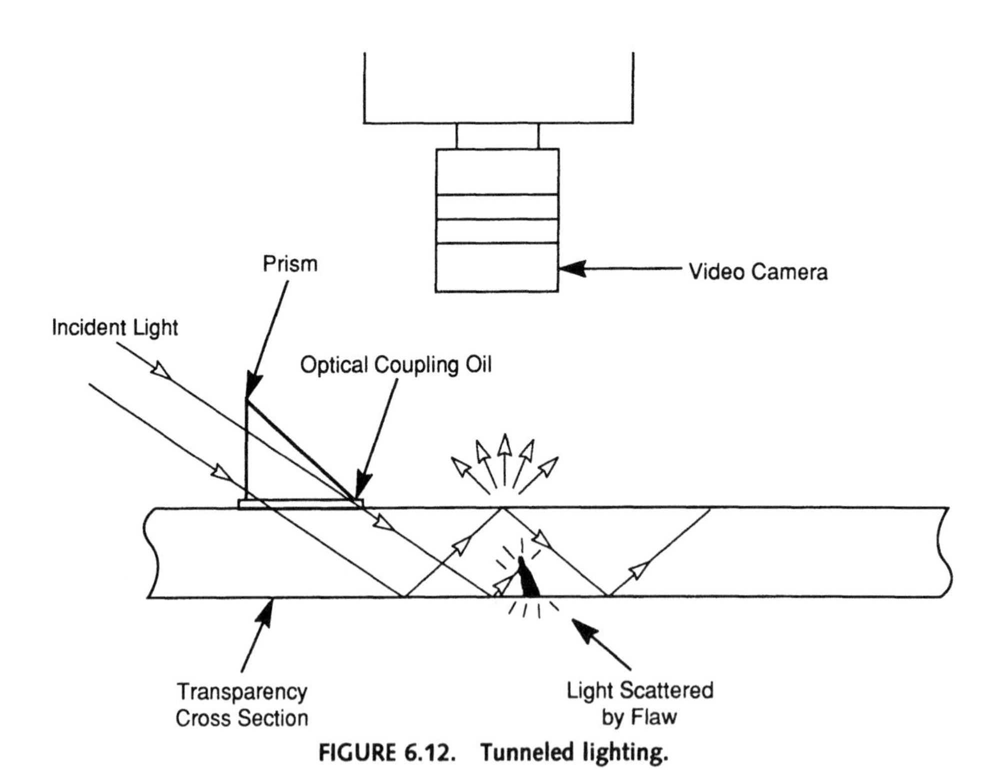

FIGURE 6.12. Tunneled lighting.

problems, the tunneling method of illumination provides exceptional images for the cameras, and can be implemented from one surface access. Damage that scatters light can appear in very high contrast to a dark-field background.

One method used to inspect helmet visors is a two-surface access form of dark-field imaging. As depicted in Figure 6.13, on one side of the canopy transparency there is a light source (preferably a laser) and a long focal length biconvex lens. This is used to prepare a source of highly parallel light rays. The high-quality beam passes through the optical material and is met by a lens which is a duplicate to the one on the source side. A small sphere or piece of wire (known as a field stop) is then used to block all light passing through the focus point of the optical train. It is only when specific defects exist in the optical path of the large beam that the rays are deflected from their parallel paths. These light rays will no longer pass through the focus point but do pass on to the video camera behind the focus point. Thus, only defects or distortion appear as light against an otherwise dark background in the image field. While this method appears ideal, it is normally performed on an optical bench or under otherwise very controlled, optically stable conditions. Implementing this scheme in an automated industrial work cell, however, would not be practical.

Another method of the general visual techniques uses high-resolution field distortion measurements. Like the dark-field methods, it is currently being used to analyze helmet visors, but on a much higher level of criticality. A system using this technique was originally developed for the detailed distortion analysis of F-111 windshields, but has found uses in a variety of areas [33, 34]. The system has a servo-driven angular deviation determiner, which is used to leave the system in position to image an area of the transparency using a lateral shearing interferometer. The interference pattern generated by the interferometer is gauged by a linear CCD array. Processing of the CCD data determines lens and distortion effects for an area to very subtle diopter changes. This method appears highly satisfactory for the gauging and analysis of transparency distortion currently performed only by humans and by relatively simple machines such as the ADAM 2 (used for measuring angular deviation of F-16 canopy transparencies [35, 36]). One disadvantage is that the transparency surface would likely have to be moved around an optical bench, in order to perform the measurements. A second consideration is that the distortion fields determined have no correspondence to which surface might be the prime contributor of the effect. Again, this technique does not provide all the information necessary to guide the transparency repair process.

A method of illumination that borrows from the tunneling method and has few of its disadvantages can be simply labeled direct illumination. Figure 6.14 shows how this method works. Using a strobe source, light is directed to the local area of interest at an angle out of the normal plane of imaging. Much of the light is reflected from the two surfaces of the transparency and is lost. Some of the light is captured within the transparency surface and tunnels there. The rest of the light

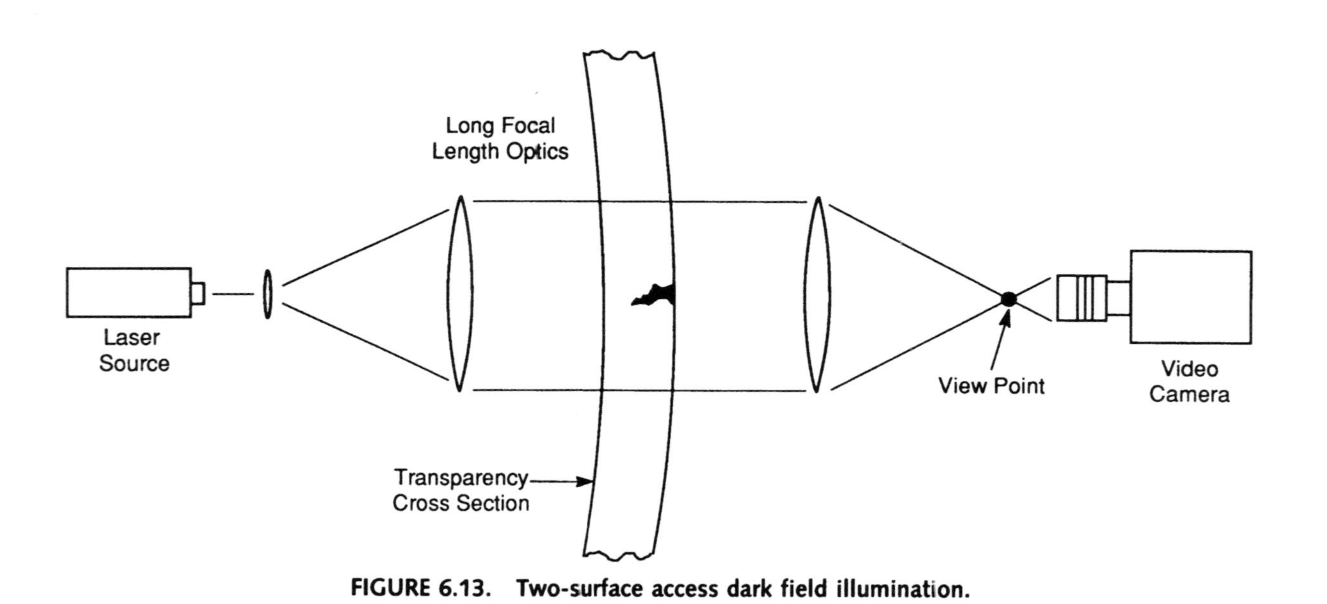

FIGURE 6.13. Two-surface access dark field illumination.

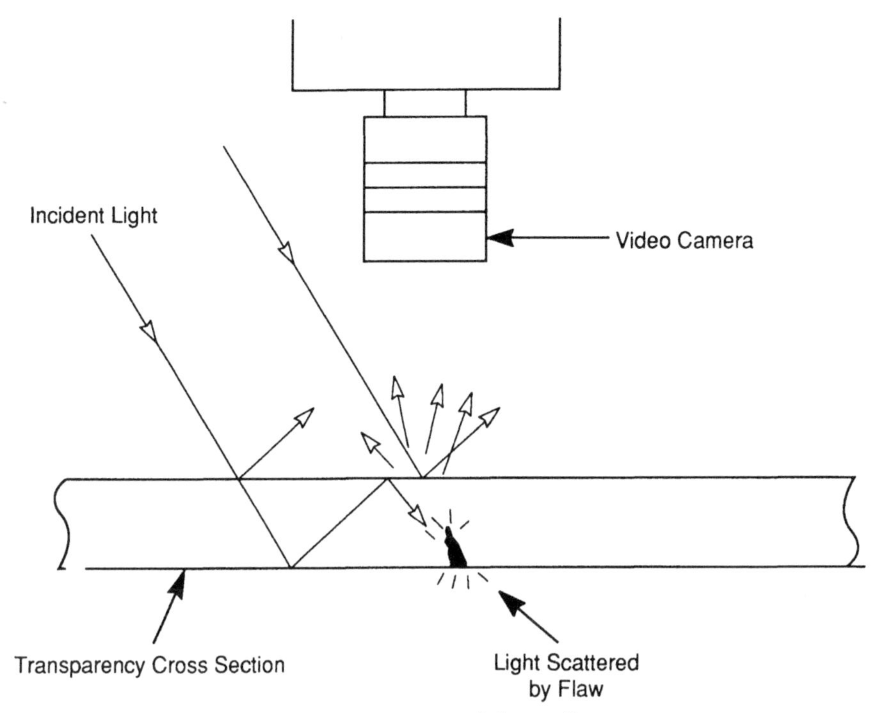

FIGURE 6.14. Non-coaxial direct illumination.

passes through the transparency material. The light passing through the surface and the light captured in the tunneling have the chance of hitting flaws and radiating in all directions. Due to the fact that much of the light is lost in this method, a very high amount of incident light must be used to get a useful flaw-to-background contrast. This can be provided by a strobe source with focusing optics. This method also lends itself well to the prospect of being unobtrusive to the imaging cameras.

Image acquisition of an abraded area presents another set of inspection problems which must be accommodated. In an ideal scenario using visual flaw detection techniques, a machine vision system would constantly take data from the current repair area and monitor that data for the detection of flaw removal. However, test images taken during an actual sanding operation were found to be unacceptably complex from an image processing standpoint. Also, the repair process devised removes a quantized layer of surface material at each step in the process to control distortion, i.e., sanding occurs for a fixed period of time before inspecting to see if the flaw has been removed. Therefore, even if a flaw were detected to have been removed at a certain instant, the current level of material being removed would have to be completed, so it is not necessary to inspect continuously.

An interesting discovery during flaw detection imaging tests was that wetted surfaces can be imaged as though they were not roughened by abrasion. Abraded surfaces offer the imaging system saturated background fields, which cannot be subsequently processed to reveal the details of the sanding pattern. Thus, a true flaw would image with excellent contrast to the abraded surrounding area. Tests run to extend the wetting period of the liquid, and to assure that the liquid coats and wets evenly with no bubbles, was the topic of another study [37]. One of the best surfactant mixtures found is simply a mixture of isopropyl alcohol and water, although water was found sufficient by itself when the time between application and imaging was subsequently reduced.

Realistically, certain types of flaws in the canopy transparency are either extremely difficult to detect or measure. Crazing, discoloration, and optical distortion are among the most difficult classes to gauge. Experiments were performed within wide bounds in an attempt to determine how well (in terms of cost, speed, and accuracy) flaw detection techniques might perform in an industrial setting. The results obtained from the tests performed implied that machine vision–based inspection of the entire transparency was not cost-effective from a time standpoint alone. Extrapolated results showed flaw determination time of over four hours for the relatively small F-4 forward canopy transparency. This time was only for detecting flaws using noncoaxial direct illumination. Optical distortion and halation were not measured, and many examples of crazing could not be detected.

It was determined that the skill of the canopy transparency repair technicians could be utilized as a cost-effective solution for inspecting the transparency surface. The human skill of finding very subtle effects would be used to provide guidance to the repair system during the initial inspection of the transparency. The inspector would judge whether the transparency is repairable and decide where to have the repairs performed. The inspector simply identifies areas to be repaired by highlighting them with opaque water-soluble markings (white tempera paint). If desired, the inspector could mark regions which had no flaws, and could omit certain flaws if they were considered to be in noncritical areas or could not be repaired without a severe distortion penalty. The decision to use human-assisted inspection also allowed for a larger field of view (FOV) since these markings were easy to distinguish in size and contrast. Thus, it was determined that the inspector would help to determine the location of the repair sites, but the remainder of the inspection operations would be automated as much as possible.

3.3. Flaw Surface Determination

The removal of a flaw on one side of the surface affects the removal of flaws in close proximity on the same side of the surface. On the other hand, a flaw within the same area of the transparency but on the other side of the surface is unaffected by the removal of flaws on the other side. However, the combined removal of two flaws in close proximity but on opposite sides of the surface greatly reduces the optical quality of the transparency in that area.

The design of proper flaw removal stratagems relies on the determination of which surface of the transparency (inside or outside) a flaw resides. The techniques available for determining the surface of a flaw include both contact and noncontact methods with the transparency surface. Several of these techniques were investigated [30].

While closeness to, or contact with, the transparency surface is considered a detriment in evaluating the suitability of an inspection process for automation, several techniques requiring this were evaluated. First, as mentioned earlier, laser backscatter analysis might provide information regarding which surface the flaws are on. However, experimentation offered no immediate indication that the surface would be distinguished from the scatter patterns. Second, the transparency surface could be touched by a delicate whisker attached to an accelerometer. The whisker would then be dragged along the transparency to record surface flaw locations. Such a tactile technique would possibly provide usable results for surface flaws of large size that have an open groove-type structure, but many flaws and crazing have no surface-detectable signatures. A third proximity method would use a vision system with a highly magnified field of view (similar to a microscope). The resulting image would have a very shallow depth of field (in the manner of confocal microscopy), so it would be possible to focus only on the flaws on the side of the surface closest to the inspection equipment. This has the disadvantage of requiring accurate surface standoff distances, as well as the need for developing good and fast automatic focusing algorithms. Also to its disadvantage, the inspection area would be extremely small.

Another series of techniques alters the way flaws are perceived, both in color and in intensity, to determine flaw surface. A wide variety of dye penetrants were tried with varying degrees of success. A very simple technique was to color the flaws on one side of the transparency one color, and those on the other, another color. The flaws did readily receive the dye materials, but not to a density suitable for image enhancement, i.e., the dyes were not clearly visible to the cameras. A second technique tried was to use what can be described as grazing incidence illumination. A strobe source was directed parallel to the surface of the transparency so that very little of the light was allowed to pass through the surface, thus illuminating the near side of the transparency in a brighter fashion than the side away from the source. In some cases, the results would be adequate for discrimination of front and back surface flaws based on intensity. Unfortunately, the results could not be achieved repeatedly, and often there was not even a detectable difference between near and far side flaws. This was probably due to light coupling into the surface material, which would require extremely critical alignment to prevent. This technique would also require inspection from both sides of the transparency surface.

Another technique for determining on which side of the surface the observable flaws reside uses a stereo imaging technique (details of the technique can be found in standard image processing texts such as [38, 39]). Two images of an area are made from different perspectives. Once the alignment of the cameras is determined and the overlapping field of view is known, the images can be processed to determine the approximate range of objects from the observation position. This process is depicted in Figure 6.15. A laser beam is used to generate the alignment information, with the rest of the geometry of the system known from previous calibration measurements. Relative lateral offset from one image to the next provides range information. The lateral offsets to the front and back surface of the transparency are known from the laser-generated fiducial dots, so absolute calculation is unnecessary.

A stereo imaging end effector as shown in Figure 6.16 was constructed for implementing and testing the stereo imaging technique [40]. The end effector had a combined camera field of view of about 50 mm by 65 mm (relatively small), a laser beam for alignment and calibration of the camera images, and a fiber-optic cable connected to a strobe head which provided the illumination. The primary problem with the method was that flaws appeared dissimilar from slightly different perspectives so that correspondence was difficult. This problem was most severe in the case of badly damaged transparencies, where several flaws might overlap one another.

Another method for distinguishing the surface of flaws utilizes marks placed by the inspector as mentioned previously. The transparency surface can be imaged one time to determine which surface the markings are on based upon some discriminator. The technique makes use of the fact that the transparency surface materials naturally absorb ultraviolet light in transmission. The discriminator is that markings on the far surface are of diminished intensity from those imaged on the near surface. This effect can be seen when using ultraviolet camera filters and strobe lamps which are conveniently bright sources of ultraviolet light. Implementation problems were encountered due to the inspector's nonrepeatable marking densities. Testing revealed that marking thicknesses and techniques produced a wide fluctuation in returned intensities from the far transparency surface. In a revised approach, discriminating the inspector's markings into front and rear surfaces involved initially finding the flaw markings, polishing one side of the transparency surface, and noting the flaw markings that had been removed. Although this technique is time-consuming, it was selected due to its robustness.

3.4. Flaw Severity Gradation

Following the decision to use machine vision techniques to locate transparency flaws, investigations regarding different ways to detect the severity (depth) of flaws were examined [30]. The earliest studies took place on cast acrylic panels and used manually induced flaws to represent those found on typical canopy

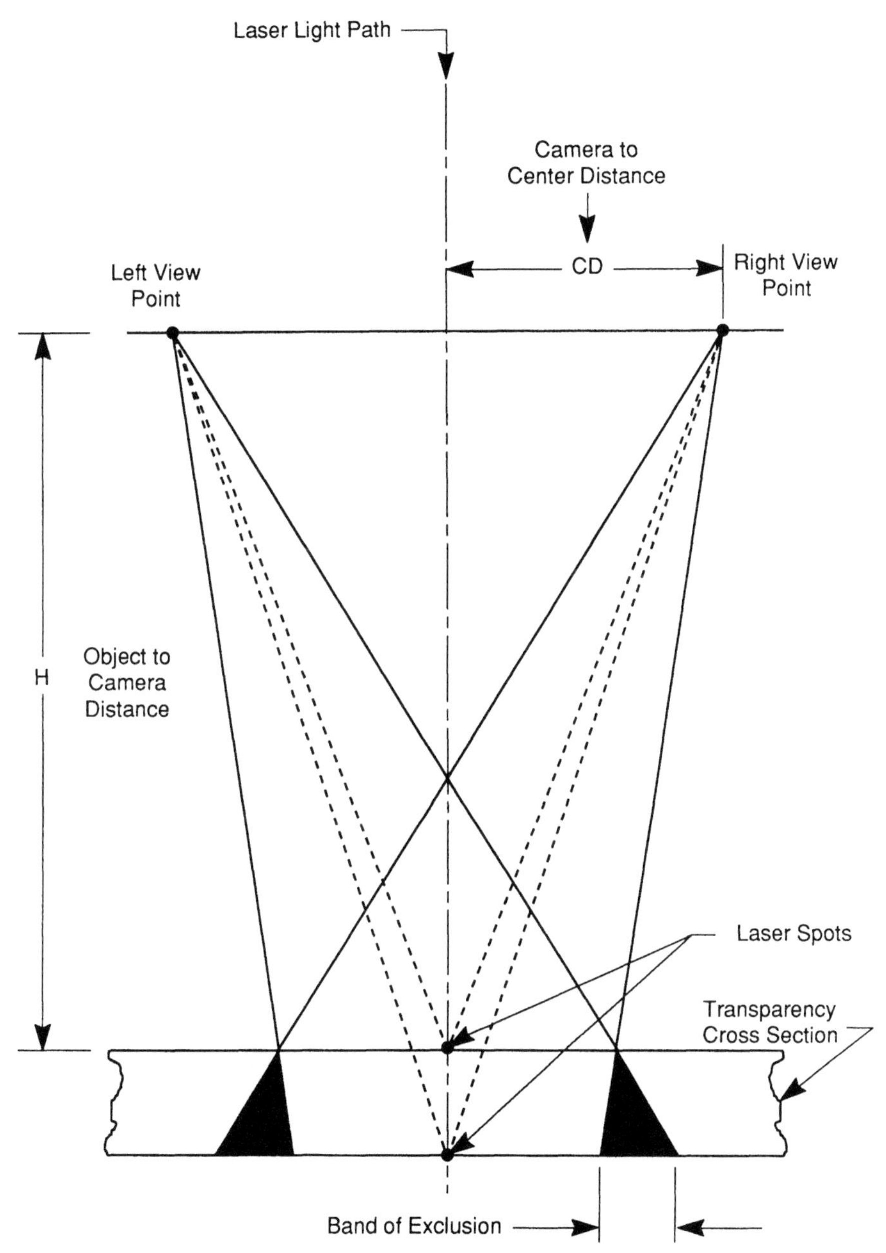

FIGURE 6.15. Stereo imaging geometry.

FIGURE 6.16. Stereo imaging end effector.

transparency surfaces. The first study used machine-milled precision grooves in the acrylic to attempt to determine if a relationship existed between flaw depth and apparent intensity, and between apparent intensity and flaw width. Groove widths of 0.787 mm, 1.19 mm, and 1.57 mm were arbitrarily chosen. The grooves ranged in depth from 0.127 mm to 0.229 mm. After a series of tests, it was apparent that there is no overall pattern relating depth and intensity of the flaws. There was, however, a trend toward the flaws being brighter as they become wider.

Determining that the milled slots did not model the transparency flaws in a reasonable way, a second series of flaw panels was created. These panels, shown in Figure 6.17, were characterized by the fact that they possessed shallow flaws similar to those existing on typical transparencies. In addition, the flaw shapes were varied by choice of flaw inducement devices typical of the objects that might cause actual service damage to a transparency. These devices were used to produce flaws of varying depth in the acrylic material by successive increase in applied force. Accurate measurements were made of the depth of the flaws using a measuring microscope with micron depth resolution. Flaw intensity was then measured and recorded. Despite the wide variety of inducement implements, a

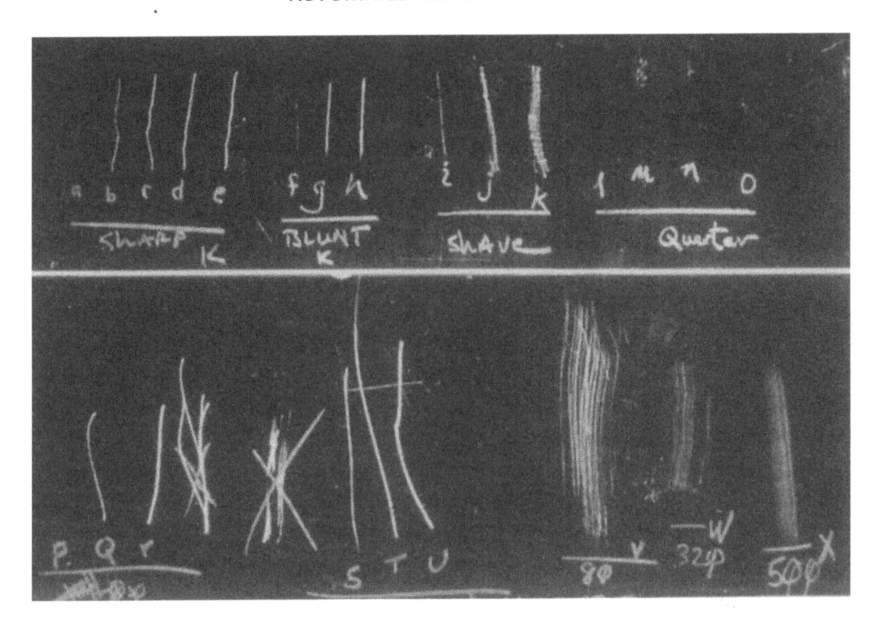

FIGURE 6.17. Panel with test flaw marks.

significant curve emerged indicating that, at least for shallow flaws, there was a relationship between depth and intensity. The relationship was not linear, but it approached linearity at depths greater than 0.02 mm. This implied that an attempt to rate flaws based on their apparent intensity was possible for shallow flaws. It was also discovered that certain implements had created brighter flaws for their depth than others. Specifically, one object marred a wide shallow trough in the plastic material which was highly reflective. The width of the flaws was then determined to be a key, relating back to the possible link drawn from the first experiments associating width and intensity in a general way.

Further data was acquired from actual (rejected and unpolished) transparency flaws identified at random. From this data, a relationship still appeared to exist between intensity of the flaw and its depth. The inability to detect small flaws during these tests was explained by the fact that the fine scratches had received a coat of wax since their creation, which is a common technique for their masking in the field between rework. Tests revealed that those flaws were of a size to be removed by normal polishing procedures, and need not be detected.

It was concluded that the reflected light from small flaws is proportional to their depth in the transparency surface, and a tertiary flaw gradation scheme was developed to coordinate with the needs of the repair process [30]. First, below a certain intensity level, flaws will be ignored resulting in a *clean* rating. This type of flaw is very slight and does not require sanding for repair since overall polishing will remove it. Second, just above the clean level, is the *minor* flaw level. The maximum depth of these flaws is in the 0.08 mm to 0.10 mm range. They are characteristically removed by lighter sanding operations. Flaws with intensity levels corresponding to depths of more than 0.10 mm, and those that saturate the camera sensor, are considered to be *major* flaws. While this scheme works well for single flaws and pits, badly scuffed areas not worked out during initial overall polishing operations can cause a major flaw grading where only a minor would be appropriate. Such abrasion-induced miscategorizations are not critical, however, because the ability to perform in-process inspection prevents excessive sanding.

Calibration to reference the grading must be performed on a regular basis due to the illumination effects caused by strobe aging and room lighting variations. Initially, the threshold values for differentiating between clean, minor, and major flaws were associated with a calibration procedure which imaged carefully gauged reference flaws. The means necessary to maintain the reference flaws in a dust-free condition did not seem reasonable for an industrial system. Instead, reflectance standards were acquired to correspond to the reference flaws in the perceived intensity domain. The reflectance standards allow large areas to contribute to the intensity measurements, and were found to make accurate references.

3.5. Image Acquisition

Imaging large, curved, transparent, framed, and damaged canopy surfaces presented unusual challenges. Important issues involved the segmentation of the surface into sizable regions to be imaged, the uneven illumination distribution created by the curving surface, and the possible presence of canopy frame and mounting hardware in acquired surface images. Addressing these issues involved careful consideration given to the cameras and illumination system selected. Rugged, inexpensive 240 (vertical) by 256 (horizontal) resolution CCD cameras were selected for the imaging development work. A variety of xenon arc strobe heads were available ranging from a fiber-optically coupled 4-J source, to a large commercial unit rated at 160 J. A variety of incandescent sources were also tried as illumination, but were deemed inferior to the strobes in the areas of deliverable contrast and heat.

System decisions regarding image acquisition were primarily centered around the trade-off between overall inspection performance speed and the need to resolve flaws for in-process inspection (and less importantly, gradation). FOVs from 50 mm to 150 mm in the vertical (creating resolutions of about 0.20 mm to 0.62 mm) were tested for possible use. A field-of-view of roughly 150 mm by 175 mm was the largest that could be used given the chosen cameras and desired resolution on the surface area. The noncoaxial direct illumination method was chosen for its ease of implementation and single-canopy surface access requirements.

A single-camera *inspection end effector* which was designed for the deliverable automated work cell is depicted in Figure 6.18. The primary elements of the inspection end effector are a camera, a strobe, a filter wheel for the camera, and a

FIGURE 6.18. Inspection end effector.

rugged frame to protect the end effector in the event of collision. The camera's filter wheel contains a number of neutral-density filters of varying tranmisttance. These filters are one mechanism used to control the magnitude of light available to the camera during calibration and inspection. The strobe lighting unit consists of a (160-J) xenon arc strobe bulb mounted at the top of the inspection end effector. The lighting unit is controlled by a remote power adjustment unit under computer command. The strobe unit's output can be varied over the equivalence of 4 F-stops of camera iris adjustment. A large parabolic reflector fitted with a fine honeycomb collimator is attached to the strobe bulb unit to create a more uniform distribution of light across the transparency surface. The strobe bulb unit is angled to provide a uniform field of low-incident angle light to the canopy surface without direct reflection to the camera. The strobe power settings, imaging and illumination geometry, filter selection, and minimum end effector standoff distance were design variables selected at the conclusion of experimentation and refined through system development.

In use, the inspection end effector must be moved to known locations, stopping at each position to acquire an image of that portion of the transparency. Inspection segmentation of the surface is represented by contiguous overlapping equal areas (called *inspection sectors*) mapped over the entire transparency contour. The areas are slightly rectangular in shape, matching the aspect ratio of the imaging sensor and lens combination. The segment edge points, along with the center point normal vector (used for orienting the inspection process), are generated from the surface model (described in a later section) with the area of each inspection sector corresponding to the FOV values of 175 mm horizontal by 150 mm vertical. While smaller FOVs provide better resolution, the penalty in time rapidly becomes intolerable to the process flow.

Use of an off-axis direct frontal illumination scheme against a curved surface produces intensity gradients over the surface. Because the illumination source is not coaxially aligned with the viewing vector, there is a gradient running along a line passing between the source and detector origins, with magnitude decreasing from the source to the detector. A second gradient is the result of illuminating a curved surface with a uniform illumination field. Experiments to modify the illumination distribution pattern to correct the gradients were largely unsuccessful. Useful effects can be generated for controlled geometries, but the surface curvature varies enough to make the effects of the imposed projected illumination gradient not worth the decreased intensity available. An alternative method to correct these gradients obtains an image representation of the gradients and uses a normalization scheme to perform the correction. Experiments determined that one intensity correction field was not adequate for the entire transparency surface. Therefore, correction data was acquired over the entire surface for each transparency type. This was done by spraying the outside surface of a transparency with a flat white water-based tempera paint generating something like a soft eggshell-like finish. The outside surface was subsequently imaged with suitable illumination and camera iris settings (the white surface reflects much of the incident light), and the images were stored for use as correction data. This technique was found to be highly effective, although it required an extensive amount of disk storage since a correction image was acquired for each inspection sector for each canopy transparency type.

Another problem encountered in the acquisition of images of the transparency surface was that of unwanted artifacts. Mounting elements to the sides of the canopy frames, and the entire perimeter of the transparency, represent reflective objects which tend to saturate the imaging camera. Edge masks were created to eliminate regions in inspection sector images where such reflective objects were found. When an image is acquired, these regions are made to simulate the perfect dark-field background, thus allowing them to be ignored.

3.6. Image Processing

Flaw segmentation refers to the ability to separate flaws from the background noise inherent in the images of the transparency surface. Flaws can be obscured by other flaws, filled with wax to make them less apparent, apparently dissimilar from different perspectives, or too small (or fine) to detect. There may also be areas of bad abrasion, inducing glare in the image of the transparency. While the detection of all of the information available in the surface images might seem desirable, specks of dust and related temporary conditions should not be detected.

The requirements for extracting useful information from images varies depending upon the type of inspection task being performed. During initial inspection to locate flaw marks, relatively large markings are to be identified and separated according to surface. During severity gradation, intensity information is sought within the boundaries of the inspector markings. During in-process inspection, regions within the inspector markings are searched for valid intensity data.

Initial inspection identifies the inspector markings on both surfaces of the transparency. The markings are relatively large in the FOV, and the same mark will often appear in several different inspection sector images (a long scratch for instance). The 8-bit gray-scale image is acquired, edge masked (if necessary), intensity corrected, and thresholded to a binary image (based upon calibration data). Connectivity analysis is performed on the binary image to identify connected objects, their dimensions, and their hierarchy. Subsequent remapping of the identified objects, known as *blobs,* is performed in the intensity domain such that each may be characterized separately. In this manner, a form of blob mass filter can be built. The mass filter can be designed to reject objects having more or less than some number of connected pixels. The resultant image can be stored for processing in a compressed format using run length encoding.

Any operation required of the inspection system after the first pass of acquisition uses template information generated from valid flaws. The template is essentially a window of the exact shape as the original inspector marking. The blob outline is eroded or dilated depending upon the operation, but provides considerable isolation from other optically relevant events on the transparency surface. Masks, corrections, thresholds, and mass filters can all be applied within these templates.

Each transparency image is identified in a transparency-specific database by the index indicating the inspection sector from which the image was acquired. Within each image, the important data is stored in pixel coordinates (throughout the processing sequence on the image processor, pixel coordinates are used). Given the camera FOV and given the position and orientation of the camera for acquiring images at each inspection sector, the pixel locations can be mapped directly to surface coordinates using the projection operations described in a later section.

Producing and decomposing global perspectives is done through the scaling and appropriate merging of individual inspection sector images. The quality of the merged images is an indication of the accuracy of the surface model and the camera positioning and alignment, where flaw marks coexisting in neighboring inspection sector images are the gauge. When the discrete images are shifted, scaled, and placed in a composite image, markings on the borders either will appear smooth and continuous or will appear offset, depending on the curvature about a particular surface region since it affects the accuracy of calculations.

3.7. Ideal Removal Patterns and Flaw Clustering

Automated surface inspection provides information about the position, shape, surface, and severity of each flaw on the transparency. The position and shape of the flaw are stored as the pixel coordinates (associated with an inspection sector) in the acquired image of the white paint flaw marking which covers the flaw and the severity of the flaw is either major or minor. In considering the distribution of flaws over the interior and exterior transparency surfaces, it is possible for some of the flaws to be small and circular, while other flaws may be long, wide, or otherwise irregularly shaped. Furthermore, some flaws may appear in groups (relatively close to one another) and in mixed combinations of major and minor severity, while other flaws may span more than one inspection sector, thus causing flaw reconstruction to result from merging such sectors. Using this information, a strategy was developed for the optimal placement of the circular sanding patterns as described in [47]. This pattern placement strategy is driven by the requirements to completely remove the flaws, to minimize the number of patterns placed, and to minimize the distortion caused by overlapping patterns.

Since the gauging of flaw severity results in the classification of flaws into two groups, major and minor, it was determined that a single ideal removal pattern could be associated with each severity group: a *medium pattern* for major flaws and a *light pattern* for minor flaws. The characteristics of the medium (light) sanding pattern were designed to completely remove the majority of flaws classified as being of major (minor) severity. As such, the medium pattern removes material to 1.5 times the depth as that of the light pattern, and removes material over 2.25 times the surface area (1.5 times the radius). The relative depths and radii for the medium and light patterns are shown in Figure 6.23.

Since the depth and maximum radius of the medium (light) pattern does not provide enough information to determine how much area of a typical major (minor) flaw can be removed, a series of tests were performed to obtain such a number. A number of typical major flaws which were long and linear (some artificially induced) were sanded using one medium pattern placed in the center of the linear flaw. The average length of these linear major flaws that were completely removed by the application of the medium pattern was defined as the *effective cutting diameter* for that removal pattern. A similar procedure was performed to determine the effective cutting diameter for the light pattern based on the application of single light patterns to typical linear flaws of minor severity. Figure 6.24 shows the effective cutting diameter for both the medium and light patterns. The pixel intensity scale shown in the figure is for reference in the discussion to follow as the medium and light patterns are constructed in eight-bit

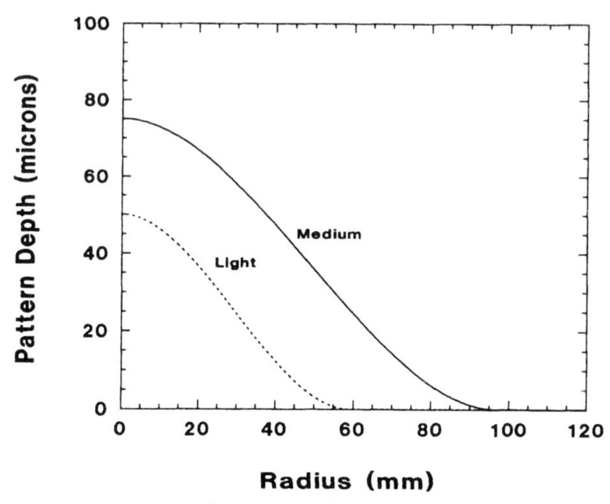

FIGURE 6.23. Removal functions for medium and light patterns.

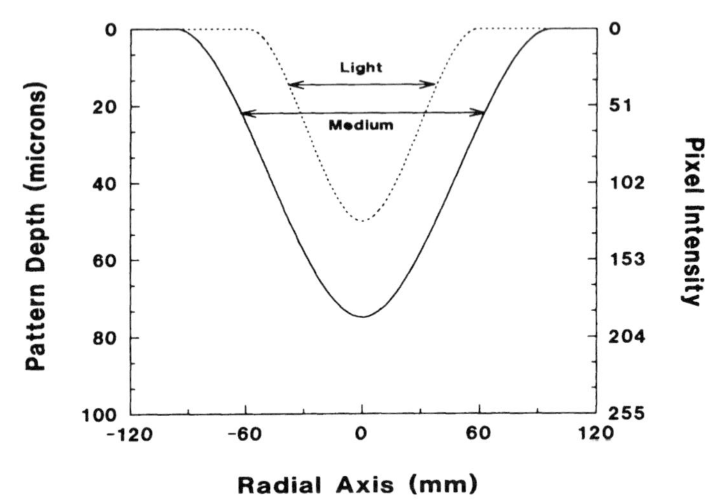

FIGURE 6.24. Effective cutting diameter for medium and light patterns.

image processing memory. The *effective cutting radius* is half of the effective cutting diameter.

Clustering refers to the process of linking flaws together into groups because taken separately, their interaction with each other during removal would have detrimental distortion effects. Similarly, *superclustering* involves grouping flaws located across neighboring inspection sectors if they are considered proximal. For the purpose of selecting locations for sanding patterns, the proximal distance measurement which is used to define the potential interaction region across each flaw, is defined to be the effective cutting radius of the pattern associated with the flaw's severity. Thus, if the edge of a flaw is within twice the effective cutting radius of any of its neighboring flaws (found in the same or neighboring inspection sectors), then the involved flaws are considered to interact with one another during their removal and consequently are merged into a supercluster. Figure 6.25 shows a map of the inspection sectors for an F-16 canopy surface with several flaw markings which cross over into multiple inspection sector images.

The process for determining superclusters begins by declaring the first flaw as the root of a supercluster. Subsequent flaws are then searched for proximity to the root or to current members of the expanding supercluster. If such flaws are within the effective cutting radius of the pattern for the root flaw of the supercluster, they are appended to the cluster and the search continues until no more additions to the cluster are made. The next unused flaw is the root of the next supercluster, and unused flaws are searched as before. The superclustering is complete when all flaws are members of a cluster for both surfaces of the canopy.

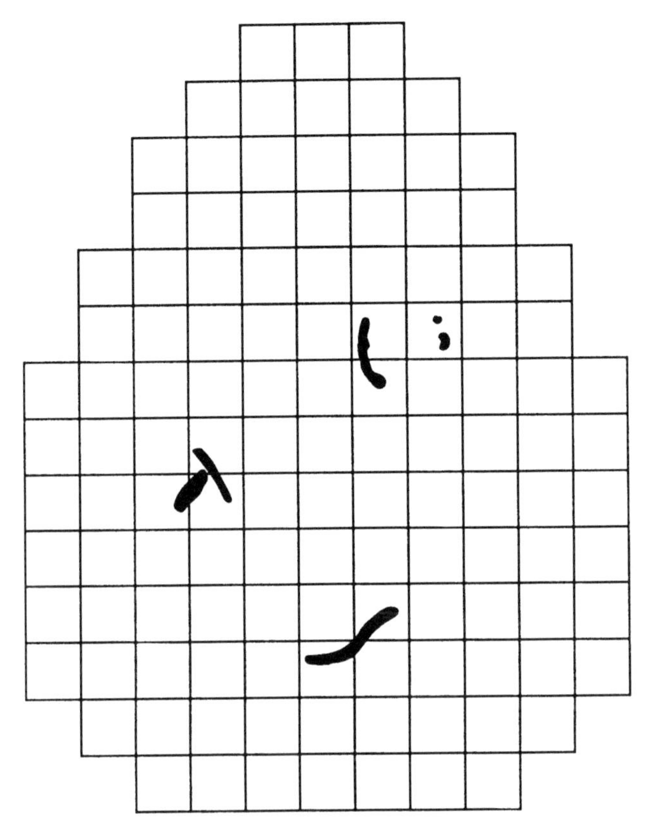

FIGURE 6.25. Inspection sectors for F-16 canopy with flaw markings.

All flaws belonging to a supercluster are gathered together from inspection sector images into a single composite image for further processing. This effort of merging the slightly overlapping inspection sector images is usually accurate, except for certain areas of the canopies where extensive curvature induces misalignment between adjacent sectors. This overlap is controlled so that it is possible to add reduced-scale flaw images together maintaining a 1:1 aspect ratio in scale (the scale factor is the same for both axes). The scale used for the reduction is determined by the maximum spread of the cluster across inspection sectors, but is not allowed to be less than two. This minimum scale factor accommodates the area of the medium pattern (recall that inspection sector images have approximately a 175-mm field of view). Figure 6.26 shows a typical flaw image representing one identified supercluster. The large mark to the left of the image covers a major flaw, while the two smaller marks in the upper right of the image cover

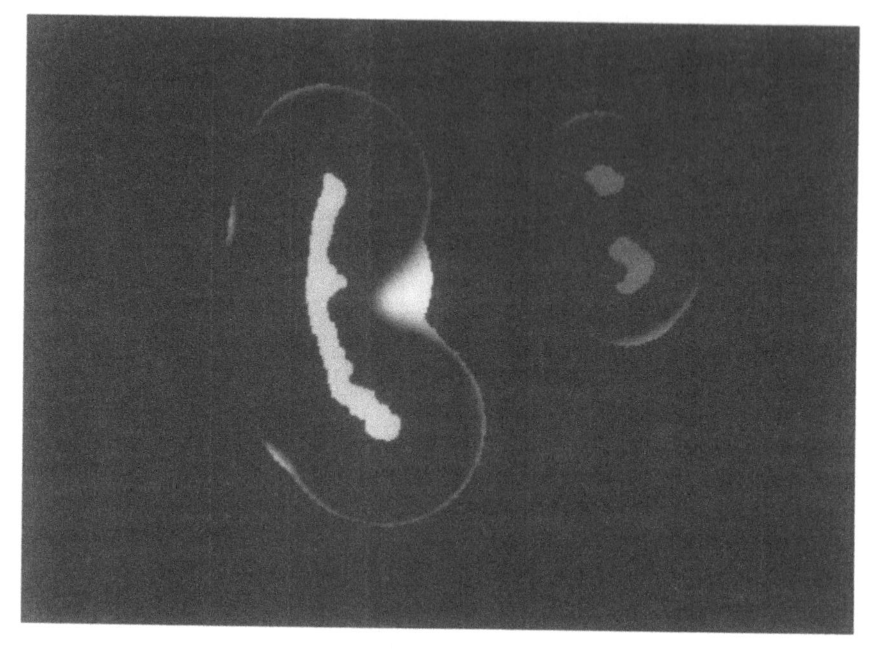

FIGURE 6.26. Typical flaw image after superclustering.

minor flaws. Such a supercluster image is ready to be processed for determining the optimal placement of sanding patterns.

3.8. Pattern Placement Algorithm

In designing a strategy for placement of sanding patterns, ideally, one would prefer that they be placed somewhere on a flaw for maximum effectiveness. Alternatively, it may be that flaws can be removed with pattern centers placed other than on the flaw, such that the effective cutting radius of each pattern sweeps over multiple flaws. While it would be possible to extend such an idea to create as large a pattern as necessary for any grouping of flaws, the problem of distortion makes this infeasible.

Superclusters with only one flaw can present simplicity to the pattern placement algorithm. If the flaw area fits within the effective cutting radius of the pattern for its stated severity, then the center of a circular sanding pattern is selected by transforming the pixel coordinates of the centroid of the flaw shape into surface model coordinates. If the flaw does not completely fit within the effective cutting radius, then it must be classed with the other, more complex forms of superclusters, i.e., those with multiple flaws, and there is a possibility of multiple inspection sector involvement. In most cases, the strategy for placing sanding patterns about a supercluster involves a more complicated algorithm.

The first step in the algorithm for complex superclusters involves construction of the *effective cutting area image* for the supercluster image. For groups of pixels in the image representing a major flaw, their pixel intensity values are assigned the value shown in Figure 6.24, which corresponds to the depth for the effective cutting diameter for the medium pattern. For groups of pixels in the image representing a minor flaw, their pixel intensity values are assigned according to the depth for the effective cutting diameter for the light pattern. After the flaw markings in the image have been colored, their shapes are extended by the effective cutting radius of the associated pattern (the effective cutting radius is scaled according to the field of view and the scaling factor required to merge the inspection sectors into a particular supercluster image). When extended shapes overlap, the maximum intensity value of overlapping pixels is chosen. The effective cutting area image constructed for the supercluster image in Figure 6.26 is shown in Figure 6.27. In this figure, extended markings for minor flaws have been assigned a gray level darker than that of extended markings for major flaws. Note that the two markings for minor flaws in the upper right-hand corner of the image have been merged into a larger object. This feature is very desirable for minimizing the number of sanding patterns placed.

Once a flaw image has been mapped into an effective cutting area image, the pattern placement strategy proceeds to determine an optimal combination of medium and light patterns placed about the effective cutting area image. The algorithm for pattern placement can be outlined as follows.

```
FOR Pattern = Medium THEN Light DO
   Construct Image Representation of Pattern
   REPEAT
      Find Best Pattern Match in Residual Effective Cutting Area Image
      IF Correlation Coefficient > Threshold THEN
         Define Best Match Location as Center for Pattern Placement
         Subtract Pattern From Residual Effective Cutting Area Image
      ENDIF
   UNTIL Correlation Coefficient < Threshold
END FOR
```

The algorithm begins by attempting to place the medium pattern, since it is preferable to place a larger pattern once than a smaller pattern several times. This minimizes the optical distortion induced by overlapping patterns. Furthermore, a medium pattern can be used on either major or minor flaws, while a light pattern should only be used on minor flaws or on residual major flaws (material near the center of a medium pattern placement that has not been completely removed).

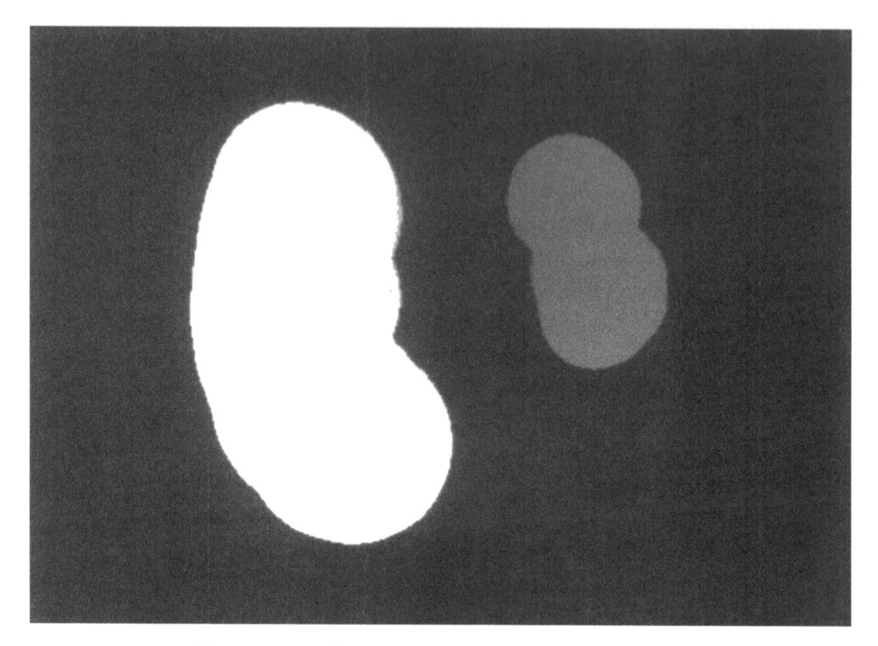

FIGURE 6.27. Effective cutting area image for flaws.

An eight-bit image representation of the medium pattern is constructed where the larger pixel intensity values correspond to deeper material removal as shown in Figure 6.24 (again, the pattern is scaled according to the field of view and scaling factor necessary to merge inspection sectors to construct the supercluster image). This image forms a template to be used for pattern matching. The first location for optimal placement of medium patterns is determined by image correlation using the medium pattern template against the effective cutting area image. Suitability of this location is determined by a threshold which is based on a fraction of perfect mismatch for a given pattern (which happens to be the total intensity for the image representation of the pattern). If the match is acceptable, the location is transformed back onto the transparency surface as the center of a sanding pattern. Then the medium pattern template is subtracted (clipping negative intensity values at zero) from the effective cutting area image resulting in a residual cutting area image. The algorithm continues to place medium patterns about the residual cutting area image and ceases when the measure of the best correlation match is no longer acceptable (based on a threshold). The algorithm then continues to place light patterns about the residual cutting area image in a similar fashion (using a different correlation threshold).

For the effective cutting area image in Figure 6.27, the result of placing four medium patterns is shown in Figure 6.28. The placement of the last medium

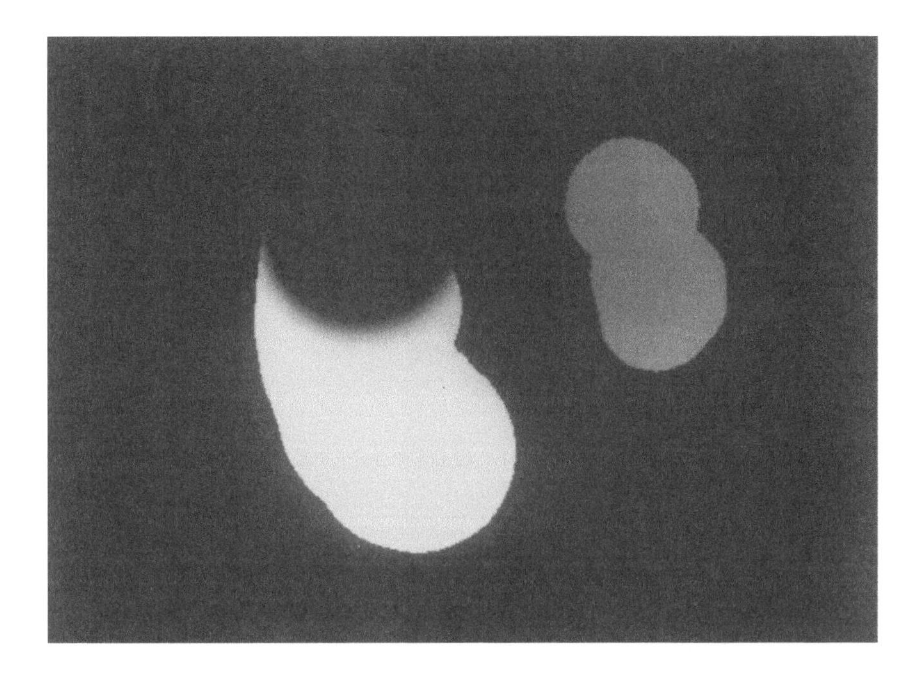

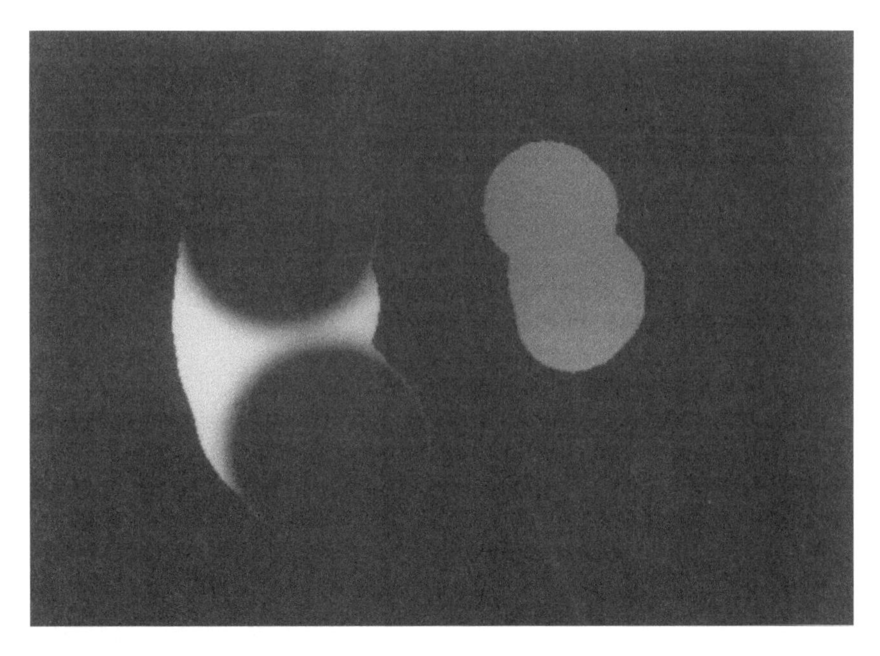

FIGURE 6.28. Results of pattern placement algorithm. After placement of (a) first medium pattern, (b) second medium pattern, (c) third medium pattern, and (d) fourth medium pattern.

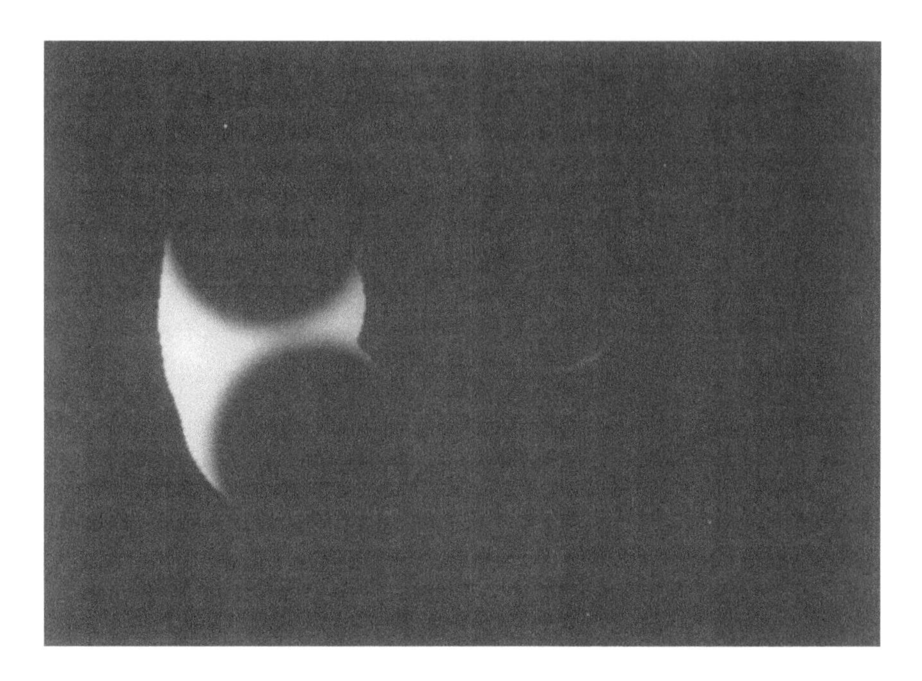

FIGURE 6.28. *(Continued)*

pattern barely exceeds the match threshold to remove the remaining medium flaw area, and what remains of the flaw areas at this point is insufficient to support the placement of light patterns. The result of placing all four medium patterns superimposed with the original flaw markings in the supercluster image is shown in Figure 6.29. Note that the requirement that none of the original flaw mark areas remain has been satisfied.

Two primary controls for the sanding pattern placement operation are the correlation match threshold and the effective cutting radius for both the medium and light patterns. The match thresholds are very critical values and have been empirically refined over time to produce repeatable quality results. The thresholds are set to minimize unnecessary patterns and yet remove all of the flaws. The effective cutting radius controls how closely each type of pattern can be placed. Although initially these effective cutting radii were based on sanding tests specifically designed to determine them, exhaustive operations revealed that these numbers could be altered slightly to produce better results.

The primary step in the algorithm for pattern placement involves matching combinations of the medium and light pattern templates against areas of effective cutting area image. This type of problem arises in many image analysis applica-

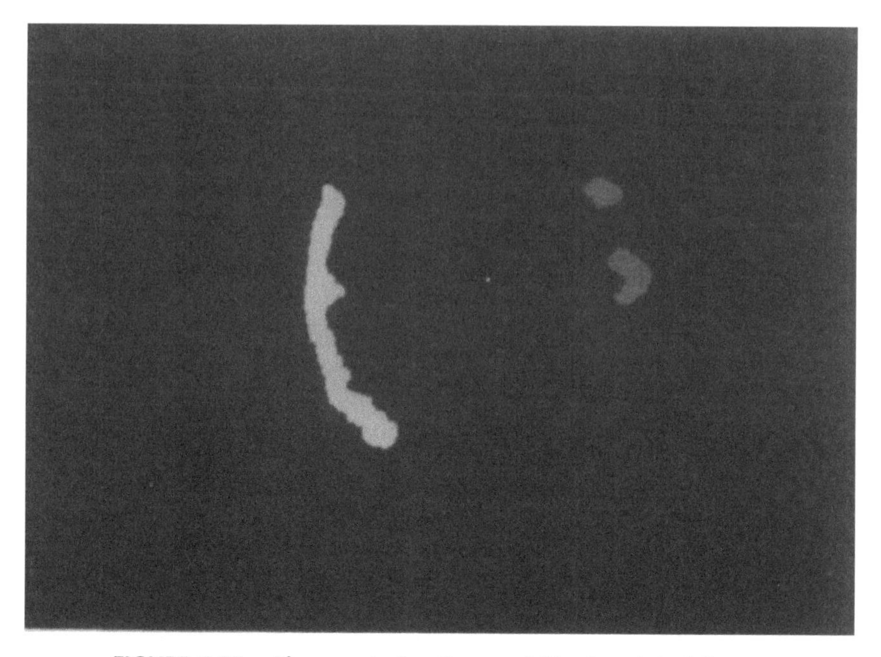

FIGURE 6.29. Placement of patterns relative to original flaws.

tions, where it is often necessary to determine whether an image contains a particular object, and if present, the exact location of that object. Provided that the appearance of this feature in the image is known accurately, then object detection can be performed by a template matching operation where the object being searched is the template that is compared to the image under investigation. A similarity measure is computed which reflects how well the image data match the template for each possible template location. The point of maximal match can be selected as the location of the feature if the similarity measure is acceptable.

A variety of measures of match or mismatch can be used in template matching algorithms as found in [38, 39]; however, the actual measure chosen depends on characteristics of the image and the template. For the purpose of locating medium and light pattern templates within the effective and residual cutting area images, the *absolute difference mismatch* measure was found to be most effective.

The template matching calculations can be visualized by imaging the template being shifted across the image to different offsets. The template is shifted to offsets in the image such that the center of the template is positioned over a point in the image which is valid for pattern placement (within the original effective cutting area). Note that when the template is shifted to an offset that is near the boundary of the image, the template shift is allowed to wrap around in the image instead of being clipped at the boundary. The sum of the absolute differences between the superimposed intensity values within the area of the template is computed, and the offset where a minimum sum occurs is the location of the best match.

Alternatively, the template matching calculations can be visualized by imagining the template in the center of some image, and the source image being shifted across the template image by different offsets. The source image is shifted by offsets to the center of the template image, where the offsets are determined by every point in the source image which is valid for pattern placement (within the original effective cutting area). Note again that when the source image is shifted by an offset determined by a point near the boundary of an image, the source image shift is allowed to wrap around. The template image has a corresponding mask image which contains an intensity value of one for every nonzero intensity value in the template image. An absolute difference image is constructed from the absolute differences between the superimposed intensity values in the template image and the shifted source image. The sum of the absolute differences within the area of the template is computed by the dot product of the mask image with the absolute difference image. Figure 6.30 shows the result of the absolute difference between a template image and a shifted source image, where the template image represents the medium pattern and the source image is the original effective cutting area image.

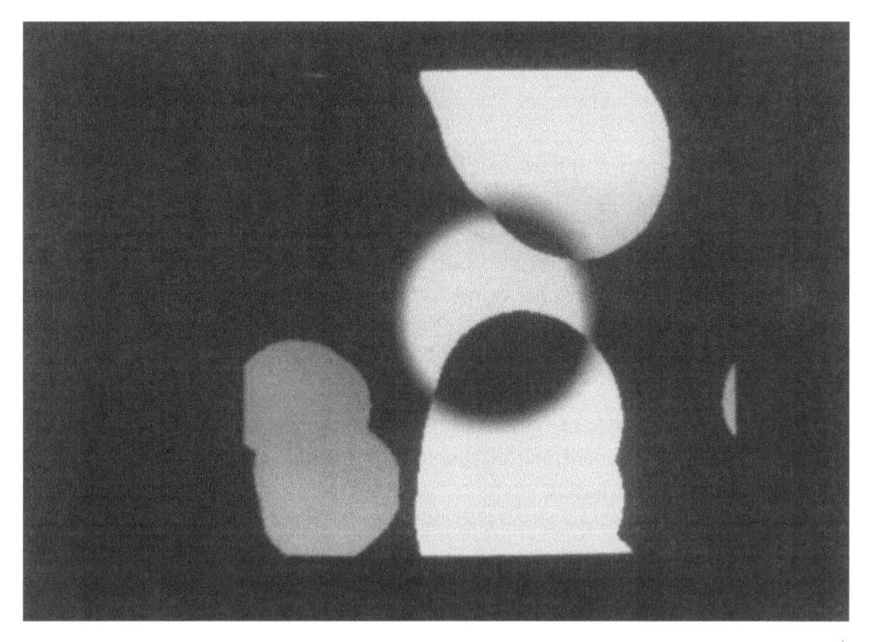

FIGURE 6.30. Absolute difference between template and shifted image.

4. SURFACE MODELING

4.1. Bi-Parametric Surface Definition

The initial canopy transparency surfaces for automated repair were the F-4 forward canopy, the F-4 aft canopy, and the canopy for the single-seat version of the F-16 (these canopies were shown in Figure 6.2). Each canopy frame with transparency is mounted to a unique fixture base which allows the canopy to be repeatedly mounted in only one orientation. The fixture bases, with mounted canopies, can be loaded onto a two-axis positioner which can rotate the fixture base about its center and translate the fixture base center along a single straight line. This positioner can be used for presenting specific regions of a fixtured-canopy surface to be within the work envelope of a robot. The upright orientation of the canopies when mounted to a fixture was chosen to facilitate drainage of liquids applied to the transparency surface during rework, to maximize use of the robot work envelope, and to minimize work cell floor space requirements.

Since the canopies can be rotated about the center of the fixture base, it is logical to impose the cylindrical coordinate system (radius r, angle θ, and height Z) when describing points on the transparency surface. The Z-axis is defined to be the axis of rotation (a vertical line through the fixture base center and perpendicular to the base) where the base is at $Z = -1000$ mm and the $+Z$ direction is up. The angle $\theta = 0°$ is the same for all three fixture bases and occurs at the angle at which the line of symmetry for the transparency surface is found (the line of symmetry is a curve from the top to bottom of a mounted canopy which divides the surface into two symmetrically identical halves; each of the three canopy surfaces mentioned has this symmetry line directly above the pilot's head). Figure 6.19 shows the F-4 aft canopy transparency in cylindrical coordinates. Three-dimensional Cartesian coordinates will be represented by $[x\ y\ z]^T$ and cylindrical coordinates will be represented by (r, θ, Z). The points on the canopy surface can be converted from the cylindrical coordinate system into the Cartesian coordinate system by the following simple transformation: $x = r\cos\theta$, $y = r \sin\theta$, $z = Z$.

The geometric nature of typical fighter aircraft transparencies (as evidenced in the geometries of the mentioned transparencies) induces certain assumptions that can be made in constructing surface models. One assumption is that the outside surface of a canopy transparency is convex. In other words, if $\vec{n}$ is the normal vector in Cartesian coordinates at any point (r, θ, Z) (cylindrical coordinates) on the outside surface and if $\vec{v}$ is a vector in Cartesian coordinates from the $(0, 0, Z)$ point (cylindrical coordinates) to that surface point, then $\vec{n} \cdot \vec{v} > 0$, i.e., the angle between the two vectors must be acute. Based on this assumption, it can further be assumed that in the cylindrical coordinate system, the radius value r can be uniquely determined at any angle θ and height Z (on the surface) and is given by the function $r = f(\theta, Z)$. Finally, it is assumed that the surface has low curvature (slow gradients).

From the description for the radius as a function of the angle and height as given in the assumption above, the transformation from cylindrical coordinates to Cartesian coordinates can be rewritten as

$$\begin{bmatrix} x \\ y \\ z \end{bmatrix} = \begin{bmatrix} f(\theta, Z)\cos\theta \\ f(\theta, Z)\sin\theta \\ Z \end{bmatrix}$$

The Cartesian coordinate values are now only a function of the angle θ and height Z thus giving

$$x = S_x(\theta, Z), \qquad y = S_y(\theta, Z), \qquad z = S_z(\theta, Z)$$

which is the mathematical representation for the geometric model of a continuous, two-parameter surface function $S(\theta, Z)$. Parametric surface equations are used because they allow solid models to be developed which are intrinsically independent of any coordinate system, which are bounded and nonplanar, which have defined tangent planes with respect to any coordinate system, which easily

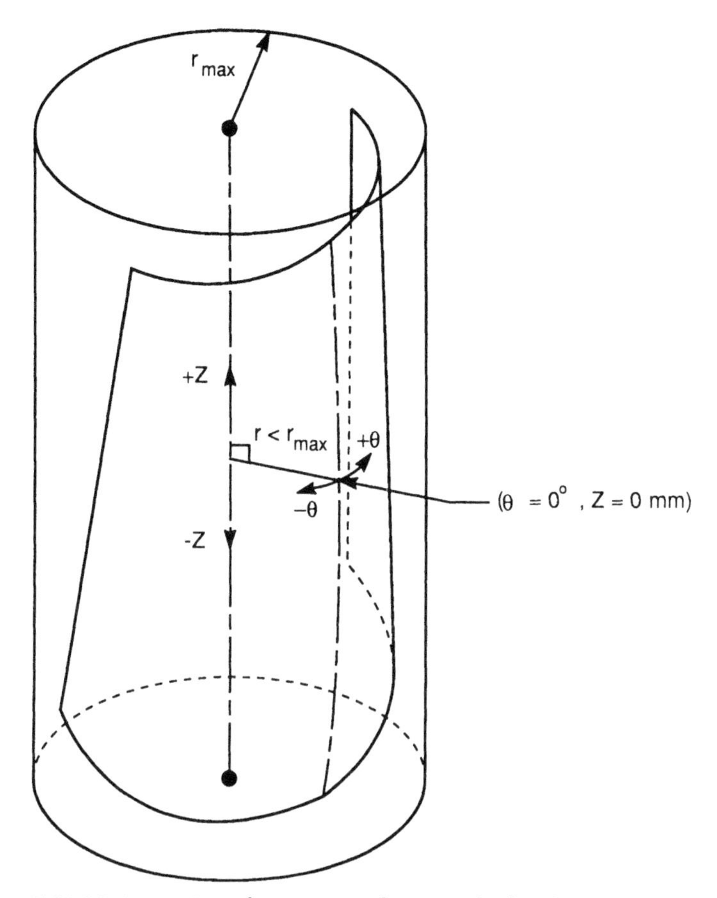

FIGURE 6.19. F-4 aft canopy surface in cylindrical coordinates.

can be divided into smaller regions (patches) through parameterization schemes, and which are numerically simple to compute.

The entire canopy surface is bounded in the parameter space such that $\theta_{min} \leq \theta \leq \theta_{max}$ and $Z_{min} \leq Z \leq Z_{max}$. The outside surface of the F-4 forward canopy has parameter space limits of $-99° \leq \theta \leq 99°$ and $-800 \leq Z \leq 350$ mm, while the outside surface of the F-4 aft canopy has parameter space limits of $-96° \leq \theta \leq 96°$ and $-775 \leq Z \leq 325$ mm, and the outside surface of the F-16A canopy has parameter space limits of $-89° \leq \theta \leq 89°$ and $-920 \leq Z \leq 1335$ mm. The edge of the transparency surface (where the surface meets the canopy frame) is defined by four parametric curves. The curves $\text{Edge}_{\theta_{min}}(Z)$ and $\text{Edge}_{\theta_{max}}(Z)$ give the minimum and maximum angles, respectively, at a given height Z. The curves

$\text{Edge}_{Z_{min}}(\theta)$ and $\text{edge}_{Z_{max}}(\theta)$ give the minimum and maximum height, respectively, at a given angle θ.

Since the transparency surface for each of the canopies has some physical thickness (10 mm for the F-4 transparencies and 15 mm for the F-16A transparency), and since the inside canopy frames have obstructions near certain edges, a separate geometric model, parameter space limits, and edge curves are defined for the inside transparency surface. The parametric surface equation for the inside surface, however, can be derived by moving points on the outside surface along their corresponding normal vectors the distance of the transparency thickness. Due to the convex nature of the transparency, the surface area for the inside model typically is less than that for the outside model.

4.2. Geometric Surface Models

A number of geometric modeling techniques exist for representing surfaces using biparametric equations as can be found in [41, 42]. These techniques vary in mathematical complexity, geometric accuracy, intensity of numerical computation, and amount of storage needed for coefficients. The chosen method should be an accurate representation of the surface (both points and normals), should not be mathematically complex to the extent that numerical computations are time-consuming, and should not require a large amount of storage for equation coefficients. The surface equation for the chosen method is determined by fitting the equation coefficients to a set of sample surface points.

The simplest technique for representing a surface is to use a connected array of piecewise flat panels where each panel is constructed between neighboring surface sample points. Points on the flat panel are linearly interpolated and the normal is constant across the panel. Although this technique is computationally simple, the geometric accuracy is sacrificed by the linear interpolation and the undefined normals at the boundaries of neighboring flat panels. Closer surface samples could be acquired, thus improving the geometric accuracy; however, the memory needed to store these points would be substantial.

An improvement on linear interpolation is to use the common technique of representing data by a least-squares fit to a cubic polynomial function. To model surfaces, a cubic polynomial fit is performed along each row of data (choose one of the two parametric variables) to produce a new column of data. This new column is now only parameterized in the other of the two variables and so a cubic polynomial fit is performed on this new column of data. Although this bicubic polynomial surface modeling technique provides better interpolation, it is unacceptable because it performs global fitting, thereby sacrificing geometric accuracy.

Since improved geometric accuracy is desired, the technique for representing surface curves using cubic splines may be considered. To model surfaces, bicubic splines can be used in a manner similar to the modeling of surfaces using bicubic polynomials. Cubic spline curves can be represented by piecewise parametric cubic curves which pass through the sample points [42]. Cubic spline curves are geometrically accurate, yet they require much storage for the coefficients of each piecewise parametric cubic curve.

The method finally chosen for modeling transparency surfaces involves a composite surface of overlapping bicubic patches where each patch is of the 16-point form [22]. A bicubic patch is a curve-bounded collection of points whose coordinates are given by continuous, two-parameter, single-valued cubic polynomial functions. The 16-point form of a bicubic patch [42] shown in Figure 6.20 defines a surface patch that passes through 16 sample points, where the simplest of the 16-point form models is when the samples are equally spaced in both parameters. All the surface sample points are used to define the overlapping bicubic surface patches where a bicubic surface patch is made from every possible 4 × 4 set of neighboring points, thus yielding overlapping patches with common points.

The algebraic form of any parametric bicubic patch is given by

$$p(u, w) = \sum_{j=0}^{3} \sum_{i=0}^{3} a_{ij} u^i w^j, \qquad u, w \in [0, 1]$$

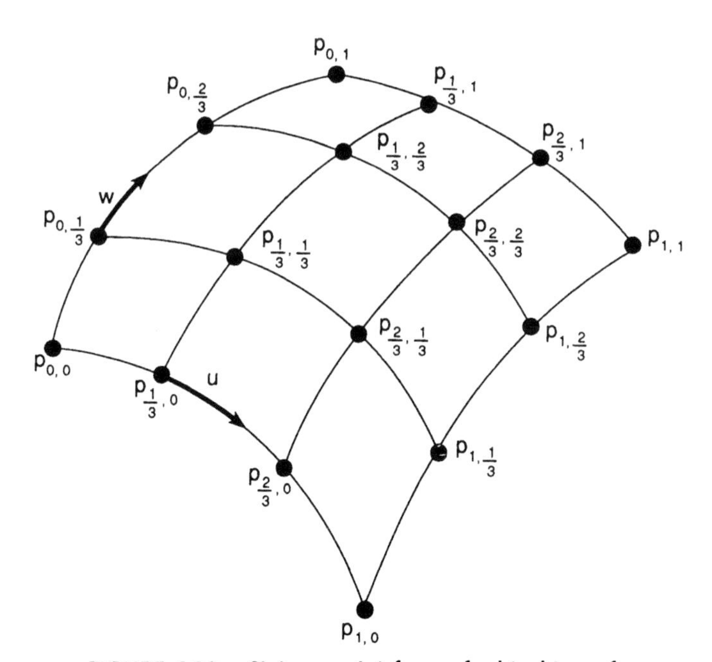

FIGURE 6.20. Sixteen-point form of a bicubic surface.

which can be rewritten as

$$p(u, w) = \hat{a}_3 w^3 + \hat{a}_2 w^2 + \hat{a}_1 w + \hat{a}_0$$

where each

$$\hat{a}_j = a_{3j} u^3 + a_{2j} u^2 + a_{1j} u + a_{0j}$$

represents a parametric cubic curve. Each bicubic patch can be evaluated using the parameter u to obtain a point, and the resulting four points can be evaluated using the parameter w to obtain a point on the bicubic patch. Thus, the 16-point form of a bicubic surface patch can be evaluated via the four-point form of a parametric cubic curve. Using the four-point form, a parametric cubic curve $p(u)$, $u \epsilon [0, 1]$ is defined in terms of four points

$$p(0) = p_1, \quad p(1/3) = p_2, \quad p(2/3) = p_3, \quad p(1) = p_4$$

which are equally spaced in the parametric variable. As shown in [42], the function $p(u)$ can be written as

$$p(u) = G_1 p_1 + G_2 p_2 + G_3 p_3 + G_4 p_4$$

where

$$\begin{aligned} G_1 &= -4.5u^3 + 9u^2 - 5.5u + 1 \\ G_2 &= 13.5u^3 - 22.5u^2 + 9u \\ G_3 &= -13.5u^3 + 18u^2 - 4.5u \\ G_4 &= 4.5u^3 - 4.5u^2 + u \end{aligned}$$

The normal vector is the cross-product of the two tangent vectors obtained by differentiating the surface function $p(u, w)$ in each of the two parameter variables.

A point on the transparency surface, referenced by the parameter values of θ and Z, is determined from the bicubic surface patch where these two parameter values are nearest the center of the patch. The method of constructing the composite surface from overlapping patches guarantees C^0 continuity; i.e., there are no gaps or breaks between beginning and ending points on any curve on this composite surface. This overlapping patch method, however, fails to guarantee C^1 continuity; i.e., the tangent vector for a surface point may not be the same for all bicubic surface patches that contain the point. If the spacing between the samples in the two parameters is small, and since the transparency surface is fairly smooth with no discontinuities, the normal is nearly approximated by the normal at the point on the centered bicubic surface patch. A better approximation is obtained from the average of the normals on all overlapping bicubic surface patches that contains the point. This surface modeling technique has acceptable geometric accuracy, requires a minimal amount of storage for coefficients, and is computationally simple.

Another technique for modeling surfaces involves the application of Bezier patches. Bezier patches are composite bicubic surface patches using blending functions applied to the control vertices of its characteristic polyhedron [42]. Bezier patches are described globally but remain sensitive to local data and have been utilized for the correction of image deformation from lens distortion [43]. Using this technique, the control vertices of a Bezier patch are determined to fit the sample set of data. Although this technique has all the advantages desired, its was not investigated further since the chosen method proved to be effective in modeling the surfaces of transparencies.

4.3. Sampling the Surface Contour

Three-dimensional sample points of the transparency surface to be modeled must be obtained in order to derive the parametric surface equation for the modeling technique chosen. For the manufacturer of a canopy transparency, certification of the transparency shape involves the use of a check fixture, but it is believed that CAD (computer-aided design) models exist for some or all of the canopies under consideration. However, this CAD data was difficult to locate and, in some cases, considered suspect, therefore this avenue for collecting surface data was not pursued. Instead, the needed surface sample data could be obtained from the actual canopy transparencies using any of the numerous techniques available for three-dimensional sensing.

Many three-dimensional sensing techniques using electro-optic sensors have been developed in the past and can be divided into several categories: time-of-flight direct sensing, shape-from techniques, stereo photogrammetry, and structured lighting. A survey of these methods can be found in [44, 45]. The time-of-flight direct sensing is an active method where a laser range finder which can sense the depth to any surface point within its field of view is typically used. The shape-from techniques are passive monocular approaches which recover relative depth from such information as texture, shading, contours, and motion. Stereo photogrammetry is also a passive technique which uses multiple visual sensors (usually two) to estimate disparity thus yielding depth information (solving the correspondence problem, however, is difficult). Structured lighting is an active technique which uses coded illumination patterns such as grids, stripes, or a laser-generated array of dots [46]. These techniques were not used since they were costly to implement.

A simpler and more direct approach to obtaining the surface sample points can be performed by a coordinate measurement machine (CMM). A CMM is a universal measuring device that can be programmed to traverse the three-dimensional contour of the surface with a hard, electronic, or noncontact probe to sample the location of surface points. The probes are designed to measure the surface points with ultimate accuracy. Noncontact probes are similar to those electro-optic sensing techniques described above. The electronic probe is also known as a touch probe with a switching mechanism that is activated when the probe is in contact with the surface. However, a coordinate measurement machine would provide much greater accuracy than required for the transparency surface model, and may not be readily available (in a size capable of sampling the surface of an F-16 canopy transparency).

The method selected for acquiring samples of the transparency surfaces utilizes a robot with a simple touch probe end effector, a fixture base positioner and custom control software. The touch probe is similar in concept to that used with a coordinate measurement machine, and the fixture base positioner provides the rotation of the mounted transparency so that the entire surface can be reached by the robot. An imaginary enclosing cylinder is imposed on the transparency surface where the height is given by the parameter space limits for the variable Z and the radius is larger than the radius of any point on the surface. Samples can only be acquired from those cylindrical surface positions where θ and Z are within the parameter space limits and the surface boundary defined by the edge curves.

In order to acquire sample points from the surface using this technique, the touch probe is positioned at valid locations along the surface of the imaginary cylinder at equally spaced intervals for each of the two parameters θ and Z. The probe is oriented such that it is normal to the imaginary cylinder surface, i.e., perpendicular to the Z-axis of rotation. The touch probe is moved to decreasing radii until it comes in contact with the surface. The cylindrical coordinates of the contact point are converted to Cartesian coordinates and then stored as a surface sample point indexed by the particular θ and Z parameters. Once all the possible surface samples points from valid parametric locations (within the bounds defined by the edge curves) have been acquired, sample points that were not acquired at the parametric locations within the parameter space limits (because these points were outside the edge curve boundary) are generated by extrapolating from the acquired points. The resulting rectangular array of sample points is then used to determine the parametric surface equation for the modeling technique chosen. This method of acquiring surface sample points by moving inward on an imaginary cylinder is based on the earlier assumption that the outside surface to be modeled is convex.

The transparency surface for the F-4 aft canopy section was sampled at interval spacings of $\theta_{step} = 16°$ and $Z_{step} = 8.46$ mm, while the transparency surface for the F-4 forward canopy section was sampled at interval spacings of $\theta_{step} = 16.5°$ and $Z_{step} = 115$ mm, and the transparency surface for the F-16A canopy was sampled at interval spacings of $\theta_{step} = 14.8°$ and $Z_{step} = 141$ mm. These spacing intervals were chosen arbitrarily based on the observed curvature of the transparency surfaces. The position and normal vector accuracy of the modeling technique with the chosen sample intervals was then tested. Positional accuracy was verified by measuring actual distances across the transparency surface between two reference points and correlating that with the distance measure computed by the model. The accuracy of the normal vectors computed by the model was verified by using a robot to position a laser beam coaxial with the computed normal vector, and observing the reflected beam from the surface which should be aligned with the incident beam.

4.4. Operations Using a Geometric Surface Model

Once the geometric model of a transparency surface has been constructed, a number of operations using the surface model can be performed. These operations provide a foundation upon which more complex surface operations or mapping algorithms can be defined. In particular, these operations consist of the construction of the tangent plane at any point on the surface, the specification of curve segments originating from any point on the surface, and the measurement of distances along a surface curve segment.

Let T be the XY-plane of three-dimensional points in Cartesian coordinates at $Z = 0$ and with normal vector $[0\ 0\ 1]^T$. Let $\vec{p}$ be a point on a three-dimensional surface S with normal vector $\vec{n}$ (this development is strictly for the outside transparency surface model; for the inside surface, the normal vector used is $-\vec{n}$). The plane T can be positioned such that the origin of the plane is coincident with the point $\vec{p}$ and the normal vector of the plane T is coincident with the surface normal vector $\vec{n}$ ($-\vec{n}$ for the inside surface). A geometric representation of this is shown in Figure 6.21. As shown, the vectors $\vec{p}^u$ and $\vec{p}^w$ on the plane T correspond to the $+X$ and $+Y$ planar axes, respectively.

A vector $\vec{q}$, defined relative to the original XY-plane, can become the vector $\vec{q}_{\text{new}}$ which is relative to the new position and orientation of the plane T. This transformation is given by $\vec{q}_{\text{new}} = M \cdot \vec{q}$ where in homogeneous coordinates

$$M = \begin{bmatrix} 1 & 0 & 0 & p_x \\ 0 & 1 & 0 & p_y \\ 0 & 0 & 1 & p_z \\ 0 & 0 & 0 & 1 \end{bmatrix} \begin{bmatrix} \cos\gamma & -\sin\gamma & 0 & 0 \\ \sin\gamma & \cos\gamma & 0 & 0 \\ 0 & 0 & 1 & 0 \\ 0 & 0 & 0 & 1 \end{bmatrix} \begin{bmatrix} 1 & 0 & 0 & 0 \\ 0 & \cos\alpha & -\sin\alpha & 0 \\ 0 & \sin\alpha & \cos\alpha & 0 \\ 0 & 0 & 0 & 1 \end{bmatrix}$$

The first transformation (rightmost matrix) is a rotation about the X-axis, the second transformation (middle matrix) is a rotation about the Z-axis, and the final transformation (leftmost matrix) is a translation by the vector $\vec{p}$. The original

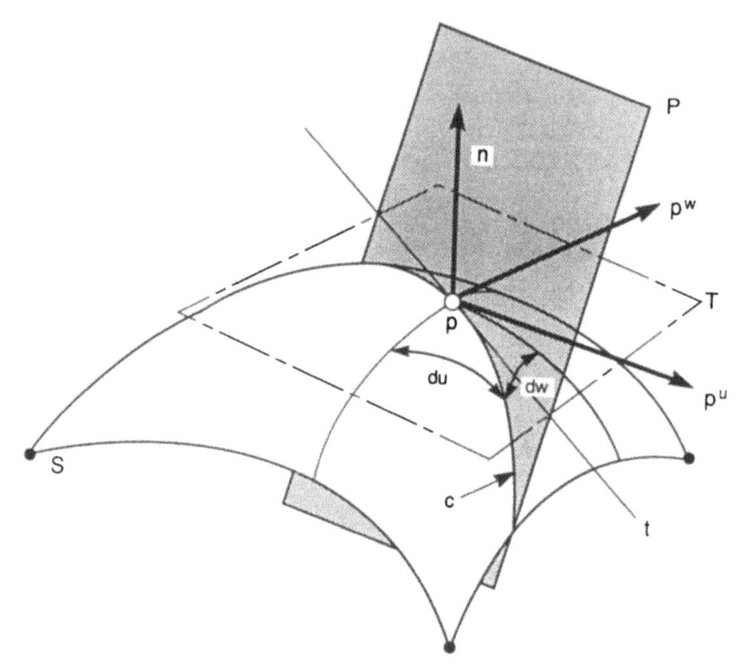

FIGURE 6.21. Tangent plane at surface point.

XY-plane is rotated about the X-axis so that the original normal vector to the plane ($[0\ 0\ 1]^T$) is coincident with the vector

$$\vec{v} = [0\ -\sqrt{1 - n_z^2}\ n_z]^T$$

where all normal vectors are given by direction cosine values. Thus, the rotation about the X-axis is given by the angle

$$\alpha = \cos^{-1}([0\ 0\ 1]^T \cdot \vec{v}) = \cos^{-1}(v_z) = \cos^{-1}(n_z)$$

The rotation about the Z-axis serves to make the vector $\vec{v}$ coincident with the normal vector $\vec{n}$. Thus, the angle of rotation is given by the angle

$$\gamma = \text{atan2}(n_y, n_x) - \text{atan2}(v_y, v_x)$$

where atan2(y, x) returns $\tan^{-1}(y/x)$ corrected for the appropriate quadrant based on the signs of the values x, y.

Again referring to Figure 6.21, let t be any line on the tangent plane T and through the point $\vec{p}$. Note that the line t can be defined in terms of the parametric variables u, w (θ, Z). Let P be a plane intersecting the tangent plane T that also contains the line t. Although there are an infinite number of such intersecting planes, only the one plane P that also contains the surface normal vector $\vec{n}$ is selected. The intersection of the plane P and the surface S defines a curve c. Similar to the line t, the curve c can be referenced by the parametric variables (θ, Z).

The distance along the surface curve segment c between two points referenced by (θ_0, Z_0) and (θ_1, Z_1), is computed using a type of adaptive quadrature numerical technique. Note that the reference (θ_0, Z_0) can be reparameterized as $(u, w) = (0, 0)$ and the reference (θ_1, Z_1) can be reparameterized as $(u, w) = (1, 1)$. The algorithm first computes the Euclidean distance between the two specified points. The curve segment between these two points is subdivided into four segments equally spaced in the parameters (θ, Z) where segment endpoints are given by

$$\begin{aligned} p_0 &= (\theta_0, Z_0) \\ p_1 &= S(\theta_{0.25}, Z_{0.25}) \\ p_2 &= S(\theta_{0.5}, Z_{0.5}) \\ p_3 &= S(\theta_{0.75}, Z_{0.75}) \\ p_4 &= S(\theta_1, Z_1) \end{aligned}$$

The sum of the Euclidean distances between the endpoints for the four segments is computed by

$$\sum_{i=0}^{3} |p_{i+1} - p_i|$$

and is compared to the Euclidean distance between the two original endpoints. If the difference is within an acceptable tolerance (e.g., $\epsilon = 0.5$ mm), the length of the curve segment is the distance between the two original endpoints. Otherwise, the length of the curve segment is the sum of the distances returned by this algorithm applied to each of the four segments, in a recursive manner, until the tolerance is acceptable for each segment of all subdivisions. This technique approximates the arc length using a summation of linear distances, where the curve is recursively subdivided into segments until the curvature of each segment approaches a line.

4.5. Surface Projections

The need to project planar data points onto the transparency surface model arises from the methods developed for automating the surface repair and inspection processes. In one application of the projection operation, the generalized planar sanding pattern can be applied at arbitrary surface locations for any transparency type. In another application, the projection operation can be used to associate surface locations with pixel coordinates in images acquired from different inspection sectors.

The operation of projecting sanding patterns onto the transparency surface molds the planar pattern to the surface contour (the pattern is represented by a series of points along concentric circles). The goal is to project neighboring points onto the curved surface, maintaining the separation distance of the corresponding points on the plane. The center of the sanding pattern will be at a point $\vec{p}$ on the surface S as was shown in Figure 6.21. The planar set of points representing the general sanding pattern are defined on the plane T, which is transformed to the position shown in Figure 6.21.

The two-dimensional Cartesian coordinates $[x, y]^T$ of a planar point for the sanding pattern are converted to the polar coordinates (d, ϕ). The angle ϕ determines the line t on the plane T in Figure 6.21 and the radius d is the requested distance to a point along the corresponding curve segment c from the point $\vec{p}$ necessary to mold the planar pattern to the curve. In this technique, the three-dimensional Cartesian coordinates of the pattern point on the transformed plane T are first converted to cylindrical coordinates. The resulting (θ, Z) values are then used to obtain a point $\vec{q}$ on the surface. The requested distance d is compared to the distance between the points $\vec{p}$ and $\vec{q}$. If the difference is within an acceptable tolerance (e.g., $\epsilon = 0.5$ mm), then the point $\vec{q}$ is taken as the projected point. Otherwise, the algorithm is repeated iteratively with the point $\vec{q}$ as the next location for the tangent plane placement and the two-dimensional coordinates $[x\ y]^T$ of the pattern point scaled by the value $1 - |\vec{p} - \vec{q}|\ /d$.

The application of this technique depends on the relatively small sanding pattern radii (not more than 100 mm) and on the geometric nature of the outside surface of transparencies (e.g., convex and slow gradients). The technique has been described for the outside surface; for the inside surface, the normal vector $\vec{n}$ at the point $\vec{p}$ is negated prior to beginning the procedure, and surface distance computations are performed using the geometric model for the inside surface. The maximum possible distance d from the surface point $\vec{p}$ along the projected curve segment c is the distance to the point at which the curve segment intersects with the edge curves that define the surface boundary.

The operation of projecting image pixels onto the transparency surface warps the planar pixel points onto the surface in an approximation to the perspective viewing. From the known camera geometry (e.g., position, orientation and field of view), image pixel coordinates are transformed into actual planar dimensions. The center of this plane of points is positioned at the point $\vec{p}$ on the surface S (the camera is aimed at the point $\vec{p}$ and the viewing direction is coincident with the negative of the normal vector $\vec{n}$ at that point).

4.6. Surface Area Partitioning

The need to partition the surface into nearly equal area patches is related to the mechanics of operating about that surface. For inspection, the surface is divided into patches determined by a constant field of view. For repair, the surface is divided into equally spaced points (corners of patches) determined by overall surface polishing requirements for constant traversal velocity. The method for partitioning the surface utilizes the surface projection operation described in the previous section.

The technique for surface area partitioning requires two parameters: x_{step}, which corresponds to the desired distance between points in the direction of the θ parameter while holding the parameter Z constant, and y_{step}, which corresponds to the desired distance in the direction of the Z parameter while holding the parameter θ constant. With the upright orientation of the transparency surface model, the x_{step} parameter corresponds to horizontal distances around the surface, while the y_{step} parameter corresponds to vertical distances up and down the surface. In order to maintain these desired distances over the entire surface, x_{step} and y_{step} should both be relatively small, since large step sizes result in a coarse appproximation for nonplanar surfaces.

The method begins by locating the center surface point in the model which is referenced by the parameters

$$\frac{1}{2}\,(\theta_{min} + \theta_{max}, Z_{min} + Z_{max})$$

This center model point is defined to be the partition point $p_{0,0}$. In general, the technique propagates the generation of partition points from the center point outward to the transparency surface edges using the directional x_{step}, y_{step} incremental distance values specified. In an iterative manner, for level $k = 1, 2, \ldots, N$, where at level $k = N$ no more partition points are found that are on the surface, define the partition points $p_{i,j}$ for $-k \le i,j \le k$ and $|i| + |j| = k$ by looping over the j indexes for each i index. The map of points being generated at each level is shown in Figure 6.22 (this figure shows increments in x_{step}, y_{step} and is not necessarily spaced equally in the parametric variables θ, Z).

Each point p_{ij} being defined at the current level k of the iteration is possibly determined by as many as two neighboring points from the previous level $k - 1$ of iteration. Let q_1 be the neighboring point from the previous level, which is immediately to the left or right of the point p_{ij}; and let q_2 also be a point from the previous level, which is immediately above or below the point p_{ij}. Computationally, the point p_{ij} can be determined from the points q_1, q_2 by considering the following four cases:

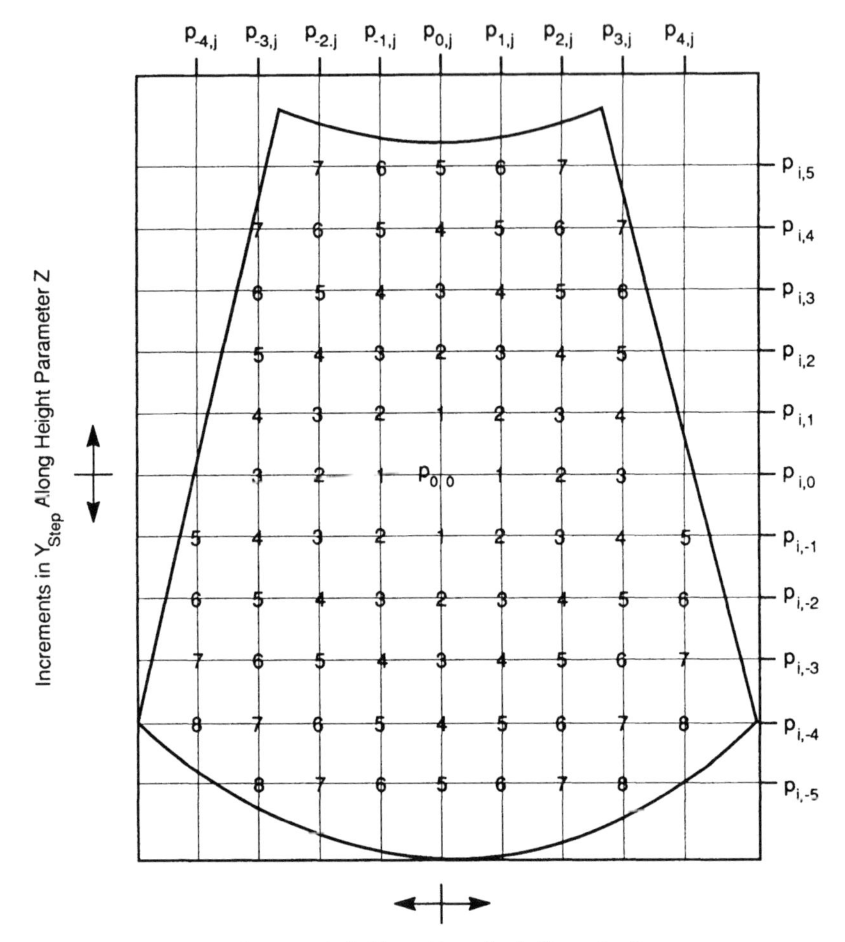

FIGURE 6.22. Indexing scheme for surface area partitioning.

1. If both points q_1, q_2 have already been defined, then the point p_{ij} is determined by the θ parameter of a point resulting from projecting the planar point $[x_{\text{step}}\ 0]^T$ (if $i > 0$, otherwise $[-x_{\text{step}}\ 0]^T$) from q_1, and by the Z parameter of a point resulting from projecting the planar point $[0\ y_{\text{step}}]^T$ (if $j > 0$, otherwise $[0\ -y_{\text{step}}]^T$) from q_2.
2. If the point q_1 has not been defined yet and the point q_2 has, then the point p_{ij} is the point resulting from projecting the planar point $[0\ y_{\text{step}}]^T$ (if $j > 0$, otherwise $[0\ -y_{\text{step}}]^T$) from q_2.
3. If the point q_2 has not been defined yet and the point q_1 has, then the point p_{ij}

is the point resulting from projecting the planar point $[x_{\text{step}}\ 0]^T$ (if $i > 0$, otherwise $[-x_{\text{step}}\ 0]^T$) from q_1.

4. If both points q_1, q_2 have not been defined yet, then the point p_{ij} cannot be defined.

The accuracy of this method of partitioning the transparency surface into equal area patches has been verified by mapping these partition points onto an actual transparency surface using a robot for positional placement and measuring the desired distances and areas.

5. THE ROBOTIC CANOPY POLISHING SYSTEM

5.1. System Components

The Robotic Canopy Polishing System (RCPS) is a three-robot work cell which automatically locates surface blemishes in acrylic transparencies and then polishes the material to remove the flaws and restore optical clarity. [48] Two of the robots have polishing end effectors to be used for the rework operations which include flaw sanding, surface polishing, washing, and drying. The third robot has an inspection end effector to be used for the machine vision inspection tasks of locating and grading flaws and in-process inspection. Each of the two polishing robots operates simultaneously and asynchronously on individual canopies, sharing the services of the inspection robot. The inspection robot is located between the polishing robots so that it can inspect either of the two canopies being reworked.

Each of the three robots is a Westinghouse Unimation[6] PUMA[7] 761 industrial robot. The robot arm is electrically driven and has six degrees of freedom (DOF), a 1600-mm reach, and a 10-kg payload capacity. Each robot has an associated controller which provides coordinated motion by issuing the appropriate signals to the servo motors in the arms. The robot controller hardware consists of an LSI-11/73 CPU executing the VAL-II[8] (Versatile Automation Language) control system, a serial communications port for supervisory control, and a parallel I/O interface. This robot has suitable accuracy and repeatability for positioning the inspection end effector for flaw inspection, and for positioning the polishing end effector at the rework location and manipulating it while withstanding the polishing forces. Robot wrists are sealed to prevent slurry contamination of the mechanical elements.

---

[6] Unimation is a registered trademark of Westinghouse Electric Corporation.

[7] PUMA is a registered trademark of Westinghouse Electric Corporation.

[8] VAL-II is a trademark of Unimation, Incorporated.

Each polishing robot has an associated fixture positioner to which the canopies are mounted for rework. These are located directly in front of the robots. The canopies are mounted vertically in fixtures which bolt to the positioner. Each positioner has two axes of motion, providing 460 mm of translation and 540 degrees of rotation. The positioner is used to present the transparency area to be reworked or inspected to the appropriate robot. The positioner also allows the fixture to be moved outside of the robot work envelope for loading and unloading of canopies. Rotational and translational control of the two canopy fixture positioners was implemented using a Berkeley Axis Machine servo controller (BAM-4)[9]. This controller contains a custom CPU board executing under the Parasol II[10] programmable command language. The controller is implemented with supervisory control via a serial communications port.

Each polishing robot has an associated pad table located adjacent to the robot. The pad table holds the sanding and polishing pads required for rework. These pads include 75-mm diameter sanding pads with 30-μm, 15-μm, and 9-μm abrasive sandpaper, and 100-mm-diameter pads for slurry polishing and washing. The polishing robots automatically replace the pads as the abrasive paper becomes worn (based on a time-in-use criteria), picking up new pads from those supplied on the polishing pad table. The operator must make certain that the tables have an adequate supply of pads with fresh abrasives (manually attached to the backing pads) prior to the rework operation.

Work cell control and high-level process coordination are provided by a Masscomp 5400 minicomputer executing under the RTU[12] operating system, Masscomp's real-time version of UNIX[13]. The cell controller has overall responsibility for the work cell, and all other devices in the work cell are slaves to the cell controller. Machine vision operations are provided by an International Robomation Intelligence (IRI) D-256 computer system executing under the Regulus[11] operating system. The image processing hardware includes camera interfaces, an image digitizer, image memory for four 256 by 256 eight-bit frames, and dedicated frame processors.

The RCPS was designed so that each polishing robot and positioner pair acts as an autonomous unit (a *polishing station*), operating independently of the other. This is accomplished by cloning a copy of the control software for each polishing station. The two tasks execute concurrently on the cell control computer, each task controlling the operation of one polishing station. Within each task, process-

[9] BAM-4 is a trademark of Berkeley Process Control.

[10] Parasol II is a trademark of Berkeley Process Control.

[11] Regulus is a registered trademark of Alcyon Corporation.

[12] RTU is a registered trademark of Massachusetts Computer Corporation (MASSCOMP).

[13] UNIX is a registered trademark of AT&T.

ing is distributed between the cell controller and the robot controllers to lessen the computational load on the cell controller and to allow each robot controller direct functioning of its end effector. The cell controller is responsible for scheduling and monitoring the overall sequence of transparency rework operations and for positioning the robots and fixtures. Once a canopy is in position for rework, each robot controller is responsible for control of the polishing end effector and execution of the polishing motions, based on motion instructions received from the cell controller. The robot controllers independently execute the pad change operation. Thus by means of multitasking and distributed processing, the cell controller regulates the two polishing operations occurring simultaneously and asynchronously.

The cell controller is also responsible for the allocation of shared work cell resources. For example, the dual polishing station arrangement necessitates a prioritization scheme for handling requests to the single inspection system resource. Certain inspection system operations have a long duration and are given a low priority, while other operations have a short duration and are given a higher priority. The cell controller determines when a task may be interrupted to handle a higher-priority request, and subsequently directs the movement of the inspection robot during the interruption.

5.2. Typical Rework Procedure

Before a canopy transparency can be reworked by the RCPS, a minimal amount of preparation is required. The canopy frame must be cleaned to remove accumulated dirt which could become loosened during the rework process and get trapped between a polishing pad and the surface, thus causing unwanted abrasions. Protruding hardware such as mirrors, mirror mounting brackets, and various handles are removed since they may obstruct the polishing robot motions. Finally, the canopy must be loaded into the appropriate fixture which rigidly holds the canopy in place during the rework process.

After the canopy has been prepared, it is manually inspected to locate scratches, pits, crazing, and various other types of flaws. Using white water-soluble paint, the flaws are marked (completely painted) on the surface where they appear. Flaws which can be removed by polishing and flaws that are close to the canopy frame are not marked. As indicated earlier, the manual inspection and marking of flaws significantly reduce the amount of time required to locate the flaws automatically. The polishing system will remove the marking paint and will determine the severity of each marked flaw. At a minimum, a single light pattern will be ground at each marked location.

The automated rework process begins with the initial overall inspection. During this step, the inspection robot is moved around the entire outside canopy surface and acquires images of the surface at each inspection sector. From these images, the machine vision system determines the location of flaws which have been marked for rework. While the images are being processed by the machine vision computer, the polishing robot performs an initial polish in the inside canopy surface, followed by washing to remove any residual slurry, and then drying with forced air. This initial polishing step removes small scratches from the inside surface, and also serves to remove the paint used to mark the locations of the inside surface flaws.

Once the inside surface has been polished (and the flaw markings on the inside surface have been removed), initial local inspection is performed. The inspection robot is moved to the inspection sectors where flaw markings were found in acquired images, and a second image is acquired. The remaining flaw markings found in these images are verified to be on the outside surface, and consequently, the missing flaw markings are determined to be on the inside surface. The machine vision system separates the inside and outside surface flaw markings from the inspection sector images, and while the images are being processed, the polishing robot performs the initial polish, wash, and dry of the outside canopy surface.

At this point, the inside and outside surfaces have been polished, and the flaws that were covered by the paint are exposed. The inspection robot again is moved to each inspection sector where flaw markings were found, and a third and possibly fourth image is acquired depending upon which surfaces of the inspection sector contained flaw markings. These images are processed to determine the severity of each flaw. The location, shape, surface, and severity of each flaw is recorded, and this information is used to compute the type and location of initial sanding patterns necessary to remove the flaws.

Flaw rework consists of sanding at the individual rework sites to remove the flaws with intermittent inspection to monitor progress. Flaw removal is primarily achieved through coarse sanding with 30-μm abrasive. The defined levels of coarse sanding available are light, medium, and heavy (only light and medium are used for initial sanding and heavy patterns are medium patterns with larger radii). Upon completion of the initial sanding, automated inspection of the flaw site determines if the flaw has been removed. If not, the next level of sanding severity is performed, again followed by inspection. This procedure continues until the flaw is removed or the highest level of sanding is completed. If the flaw still remains after heavy sanding, the operator is queried for instruction to either continue or stop sanding.

The coarse sanding is performed for all flaws marked for rework on the inside surface. The sanded areas are then refinished first with 15-μm abrasive and then with 9-μm abrasive. Once all flaws have been reworked on the inside surface, a final overall polish is performed to complete the restoration of that surface. The flaw rework and final polish sequences are then repeated for the outside surface.

5.3. System Performance

According to the RCPS operators at Hill AFB, the average rework time for an F-4 canopy transparency (either forward or aft section) by the RCPS is about eight hours, while for an entire F-16 canopy transparency the average rework time is about 12 hours. This time is variable and depends on the condition of the canopy and the exact rework procedure required. By comparison, the amount of time required for a human to rework a transparency manually is about 24 hours for either of the F-4 canopies, while the amount of time to rework an F-16 canopy transparency manually is estimated at approximately 33 hours [49].

The most important measure of the performance of the RCPS is the quality of the transparency after rework. Although the potential to introduce optical distortion exists, adequate process direction and control by experienced canopy transparency repair technicians and system operators ensures that transparency surface integrity is maintained. The inspector tailors the flaw removal procedures to be used for each specific flaw type by careful marking, which takes into account transparency optical requirements which vary with location zones on the transparency. Canopy transparency repair experts have judged and accepted the resulting rework quality by viewing a grid board through the transparency surface. The final decision on the quality of a reworked transparency, however, is the judgment of the pilot who can reject a reworked transparency if the amount of distortion is considered unacceptable. No pilot has yet rejected a transparency that was returned to service after rework by the RCPS.

6. CONCLUSION

The complete development cycle of a robotic work cell for automated rework of aircraft canopy transparencies has been presented. The major components of the work have been addressed with adequate depth to describe the actual workings of the implemented system. Although many of the details have been treated superficially or omitted, the key results of the research have been adequately documented and further information may be found in the references. An attempt was made to provide a general background of the transparency rework problems faced by industry and the essence of the developmental problems faced by the researchers. Given the subjective nature of the transparency rework art, details of the inspection and repair techniques described are subject to debate by those skilled in the art, but an honest effort was made to present a generally accepted viewpoint.

Although the specific application of transparency polishing was presented here, the techniques and methodology described have application in many areas of robotics and machine vision. For instance, the existing system could be modified slightly to perform sanding of fighter aircraft randomes for the purpose of removing paint without damaging the underlying composite structure (which possesses dimensional properties critical to proper operation of the radar). Related processes include grinding and finishing of stainless steel sheet metal fabrications and application of thickness-critical surface coatings on complex contours.

ACKNOWLEDGMENTS

This research and development part of the Robotic Canopy Polishing System delivered to Hill Air Force Base (Ogden, Utah), under U.S. Air Force Contract No. F42650-86-C-3276. The authors also want to acknowledge the contributions made by Douglas L. Michalsky, Stanley D. Young, S. Bruce Farmer, and Bruce C. Mather, who were the other primary participants in this research and development effort.

REFERENCES

1. G.A. Grabits, "Polishing aerospace transparencies," *Proc. 1986 SME Robotic Solutions in Aerospace Manufacturing Conference,* 1986.
2. R.E. Hopkins and C. Munnerlyn, "Optical finishing," *Optical Instruments and Techniques Proc.*, 1969, pp. 85–92.
3. D.F. Horne, "Loose abrasives, impregnated diamonds and electro-plated diamonds for glass surfacing," *SPIE Proc. Advances in Optical Production Technology,* vol. 109, 1977, pp. 12–18.
4. Y. Kihira, Y. Kaneko, K. Sugiyama, and H. Yamamura, "Effects of the grinding on some physical properties of polycarbonate," *Japan Congress of Materials Research Proceedings,* pp. 259–263, Sept. 1978.
5. H. Poehlmann, "A requiem for the aircraft canopy," *Maintenance Magazine,* pp. 12–15, July/Sept. 1986.
6. U.S. Military Surface Quality Specification, MIL-O-13830A, United State Military.
7. *Standard Practice for Stress Crazing of Transparent Plastics* (ASTM Designation F791-82). Philadelphia: American Society for Testing and Materials, 1982, pp. 852–855.
8. *Standard Practice for Optical Distortion and Deviation of Transparent Parts Using the Double-Exposure Method* (ASTM Designation F733-81). Philadelphia: American Society for Testing and Materials, 1981, pp. 808–813.
9. *Standard Test Method for Measuring Binocular Disparity in Transparent Parts* (ASTM draft submitted for approval). Philadelphia: American Society for Testing and Materials, September 1989.
10. *Standard Test Method for Measuring Halation of Transparent Parts* (ASTM Designation F943-85). Philadelphia: American Society for Testing and Materials, 1985, pp. 908–913.

11. H.S. Merkel, "An automated method for measuring transparency optical distortion," Wright-Patterson Air Force Base, OH, Internal Document, 1989.
12. USAF Technical Order 1F-4C-3-1-4, United States Air Force.
13. NAVAIR 01-245FDA-3-1-4, United States Navy.
14. "Critical item development specification: F-16 transparencies," Specification No. 16ZK002D, Code Identification No. 81755, General Dynamics, Forth Worth, TX, 1978.
15. "Transparency crack criteria and repair procedures," Bulletin No. 12C-027, General Dynamics, Fort Worth, TX, 1985.
16. "Optical surface restoration of acrylic transparencies," McDonnel Douglass, St. Louis, MO, Process Specification 20029, FSCM No. 76301, June 1981.
17. J.A. Luckemeyer, "Abrasives study," Southwest Research Institute, San Antonio, TX, Automated (Robot) Canopy Polishing System: Final Report, Appendix G, USAF Contract No. F42650-86-C-3276, May 1987.
18. K. Ewald, "Acrylic window crazing," in *ASTM Subcommittee F7.08 Meeting*, 1986.
19. J.A. Luckemeyer and S.B. Farmer, "Canopy polishing tests," Southwest Research Institute, San Antonio, Tx, Automated (Robot) Canopy Polishing System: Final Report, Appendix B, USAF Contract No. F42650-86-C-3276, May 1987.
20. V.R. Sturdivant, "Canopy polishing studies," Southwest Research Institute, San Antonio, TX, Automated (Robot) Canopy Polishing System: Final Report, Appendix A, USAF Contract No. F42650-86-C-3276, May 1987.
21. J.A. Luckemeyer, "Polishing end effector design," Southwest Research Institute, San Antonio, TX, Automated (Robot) Canopy Polishing System: Final Report, Appendix F, USAF Contract No. F42650-86-C-3276, May 1987.
22. D.J. Wenzel, S.B. Farmer, D.S. McFalls, B.C. Mather, and S.D. Young, "Programmer's reference," Southwest Research Institute, San Antonio, TX, Robotic Canopy Polishing System (RCPS): Operation and Maintenance Manual, Volume IV, USAF Contract No. F42650-86-C-3276, Amendment P00001, Aug. 1988.
23. *Standard Test Method for Intensity of Scratches on Aerospace Transparent Plastics* (ASTM Designation F548-81). Philadelphia: American Society for Testing and Materials, 1981, pp. 779–780.
24. *Standard Test Method for Measuring Optical Reflections from Transparent Materials* (ASTM draft submitted for approval). Philadelphia: American Society for Testing and Materials, June 1987.
25. *Standard Test Method for Deviation of Line of Sight through Transparent Plastics* (ASTM Designation D881-48). Philadelphia: American Society for Testing and Materials, 1948, pp. 460–462.
26. *Standard Method for Measuring Optical Angular Deviation of Transparent Parts* (ASTM Designation F801-83). Philadelphia: American Society for Testing and Materials, 1983, pp. 871–878.
27. *Standard Test Method for Measuring the Angular Displacement of Multiple Images in Transparent Parts* (ASTM Designation F1165-88). Philadelphia: American Society for Testing and Materials, 1988.
28. H. Kwun, "Feasibility of measuring thickness of canopy layers using ultrasonics," Southwest Research Institute, San Antonio, TX, Automated (Robot) Canopy Polishing System: Final Report, Appendix E, USAF Contract No. F42650-86-C-3276, Dec. 1986.

29. C.M. Teller, C.R. Ursell, D.V. Smith, and S.A. Cerwin, "Development of nondestructive inspection procedures and equipment for F-5 canopy acrylic panels," Southwest Research Institute, San Antonio, TX, Final Report, SwRI Project No. 03-5260, USAF Contract No. F41608-78-G-0010, June 1978.
30. D.S. McFalls, "Canopy inspection investigations," Southwest Research Institute, San Antonio, TX, Automated (Robot) Canopy Polishing System: Final Report, Appendix C, USAF Contract No. F42650-86-C-3276, May 1987.
31. J. Torma, "Engineering support for the fluorescent penetrant inspection module (FPIM) of the integrated blade inspection system (IBIS)," Southwest Research Institute, San Antonio, TX, Final Report, SwRI Project No. 14-8373, May 1985.
32. "Automated aircraft canopy inspection system," P-InTA-87-43, International Technical Associates, Santa Clara, California, 1987.
33. J. Taboada and A.J. Duelm, "Shearing interferometer with scanned photodiode array and microcomputer for automated transparency distortion measurements," *Proc. SPIE Conference on Advances in Optical Metrology,* Aug. 1978, pp. 139–145.
34. J. Taboada, "Coherent optical methods for applications in robot visual sensing," *SPIE 3-D Machine Perception,* vol. 283, pp. 25–29, 1981.
35. L.V. Genco and H.L. Task, "Aircraft transparency optical quality: New methods of measurement," Wright-Patterson Air Force Base, OH, Air Force Aerospace Medical Research Laboratory, Report No. AFAMRL-TR-81-21, Feb. 1981.
36. L.V. Genco, "The measurement of angular deviation and its relation to weapons sighting accuracy in F-16 canopies," Wright-Patterson Air Force Base, OH, Air Force Aerospace Medical Research Laboratory Report No. AFAMRL-TR-82-8, 1982.
37. J.A. Luckemeyer and S.B. Farmer, "In-process surfactant research," Southwest Research Institute, San Antonio, TX, Automated (Robot) Canopy Polishing System: Final Report, Appendix D, USAF Contract No. F42650-86-C-3276, May 1987.
38. D.H. Ballard and C.M. Brown, *Computer Vision.* Englewood Cliffs, NJ: Prentice-Hall, Inc., 1982.
39. A. Rosenfeld and A.C. Kak, *Digital Picture Processing,* 2nd ed. Orlando, FL: Academic Press, 1982.
40. D.L. Michalsky, S.B. Farmer, J.A. Luckemeyer, D.S. McFalls, A. Rizo-Patron, V.R. Sturdivant, and S.D. Young, "Automated (Robot) Canopy Polishing System: Final Report," Southwest Research Institute, San Antonio, TX, USAF Contract No. F42650-86-C-3276, SwRI Project No. 14-1044, June 1987.
41. B.A. Barsky, "A description and evaluation of 3D models," *IEEE Computer Graphics and Applications,* vol. 4, pp. 38–52, July 1984.
42. M.E. Mortenson, *Geometric Modeling.* New York: John Wiley & Sons, 1985.
43. A. Goshtasby, "Correction of image deformation from lens distortion using Bezier patches," *Computer Vision, Graphics, and Image Processing,* vol. 47, pp. 385–394, Oct. 1989.
44. R.A. Jarvis, "A perspective on range finding techniques for computer vision," *IEEE Pattern Analysis and Machine Intelligence,* vol. PAMI-5, pp. 122–139, March 1983.
45. D. Nitzan, "Three-dimensional vision structure for robot applications," *IEEE Pattern Analysis and Machine Intelligence,* vol. 10, pp. 291–309, May 1988.

46. M.D. Altschuler, B.R. Altschuler, and J. Taboada, "Laser electro-optical system for rapid three-dimensional (3-D) topographic mapping of surfaces," *Optical Engineering*, vol. 20, pp. 953–961, Nov.-Dec. 1981.
47. D.J. Wenzel and D.S. McFalls, "An optimal material removal strategy for automated repair of aircraft canopies," *Proc. 1989 IEEE Int. Conf. on Robotics and Automation*, May 1989, pp. 370–377.
48. S.D. Young and D.L. Michalsky, "The robotic canopy polishing system," *Proc. 1989 SME Robots in Aerospace Manufacturing Conference*, Feb. 1989.
49. D.L. Michalsky, "Feasibility acceptance test report," Southwest Research Institute, San Antonio, TX, Automated (Robot) Canopy Polishing System: Final Report, Appendix L, USAF Contract No. F42650-86-C-3276, June 1987.

7

Geometric Modeling for Robot Task Planning

Claudio Mirolo
Dip. di Matematica e Informatica dell'Università di Udine,
Via delle Science 206, I-33100 Udine, Italy

Enrico Pagello
Ist. LADSEB del CNR, Corso Stati Uniti 4, I-35020 Padova, Italy
& Dip. di Elettronica e Informatica, Università di Padova

1. INTRODUCTION

Geometry, in its quantitative connotation, is a favored means of analysis to improve the technology of robots in the assembly process. This is because the objects are usually supposed to be rigid and the corresponding measures invariant, but other implicit assumptions are hidden in this sort of description of the robot's world, which limit what we can expect from a system. Later, we will discuss these assumptions and their consequences. Although it is clear that the geometric view is not sufficient, perhaps even unsuitable for advanced applications in unstructured environments, here we restrict our attention to those problems that can be approached by means of accurate geometric information.

Most of the available systems aimed at constructing geometric models of objects and environments have been developed for computer-aided design (CAD) purposes. Before the researchers in robotics felt the need for a geometric model, a few techniques for representing rigid bodies and kinematic chains were already introduced [1]. However, a standard CAD system is only helpful while simulating the robot off-line, to check its behavior without affecting the physical world and protect it from possible damages. In this respect, the programmer is supported in the debugging activity, but no further tools are provided to help reasoning about a changing world. Unfortunately, a CAD system is unsatisfactory for applications in robotics because the internal representation is not suitable for geometric reasoning, whereas several geometric problems need an efficient solution to overcome the complexity of the tasks.

In the past the geometry of objects and the geometry of motion planning were studied independently. For example, using the same system both for representing the robot world and trying to find a collision-free path is still a hard task. In our opinion, any progress toward filling the gap between object modeling and motion planning should start from a deep analysis of the computational costs. We agree with Pertin-Troccaz:

> Unfortunately computational geometry is still a bottleneck for most of the topics related to CAD-based system development (in robotics, graphics, computer vision . . .); even for well-known problems, it happens that implementing some theoretical results (e.g., the configuration space approach for spatial planning) is a very difficult and sometimes tedious task for two major reasons: (1) geometry is very combinatorial and (2) geometry is complex. . . . [2].

Meanwhile, the design of highly structured representations and refined techniques to access geometric information has grown more and more important, at least in principle. This is attested to, for example, by the presence of a section devoted to computational geometry in the recent IEEE International Conference on Robotics and Automation, held in San Diego, May 1994.[1] So, not only does robotics remain a fertile source of new problems for people working in computational geometry, but also robotics is expected to benefit from any advances in the more abstract area of computational geometry.

We tried to take a first step toward the application of a geometric modeler to motion planning with the World Modeler (WM) project [3], developed at the Institute LADSEB of the National Research Council in Padua and at the Department of mathematics and computer science of the University of Udine. The earliest version of the system was mainly intended for simulation [4], and some ideas (generalized cones, subpart trees, variable affixments) were taken from ACRONYM [5]. Later, we realized that the analysis of contact configurations involving the robot and other objects in its environment turns out to be a versatile tool for several applications, ranging from motion planning to assembly and grasp planning. In this respect, we considered a few elementary operations on the geometric model, each allowing an efficient implementation.

The chapter is organized as follows. Section 2 is a short discussion of some geometric issues in task planning, particularly assembly and grasp planning, and of the role of uncertainty. The focus of Section 3 is on motion planning: After a brief survey, a few sample approaches are reviewed, then, we introduce a set of methodological criteria to be used as a means of comparison. The use of a geometric modeler for motion planning and other problems arising in robotics is discussed in Section 4. More specifically, we are interested in the development of a set of simple and efficient primitive operations to access the model. Finally, in Section 5 we apply these primitive operations and propose a cell decomposition approach to characterize the configuration space. We conclude the section by

---

[1] At the time of this writing, this was the most recent IEEE conference on robotics and automation.

considering the possibilities of exploiting monotone subdivisions and planning in the space time.

Although several references are included at the end of the chapter, the authors do not want to pretend that it is at all complete. Rather, both the organization of the material and the referred literature reflect the authors' views. Among the other papers, collections of papers, and books presenting a survey or investigation into the relationships between geometry and robotics, the following can be mentioned: [6–16].

2. GEOMETRIC ISSUES IN ROBOTICS

We said in the introduction that the "accurate geometry" view hides some implicit assumptions that limit the range of applications. First of all, we mention these assumptions and suggest that the reader reflect about their meaning in what follows. Here are the main points:

- The considered models apply to the standard robots, usually manipulators for mechanical assemblies, such that positions and configurations can be measured with high precision.
- A unique centralized planning control is supposed, i.e., there is only one unit where decisions are made. Systems of several manipulators are possible provided all of them are coordinated at a higher level.
- A structured environment is supposed and its geometry can be measured with the desired precision.
- Any change in the environment must be the effect of an action by some robot, under the centralized control, and all the consequences of an action are supposed to be known as well.

2.1. Task Planning

The problem of automating complex tasks has been studied for several years, see for example [17]. The first attempts to develop high-level languages for robot programming came in the late 1970s. A few languages were proposed at joint level [18], manipulator level [19], and task level [20, 21], but they were not satisfactory [22].

In the framework of mechanical assembly, a task can be conveniently decomposed into a sequence of actions, each involving the movement of an object from a given configuration to a target configuration, or some auxiliary activities, such as screwing, welding, and so on. For the significance of geometry, our focus falls on these movements, which can be decomposed into the following finer steps:

1. Approaching gross motion
2. Pregrasping fine motion
3. Grasping
4. Detachment fine motion
5. Object transfer gross motion
6. Part mating (fine motion)
7. Ungrasping
8. Leaving fine motion
9. Leaving gross motion.

However, the proposed decomposition cannot be interpreted as a usual problem-solving strategy. Since the mid 1980s Lozano-Pérez and his collaborators realized that a simple *divide and conquer* approach does not apply because of intrinsic nonlocal features, and their conclusion is emphasized in [23]. For example, the transfer motion trajectory of an object depends on how the object is grasped, but to choose a suitable grasp we must take into account where and how the object has to be placed, and so on. So, rather than computing a single solution for every step, it is better to determine a set of locally consistent solutions (in other words, to reduce the set of possible solutions).

For instance, the experimental system TWAIN, designed by Lozano-Pérez and Brooks [23], was based on three loosely connected modules aimed at solving problems of grasping, fine motion, and gross motion. (Each of the subproblems listed above falls into one of these three classes.) With such a system, Lozano-Pérez and Brooks tried to integrate several tools already available for specific purposes. As a noticeable result, they proposed a *constraint propagation* technique as a control mechanism suitable to merge the partial solutions computed by different modules (see also [24, 25]).

Since TWAIN, other related approaches have come into a modular subdivision and a corresponding design of a system supporting the necessary activities. It is worth mentioning the system SHARP, proposed by Laugier and Pertin-Troccaz [26] (see also [9]). Moreover, the ideas underlying TWAIN were then developed in the project Handey [27, 28].

In the past, we also tried to contribute to the design of a robot programming system. A first attempt was an off-line programming environment with a few rich modules communicating by message passing [4]. One such module, for example, integrated the geometric modeler WM. Another implemented the Virtual Machine Language (VML) for robotized mechanical assembly [29]. Later, we cooperated within a European project [30].

The rest of this section considers the geometric issues in assembly and grasp planning and the role of uncertainty. The problem of motion planning, where geometry is prominent, is discussed in the next section.

2.1.1. Assembly.

Automating assembly and disassembly is a main goal to improve robot technology.[2] It is also a challenging topic because of the variety of aspects to be considered, not all of them being geometric. Indeed, in order to plan assemblies and disassemblies, we have to model the setting (e.g., parts, tools, layout), the robot (sensors, kinematics) and the task (grasps, part mating, movements). It can be noticed that a symbolic planner, which usually represents the tasks in a logic-oriented language, is unsuitable to dealing with the geometry of the objects and the environment. The geometric model should be queried for the following purposes (in this respect, the assembly and disassembly strategies are perfectly symmetric):

- Determining the features of two parts which are in contact
- Finding the directions such that a given part can move freely away from its contact configuration (*spanning vectors,* according to [31])
- Avoiding collisions with other parts not in contact (since the spanning vectors depend only on the contact features; in [31] this problem is solved by trial and error).

The more direct approaches to assembly planning consider all feasible combinations of parts [32–38]. Methods of this kind, based on the "liaison models," [32], or and/or graphs representing stable subassemblies, suffer from the combinatorial explosion when the number of parts increases. In fact, generating non-monotone assembly plans is *PSPACE-hard* already in two dimensions [39], whereas a class of monotone assembly problems involving polyhedra is *NP-complete* [40].[3]

Some refined strategies have been proposed to reduce the number of candidate subassemblies [39, 41–47]. Wolter claimed that a good assembly plan should consider several constraints, such as directionality, locality, complexity of the fixture, manipulability, and changes of tools [39], and later he also proposed the use of intermediate configurations given in input [47]. As further examples, in [41] the assembly directions are analyzed, whereas other features (e.g., fixture complexity and change of direction) are introduced in [42]; the CAD description of a product taken from the world database is used in [46].

An interesting geometric problem, the partitioning problem, is solved in [28].

---

[2] This will reduce the costs of the products on the one hand and the costs of retrieving components from the old products on the other. The latter would result in a reduction of pollution, as emerged, for example, during the 3rd International Workshop on Robotics in the Alpe-Adria Region, Bled, Slovenia, 1994.

[3] By monotone assembly we mean an assembly such that there exists a sequential plan without intermediate park positions for some of its parts.

It can be stated as follows: Recognize the subset of parts that can be moved to infinity by a straight motion avoiding conflicts with other parts. Moreover, the generalization to the cases where nonstraight motions are needed is discussed in [48]. Good results can be obtained in two dimensions when the disassembly direction is fixed in advance and each component can be separated by a single translation. For example, $O(Mn + M \log M)$ time is sufficient to compute a solution in the case of M simple polygons with n sides [49] (which is also a reference for other applications of the tools of computational geometry to disassembly planning).

The *nondirectional blocking graph,* introduced in [50], describes the internal structure of the assembly and represents how each part prevents the others from being moved in some direction. The authors show that the size of the graph is polynomial in the number of parts and can be computed in polynomial time, given the geometric model of the assembly (see also [40]). Finally, Nnaji fully describes a system where assemblies are defined as sets containing parts which satisfy spatial relationships [31].

Recently, we tried to study in more depth the relation between collision detection and disassembly plans [51]. To this aim, we combined an explicit representation of the constraints to the motions of the parts and the concept of spanning vectors. Basically, the method iterates the computation of the spanning vector sets whenever a collision of the moving parts is found. It is not complete, but works in real-world examples, like that in Figure 7.1, where the geometry is not very complex.

Other characteristics, beyond the geometric issues, are referred to as specific task constraints in [30]. For instance, since the parts that are loosely connected to each other should be disassembled in an earlier step, knowledge about the tightness of the liaison can be profitably used [52]. Key items of information to generate assembly sequences are the liaison types and the assembly directions. From these data several important geometrical, topological, and technological constraints can be inferred and exploited to reduce the computational costs of

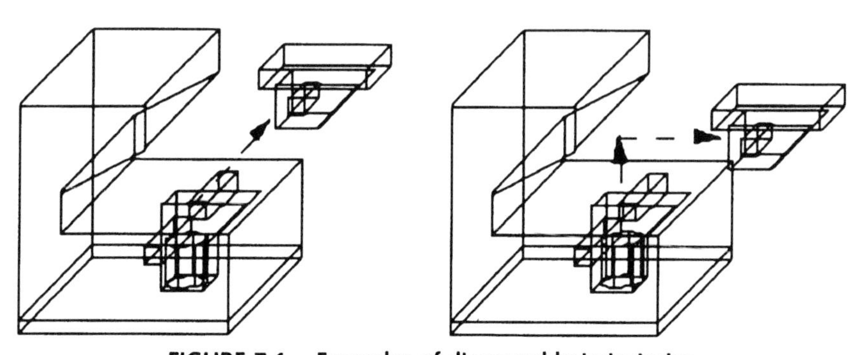

FIGURE 7.1. Examples of disassembly trajectories.

planning, so that the planner can generate optimal assembly plans in time roughly proportional to the number of parts.

2.1.2. Grasping.

Researchers from different areas have been concerned with the problem of grasping objects. As pointed out in [53], their work can be classified into two broad categories: design of manipulators and planning. The grasp planning problem is the problem of choosing a suitable grasp for a given task and environment. The grasp planner is usually thought of as part of a more general system, the task planner, and interacts with other modules, e.g., the path planner and the fine-motion planner.

For dexterous hands most of the work is based on the analysis of human prehension [54–58]. These methods contributed to a better understanding of the problem, giving rules of thumb for the design of artificial manipulators, but do not seem sufficient to solve the whole problem. Contact forces and friction are studied in [59–61]. In [62], the objects are represented in the force domain. Moreover, several criteria for evaluating the goodness of a grasp have been proposed, e.g., maximizing stability [63], minimizing contact forces [64], task-oriented measures [65, 66].

Geometric algorithms for computing the fingertip contacts and the exerted forces on the surface of the object can be found in [64, 67–70] for the planar case, and in [71–73] for three dimensions. However, the problem of finding a grasp configuration that is feasible by a multifingered hand has not been much investigated. To this aim, the physical and geometric conditions of stability and accessibility have to be integrated with other task-related conditions.

From a geometric viewpoint, the analysis cannot be restricted to the kinematic of grasp planning, concerning how the fingers approach the target surfaces, but should also take into account the exerted forces. In other words, both the *reachability* and *stability* problems have to be solved to determine a suitable set of contact points and to ensure that all the involved forces and torques be balanced. To schematize, the geometric issues are:

- Geometry of prehension, determined by the shapes of the gripper, the object to be grasped and the surrounding objects (*local* and *static* reachability, arising at pickup and at put-down)
- Geometry of the motions of the gripper, possibly carrying an object, near the obstacles and far from them (*dynamic* reachability, the hand and the grasped object act as a single rigid body)
- Geometry of the exerted forces and torques which ensure the safety of the grasp (stability).

A grasp configuration is reachable if the fingers do not collide with other objects in the environment. We can distinguish among local (constraints at the contact points, due to the specific shape, size, and configuration of the hand and the object, independently of the rest of the environment), static (conditions related to the initial and final positions of the grasped object and the other objects in the environment), and dynamic (collision avoidance during the motions) reachability. While many results exist for a gripper, mainly based on the configuration space approach, very few are available for an articulated hand. Reachability (*safety,* according to [7]) is addressed in [71]. The kinematics of grasping is discussed in [74].

Stability refers to the balance among the external forces and those exerted by the hand [75]. Each force may change during manipulation, but we want the fingertip forces to restore the equilibrium configuration for small perturbations. The stability of grasps in the particular case of end effectors with two jaws is discussed, for example, in [76]. In a more general framework, we look for a set of contact points on the object surface such that any force can be exerted in the absence of friction. This condition is generally referred to as *force closure* [77, 78]. It has been shown that any force-closure grasp can be made stable [79], hence considering force-closure grasps is sufficient to determine stable grasps.

This subject has also been studied from a more computational viewpoint. For example, in [80] it is proved that a force- and torque- closure grasp can be computed in $O(n^{3/2} \log^{1/2} n)$ for a convex polyhedron with n faces. Another geometric and combinatorial problem is the determination of an upper bound to the number of fingers necessary to accomplish a grasp in particular situations [81].

However, a lot of information must be considered to perform a grasp. Pertin-Troccaz proposed an interesting strategy, actually under some restrictive hypotheses, to capture the information relevant to a specified task [2]. Her technique is based on a symbolic analysis, followed by the numerical analysis of a small set of candidate grasps involving local geometric properties. It is worth noting that in the conclusion of the cited paper Pertin-Troccaz emphasizes the distinction of two levels of geometric reasoning, symbolic and numerical, as a general paradigm for robot programming.

There is one further aspect to be considered, showing that the geometry of grasps cannot be analyzed out of a more general context without losing useful information. Human behavior suggests that the choice of a grasp should also depend on the action to be performed. This problem of *functionality* rests on the existence of a cognitive model to answer such questions as: "What is a shape for?" and "How can an object be used?" [82]. Moreover, function-to-form information is also useful in identification problems [83, 84].

In the framework outlined above, we proposed algorithms for computing stable grasps [30, 85, 86]. Sometimes, the functionality of the grasped object can be simply related to points and directions. The point where the force (or torque) is applied, the direction of motion, or the suitable orientation of the object are natural examples of points and directions that can be chosen to drive heuristic rules and determine a grasp configuration [86]. The model of functionality in

[85] is based on the primitives *push, turn, swing,* and *lever,* which can be combined to represent more complex functions, e.g., screw is the result of push and turn. From a geometric viewpoint, the application of each such primitive operation to a particular task is instantiated by giving three vectors: for the object (its axis), the action, and the applied force, see Figure 7.2.

2.2. Uncertainty

An important question in task planning is how to deal with uncertainty. In particular, this problem arises in realistic part-mating strategies and may require the use of sensory feedback. At least two different kinds of uncertainty have been recognized, i.e., the possible lack of knowledge about the environment and the robot control (see also [87]). The first arises when the robot does not know its environment with the necessary precision, or when it cannot know its environment because of the sensor limits. The second depends on the fact that the actual trajectory may be slightly different from the expected one due to actuator limits.

In order to deal with uncertain knowledge about the environment, a plan must encapsulate new operations such as calibration, location of relevant objects, compliant motion control, error detection, and recovery. If we refer to the limited accuracy of any geometric measure, it is clear that uncertainty is unavoidable and it affects in particular the fine-motion strategies. As a matter of fact, inaccuracies are introduced also because of the finite precision arithmetic [88].

While planning some movement, we can follow two main approaches:

- We can provide a complete plan, or skeleton, under the assumption of infinite precision, and only in a second step perturb such a skeleton to take into account the uncertainty.

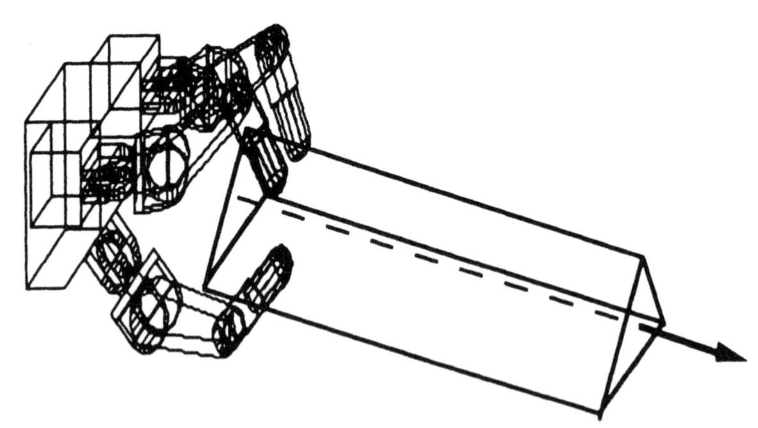

FIGURE 7.2. Selection of the approach direction for a screwing operation.

- Otherwise, we can think of the uncertainty as an intrinsic part of the task, affecting every decision in the planning process.

The former approach gave rise to different techniques based on the instantiation of plan skeletons representing motion strategies, or on the ability of replanning a motion which is going to fail because of imprecise knowledge about the environment. Typically, geometric uncertainty is analyzed to provide useful models of probabilistic distribution, e.g. [89]. A classic example of the latter approach is the *preimage back-chaining* technique, introduced in [90], and developed in [91, 92]. The technique was also developed by other researchers mentioned in the survey [87] (see also [93]).

Further difficulties arise because the amount of uncertainty is not a static property. In particular, the accuracy of the trajectory of a robot does not only depend on the corresponding kinematic, but must also take into account the exerted forces. Some authors have developed a representation of the contact space of parts [94], which allows to treat the uncertainty on the model, configuration, and control in a suitable way [95].

It is also interesting to investigate the relationship between grasp planning and uncertainty. As pointed out in [7], "If the initial configuration of the target object is subject to substantial uncertainty, . . . the grasp motion should reduce the uncertainty." Compliance can be used to reduce the uncertainty about the reached configuration, e.g., [96]. However, it should be noticed that the problem of controlling compliant motions has been proved nondeterministic exponential time hard [97].

A more general interpretation of uncertainty includes the case of incomplete knowledge. Path planning with partial information is a typical problem for autonomous robots, which often operate in unstructured environments, and has been tackled by means of heuristic approaches [98, 99]. However, nonheuristic algorithms have also been developed by Lumelsky in the case of dynamic planning for sensory robots, e.g. [100], and his analysis is particularly interesting from the geometric viewpoint.

However, the role of uncertainty cannot be wholly understood from a purely geometric viewpoint. Dean considers this problem together with time passing as a milestone to be able to design real autonomous machines acting in a real world [101]. In the unstructured environments there is a lack of a priori knowledge, involving both the location of the objects at a given time and the possibility that a configuration changes because of unpredictable events (see, for example, [102]). Moreover, it is the temporal evolution which allows the acquisition of new information to increase the knowledge about the world. In this respect, some different degrees of uncertainty can be distinguished: fixed and incompletely known obstacles, known and randomly movable obstacles, incompletely known and randomly movable obstacles [103].

3. MOTION PLANNING

The classical approach to robot motion planning, thoroughly described in [10], makes extensive use of geometry. On the one hand, it should be observed that these techniques principally apply to gross motion, since the solutions do not assume any sensory feedback and are subject to the restrictions pointed out at the beginning of Section 2. On the other, they contribute a clean theoretical framework where questions about performances or what we can actually expect from a planning system can be stated in precise terms.

Before mentioning some milestone works and results in motion planning and trying a brief review of a selection of basic approaches, we introduce the problem and the underlying abstract model. The problem can be stated as follows: For a movable system (a mechanical arm or a navigating body in the simplest cases) within a world of rigid obstacles, find a movement from the given configuration to a target configuration.

In the case of navigation the solution consists of a suitable collision-free path, possibly with rotations. More generally, we are looking for a coordinated and safe motion of a set of parts, such as the links of a manipulator, constrained to each other and to the environment. However, every such problem is equivalent to a search for a collision-free trajectory in a suitable multidimensional space, the *configuration space* (C-space) [104], where the moving system is represented by a point with at least as many coordinates as the degrees of freedom, and the obstacles are mapped into the corresponding regions of nonfree configurations.

Examples of the C-space are shown in Figures 7.3–7.5. The sample problems are very simple and chosen such that both the original space and the corresponding C-space are two-dimensional. The movable systems are: (1) A translating polygon in the plane; (2) the stick model of a robot arm with two revolute joints; and (3) two robot arms, each having only one degree of freedom.

With respect to these settings it is also possible to comment on the implicit assumptions listed in Section 2. First, the hypotheses of fine control and detailed knowledge of the environment are directly related to the choice of an algebraic representation of the space, requiring precise devices and measures. Then, modeling the active components by a point in the C-space is only allowed if there exists a unique planning unit, and any computed solution is necessarily the coordinated motion of all the movable parts. Finally, the perfect predictability hypothesis guarantees that the computed solution can actually be exploited to carry out the plan.

Having chosen the C-space to model every reasonable motion planning problem, the next step is its computational characterization. The description of the free space has a twofold purpose: locating the initial and target configurations, and searching for a continuous path connecting such points. Typical characterizations are the subdivision into disjoint regions or the construction of a road map within each connected component of the free space.

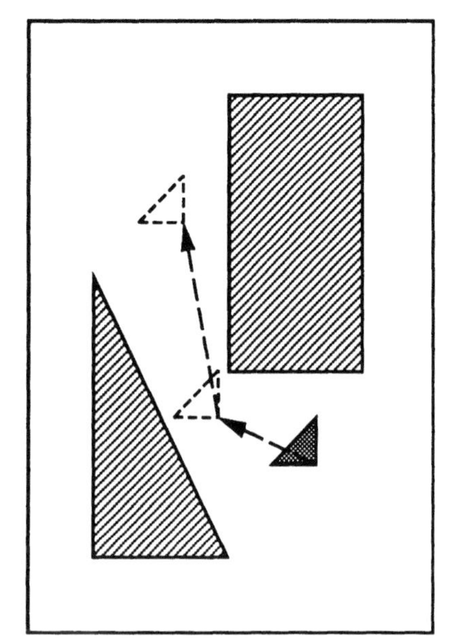

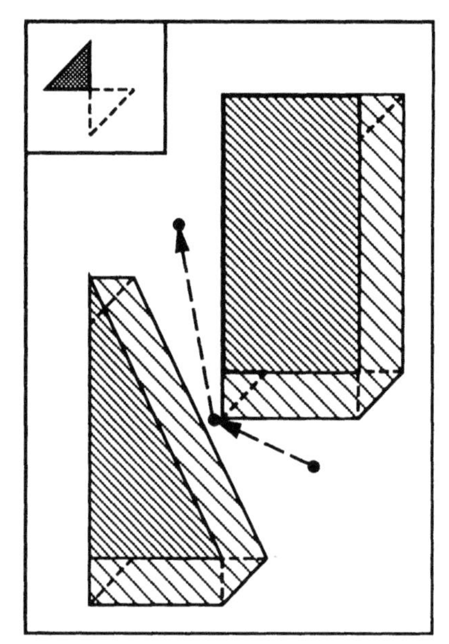

FIGURE 7.3. (a) Planning translations; and (b) interpretation in the configuration space.

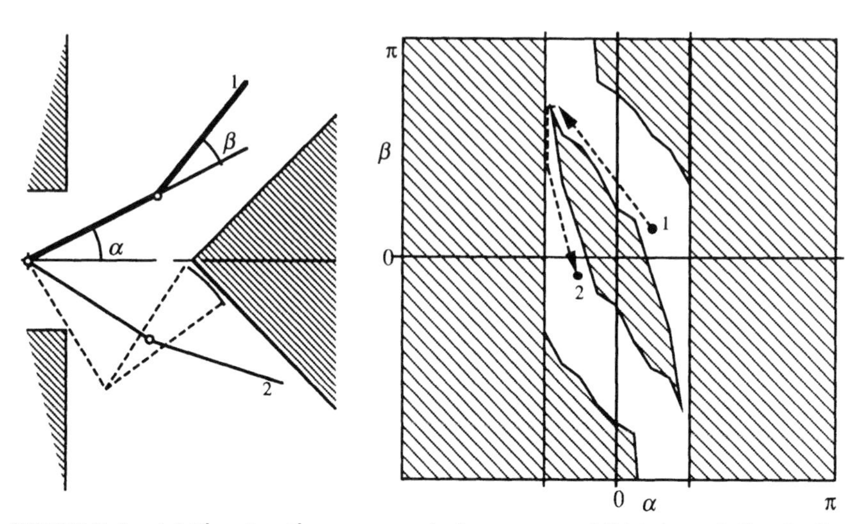

FIGURE 7.4. (a) Planning the movement of an arm; and (b) interpretation in the C-space.

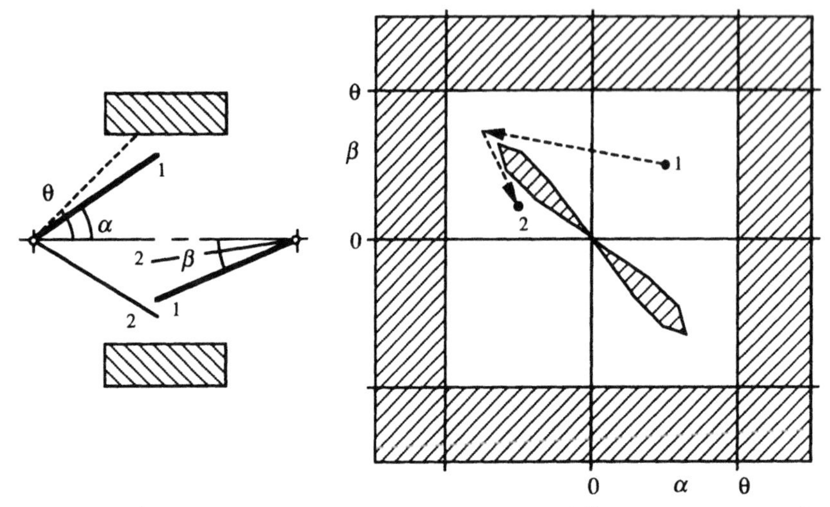

FIGURE 7.5. (a) Planning coordinated movements; and (b) interpretation in the C-space.

3.1. Essential Survey

Borrowing terminology from [10], a first distinction is between approximate and exact characterizations of the free space. Most of the approximate methods rest on the pioneering work by Lozano-Pérez and collaborators at the Massachusetts Institute of Technology (MIT) [105, 106], whose development resulted in the design of system prototypes, a noticeable one being [107], relative to a flying object with six degrees of freedom. A different characterization of the C-space has been proposed by Brooks [108], who contributed to several MIT research projects. He suggested representing the free space by a set of cones bounding *freeways*.

On the other hand, the exact techniques refer especially to the more abstract analysis of the problem preferred at the Courant Institute by Schwartz, Sharir, and others, searching in depth for the consequences of using algebraic tools. Some fundamental papers of their approach are collected in [11] and an introduction can also be found in [109]. Although some of the proposed techniques are hopelessly unsuitable for practical applications, the role of different parameters in the algebraic setting is enlightened. A further contribution of this methodology is the systematic use of judgments based on evaluation of the computational costs.

By using elegant algebraic and topological concepts, Schwartz and Sharir developed a general algebraic technique [110]. They were able to apply some previous works by Tarski [111] and Collins [112] on the decidability of a particular subset of first-order logic with an interesting geometric interpretation. (A less technical introduction can be found in [113, 114]). Indeed, Collins had developed a general algorithm for cellular algebraic decomposition as a fundamental step for eliminating the quantifiers in the considered formulas. This allowed a suitable characterization of the space, provided the problem could be stated in terms of semialgebraic sets, which is not at all restrictive. Schwartz and Sharir admit that such a general-purpose technique is only interesting from a theoretical viewpoint because of their "general but catastrophically inefficient algorithm," [115]. However it should be considered an important step toward a deeper understanding of the motion planning problem. There are two key points:

- In principle, every conceivable problem of (purely geometric) motion planning can be attempted by means of algebraic techniques, under unrestrictive hypotheses.
- The upper bounds to the computational costs depend polynomially on the complexity of the constraints imposed to the motion, but exponentially on the number of degrees of freedom.

The commonsense principle stating that a too general technique cannot be an efficient one has been formally proved by Reif [116], who showed that the general motion planning problem is computationally intractable since the time required for the solution depends exponentially on the number of degrees of freedom. Moreover, similar lower bounds to the computational costs are discussed in [117] and, more concisely, in [109]. In light of Reif's results, no substantial improvements can be expected for a general-purpose technique. However, Canny improved the upper bound and showed that the doubly exponential growth of the computational costs estimated by Schwartz and Sharir could be reduced to simply exponential [118]. In his work, he used a road map instead of a cell decomposition of the free space.

A selection of promising results obtained with exact techniques follow. The motion of a convex polygon among polygonal obstacles can be planned within $O(n\ \lambda_s(n) \log n)$ time, where $\lambda_s(n)$ is a function almost linear in the number of sides n [119]; for pure translations the upper bound is reduced to $O(n \log n)$ [120, 121]. The cost of the shortest collision-free path for a moving point and polygonal obstacles is $O(n^2 \log n)$ [122]. About shortest paths in a three-dimensional polyhedral world, we know that the problem is *NP-hard* [97], and can be solved in $n^{O(\log n)}$ parallel time by providing a considerable amount of processors [123]. If the point is constrained to slide on the surface of a polyhedron with n faces, then a plan requires $O(n^2)$ time [124]; other interesting upper bounds are proved in [125].

A few systems try to exploit the efficient techniques for computing proximity measures and properties made available by researchers in computational geometry. In particular, algorithms for collision detection are used to plan a collision-free path by iterating a guess-and-modify process. At each step of the process a trajectory is estimated, then it is tested for collisions, and possible negative results direct the choice of alternative paths. This approach is followed in [126] to develop a system integrating a geometric modeler (see also [127, 128]). One further example is the collision detection algorithm presented in [129], which can also deal with a special kind of uniform rotation.

Computing the distance of two objects may play a similar role in motion planning strategies as shown in [130, 131]. Of particular interest, both in theory and in practice, is the approach followed by Canny and Lin [132–134], where the distance between couples of bodies is sampled in order to incrementally draw a road map. This technique, which benefits by a fast algorithm to compute distances, seems to work in rather general situations.

Several applications use approaches based on potential field functions, introduced by Khatib [135], and then developed by other researchers, e.g. [136]. It is also worth mentioning the techniques for planning motions on-line, resting on the sensory information [100, 137, 138]. Finally, a few emerging lines of research are listed below:

- Motion planning with many degrees of freedom [139, 140].
- Motion planning with moving obstacles: for bounded velocity the problem is NP-hard in the plane [97], and *PSPACE-hard* in three dimensions [141]. Assuming that the velocity of the point robot is higher than that of the moving obstacles may be helpful, as shown in [142] for the planar case, where an $O(n^2 \log n)$ upper bound is obtained. See also [143, 144].
- Nonholonomic motion planning, where not only the configurations but also the velocities are constrained [145–147].

3.2. Review of Some Techniques

To give the reader an idea about the different approaches proposed in the early 1980s, which are the basis of the successive developments, we consider the work by Lozano-Pérez and collaborators, the technique of critical regions, as an example of exact cell decomposition, and the technique of retraction, as an example of road map. We briefly outline these techniques, which are also interesting from the applicative viewpoint.

3.2.1. Work by Lozano-Pérez.

Lozano-Pérez first realized that some available techniques solving navigation problems could be profitably used provided the moving body was sufficiently small, as it is in the case of the configuration space. Then, he tried to extend this approach to many different situations, involving more degrees of freedom (rotations, arm joints, and so on). Finally, to overcome the difficulties arising because of the increased degrees of freedom, he developed a cell decomposition technique. Some noticeable references are [23, 27, 104, 106, 148–150].

In one of his earliest works [148], Lozano-Pérez considers the translational motion of a polygon among polygonal obstacles in the plane. In the corresponding two-dimensional C-space the grown obstacles are still polygons, as illustrated in Figure 7.3. The C-space is characterized by the visibility graph, in fact a road map connecting by a segment every couple of visible vertices (two points are visible if the linking segment does not intersect any grown obstacle), where also the starting and goal configurations are treated as vertices. An example is shown in Figure 7.6. The optimal collision-free path can be computed by the graph-searching algorithm VGRAPH, already designed for navigation problems.

Extending this strategy to three or more dimensions is not straightforward. Consider for example the case of a polyhedron translating among polyhedral obstacles. Although the grown obstacles are still polyhedra, this time a solution path is no longer guaranteed to pass through the vertices, but all the points in each edge must be taken into account, giving rise to a noncombinatorial problem. Then, some approximations are needed and the computational costs increase. Further complications are introduced by rotations, since the obstacles mapped in the C-space have curved faces and edges.

In principle, the polygonal (polyhedral) description of the C-space, suitable for purely translational problems, can be seen as a particular planar (hyperplanar) section of a more general C-space. A finite set of slices, i.e., packs of thin sections, can be used to characterize the C-space by describing the projections of each slice. Such a characterization can be easily understood in the case of planar navigation with rotations (see Figures 7.7–7.8). Every section is a planar section

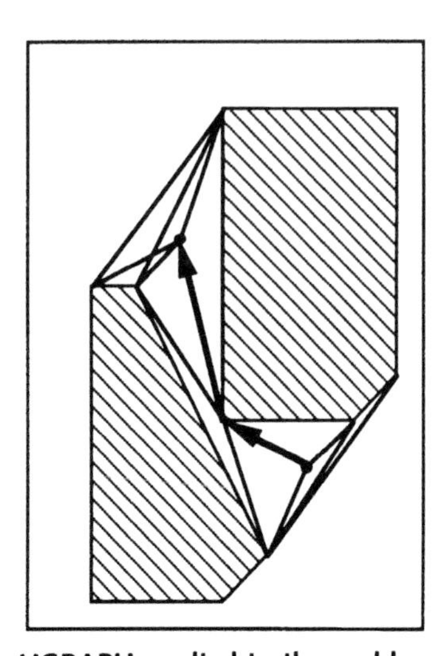

FIGURE 7.6. VGRAPH applied to the problem of Figure 7.3.

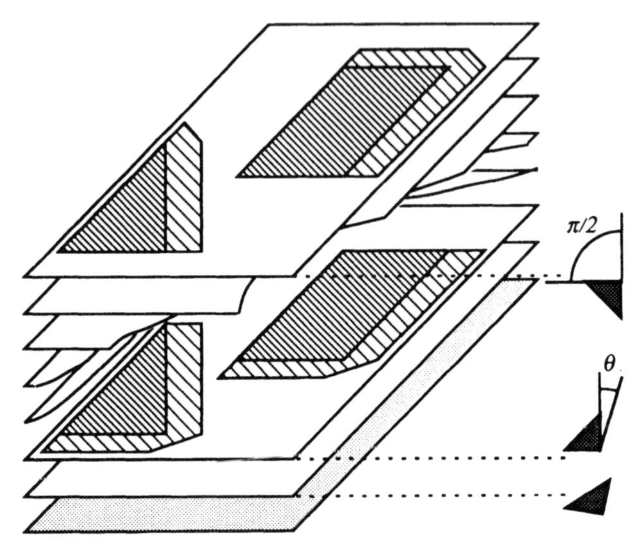

FIGURE 7.7. Sections of the C-space corresponding to different orientations.

corresponding to a fixed orientation of the moving body. The perpendicular projection of a pack of sections determines a slice projection, where a collision-free path can be found by building an approximated visibility graph and using VGRAPH [104]. The slice projections are also a good tool for separating the translational and rotational parameters in C-spaces with many dimensions.

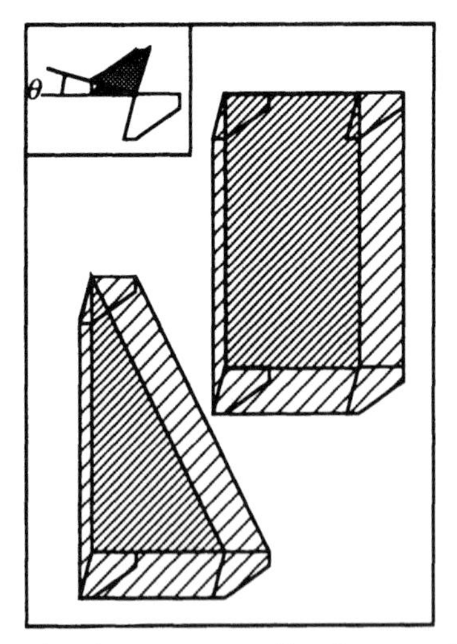

FIGURE 7.8. Slice projection corresponding to a small set of orientations.

In order to characterize more complex C-spaces, such as those relative to manipulators with revolute joints, Brooks and Lozano-Pérez developed a cell subdivision technique [106]. They use a hierarchical strategy and decompose the C-space into cells bounded by isothetic hyperplanes. All the cells completely inside the C-space obstacles are discarded; the others are labeled either "empty" or "mixed," depending on whether we know that they are free from C-space obstacles or not (see Figure 7.9). While searching for a collision-free path, some mixed cells are further decomposed only if no trajectory through empty cells can be found.

However, a few degrees of freedom are sufficient to lower the performances of any real motion planning system. Therefore, some heuristic strategies are necessary to drive the search for a collision-free trajectory, as interesting research by Donald shows [107]. Other examples of systems based on a similar approach were proposed by Faverjon, who uses octrees to represent the C-space [151–153]. It should also be observed that the time spent computing a solution in the mentioned planning systems is usually too high to allow real-time applications. This is because these techniques deal with overly detailed information.

3.2.2. Critical Regions.

The method of *critical regions* is based on a cell decomposition of a suitable projection of the C-space. A new feature is that the shape of the cells is not predetermined but depends on the actual properties of the free space. Another

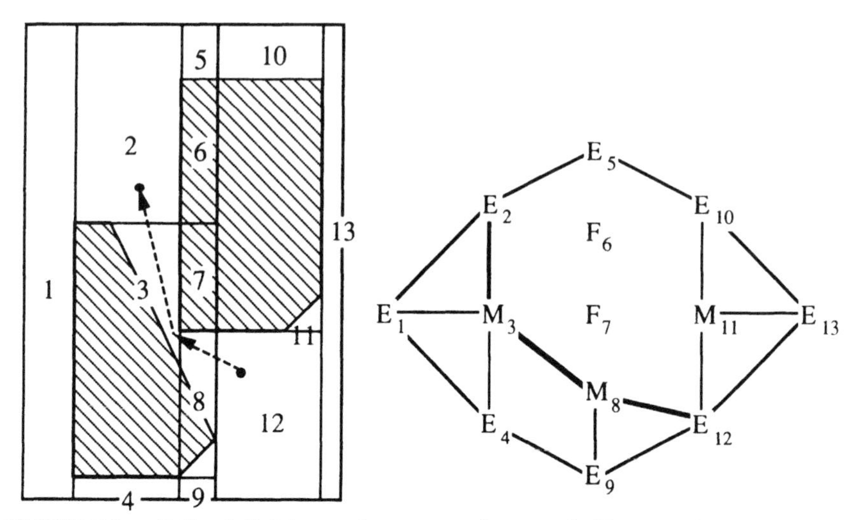

FIGURE 7.9. Cell subdivision and corresponding graph (E = empty, M = mixed, F = full). The solution of Figure 7.3. requires a finer subdivision involving the mixed cells 3 and 8.

feature is the role of topological concepts: In fact the noncritical regions are regions where appropriate topological properties are invariant. Some sample examples are discussed in [154]. It is not a systematic approach, since the imagination of the authors is relevant for applying it to different problems. However, we can recognize some important points:

- The cell subdivision is based on analysis of a convenient reference space whose dimension is less than the dimension of the whole C-space; this also implies the separation of different classes of configuration parameters.
- The reference space is decomposed into noncritical regions, bounded by a set of critical regions representing crucial configurations of the reference point of the moving object.
- The topological properties of the portions of the C-space related to different noncritical regions may be different. In other words, the topology of the free space is likely to change if the projection of the reference point of the moving object crosses a critical region.

The case of a rod moving in a plane [154] is a good example to understand the spirit of this method. Figures 7.10–7.12 show the reference space with the critical curves, a characterization of the topology corresponding to noncritical regions, and the relation with the three-dimensional C-space; the simple problem has the only purpose of giving an intuitive idea of what is going on. To search for a collision-free path, the connected components of the free space are partitioned into cells according to the corresponding noncritical regions, and then the adja-

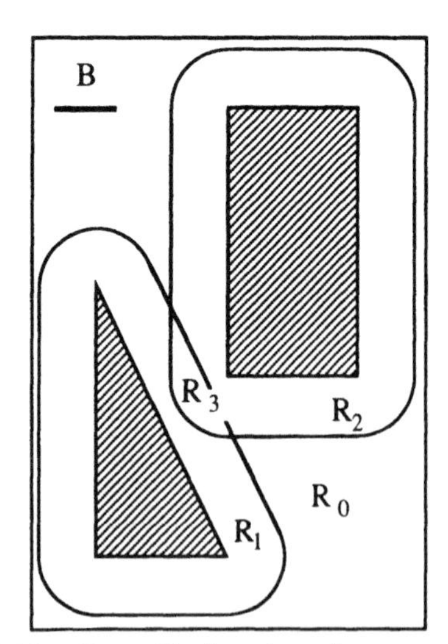

FIGURE 7.10. Noncritical regions for a moving rod B.

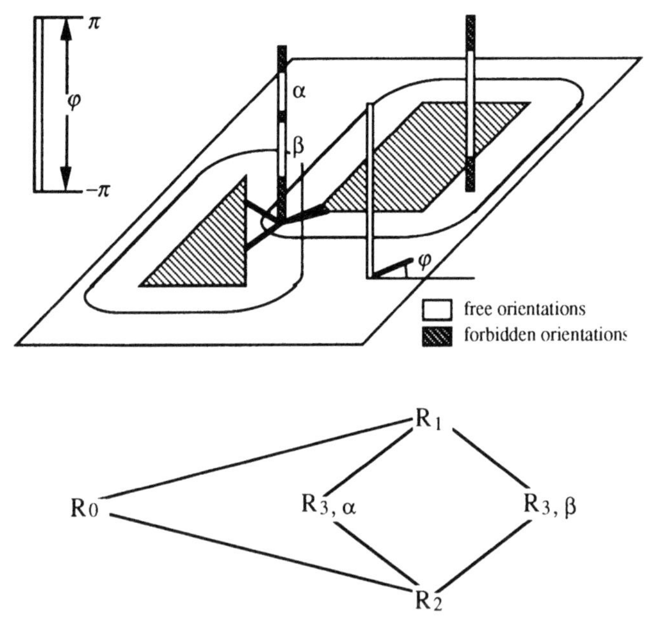

FIGURE 7.11. Free orientations for different noncritical regions and adjacency graph.

cent cells belonging to the same connected component are linked together. This defines a graph, where a standard searching algorithm can be applied. The bars in Figure 7.11 (the fibers in the terminology of [109]) represent the connectivity properties of the portion of C-space whose projection falls in a particular non-

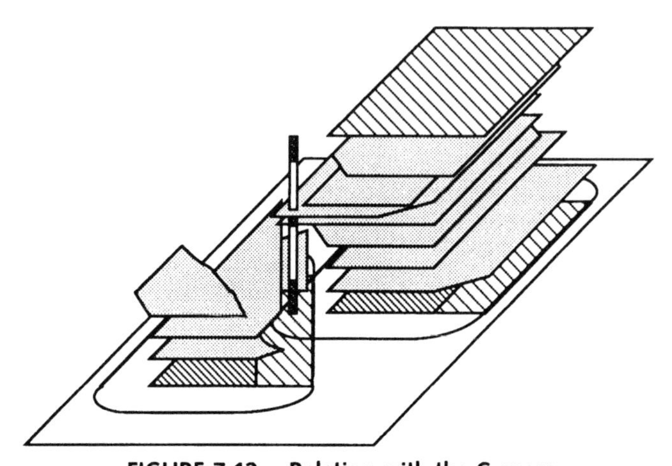

FIGURE 7.12. Relation with the C-space.

critical region. For example, two disconnected cells are determined above R3. For simplicity, the nodes in the graph are labeled only with the name of the region when there is a single cell corresponding to a noncritical region.

The critical curves shown in Figures 7.10–7.12 are also the contours of the projections of the C-space obstacles in the reference space; out of these contours any orientation of the rod is safe. But it should be noticed that there may be other kinds of critical curves representing points where some important changes in the set of allowed orientations arise. For instance, because of the simplicity of the example, there are no critical curves due to contacts of the rod with a corner and a wall at the same time.

Let us briefly consider the three-dimensional extension of the rod problem according to [115]. In this case the reference space is the surface of a sphere representing all possible orientations of the rod, and the critical curves are built in that surface. A point in the reference space corresponds to a three-dimensional section of the C-space which is the configuration space of a translational problem and then maps polyhedra into grown polyhedra. So, any noncritical region defines a class of topological equivalence among sections: Two equivalent sections are such that for any possible collision-free path in the former, there exists a qualitatively similar path in the latter.

Finally, to give an idea of the computational costs of this technique, it is worth mentioning that the general motion planning problem for a rod in the plane requires $O(n^5 \log n)$ time in the worst case, where n is the number of straight line segments bounding the environment [154]. However, in the opinion of the authors, better performances can be expected in practice. More generally, the algorithms based on this approach, both for rods and polygons, are polynomial time with degree depending on the type of critical curves.

3.2.3. Retraction.

The method of *retraction* bases the characterization of the C-space on a road map which is a generalized Voronoi diagram [155, 156]. The following points, relative to the case of a moving disk, can be noted:

- The road map is a particularly cheap description of the C-space.
- Since the road map is a Voronoi diagram (built of straight line segments and arcs of parabola), the planned motions are as far as possible from the obstacles.
- The road map is independent of the radius of the disk.
- Updating the Voronoi diagram after a local perturbation is easier than recomputing the road map from scratch.

A problem for a disk is shown in the example of Figure 7.13 (see also [156]). Under unrestrictive hypotheses, given any set of obstacles bounded by n straight walls, planning the motion of a disk in the plane requires $O(n)$ storage and $O(n \log n)$ time. A new solution can be computed in $O(n)$ time if only the disk

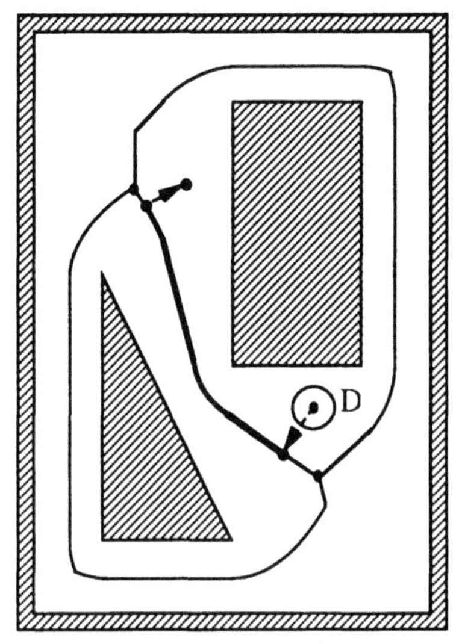

FIGURE 7.13. Road map to plan the motion of disk D.

and the initial and target configurations are changed. Moreover, the Voronoi diagram can be updated in $O(n)$ time if only one wall is introduced or removed, i.e., for a local perturbation of the environment.

3.3. Geometric vs. Combinatorial Analysis

So far, we have outlined a few different approaches to the motion planning problems. By comparing such methods we note that:

- We can recognize two main steps, at least in principle: (1) A geometric step, measuring the relevant features of the C-space; and (2) a combinatorial step, usually the search for a path in a graph.
- It is usually necessary to organize the configuration parameters into two classes, e.g., translational and rotational, and deal with such classes separately in order to reduce the dimensions of the space.

As a basis for a judgment about the techniques to characterize the configuration space and find collision-free paths we suggest considering the properties listed schematically below:

- *Completeness,* meaning that a solution can be eventually found every time solutions exist (exact approaches, according to [10]). The advantages of complete planning strategies are obvious, but there is a price to be paid in terms of computational costs, e.g., because of the proliferation of small cells. Examples of general complete strategies can be found in [11, 118, 134].

- *Flexibility* in the choice of a particular collision-free path, in order to satisfy other constraints. This is a feature of the cell decompositions as opposed to road maps, since all the information about the free space is potentially available.
- *Variable granularity,* meaning that the free space is not represented with the same detail everywhere, which is typical of the approximated cell decompositions, e.g. [106, 151]. The advantage is that fine grain descriptions are confined to the zones where a higher resolution is really needed.
- *Selectivity,* meaning that every item of geometric information should be computed only if necessary. This presumes a high organization of the geometric information and is poorly addressed in the standard approaches, which tend either to assume that the entire free space must be explicitly computed or to treat these kinds of problems by heuristic means; a noticeable exception is [132, 134].
- *Modularity* of the description of different regions of the C-space and of the refinement operations, which may be also exploited to apply parallel architectures.
- *Efficiency* of the primitive operators, which are applied several times and then should achieve low query time. Interestingly enough, collision detection, intersection detection, distance, and depth of collision, although intensely studied (e.g. [127–129, 133, 157–162]) are hardly ever considered in connection with cell decompositions.

Specifically for cell decompositions:

- *Adherence* of the cell shape to the shape of the free space (e.g., critical regions and general algebraic technique). The advantage is a reduced number of cells to describe the free space with a given accuracy, at the price of an increased complexity of the cell descriptions.

We have chosen the above properties to establish some methodological criteria and direct our research. Indeed, we think that a careful consideration of the properties of granularity, selectivity, modularity, efficiency, and adherence should allow the development of motion planning techniques (for some restricted class of problems) which are computationally tractable without losing completeness.

4. DESIGN OF A GEOMETRIC MODELER FOR ROBOTICS

Now the question to be answered is: How could a geometric modeler help to solve motion planning problems? We think that a geometric modeler suitable for applications in robotics should make available a set of efficient operations to deal with the configurations of the objects in contact, which are related to the boundary of the free space. Such operations should provide all the information necessary for the geometric step mentioned before. On the other hand, the symbolic reasoning needed to complete a planning task can be performed in a more general system, where the geometric modeler is a particular tool. As Hopcroft and Kraft point out in [163], "to describe a physical process or transformation, [such] a system must have representations for both the objects involved and the task itself."

The abstract properties of several representation schemes have been analyzed in some surveys that appeared in the last decade [1, 164–167]. On the other hand, in [163] Hopcroft and Kraft suggest founding a new science for studying how to model and reason about physical objects, not only from the geometric viewpoint.

The design of efficient techniques to compute proximity measures and properties is also of great interest in its own right, since it requires nontrivial representations of the problems. Efficient algorithms solving several proximity problems (intersection, distance, depth of collision) have been proposed for hierarchical representations [157, 158, 168]. Specific techniques for intersection detection can be found in [169, 170]. Intersection between convex polyhedra can be detected efficiently via duality transformations also in more than three dimensions [171], but the representation of polyhedra is not invariant by rotation. Moreover, depth of collision and other proximity problems with applications to motion planning are discussed in [159, 160, 172], where the proposed algorithms, although considerably less efficient (linear time), do not rest on any particular property of the representation. The best algorithms, among the mentioned ones, solve the proximity problems in $O(\log n)$ time for convex polygons in the plane and $O(\log^2 n)$ time for convex polyhedra in three dimensions using $O(n)$ storage, where n is the total number of vertices. Techniques for nonconvex bodies are proposed in [173].

As mentioned in the previous section, collision (or intersection) detection algorithms are often used in those planning systems where the desired path is found by iterating a guess-and-modify process [126–129]. Computing the distance of two objects may play a similar role, as shown in [130, 131, 161, 162, 174]. In this respect, the proposals in [133, 175] are of particular interest, both from a theoretical and an applicative point of view.

4.1. Purposes and Related Problems

What we expect from a geometric modeler is a set of geometric tools to build new objects, observe the state of the model, and change it consistently with respect to the geometric constraints. However, the important point is determining the spatial relationships among the objects. Unfortunately, a major drawback of the standard CAD systems is the lack of primitive operations aimed at recognizing the state of the model and guaranteeing consistency.

In our opinion, detecting collisions between couples of objects, under translation or rotation, seems to be a powerful tool. A sample set of primitive operations may solve the following problems:

Problem 1. Given two objects P and Q, and a direction d, find out the shortest translation of P such that P and Q share a point and a support line (plane) perpendicular to d.

Problem 2. Given two objects P and Q, and a direction d, find the collision configuration for P translating along d.

Problem 3. Given two objects P and Q, and a vector v, find out the collision configuration for P rotating around v.

The problems stated above are at the basis of the approach we will outline in the next section. An instance of any such problem, either in two- or three-dimensional space, is solved by providing a contact configuration. For our purposes, the objects are convex polygons (convex polyhedra in three dimensions), disks (spheres), or rods. The restriction to convex bodies is motivated by the need of designing efficient algorithms, since lots of collisions have to be detected in order to characterize a sufficient portion of the free space.

A key observation is that Problems 1–3 and their solutions can be interpreted in the configuration space. Think of Q as a fixed obstacle constraining the motion of P. The original obstacle Q is mapped into the region Q' that represents the forbidden configurations of P because of interference with Q. If the configurations are positions of the reference point p of a translating body P, then Q' is still a convex body, whose boundary can be probed by solving the Problems 1 and 2. On the other hand, if the coordinates of the configuration space represent rotations around a fixed joint, Q' is a monotone region [176, 177], whose boundary can be explored by solving Problem 3. Thus, an implicit representation of the C-space obstacles is available and ready to be accessed. Actually, for a certain class of planning problems, including those mentioned above, the boundaries of the C-space obstacles give rise to a monotone subdivision which can be efficiently maintained and updated by standard techniques of computational geometry [178–186] (techniques in [185, 186] are spatial subdivisions; the others are planar).

Based on the previous remarks, the analysis of contact configurations directs a cell decomposition with the nice property that, to find a solution, it is not necessary to explicitly compute the whole boundary of the C-space obstacles. Thus, we investigate the possibility of incremental characterization of the configuration space and low-cost updates for small changes in the environment, such as adding or removing obstacles. It is also interesting to notice that, although the necessity of local refinements of the configuration space is usually considered in all practical applications to avoid wasting time and storage, little work has been done to study how the update costs could be reduced.

4.2. Representation and Algorithms

The effectiveness of a planning system based on operators for detecting collisions rests on the efficiency of the algorithms implementing such operators. Therefore, a major goal of our work is the design of efficient collision detection algorithms to be applied to a world of convex polyhedra [3, 187–189]; an instance of such a problem is illustrated in Figure 7.14.

Here we spend a few words on our solution to Problem 2 of the previous subsection for two convex polyhedra P and Q. For two polygons in the plane, a simple logarithmic time solution is developed in [3]. The extension to three dimensions proposed there works under restrictive hypotheses. The approach has been successively improved leading to a general technique such that collision detection for a couple of polyhedra is reduced to collision detection for couples of planar sections and minimization of a convex function [190–192].

To develop fast algorithms in three dimensions, we exploited the properties of drum-based polyhedron representations chosen for our geometric modeler (and

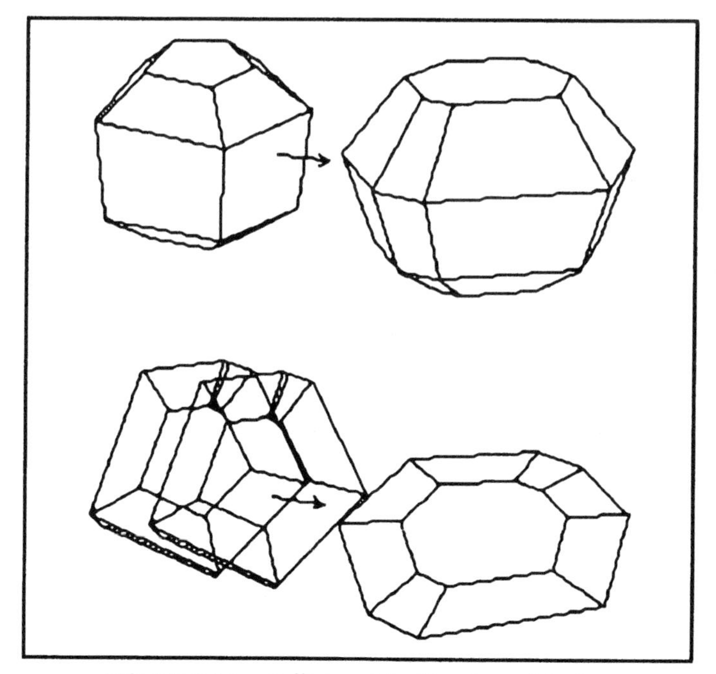

FIGURE 7.14. Collision detection for polyhedra.

already used, e.g., in [170]). Roughly speaking, a primitive solid is a preprocessed convex polyhedron whose representation is characterized by a set of parallel sections; each portion lying between two consecutive sections is a drum. (This representation is discussed and compared with different schemes in [3]). The algorithms for collision detection require $O(\log n)$ time for polygons in the plane and $O(\log^2 n)$ average time for polyhedra in the ordinary space, where n is the total number of vertices as usual. To detect collisions in polylog time, the following points are important:

- The set of primitive solids is restricted to the convex polyhedra.
- The representations are preprocessed.
- All the items of information are accessed or computed only if really necessary.

A key role is played by the binary search paradigm, also in a suitable two-dimensional version which is connected to the standard convex minimization techniques [193]. The mechanism is very simple and effective: By comparing two geometric items (e.g., two polygon sides) we try to get as much information as we can about what items are no longer worth considering to find the solution. If we carefully choose the items to be compared, then about half of the candidate items can be discarded each time. Moreover, the problems in three dimensions are reduced to simpler subproblems in the plane, whose instances are solved by binary searches in their turn. We refer the reader to the appendix for some more comments on this subject.

5. APPLICATIONS

In this section, we outline a cell decomposition approach to robot motion planning based on the availability of primitive operations to detect collision configurations for simple motions. The framework we assume is the best known one [10]. In theory, the motion planning problem is quite well understood, both because of the existence of general algebraic techniques [110, 158], which in principle apply to every conceivable situation, and because of the negative results that prove the practical intractability of the problem in its whole generality [116]. Therefore, research addresses limited instances of the problem that can be made computationally attractive.

Because of the role of local geometric properties, our work in progress is partly in the spirit of [132, 133], although we preferred a cell decomposition to a road map approach. For some problems, our treatment of cells and rotations is strongly related to the approach based on critical regions [110], in that parts of the cell boundaries are critical curves and the constraints imposed upon free orientations have a topological rather than geometric representation.

Our methods have been extended to deal with both spatial and temporal constraints for the problem of finding a free path in an environment cluttered with moving obstacles. Following the approach of [194], that decomposes the problem into path planning among stationary obstacles and velocity planning along a fixed path, we applied our techniques to both the steps [195]. (The knowledge about temporal constraints is also studied independently in [196]).

In the rest of the section we outline the grounds of our present work in motion planning. First, we define the cells and sketch the decomposition based on collision detection. Then, we discuss the property of monotonicity. Finally, we consider the problem in dynamic environments.

5.1. Cell Subdivision

We discuss the cases of translations in the plane, translations in the space, translations and rotations in the plane.[4] To this aim, recall that the solutions of the collision problems can be interpreted either in the real world or in the configuration space, according to our goals.

5.1.1. Free Translations in the Plane.

We first consider a simple navigation problem: Given a set of polygonal obstacles, compute a connected region of collision-free configurations containing both the initial and target configurations of a translating convex polygon.

Let P be the moving convex polygon and p be its reference point. Without losing generality, we represent the obstacles as a set of convex components Q_i, for $i \in \text{Obs}$. The corresponding configuration space has two dimensions, since every configuration is described by the position of p, and the transformed obstacles are convex. A cellular description of the free space is computed by first localizing each convex component Q_i' of the configuration space obstacles in a strip (instances of the *strip problem*) and then iterating a refinement step to cut off free portions of the strips while looking for a free channel (instances of the *boundary problem*). The basic problems are stated below:

- Strip problem—Build the thinnest horizontal strip containing a given configuration space obstacle Q_i'.
- Boundary problem—Find the point hit by p on the boundary of Q_i' by moving horizontally in direction x or $-x$, and build a corresponding support line.

These problems are immediately reduced to Problems 1 and 2, respectively. We can analyze the connectivity properties of the free space by decomposing it into a set of trapezoidal cells:

---

[4] In the case of translations and rotations in the plane, the state of the art is a partial characterization of the configuration space. The search for a complete technique satisfying the methodological criteria of Section 3.3 is still in progress.

Definition 1. A cell c in two dimensions is a trapezium with horizontal bases u_1u_2 and v_1v_2. The uncertainty about the actual placement of each Q_i' is represented by a couple of sets Ω_L, $\Omega_R \subseteq$ Obs.

Intuitively, the more items that are contained in Ω_L and Ω_R, the less we know about the free space in cell c. Suppose there exists a free position of p in c, then $i \in \Omega_L$ (Ω_R) if, on the basis of the available information, p might hit against Q_i' by moving left to right (right to left). From the above definition, we know that a cell is free if $\Omega_L = \Omega_R = \emptyset$. Otherwise we can apply one of two refinement operations relative to each Q_i' such that $i \in \Omega_L \cup \Omega_R$.

If $i \in \Omega_L \cap \Omega_R$, i.e., we do not know anything about the placement of Q_i' relative to c, we solve the strip problem. The two support points are the ends of a segment s completely contained in Q_i' (by convexity) and cutting the strip into two disjoint portions. Such a segment cannot be traversed by a collision-free path and gives rise to a natural decomposition of c into four cells as shown in Figure 7.15. The reduced uncertainty is represented by the updated couples corresponding to every new cell: $<\Omega_L - \{i\}, \Omega_R - \{i\}>$ for c_1 and c_2, $<\Omega_L, \Omega_R - \{i\}>$ for c_3, and $<\Omega_L - \{i\}, \Omega_R>$ for c_4.[5]

If on the other hand $i \in \Omega_L - \Omega_R$ ($i \in \Omega_R - \Omega_L$), i.e., p may hit against Q_i' by moving from left to right (from right to left), then we can solve the boundary problem with respect to direction x ($-x$) and find the side of the left (right) boundary of Q_i' intersecting a given horizontal line. The resulting decomposition of c into three cells is shown in Figure 7.16. The new couples are $<\Omega_L - \{i\}, \Omega_R>$ for c_1 and $<\Omega_L, \Omega_R>$ for c_2 and c_3.

The outline of an algorithm for planning translations in two dimensions is straightforward; its practical efficiency depends on the techniques used to find a channel and to choose the cells to be decomposed:

1. Let the only active cell c_0 be an isothetic rectangle with $\Omega_L = \Omega_R =$ Obs.
2. Look for a channel $\mathscr{C} = \{c_j \mid j \in J\}$ of active cells connecting the initial and target positions; if there is no such channel, then collision-free paths do not exist, otherwise go to Step 3.
3. If all the cells of $\mathscr{C}$ are such that $\Omega_L = \Omega_R = \emptyset$, then $\mathscr{C}$ is a solution; otherwise select and decompose one or more of the cells of $\mathscr{C}$ with either $\Omega_L \neq \emptyset$ or $\Omega_R \neq \emptyset$—the new cells are all marked active, whereas the decomposed ones are no longer active—then go to Step 2.

If the moving object P is a disk centered at p, then the grown obstacles Q_i' are still convex but no longer polygons. In fact, their boundaries can be described by

[5] In Figure 7.15 the segment s lies within c. If s intersects the walls of the cell, we can preliminarily cut c by a straight line parallel to x through each intersection point. Similar considerations apply to the other decomposition operations.

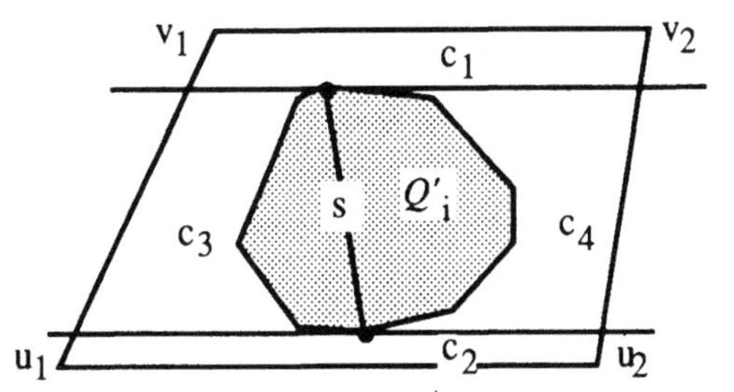

FIGURE 7.15. Cell refinement by solving the strip problem.

a finite set of straight line segments and circular arcs. Thus, we can plan free translations as before, provided we generalize the class of cells by allowing circular left and right sides. The generalization of the strip and boundary problems needs only some technical refinements.

5.1.2. Free Translations in the Space.

The properties of contacts can also be exploited in three dimensions. Assume we are able to solve one further collision problem, lying between the natural extensions of the strip and boundary problems, as required by the refinement steps mentioned below. The aim is bounding the cells by six planes.

Definition 2. A cell c in three dimensions is built by prolonging in direction z a two-dimensional cell and taking the portion between two planes not parallel to z. The uncertainty about each Q_i' is represented by four sets Ω_L, Ω_R, Ω_B, Ω_T $\subseteq$ Obs.

The sets Ω_L and Ω_R describe properties of the projection of the configuration space along z as in two dimensions; Ω_B and Ω_T have an analogous interpretation with respect to z. If $i \in \Omega_B$ (Ω_T) then p might hit Q_i' while moving bottom-up along z (top-down along $-z$). Moreover, Ω_L, $\Omega_R \subseteq \Omega_B \cap \Omega_T$. A cell is free if $\Omega_B = \Omega_T = \varnothing$. Otherwise, one of three refinement operations is possible, relative to each of the obstacles Q_i such that $i \in \Omega_B \cup \Omega_T$.

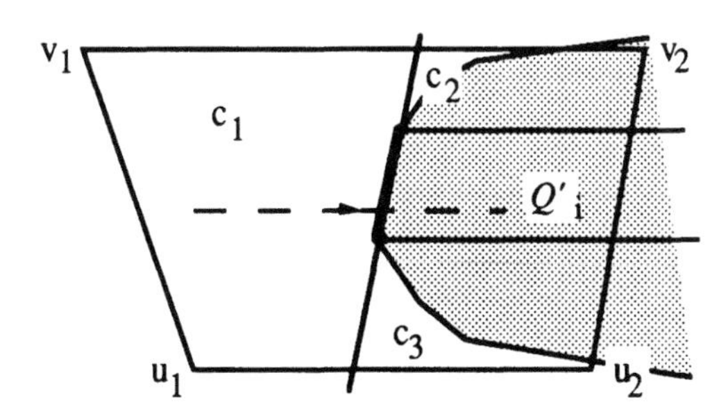

FIGURE 7.16. Cell refinement by solving the boundary problem.

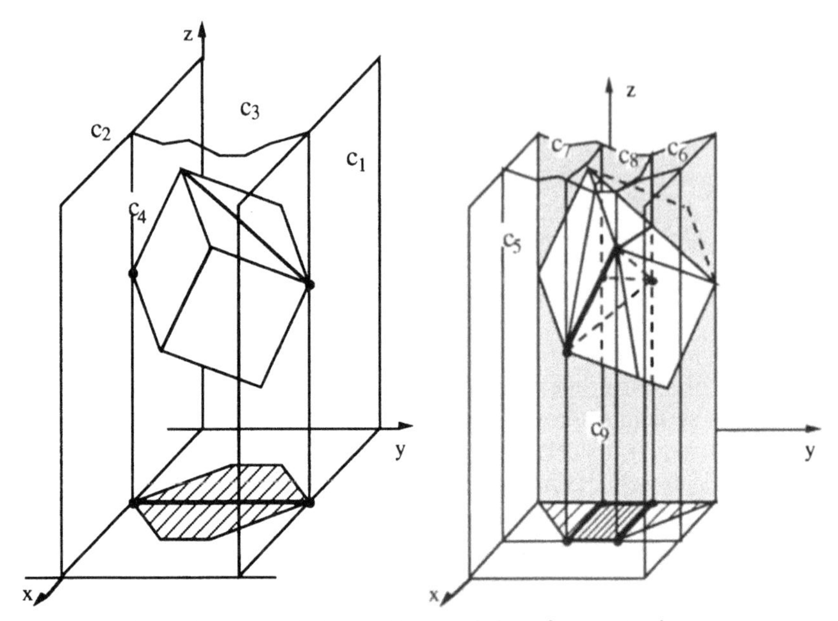

FIGURE 7.17. (a) First step; and (b) refinement of C_4.

If $i \in \Omega_L \cap \Omega_R$ we solve a strip problem for the projection into the plane $x - y$. Starting from a cell c_0 with sets of obstacles $<\Omega_L, \Omega_R, \Omega_B, \Omega_T>$, all containing i, the new cells relative to a cube Q_i' are illustrated in Figure 7.17a. Ω_L, Ω_R, Ω_B, and Ω_T are updated as follows: $<\Omega_L - \{i\}, \Omega_R - \{i\}, \Omega_B - \{i\}, \Omega_T - \{i\}>$ for c_1 and c_2, $<\Omega_L, \Omega_R - \{i\}, \Omega_B, \Omega_T>$ for c_3 and $<\Omega_L - \{i\}, \Omega_R, \Omega_B, \Omega_T>$ for c_4.

If $i \in \Omega_L - \Omega_R$ ($i \in \Omega_R - \Omega_L$) we solve a boundary problem for the projection and build planes parallel to z as shown in Figure 7.17b. That particular region whose projection lies entirely inside the projection of Q_i' is cut by suitable planes bounding a portion S of Q_i' (the convex hull of four points). S plays the role of the segment s in two dimensions. The tuples of sets are $<\Omega_L - \{i\}, \Omega_R - \{i\}, \Omega_B - \{i\}, \Omega_T - \{i\}>$ for c_5, $<\Omega_L - \{i\}, \Omega_R, \Omega_B, \Omega_T>$ for c_6 and c_7, $<\Omega_L - \{i\}, \Omega_R - \{i\}, \Omega_B - \{i\}, \Omega_T>$ for c_8 and $<\Omega_L - \{i\}, \Omega_R - \{i\}, \Omega_B, \Omega_T - \{i\}>$ for c_9.

Finally, if $i \notin \Omega_R \cup \Omega_L$ and $i \in \Omega_B - \Omega_T$ ($i \in \Omega_T - \Omega_B$) we solve the boundary problem with respect to z ($-z$). The contact face and its plane determine the decomposition of Figure 7.18. The tuples are $<\Omega_L - \{i\}, \Omega_R - \{i\}, \Omega_B - \{i\}, \Omega_T - \{i\}>$ for c_{10} and $<\Omega_L - \{i\}, \Omega_R - \{i\}, \Omega_B - \{i\}, \Omega_T>$ for c_{11}.[6]

---

[6] However, the intersection between cut planes and cell walls may require finer decompositions to preserve the geometry of the cells as established by the above definition.

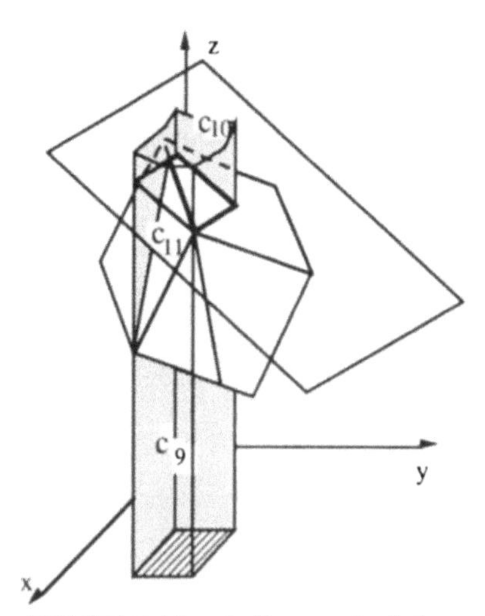

FIGURE 7.18. Refinement of C_8.

5.1.3. Rotational Information.

Consider again the planar case. In spite of their apparently translational nature, a deeper analysis of the solutions of Problems 1 and 2 adds further useful information about nonfree orientations of a moving convex polygon in the form of topological properties which are invariant within a cell. Although the configuration space is three-dimensional for the complete navigation problem, it is nevertheless possible to construct a planar cell decomposition and maintain only symbolic information on the orientations as for the critical region approach [54].

Let P be the moving convex polygon, p a reference point in its interior, D' the largest disk centered at p contained in P, and D'' the smallest disk centered at p containing P. In principle, we can think of building a characterization of the free space relative to both D' and D'', and then overlap the resulting cell decompositions. The free positions of D'' are exactly the positions where P is allowed to take every possible orientation, whereas the forbidden positions of D' are also forbidden for P. In fact, we can interleave the refinement processes relative to the two disks (we do not discuss the details here).

Moreover, call *rod* a radius of D'' connecting p with one furthest point q on the boundary of P. Of course, any forbidden configuration of pq corresponds to a forbidden configuration of P. For a given position of p, the free configurations of pq define a set of open intervals and pq cannot move continuously from one orientation to another unless the two orientations belong to the same interval. This provides information on unsafe rotations of P. On this basis, we can refine the decomposition so that in each cell the number and order of sample orientations which are nonfree for pq, and then P, are invariant. Since the actual values of nonfree orientations change at every point, such orientations have a symbolic label representing one obstacle side or vertex which is responsible for the interference with *pq*. The most important information related to the analysis of rod configurations concerns spatial adjacency. Indeed, crossing the boundary shared by two contiguous cells constrains the way in which the rod orientation can change. Partial information about nonfree orientations and spatial adjacency is implicit in the solution of Problems 1 and 2.

Of course, the properties of contacts can be analogously exploited in three dimensions, but they seem to be less informative because of the augmented number of dimensions to represent orientations.

5.2. Monotone Cell Subdivisions

A new question arises: What about different motion planning problems, such as planning the movements of a manipulator with revolute joints? We are just beginning to see a possible answer. At first glance, most of the nice features of the cell decomposition process seem to rest on convexity properties, which do not hold in the case of revolute manipulators. However, to some extent the refinement steps are based on two weaker properties, i.e., connectivity and monotonicity, meaning that the intersection of a given region with any horizontal line is either empty or a segment. Monotonicity of a configuration space obstacle intuitively means that any motion of a particular type in the original space is such that the obstacle is "entered" and "exited" at most once. Moreover, monotone regions in a space give rise to monotone decompositions, suitable to be represented by dynamic structures which allow efficient point location and updates [180–186].

Consider, for example, the model of a planar robot consisting of 2 stick links connected to each other and to the environment by revolute joints at their end points. Every possible configuration is completely described by the couple (θ_1,θ_2) of joint angles or related measures. It can be easily proved that, if the second link is not longer than the first, then every convex obstacle is mapped into a configuration space obstacle which is monotone with respect to the horizontal direction $d = (0, 1)$. Moreover, under some further weak hypotheses, the transformed obstacles are also connected.

So, there is only one thinnest strip parallel to *d*, which contains a connected and monotone configuration space obstacle Q'. By analogy with the case of translations, (1) The half-planes above and below the strip are free from Q'; and (2) there are two disjoint regions of the strip, such that Q' is hit by moving in direction *d* starting from one of them and by moving in the opposite direction starting from the other.

We finally illustrate how the strip and boundary problems generalize to the case of an arm model consisting of two stick links. Figures 7.19–7.21 show a

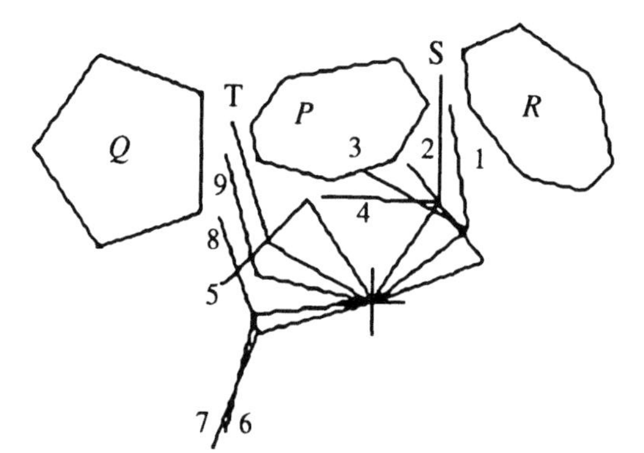

FIGURE 7.19. Planning the motion of a revolute manipulator.

sample motion planning problem and the characterization of the corresponding configuration space. The configuration where α is the angle from the vertical axis to the first link and β the angle from the ray prolonging the first link to the second link is represented by the point $(\tan(\beta/2), \tan(\alpha/2))$. All the curves drawn in Figures 7.20 and 7.21 are algebraic in the variables $x = \tan(\beta/2)$ and $y = \tan(\alpha/2)$, each variable having degrees less than or equal to 2.

Referring to Figure 7.19, we have to plan the motion of a planar manipulator from a starting configuration (S) to a target one (T) avoiding collision with three obstacles P, Q, and R. The subdivision of Figure 7.20 is the result of applying a suitable version of the strip problem to P, Q, and R. The broken curves play the same role as the segment s in Figure 7.15, whereas the points represent the configurations S and T. The partial characterization of Figure 7.21 is sufficient to provide the free motion whose intermediate configurations are numbered 1 to 9 and correspond to the real motion depicted in Figure 7.19. The new curves drawn

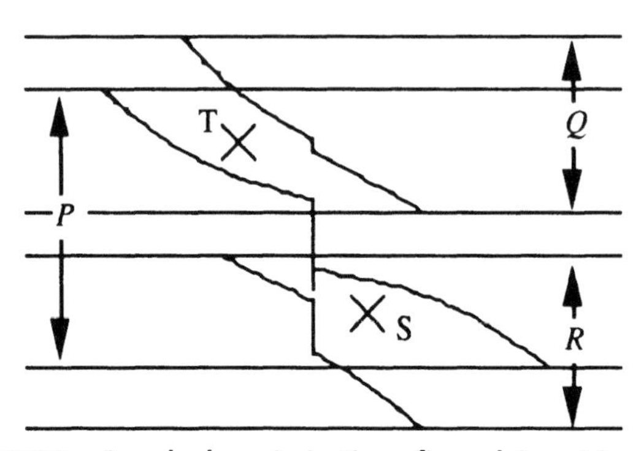

FIGURE 7.20. Rough characterization after solving strip problems.

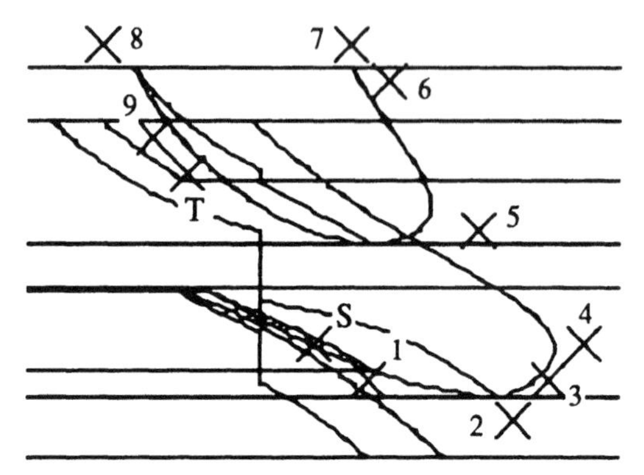

FIGURE 7.21. Partial characterization of the C-space.

in Figure 7.21 result from the analysis of the constraints imposed by sides or vertices of the obstacles after solving boundary problems based on Problem 3. Actually, only five sides and one vertex are considered.

5.3. Planning in the Space Time

Motion planning in dynamic environments has recently attracted great attention, e.g. [142, 144]. In order to try some experiments with plans in space and time, in [195] we have referred to the velocity tuning method [194], which looks very nice because of the separation between space and time information. The proposed approach is a noncomplete two-step technique for a robot translating in the plane, where there are both fixed and moving obstacles (straight motion, constant velocity): First, a few free paths are computed, relative to the fixed obstacles; then for such paths the velocity planning problem is solved to avoid collisions with the moving objects.

Collision detection can be exploited in the first step. Then, the paths avoiding collisions with the fixed obstacles are analyzed in a segment-by-segment manner to test all collisions with boxes embedding the moving obstacles. Thus, we obtain the same space time diagram as in [194]. Moreover, the velocity tuning problem is similar to the standard path planning in two dimensions (displacement and time), except that we must take care of two peculiar constraints: the irreversibility of time and the limited velocity of the robot. This constraints prevent the point robot from following arbitrary trajectories in the space time.

Different solutions for the navigation among fixed obstacles give rise to different solutions for the velocity tuning that can be compared in order to improve some features of the planned motion. Of course, we have to be careful about the main objection to the approach [194], i.e., whenever the velocity planning phase fails, the planner does not have a systematic way to generate another path among the stationary obstacles.

6. CONCLUSIONS AND PERSPECTIVES

After surveying several works involving the application of geometric reasoning to robotics, we have outlined a research in progress aimed at developing an approach based on collision detection to a class of motion planning problems. In particular, we have discussed how a few simple primitive operations allow us to get significant geometric information in order to characterize the configuration spaces of motion planning problems. The elementary problems under consideration are suitable for efficient algorithmic solutions.

Although we cannot expect to apply such an approach to problems with many degrees of freedom, some advantages are related to the possible use of dynamic structures, specifically: (1) Partial characterization of the configuration space; and (2) fast updates for local changes. Meanwhile, we are also testing the effectiveness of answering collision detection queries in heuristic strategies and for other problems such as velocity, grasp, and assembly planning.

A careful analysis concerning what can be done by symbolic computations, and what really need detailed geometric (or numeric) computations, is crucial to making further progress. However, it should be clear that geometry is not sufficient to plan the behavior of a robot at the level of task, but each subproblem should be explored from different viewpoints independently.

ACKNOWLEDGMENTS

The research described in this paper has been supported by the Italian government (CNR and MURST projects on Robotics and Automated Planning). Several people have contributed, and special thanks are due to P. Bison, C. Ferrari, E. Modolo, and L. Stocchiero.

APPENDIX. ALGORITHM IN THREE DIMENSIONS

Roughly speaking, to efficiently detect collisions between convex polyhedra, we reduce the problem to collisions of planar sections and minimization of a convex function of arity two. The representation of convex polyhedra is characterized by a list of parallel sections, such that each vertex belongs to one section and the portion between two sections defines a drum.

The bivariate convex function can be defined as follows for a couple of convex objects. Independently for each object, choose a reference point and direction and consider all the planar sections perpendicular to that direction. Then it can be easily proved that the collision translation for couples of parallel sections is a convex function of the couples of distances between the planar sections and the corresponding reference points.

The minimum of the function is the length of the collision translation, whereas the point of minimum determines the couple of sections, one for each polyhedron, containing the contact points (provided a collision configuration exists). To find the point of minimum in a finite number of steps, we observe that the topology of two parallel sections is different only if there are polyhedron vertices between them. So we call "critical" the distances from the reference point to section planes with some vertex. By drawing a horizontal line for each critical distance relative to one polyhedron and a vertical line for each critical distance relative to the other, we define a rectangular grid which partitions thc domain of the convex function. Thus, a discretized version of the minimization problem asks for the grid rectangle containing the point of minimum of the convex function. Since two sections corresponding to a point in the interior of a given grid rectangle do not contain vertices of the polyhedra, minimizing the convex function in the rectangle is equivalent to finding a collision configuration for two drums. As a consequence, the collision problem for two polyhedra is reduced to a search for the grid rectangle with the point of minimum, plus collision detection for two drums.

Most of the work aimed at finding the solution is done while searching the grid. At each step of the search, to discard a portion of the grid as large as possible we select either a pair of planar sections or a section and a drum and detect collisions relative to the selected items. On the basis of the contact configuration if the items actually collide, or of some suitable separation information if they do not, we are able to cut off a portion of the grid by a straight line. The search algorithm uses a standard minimization technique, suitably modified to take into account the discrete nature of the grid.

Thus, the three-dimensional algorithm is characterized by a planar minimization which needs $O(\log n)$ steps in the average and at each step asks for a collision configuration of two sections. The latter problem is equivalent to a planar problem and solved by an ordinary binary search. This gives the cost:

$$\log n \text{ (expected time for minimization)} \times \log n \text{ (collision detection for planar sections)} = \log^2 n \text{ (expected time)}$$

By using layered DAGs [177], the preprocessed representations of the polyhedra require only $O(n)$ storage. A detailed description of the algorithm and the proof of correctness are in [191].

REFERENCES

1. A.A.G. Requicha, "Representation for rigid solids: Theory, methods and systems," *ACM Computing Surveys*, vol. 12, No. 4, pp. 437–464, Dec. 1980.
2. J. Pertin-Troccaz, "Geometric reasoning for grasping: A computational point of view," in *CAD Based Programming for Sensory Robots*, B. Ravani Ed., *NATO-ASI Series: Computer and System Science*, Vol. 50. Berlin: Springer-Verlag, 1988, pp. 397–423.
3. C. Mirolo and E. Pagello, "A solid modeling system for robot action planning," *IEEE-CG&A*, vol. 9, No. 1, pp. 55–69, Jan. 1989.
4. E. Pagello, P. Bison, C. Mirolo, G. Perini, and G. Trainito, "A message passing approach to robot programming," *Computers in Industry*, vol. 7, No. 3, pp. 237–247, June 1986.
5. R.A. Brooks, "Symbolic reasoning among 3-D models and 2-D images," *Artificial Intelligence*, vol. 17, No. 1–3, pp. 285–348, Aug. 1981.
6. B.R. Donald, Ed., "Special issue—Computational robotics: The geometric theory of manipulation, planning, and control," *Algorithmica*, vol. 10, No. 2/3/4, pp. 91–352, Aug./Sept./Oct. 1993.
7. T. Lozano-Pérez and R.H. Taylor, "Geometric issues in planning robot tasks," in *Robotics Science*, M. Brady, Ed. Cambridge, MA: The MIT Press, 1989, pp. 227–259.
8. J.T. Schwartz and C.K. Yap, Eds., *Algorithmic and Geometric Aspects of Robotics*, vol. 1. Hillsdale, NJ: Lawrence Erlbaum Associates, 1987.
9. C. Laugier, "Geometric reasoning in motion planning," in *Geometry and Robotics*, J.D. Boissonnat and J.P. Laumond, Eds. (LCNS 391), Berlin: Springer-Verlag, 1989, pp. 377–413.
10. J.C. Latombe, *Robot Motion Planning*. Norwell, MA: Kluwer Academic Publishers, 1991.
11. J.T. Schwartz, M. Sharir, and J. Hopcroft, Eds., *Planning, Geometry, and Complexity of Robot Motion*. Norwood, NJ: Ablex Publishing Corporation, 1987.
12. J.T. Schwartz and M. Sharir, "Algorithmic motion planning in robotics," in *Algorithms and Complexity*, J. van Leeuwen, Ed., *Handbook of Theoretical Computer Science*, vol. A. Amsterdam: Elsevier, 1990. pp. 391–430.
13. M. Sharir, Ed., "Algorithmic motion planning in robotics," *Annals of Mathematics and Artificial Intelligence*, vol. 3, No. 1, pp. 1–150, Jan. 1991.
14. D. Kapur and J.L. Mundy, Eds., "Special volume on geometric reasoning (selection from Workshop on Geometric Reasoning, Oxford, 1986)," *Artificial Intelligence*, vol. 37, No. 1–3, pp. 1–412, Dec. 1988.
15. Y. Hwang and N. Ahuja, "Gross motion planning—A survey," *ACM Comp. Surveys*, vol. 24, No. 3, pp. 219–292, Sept. 1992.
16. J.D. Boissonnat and J.P. Laumond, Eds., *Geometry and Robotics* (LNCS 391). Berlin: Springer-Verlag, 1989.
17. J. Feldman, G. Falk, G. Grape, J. Pearlman, I. Sobel, and J.M. Tanebaum, "The Stanford hand-eye project," in *Proc. of the 1st Int. Joint Conf. on Artificial Intelligence*, 1969, pp. 521–526.
18. M. Salmon, "SIGLA—The Olivetti SIGMA robot programming language," in *Proc. of the 8th ISIR*, 1978, pp. 358–363.

19. R.A. Finkel, R. Taylor, R. Bolles, R. Paul, and J. Feldman, "AL—A programming system for automation," Artificial Intelligence Lab., Stanford University, Memo AIM-213, 1974.
20. L.I. Lieberman and M.A. Wesley, "AUTOPASS: An automatic programming system for computer controlled mechanical assembly," *IBM J. Research and Development,* vol. 21, No. 4, pp. 321–333, July 1977.
21. R.J. Popplestone, A.P. Ambler, and I.M. Bellos, "An interpreter for a language for describing assemblies," *Artificial Intelligence,* vol. 14, No. 1, pp. 79–107, Aug. 1980.
22. R.A. Volz, "Report of the robot programming language working group: NATO workshop," *IEEE J. of Robotics and Automation,* vol. 4, No. 1, pp. 86–90, Feb. 1988.
23. T. Lozano-Pérez and R.A. Brooks, "An approach to automatic robot programming," Massachusetts Institute of Technology AI Lab., Cambridge, MA, AI Memo No. 842. April 1985.
24. B. Frommherz and K. Hörmann, "A concept for a robot action planning system," in *NATO ASI Series,* vol. F29, *Languages for Sensor-Based Control in Robotics,* U. Rembold and K. Hörmann, Eds. Berlin, Heidelberg: Springer-Verlag, 1987, pp. 125–145.
25. C. Laugier and J. Pertin-Troccaz, "SHARP: A system for automatic programming of manipulation robots, in *3rd Int. Symp. on Robotics Research,* 1985, pp. 125–132.
26. C. Laugier and J. Pertin-Troccaz, "SHARP: A system for automatic programming of manipulation robots, in *Robotics Research, The Third Int. Symposium,* O.D. Faugera & G. Giralt, Eds. Cambridge, MA: The MIT Press, 1986, pp. 125–132.
27. T. Lozano-Pérez, J. Jones, E. Mazer, P. O'Donnell, E. Grimson, P. Tournassoud, and A. Lanusse, "Handey: A robot system that recognizes, plans, and manipulates," *Proc. of the IEEE Int. Conf. on Robotics and Automation,* 1987, pp. 843–849.
28. T. Lozano-Pérez, J.L. Jones, E. Mazer, and P.A. O'Donnell, *HANDEY—A Robot Task Planner.* Cambridge, MA: The MIT Press, 1992.
29. P. Bison, E. Pagello, and G. Trainito, "VML: An intermediate language for robot programming," *Robotics and Computer Integrated Manufacturing,* vol. 5, No. 1, pp. 11–19, Jan. 1989.
30. K. Hörmann, G. Werling, B. Frommherz, J. Hornberger, A. Pezzinga, R. Gallerini, P. Bison, C. Mirolo, E. Pagello, and L. Stocchiero, "Task level programming," in *Integration of Robots into CIM,* R. Bernhardt, R. Dillman, K. Hörmann, and K. Tierney, Eds. New York: Chapman & Hall, 1992, pp. 122–167.
31. B.O. Nnaji, *Theory of Automatic Robot Assembly and Programming.* New York: Chapman & Hall, 1993.
32. A. Bourjault, "Contribution a une approche methodologique de l'assemblage automatisé: Elaboration automatique des sequences operatoires," Université de Besançon, Besançon, France, Ph. D. Thesis, Nov. 1984.
33. T. De Fazio and D.E. Whitney, "Simplified generation of all mechanical assembly sequences," *IEEE J. of Robotics and Automation,* vol. 3, No. 6, pp. 640–658, Dec. 1987. (errata vol RA-4, No. 6, pp. 705–708, Dec. 1988.
34. A. Delchambre and P. Gaspart, "KBAP: An industrial prototype of knowledge-

based assembly planner," in *Proc. of the IEEE Int. Conf. on Robotics and Automation,* 1992, pp. 2404–2409.

35. L.S. Homen de Mello and A.C. Sanderson, "AND/OR graph representation of assembly plans," *IEEE Trans. on Robotics and Automation,* vol. 6, No. 2, pp. 188–199, Apr. 1990.
36. L.S. Homen de Mello and A.C. Sanderson, "Representations of mechanical assembly sequences," *IEEE Trans. on Robotics and Automation,* vol. 7, No. 2, pp. 211–227, Apr. 1991.
37. J. Hornberger and B. Frommherz, "Automatic generation of precedence graphs," in *Proc. of the 18th ISIR,* 1988, pp. 453–466.
38. H. Sekiguchi, K. Kojima, and K. Inoue, "Study on automatic determination of assembly sequences," *Annals of the CIRP,* vol. 32, No. 1, pp. 371–374, 1983.
39. J.D. Wolter, "On the automatic generation of assembly plans," in *Proc. of the IEEE Int. Conf. on Robotics and Automation,* 1989, pp. 62–68.
40. R.H. Wilson, J.C. Latombe, and T. Lozano-Pérez, "On the complexity of partitioning an assembly," Stanford University, Stanford, CA, Stanford Technical Report STAN-CS-92-1458, Dec. 1992; also presented at the Workshop on Assembly and Task Planning, IEEE Conf., Atlanta, GA, 1993.
41. D.F. Baldwin, "Algorithmic methods and software tools for the generation of mechanical assembly sequences," Massachusetts Institute of Technology, Cambridge, MA, S.M. Thesis, May 1988.
42. K.I. Huang, "Development of an assembly planner using decomposition approach," in *Proc. of the IEEE Int. Conf. on Robotics and Automation,* 1993, pp. 63–68.
43. S. Lee and Y.G. Shin, "Assembly planning based on subassembly extraction," in *Proc. of the IEEE Int. Conf. on Robotics and Automation,* 1990, pp. 1606–1611.
44. S. Lee, "Backward assembly planning with assembly cost analysis," in *Proc. of the IEEE Int. Conf. on Robotics and Automation,* 1992, pp. 2382–2391.
45. S. Lee and F.C. Wang, "Physical reasoning of interconnection forces for efficient assembly planning, in *Proc. of the IEEE Int. Conf. on Robotics and Automation,* 1993, pp. 307–313.
46. Y.F. Huang and C.S.G. Lee, "An automatic assembly planning system," in *Proc. of the IEEE Int. Conf. on Robotics and Automation,* 1990, pp. 1594–1599.
47. J. Tsao and J.D. Wolter, "On the automatic generation of assembly plans," in *Proc. of the IEEE Int. Conf. on Robotics and Automation,* 1993, pp. 71–76.
48. T. Lozano-Pérez and R.H. Wilson, "Assembly sequencing for arbitrary motions," in *Proc. of the IEEE Int. Conf. on Robotics and Automation,* 1993, pp. 527–532.
49. D. Nussbaum and J.R. Sack, "Disassembling two-dimensional composite parts via translations," *Int. J. of Computational Geometry and Applications,* vol. 3, No. 1, pp. 71–84, 1993.
50. R.H. Wilson and J.C. Latombe, "On the qualitative structure of a mechanical assembly," in *Proc. of the 10th AAAI Conf.,* 1992, pp. 697–702.
51. P. Bison, C. Ferrari, E. Pagello, and L. Stocchiero, "Mixing action planning and motion planning in building representations for assemblies," in *Proc. of the 24th Int. Symp. on Industrial Robots (ISIR),* 1993, pp. 137–144.
52. W.H. Qian and E. Pagello, "On the scenario and heuristics of disassembly," in *Proc. of the IEEE Int. Conf. on Robotics and Automation,* 1994, pp. 264–271.

53. J. Pertin-Troccaz, "Grasping: A state of the art," in *The Robotics Review 1*, O. Khatib, J.J. Craig, T. Lozano-Pérez, Eds. 1989. Cambridge, MA: The MIT Press.
54. M.R. Cutkosky, "On grasp choice, grasp models, and the design of hands for manufacturing tasks," *IEEE Trans. on Robotics and Automation*, vol. 5, No. 3, pp. 269–279, June 1989.
55. T. Iberall, "The nature of human prehension: Three dexterous hands in one," in *Proc. of the IEEE Conf. on Robotics and Automation*, 1987, pp. 396–401.
56. H. Liu, T. Iberall, and G.A. Bekey, "The multi-dimensional quality of task requirements for dexterous robot hand control," in *Proc. of the IEEE Int. Conf. on Robotics and Automation*, 1989, pp. 452–457.
57. D.M. Lyons, "A simple set of grasps for a dexterous hand," in *Proc. of the IEEE Int. Conf. on Robotics and Automation*, 1985, pp. 588–593.
58. T.N. Nguyen and H.E. Stephanou, "A topological algorithm for continuous grasp planning," in *Proc. of the IEEE Int. Conf. on Robotics and Automation*, 1990, pp. 670–675.
59. M.T. Mason and J.K. Salisbury, *Robot Hand and the Mechanics of Manipulation*, Cambridge, MA: The MIT Press, 1985.
60. J.R. Kerr and B. Roth, "Analysis of multifingered hands," *The Int. J. of Robotics Research*, vol. 4, No. 4, pp. 3–17, Winter 1986.
61. Y. Nakamura, K. Nagai, and T. Yoshikawa, "Mechanics of coordinate manipulation by multiple robotic mechanisms," in *Proc. of the IEEE Int. Conf. on Robotics and Automation*, 1987, pp. 991–998.
62. A.R. Grupen and R.S. Weiss, "Force domain models for multifingered grasp control," in *Proc. of the IEEE Int. Conf. on Robotics and Automation*, 1991, pp. 418–423.
63. H. Hanafusa and H. Asada, "Stable prehension by a robot hand with elastic fingers," in *Proc. of the 7th Int. Symp. on Industrial Robots*, 1977, pp. 361–368.
64. X. Markenscoff and C. Papadimitriou, "Optimum grip of a polygon," *The Int. J. of Robotics Research*, vol. 8, No. 2, pp. 17–29, Apr. 1989.
65. Z. Li and S. Sastry, "Task-oriented optimal grasping by multifingered robot hands," *IEEE Journal of Robotics and Automation*, vol. 4, No. 1, pp. 32–44, Feb. 1988.
66. Y. Park and G. Starr, "Optimal grasping using a multifingered robot hand," in *Proc. of the IEEE Int. Conf. on Robotics and Automation*, 1990, pp. 689–694.
67. T. Omata, "Fingertip positions of a multifingered hand," in *Proc. of the IEEE Int. Conf. on Robotics and Automation*, 1990, pp. 1562–1567.
68. Y. Park and G. Starr, "Grasp synthesis of polygonal objects," in *Proc. of the IEEE Int. Conf. on Robotics and Automation*, 1990, 1990, pp. 1574–1580.
69. J. Ponce and B. Faverjon, "On computing three-finger force-closure grasps of polygonal objects," in *Proc. of the 5th Int. Conf. on Advanced Robotics*, 1991, pp. 1018–1023.
70. J. Canny and C. Ferrari, "Planning optimal grasps," in *Proc. of the IEEE Int. Conf. on Robotics and Automation*, 1992, pp. 2290–2295.
71. N.S. Pollard and T. Lozano-Pérez, "Grasp stability and feasibility for an arm with an articulated hand," in *Proc. of the IEEE Int. Conf. on Robotics and Automation*, 1990, pp. 1581–1585.

72. L. Romdhane and J. Duffy, "Optimum grasp for multi-fingered hands with point contact with friction using the modified singular value decomposition," in *Proc. of the IEEE Int. Conf. on Robotics and Automation,* 1991, pp. 1614–1617.
73. T. Omata, "An algorithm for computing fingertip positions of a spatial equilibrium grasp with a multifingered hand," in *Proc. of the IEEE/RSJ-IROS'91,* 1991, pp. 711–716.
74. K. Roberts, "Coordinating a robot arm and multi-finger hand using the quaternion representation," in *Proc. of the IEEE Int. Conf. on Robotics and Automation,* 1990, pp. 1252–1257.
75. B.S. Baker, S. Fortune, and E. Grosse, "Stable prehension with three fingers," in *Proc. of the 17th Annual Symposium on Theory of Computing, SIGACT,* 1985, pp. 114–120.
76. J. Barber, R. Volz, R. Desai, R. Runinfeld, B. Schipper, and J. Wolter, "Automatic two-fingered grip selection," in *Proc. of the IEEE Int. Conf. on Robotics and Automation,* 1986, pp. 890–896.
77. V.D. Nguyen, "Constructing force-closure grasps in 3D," in *Proc. of the IEEE Int. Conf. on Robotics and Automation,* 1987, pp. 240–245.
78. V.D. Nguyen, "Constructing force-closure grasps," *The Int. Journal of Robotics Research,* vol. 7, No. 3, pp. 3–16, June 1988.
79. V.D. Nguyen, "Constructing stable grasps in 3D," in *Proc. of the IEEE Int. Conf. on Robotics and Automation,* 1987, pp. 234–239.
80. W. Meyer, "Seven fingers allow force-torque closure grasps on any convex polyhedron," *Algorithmica,* vol. 9, No. 3, pp. 278-292, Mar. 1993.
81. B. Mishra, J.T. Schwartz, and M. Sharir, "On the existence of multifinger positive grips," Courant Institute of Mathematical Sciences, New York, Dept. Computer Sci., Tech. Rep. No. 259, Nov. 1989.
82. M. Brady, "Artificial intelligence and robotics," *Artificial Intelligence,* vol. 26, No. 1, pp. 79–121, Apr. 1985.
83. P.H. Winston, T.O. Binford, B. Katz, and M. Lowry, "Learning physical descriptions from functional definitions, examples, and precedents," in *Robotics Research: The First Int. Symposium,* M. Brady and R. Paul Eds. Cambridge, MA: The MIT Press, 1984, pp. 117–135.
84. M. Di Manzo, E. Trucco, F. Giunchiglia, and F. Ricci, "FUR: Understanding functional reasoning," *Int. Journal of Intelligent Systems,* vol. 4, No. 4, pp. 431–457, Winter 1989.
85. P. Bison, C. Ferrari, E. Pagello, and L. Stocchiero, "A heuristic approach to automatic grasp planning for a 3-fingered hand, *Journal of Intelligent and Robotic Systems,* to appear.
86. P. Bison, C. Ferrari, E. Pagello, and L. Stocchiero, "Using sensor data and a priori knowledge for planning and monitoring multifingered hand grasping," in *Proc. of the Int. Workshop on Sensorial Integration for Industrial Robots,* 1989, pp. 156–161.
87. J.C. Latombe, "Motion planning with uncertainty: On the preimage backchaining approach," in *The Robotics Review 1,* O. Khatib, J.J. Craig, and T. Lozano-Pérez, Eds. Cambridge, MA: The MIT Press, 1989, pp. 53–69.
88. C.M. Hoffmann, "The problems of accuracy and robustness in geometric computation," *IEEE Computer,* vol. 22, No. 3, pp. 31–41, March 1989.

89. I. Mazon, "Modelling positioning uncertainties," in *Geometry and Robotics*, J.D. Boissonnat and J.P. Laumond, Eds. Berlin: Springer-Verlag, 1989, pp. 336–360.
90. T. Lozano-Pérez, M.T. Mason, and R.H. Taylor, "Automatic synthesis of fine-motion strategies for robots," the *Int. J. Robotics Research*, vol. 3, No. 1, pp. 3–24, Spring 1984.
91. B.R. Donald, *Error Detection and Recovery in Robotics* (LNCS 336). Berlin: Springer-Verlag, 1987.
92. B.R. Donald, "A geometric approach to error detection and recovery for robot motion planning with uncertainty," *Artificial Intelligence*, vol. 37, No. 1–3, pp. 222–271, Dec. 1988.
93. H.F. Durrant-Whyte, "Uncertain Geometry," in *Geometric Reasoning*, D. Kapur and J.L. Mundy, Eds. Cambridge, MA: The MIT Press, 1989, pp. 447–481.
94. C. Laugier, "Planning fine motion strategies by reasoning in the contact space," in *Proc. of the IEEE Int. Conf. on Robotics and Automation*, 1989, pp. 653–659.
95. S.N. Gottschlich and A.C. Kak, "AMP-CAD: An assembly motion planning system," in *Proc. of the IEEE Int. Conf. on Robotics and Automation*, 1992, pp. 2355–2360.
96. K.Y. Goldberg, "Orienting polygonal parts without sensors," *Algorithmica*, vol. 10, No. 2/3/4, pp. 201–225, Aug./Sept./Oct. 1993.
97. J.F. Canny and J. Reif, "New lower bound techniques for robot motion planning problems," in *Proc. of the 28th IEEE Symp. on Foundations of Computer Science*, 1987, pp. 49–60.
98. J.L. Crowley, "Coordination of action and perception in a surveillance robot," *IEEE Expert, Intelligent Systems and their Applications*, vol. 2, No. 4, pp. 32–43, Winter 1987.
99. H. Moravec, "Sensor fusion in certainty grids for mobile robots," *AI Magazine*, vol. 9, No. 2, pp. 61–74, Summer 1988.
100. V.J. Lumelsky, "Effect of kinematics on motion planning for planar robot arms moving amidst unknown obstacles," *IEEE Journal of Robotics and Automation*, vol. RA-3, No. 3, pp. 207–223, June 1987.
101. T. Dean, "Tradeoffs in spatial reasoning," presented at the Workshop on the Integration of AI into Robotics, IEEE Int. Conf. on Robotics and Automation, Phoenix, AZ, May 1989.
102. R.A. Brooks, "A robust layered control system for mobile robot," *IEEE Journal of Robotics and Automation*, vol. 2, No. 1, pp. 14–23, March 1986.
103. S. Badaloni, E. Pagello, and C. Sossai, "An autonomous mobile robot in a temporally rich domain: Some considerations from the standpoint of AI," in *Proc. of the 2nd Int. Conf. on Intelligent Autonomous Systems*, 1989, pp. 672–682.
104. T. Lozano-Pérez, "Spatial planning: A configuration space approach," *IEEE Trans. on Computers*, vol. C-32, No. 2, pp. 108–120, Feb. 1983.
105. T. Lozano-Pérez, "Automatic planning of manipulator transfer movements," *IEEE Trans. on Systems, Man, and Cybernetics*, vol. 11, No. 10, pp. 681–698, Oct. 1981.
106. R.A. Brooks and T. Lozano-Pérez, "A subdivision algorithm in configuration space for findpath with rotation," *IEEE Trans. on Systems, Man, and Cybernetics*, vol. 15, No. 2, pp. 224–233, March-April 1985.

107. B.R. Donald, "A search algorithm for motion planning with six degrees of freedom," *Artificial Intelligence*, vol. 31, No. 3, pp. 295–353, March 1987.
108. R.A. Brooks, "Solving the find-path problem by good representation of free space," *IEEE Trans. on Systems, Man, and Cybernetics*, vol. 13, No. 3, pp. 190–197, March-April 1983.
109. C.K. Yap, "Algorithmic motion planning," in *Algorithmic and Geometric Aspects of Robotics*, vol. 1, J.T. Schwartz and C.K. Yap, Eds. Hillsdale, NJ: Lawrence Erlbaum Associates, 1987, pp. 95–143.
110. J.T. Schwartz and M. Sharir, "On the piano mover's problem II: General techniques for computing topological properties of real algebraic manifolds," *Advances in Applied Mathematics*, vol. 4(3), pp. 298–351, Sept. 1983 also in [11].
111. A. Tarski, *A Decision Method for Elementary Algebra and Geometry*. Berkeley, CA: University of California Press, 1948.
112. G.E. Collins, "Quantifier elimination for real closed fields by cylindrical algebraic decomposition," in *Proc. of the Second GI Conference on Automata Theory and Formal Languages* (LNCS 33). Berlin: Springer-Verlag, 1975, pp. 134–183.
113. D.S. Arnon, G.E. Collins, and S. McCallum, "Cylindrical algebraic decomposition I: The basic algorithm," *SIAM Journal on Computing*, vol. 13, No. 4, pp. 865–889, Nov. 1984.
114. D.S. Arnon, "Geometric reasoning with logic and algebra," *Artificial Intelligence (Special Volume on Geometric Reasoning)*, vol. 37, No. 1–3, pp. 37–60, Dec. 1988.
115. J.T. Schwartz and M. Sharir, "On the piano mover's problem V: The case of a rod moving in three dimensional space amidst polyhedral obstacles," *Communications on Pure and Applied Mathematics*, vol. 37, No. 6, pp. 815–848, 1984 also in [11].
116. J.H. Reif, "Complexity of the generalized mover's problem," in *Proc. of the 20th IEEE Symp. on Foundations of Computer Science*, 1979, pp. 421–427 also in [11].
117. J.E. Hopcroft, D. Joseph, and S. Whitesides, "Movement problems for 2-dimensional linkages," *SIAM J. on Computing*, vol. 13, No. 3, pp. 610–629, Aug. 1984.
118. J. Canny, *The Complexity of Robot Motion Planning*. Cambridge, MA: The MIT Press, 1988.
119. K. Kedem and M. Sharir, "An efficient motion-planning algorithm for a convex polygonal object in two-dimensional polygonal space," *Discrete Comput. Geom.*, vol. 5, No. 1, pp. 43–75, Feb. 1990.
120. K. Kedem, R. Livne, J. Pach, and M. Sharir, "On the union of Jordan regions and collision-free motion amidst polygonal obstacles," *Discrete Comput. Geom.*, vol. 1, No. 1, pp. 59–71, Spring 1986.
121. D. Leven and M. Sharir, "Planning a purely translational motion for a convex object in two-dimensional space using generalized Voronoi diagrams," *Discrete Comput. Geom.*, vol. 2, No. 1, pp. 9–31, Spring 1987.
122. M. Sharir and A. Schorr, "On shortest paths in polyhedral spaces," *SIAM J. on Computing*, 15, pp. 193–215, Feb. 1986.
123. J.H. Reif and J.A. Storer, *Three-Dimensional Shortest Paths in the Presence of Polyhedral Obstacles* (LNCS 324). Berlin: Springer-Verlag, 1988, pp. 85–92.
124. J. Chen and Y. Han, "Shortest paths on a polyhedron," in *Proc. of the 6th ACM Symp. on Computational Geometry*, 1990, pp. 360–369.

125. J.H. Reif and J.A. Storer, "Shortest paths in Euclidean space with polyhedral obstacles," Computer Science Department, Brandeis University, Waltham, MA, Tech. Rep. CS-85-121, 1985.
126. S. Cameron and J. Aylett, "ROBMOD: A geometric engine for robotics," in *Proc. of the IEEE Int. Conf. on Robotics and Automation*, 1988, pp. 880–885.
127. S. Cameron, "Collision detection by four dimensional intersection testing" *IEEE Trans. on Robotics and Automation,* vol. 6, No. 3, pp. 291–302, June 1990.
128. S. Bonner and R.B. Kelley, "A novel representation for planning 3-D collision-free paths," *IEEE Trans. on Systems, Man, and Cybernetics*, vol. 20, No. 6, pp. 1337–1351, Dec. 1990.
129. J. Canny, "Collision detection for moving polyhedra" *IEEE Trans. on Pattern Analysis and Machine Intelligence,* vol. 8, No. 2, pp. 200–209, March 1986.
130. E.G. Gilbert and C.P. Foo, "Computing the distance between general convex objects in three-dimensional space," *IEEE Trans. on Robotics and Automation,* vol. 6, No. 1, pp. 53–61, Feb. 1990.
131. S. Cameron and R.K. Culley, "Determining the minimum translational distance between two convex polyhedra," in *Proc. of the IEEE Int. Conf. on Robotics and Automation,* 1986, pp. 591–596.
132. J. Canny and M.C. Lin, "An opportunistic global path planner," in *Proc. of the IEEE Int. Conf. on Robotics and Automation,* 1990, pp. 1554–1559.
133. M.C. Lin and J. Canny, "A fast algorithm for incremental distance calculation," in *Proc. of the IEEE Int. Conf. on Robotics and Automation,* 1991, pp. 1008–1014.
134. J. Canny and M.C. Lin, "An opportunistic global path planner," *Algorithmica,* vol. 10, No. 2/3/4, pp. 102–120, Aug./Sept./Oct. 1993.
135. O. Khatib, "Real-time obstacle avoidance for manipulators and mobile robots," *the Int. J. Robotics Research,* vol. 5, No. 1, pp. 90–98, Spring 1986.
136. E. Rimon and D.E. Koditschek, "Exact robot navigation using artificial potential functions," *IEEE Trans. on Robotics and Automation,* vol. 8, No. 5, pp. 501–518, Oct. 1992.
137. J. Cox and C.K. Yap, "On-line motion planning: Case of a planar rod," *Annals of Math. and Artificial Intelligence,* vol. 3, No. 1, pp. 1–20, June 1991.
138. E. Rimon and J.F. Canny, "Construction of C-space roadmaps from local sensory data—What should the sensors look for?" in *Proc. of the IEEE Int. Conf. on Robotics and Automation,* 1994, pp. 117–123.
139. J. Barraquand and J.C. Latombe, "A Montecarlo algorithm for path planning with many degrees of freedom" in *Proc. of the IEEE Int. Conf. on Robotics and Automation,* 1990, pp. 1554–1559.
140. L. Kavraki and J.C. Latombe, "Randomized preprocessing of configuration space for fast path planning," in *Proc. of the IEEE Int. Conf. on Robotics and Automation,* 1994, pp. 2138–2145.
141. J.H. Reif and M. Sharir, "Motion planning in the presence of moving obstacles," in *Proc. of the 26th IEEE Symp. on Foundations of Computer Science,* 1985, pp. 144–154.
142. K. Fujimura and H. Samet, "Planning a time-minimal motion among moving obstacles," *Algorithmica,* vol. 10, No. 1, pp. 41–63, July 1993.
143. K. Fujimura, *Motion Planning in Dynamic Environments.* Berlin: Springer Verlag, 1991.

144. K.C. Fan and P.C. Lui, "Solving find path problem in mapped environments using modified A* algorithm," *IEEE Trans. on Systems, Man, and Cybernetics,* vol. 24, No. 9, pp. 1390–1396, Sept. 1994.
145. Z. Li and J.F. Canny, Eds., *Nonholonomic Motion Planning.* Norwell, MA: Kluwer Academic Publishers, 1993.
146. J.P. Laumond, "Feasible trajectories for mobile robots with kinematic and environment constraints," in *Proc. of the International Conference on Intelligent Autonomous Systems,* 1986, pp. 346–354.
147. J. Barraquand and J.C. Latombe, "Nonholonomic multibody mobile robots: Controllability and motion planning in the presence of obstacles," *Algorithmica,* vol. 10, No. 2/3/4, pp. 121–155, Aug./Sept./Oct. 1993.
148. T. Lozano-Pérez and M.A. Wesley, "An algorithm for planning collision-free paths among polyhedral obstacles," *Comm. of the ACM,* vol. 22, No. 10, pp. 560–570, Oct. 1979.
149. R.A. Brooks, "Planning collision-free motions for pick-and-place operations," The *Int. J. of Robotics Research,* vol. 2, No. 4, pp. 19–44, Winter 1983.
150. T. Lozano-Pérez, "A simple motion planning algorithm for general robot manipulators," *IEEE J. of Robotics and Automation,* vol. 3, No. 3, pp. 224–238, June 1987.
151. B. Faverjon, "Obstacle avoidance using an octree in the configuration space of a manipulator," in *Proc. of the IEEE Int. Conf. on Robotics and Automation,* 1984, pp. 504–510.
152. B. Faverjon and P. Tournassoud, "Object level programming of industrial robots," in *Proc. of the IEEE Int. Conf. on Robotics and Automation,* 1986, pp. 1406–1411.
153. B. Faverjon and P. Tournassoud, "Motion planning for manipulators in complex environments," in *Geometry and Robotics* (LNCS 391), J.D. Boissonnat and J.P. Laumond, Eds. Berlin: Springer-Verlag, 1989, pp. 87–115.
154. J.T. Schwartz and M. Sharir, "On the piano mover's problem I: The case of a polygonal body moving amidst polygonal barriers," *Communications on Pure and Applied Mathematics,* vol. 36, No. 3, pp. 345–398, May 1983 also in [11].
155. C. O'Dunlaing, M. Sharir, and C.K. Yap, "Retraction: A new approach to motion planning," in *Proc. of the 15th Symp. on the Theory of Computing,* 1983, pp. 207–220 also in [11].
156. C. O'Dunlaing and C.K. Yap, "A retraction method for planning the motion of a disk," *Journal of Algorithms,* vol. 6, No. 1, pp. 104–111, March 1985, also in [11].
157. D. Dobkin, J. Hershberger, D. Kirkpatrick, and S. Suri, "Implicitly searching convolutions and computing depth of collision," in *Proc. of SIGAL* (LNCS 450), 1990, pp. 165–180.
158. D. Dobkin, J. Hershberger, D. Kirkpatrick, and S. Suri, "Computing the intersection depth of polyhedra," *Algorithmica,* vol. 9, No. 6, pp. 518–533, June 1993.
159. N.K. Sancheti and S.S. Keerthi, "Computation of certain measures of proximity between convex polytopes: A complexity viewpoint," in *Proc. of the IEEE Int. Conf. on Robotics and Automation,* 1992, pp. 2508–2513.
160. K. Sridharan and H.E. Stephanou, "Algorithms for rapid computation of some

distance functions between objects for path planning," in *Proc. of the IEEE Int. Conf. on Robotics and Automation,* 1994, pp. 967–972.

161. E.G. Gilbert and C.J. Ong, "New distances for the separation and penetration of objects," in *Proc. of the IEEE Int. Conf. on Robotics and Automation,* 1994, pp. 579–585.
162. J.E. Bobrow, "A direct minimization approach for obtaining the distance between convex polyhedra," *The Int. J. of Robotics Research,* vol. 8, No. 3, pp. 65–76, June 1989.
163. J.E. Hopcroft and D.B. Kraft, "The challenge of robotics for computer science," in *Algorithmic and Geometric Aspects of Robotics,* vol. 1, J.T. Schwartz and C.K. Yap, Eds. Hillsdale, NJ: Lawrence Erlbaum Associates, 1987, pp. 7–42.
164. A.A.G. Requicha and H.B. Voelcker, "Solid modelling: A historical summary and contemporary assessment," *IEEE-CG&A,* vol. 2, No. 2, pp. 9–24, March 1982.
165. A.A.G. Requicha and H.B. Voelcker, "Solid modelling: Current status and research directions," *IEEE-CG&A,* vol. 3, No. 7, pp. 25–37, Oct. 1983.
166. B. Bhanu and C.C. Ho, "CAD-Based 3D Object Representation for Robot Vision," *Computer,* vol. 20, No. 8, pp. 19–35, Aug. 1987.
167. A.A.G. Requicha and H.B. Voelcker, "Solid modelling—A 1988 update," in *CAD Based Programming for Sensory Robots,* B. Ravani, Ed., *NATO-ASI Series: Computer and System Science,* vol. 50. Berlin: Springer-Verlag, 1988, pp. 3–22.
168. D.P. Dobkin and D.G. Kirkpatrick, "Determining the separation of preprocessed polyhedra—A unified approach," in *Proc. of ICALP* (LNCS 443), 1990, pp. 400–413.
169. D.P. Dobkin and D.G. Kirkpatrick, "Fast detection of polyhedral intersection," *Theoretical Computer Science,* vol. 27, No. 3, pp. 241–253, Dec. 1983.
170. B. Chazelle and D.P. Dobkin, "Intersection of convex objects in two and three dimensions," *Journal of the ACM,* vol. 34, No. 1, pp. 1–27, Jan. 1987.
171. O. Günther and E. Wong, "A dual approach to detect polyhedral intersections in arbitrary dimensions," *Bit,* vol. 31, No. 1, pp. 2–14, Spring 1991.
172. K. Sridharan, H.E. Stephanou, and S.S. Keerthi, "On computing a distance measure for path planning, in *Proc. of the IEEE Int. Conf. on Robotics and Automation,* 1993, pp. 554–559.
173. F. Thomas and C. Torras, "Interference detection between non convex polyhedra revisited with a practical aim," in *Proc. of the IEEE Int. Conf. on Robotics and Automation,* 1994, pp. 587–594.
174. E.G. Gilbert, D.W. Johnson, and S.S. Keerthi, "A fast procedure for computing the distance between complex objects in three-dimensional space," *IEEE J. of Robotics and Automation,* vol. 4, No. 1, pp. 193–203, Apr. 1988.
175. M.C. Lin, D. Manocha, and J. Canny, "Fast contact determination in dynamic environments, in *Proc. of the IEEE Int. Conf. on Robotics and Automation,* 1994, pp. 602–608.
176. F.P. Preparata, "A new approach to planar point location," *SIAM Journal on Computing,* vol. 10, pp. 473–483, Aug. 1981.
177. H. Edelsbrunner, L.J. Guibas, and J. Stolfi, "Optimal point location in a monotone subdivision," *SIAM J. on Computing,* vol. 15, No. 2, pp. 317–340, May 1986.
178. H. Edelsbrunner, *Algorithms in Combinatorial Geometry.* Berlin: Springer-Verlag, 1987.

179. F.P. Preparata and M.I. Shamos, *Computational Geometry—An Introduction.* New York: Springer-Verlag, 1985.
180. M.T. Goodrich and R. Tamassia, "Dynamic trees and dynamic point location," in *Proc. of the 23rd ACM Symp. on Theory of Computing,* 1991, pp. 523–533.
181. Y.J. Chiang and R. Tamassia, "Dynamization of the trapezoid method for planar point location in monotone subdivisions," *Int. J. of Computational Geometry and Applications,* vol. 2, No. 3, pp. 311–333, Sept. 1992.
182. S.W. Cheng and R. Janardan, "New results on dynamic planar point location," *SIAM J. on Computing,* vol. 21, No. 5, pp. 972–999, Oct. 1992.
183. R. Tamassia, "An incremental reconstruction method for dynamic planar point location," *Information Processing Letters,* vol. 37, No. 2, June 1991, pp. 79–83.
184. F.P. Preparata and R. Tamassia, "Fully dynamic point location in a monotone subdivision," *SIAM J. on Computing,* vol. 18, No. 4, pp. 811–830, Aug. 1989.
185. F.P. Preparata and R. Tamassia, *Efficient Spatial Point Location* (LNCS 382). Berlin: Springer-Verlag, pp. 3–11, 1989.
186. F.P. Preparata and R. Tamassia, "Efficient point location in a convex spatial cell complex," *SIAM J. on Computing,* vol. 21, No. 2, pp. 267–280, Apr. 1992.
187. C. Mirolo and E. Pagello, "Motion planning algorithms for mechanical assemblies," in *Preprints of IFAC Symp. on Robot Control SYROCO-88,* 1988, pp. 42.1–42.6.
188. C. Mirolo and E. Pagello, "Local geometric issues for spatial reasoning in robot motion planning," in *Proc. of the IEEE/RSJ-IROS '91,* 1991, 569–574.
189. C. Mirolo and E. Pagello, "Robot motion planning based on collision detection," in *Proc. of the 3rd International Workshop on Robotics in Alpe-Adria Region RAA'94,* July 1994, pp. 187–192.
190. C. Mirolo, "Geometric algorithms for detecting collision between convex polyhedra," Dept. of Mathematics and Computer Science, University of Udine, Udine, Italy, Tech. Rep. UDMI/01/91/RR, Jan. 1991.
191. C. Mirolo, "Polylogarithmic algorithms for collision detection based on convex minimization," Dept. of Mathematics and Computer Science, University of Udine, Udine, Italy, Tech. Rep. UDMI/01/94/RR, Jan. 1994.
192. C. Mirolo, "Convex minimization on a grid and applications," in *Proc. of the 6th Canadian Conference on Computational Geometry, CCCG'94,* Aug. 1994, pp. 314–319.
193. A.S. Nemirovsky and D.B. Yudin, *Problem Complexity and Method Efficiency in Optimization.* New York: Wiley, 1983.
194. K. Kant and S.W. Zucker, "Towards efficient trajectory planning: The path-velocity decomposition," *The Int. Journal of Robotics Research,* vol. 5, No. 3, pp. 72–89, Fall 1986.
195. E. Modolo and E. Pagello "Collision avoidance detection in space and time planning for autonomous robots," in *Proc. of the 3rd Int. Conf. on Intelligent Autonomous Systems,* 1993, pp. 216–225.
196. S. Badaloni, E. Pagello, L. Stocchiero, and A. Zanardi, "Planning temporally qualified robot actions," in *Proc. of the 6th Int. Conf. on Advanced Robotics (ICAR),* 1993, pp. 685–700.

8

Evolution of Standard Fieldbus Networks

Gianluca Cena and Luca Durante
Dip. di Automatica e Informatica,
Politecnico di Torino,
Corso Duca degli Abruzzi, 24 - 10129 Torino, Italy

Adriano Valenzano
Centro per l'Elaborazione Numerale dei Segnali

1. INTRODUCTION

Until the late 1970s communication in the factory environment was based, where it existed, on proprietary solutions; field devices could not be monitored or programmed from a remote station, different cells could not communicate with each other, and so on. In such a situation neither open solutions nor integration could be achieved.

The first attempt to define a standard protocol for field communication in the factory environment was made at the beginning of 1980s [1]. It originated in a joint effort of the International Electrotechnical Committee (IEC) and the Instrumentation Society of America (ISA). In order to facilitate progress and to ensure that the two standards were developed at the same rate, joint IEC/ISA meetings were held. The project is known as IEC 1158/ISA SP 50 Fieldbus. The slow rate at which the international standardization activity was progressing and the growing needs of the industrial users led some groups of companies operating in the industrial automation field to define their own protocols mainly for factory communication and field communication.

From among the proprietary proposals there are two which are particularly interesting: the FIP (Field Instrumentation Protocol) proposal [2], sponsored by a team of French companies, and the PROFIBUS (PROcess FIeldBUS) proposal [3], supported by a team of German companies.

The PROFIBUS and FIP teams developed components and/or products satisfying their norms and also defined a suitable protocol. This commercial strategy facilitated the diffusion of the FIP and PROFIBUS protocols which have become national standards in France and Germany, respectively. In such a way these two protocols have begun to influence IEC/ISA Fieldbus international standardization activity; at the moment fieldbus has not yet been completely standardized but many of its mechanisms are derived from FIP or PROFIBUS or both.

It is for these reasons that this chapter considers those three protocols, even though many other proposals and field communication models have been proposed during the last few years.

The chapter is structured as follows: Section 2 presents a brief history of factory communications; Section 3 describes the requirements which fieldbuses must exhibit and how they differ from other kinds of networks such as Local Area Networks (LANs). Sections 4 and 5 describe the FIP and PROFIBUS protocols, respectively, while Section 6 describes some stable aspects of the IEC/ISA Fieldbus protocol. Section 7 briefly compares the three protocols, keeping in mind the requirements explained in Section 3.

2. EVOLUTION OF THE INDUSTRIAL COMMUNICATION SYSTEMS

Interest in industrial computer networks has significantly increased in the last decade because networks are considered the primary way of simplifying the transfer of information and of achieving the degree of integration needed by computer-integrated manufacturing (CIM) systems.

The demand for communication capabilities in the industrial field has led private and public organizations, together with standardization bodies, to spend a lot of time, resources, and effort in developing suitable communication standards able to satisfy the numerous requirements typical of process control and manufacturing automation environments.

According to a widely accepted model [4, 5], communication in an industrial system can be organized as a hierarchy of three types of networks (see Figure 8.1), each one having different goals and communication capabilities. Type one networks are to be used to interconnect machines which perform tasks such as manufacturing engineering, production management, resource allocation, and so on, while type two networks are designed to be used with cell controllers, milling, inspection, and control workstations in manufacturing plants. At the lowest level in the hierarchy, type three networks (also called fieldbuses) are used to connect equipment controllers, sensors, actuators, and less intelligent devices. Attention is focused mainly on the two lower layers of the above-mentioned hierarchy, that is to say on the cell and the instrumentation (or field) levels.

A clear and stable definition of the communication architecture needed to support modern automation systems, and the subdivision of the communication capabilities according to the hierarchical model proposed, appeared only recently. In fact, a few years ago, the communications scenario in an automated factory was limited by the technology used to provide an easy and effective exchange of data and information. In particular, communication among controllers, sensors,

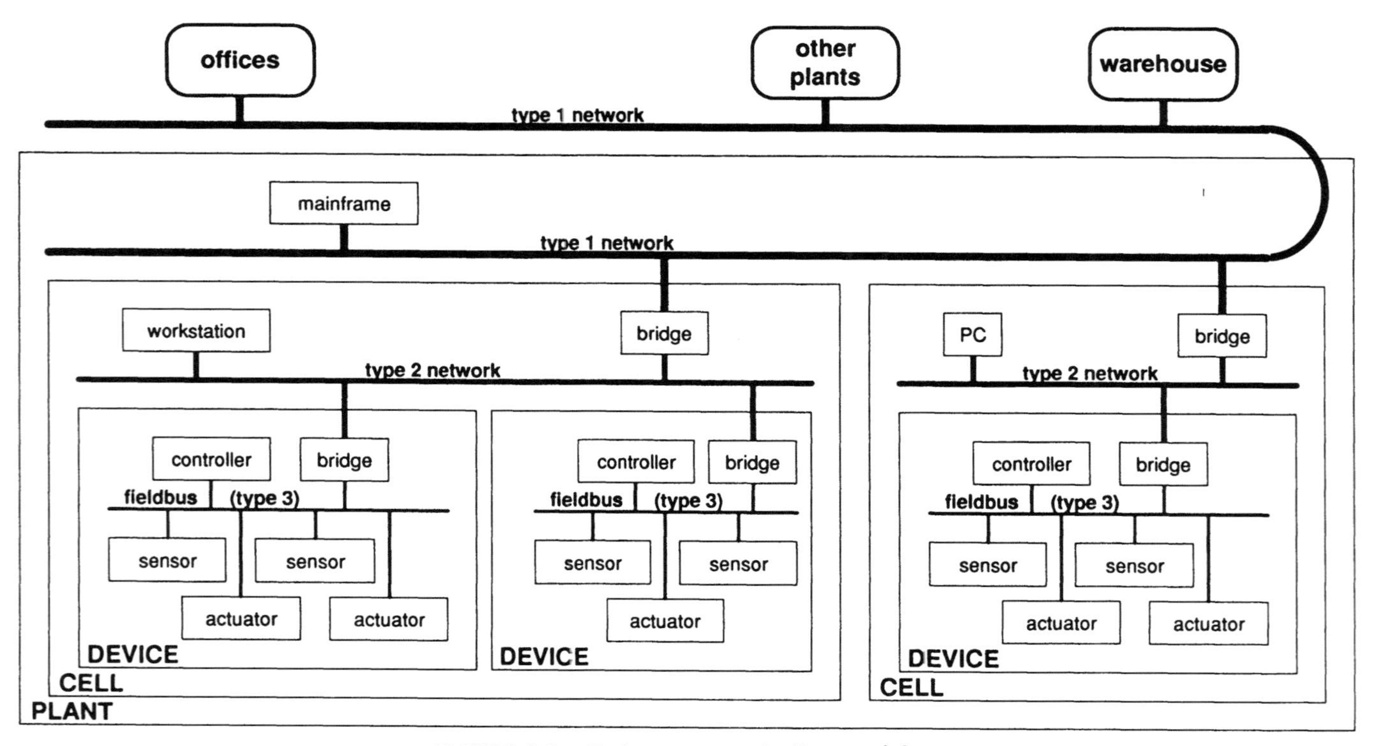

FIGURE 8.1. Factory communication model.

and actuators took place through the use of bundles of analog links. Very often ad hoc solutions were adopted, so that when a component had to be replaced it was necessary to introduce substantial modifications to the overall system.

At the cell level the situation was even worse in that information was mainly transferred manually by human operators from one automated island to the others. This kind of transfer was accomplished by actually moving the storage medium, for example punched or magnetic tapes. The advantages of using a communication network to perform such tasks was not always sufficiently clear to either the users or the manufacturers. Further, information was coded by using proprietary and incompatible formats, thus making it very difficult or impossible to exchange data between equipment manufactured by different companies. These factors led to old system architectures such as the one depicted in Figure 8.2.

The impressive evolution of digital transmission techniques and the strong conviction that their use could offer significant advantages in the factory environment led companies manufacturing devices to modify their equipment to take advantage of emerging technologies. Thus, simple digital point-to-point connections based on the well-known 20-mA current loop were quickly introduced to replace analog links. Though this kind of interface did not allow factory activity to be integrated completely, it was the first important step toward understanding the importance of adopting common and agreed-upon interfaces.

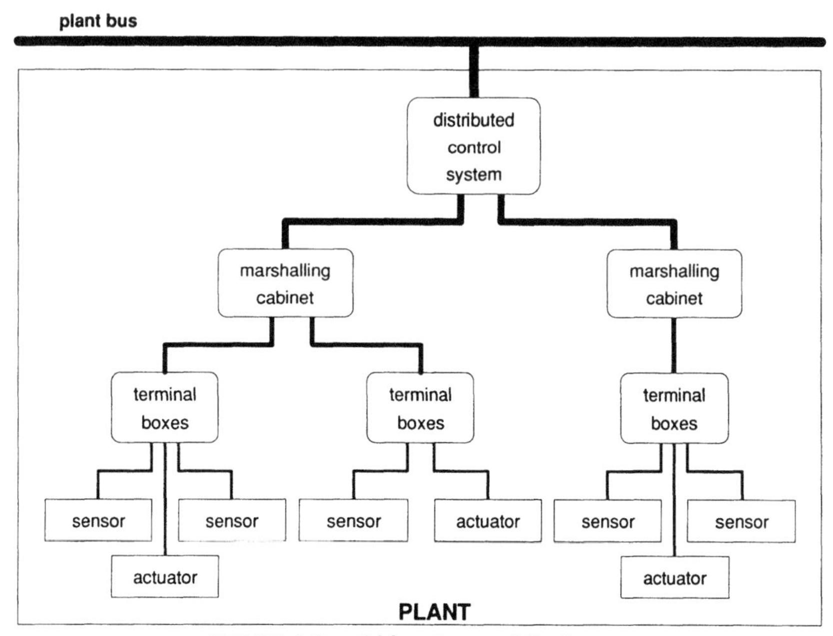

FIGURE 8.2. Old system architecture.

Another significant step forward was made at the end of the 1970s when a group of important users of automatic manufacturing systems led by General Motors developed a well-defined set of communication standards (mainly based on the ISO/OSI seven-layer model) to support applications in the factory environment. This collaboration led to the manufacturing automation protocol (MAP) [6], which can be considered a milestone in the history of the industrial communications. Although it was very ambitious, the MAP project had little success because it was not able to either manage the real-time communications needed at the shop floor level or provide a cheap interface to most of the instrumentation devices. Successive modifications introduced in the later version of the MAP protocol, such as the extended performance architecture (EPA) and the miniMAP [6] streamlined protocol profile, were not able to satisfy user needs completely.

Another factor that has affected the industrial communication scenario is the evolution of the local area networks (LANs), together with a deeper understanding of how most of the benefits expected from factory automation can be achieved only through a complete integration of the lower layers in the factory network hierarchy. This conviction has progressively led to the development of the concept of fieldbus. In practice, a fieldbus is a LAN with some enhancements added so as to satisfy requirements which are typical of a real-time system (real-time LAN, RTLAN).

The ability to satisfy the real-time requirements found at the field level makes fieldbus protocols attractive as a replacement of the bundles of analog or digital point-to-point links used to link controllers and instrumentation devices. In addition, since the architecture of fieldbuses is very similar to LANs, they can easily provide the communication backbone needed to support communications at the cell level (usually similar to those described in the MAP specifications). Fieldbuses are thus well suited to satisfying the communication requirements found at the field and cell levels, and to supporting a fully integrated system.

Two significant fieldbus protocols began to be developed in the early 1980s by two groups of European equipment manufacturers, one led by Siemens and the other by Telemecanique. These efforts resulted in two national standard specifications: the French standard FIP and the German standard PROFIBUS. The development of an international standard fieldbus was started about 10 years ago with the IEC 1158/ISA SP 50 Fieldbus project. It is well known that this proposal has not yet received consensus approval and the standardization process has made very slow progress and is still far from being completed.

At present only the physical and the data link layers are completely defined. Because there is no international standard, a number of different proprietary solutions have appeared on the market in the last few years and are now being used in a number of automated plants. Most of these solutions aim at achieving a smooth migration from the existing plants; they reuse older cables and older equipment by, for instance, interposing special adapters. A main drawback of such commercial fieldbuses is that, even though these solutions are often very similar, they are not compatible with one another. Thus, the requirement of having a common interface, both from a physical point of view (standardization of connectors, electrical signals, timing, and so on), and from the point of view of the application programs (i.e., a programming interface consisting of an agreed-upon set of service primitives and a high-level description based on a common abstract model of the manufacturing devices) is far from being satisfied.

Although the proprietary networks which are available at the moment keep many of the advantages typical of fieldbuses, they usually make it difficult for different automation cells to communicate with each other by means of a common (plant) backbone. But what is even more important, each solution requires its own interface, and this means that a field device can be connected only to the network or that kind of network for which it was designed.

Manufacturers are reluctant to promote the adoption of a single standard fieldbus; instead they prefer to design their new products based on more consolidated national standards such as WorldFIP and ISP that introduce minor changes to the French standard FIP and to the German standard PROFIBUS respectively. However, to be fair, it must be said that FIP and PROFIBUS have significantly affected the standardization activity of the IEC 1158/ISA SP50 Fieldbus in the last few years. In fact, these two fieldbus architectures are being carefully examined by international standardization committees so that their main features can be included in the resulting international standard proposal. In the following, the most significant characteristics of both FIP and PROFIBUS will be introduced so that the reader will get an idea of the mechanisms that can be found in the IEC 1158 Fieldbus.

The development of that specification is fundamental for integrating the communication functions in the plant. When the standard is ready and widely accepted, the scenario of an automated factory will probably be similar to that sketched in Figure 8.3.

2.1. Fieldbus Advantages

There is no doubt that communication plays a key role in the automated factory environment and the fieldbus approach appears to be the most suitable solution. Fieldbuses have many advantages compared to both networks used at the cell level and point-to-point links which are used to interconnect controllers to intelligent devices. The first advantage is that a single common network can be used both at the lower field level and at the intermediate cell level of the automated factory environment. This reduces both the complexity of cabling and the cost of interfacing to the communication capabilities, while at the same time it offers a very flexible architecture. Thus the so called "plug-and-play" concept can be easily reached in real systems. When fieldbuses are compared to conventional analog and digital point-to-point links the following advantages can be found:

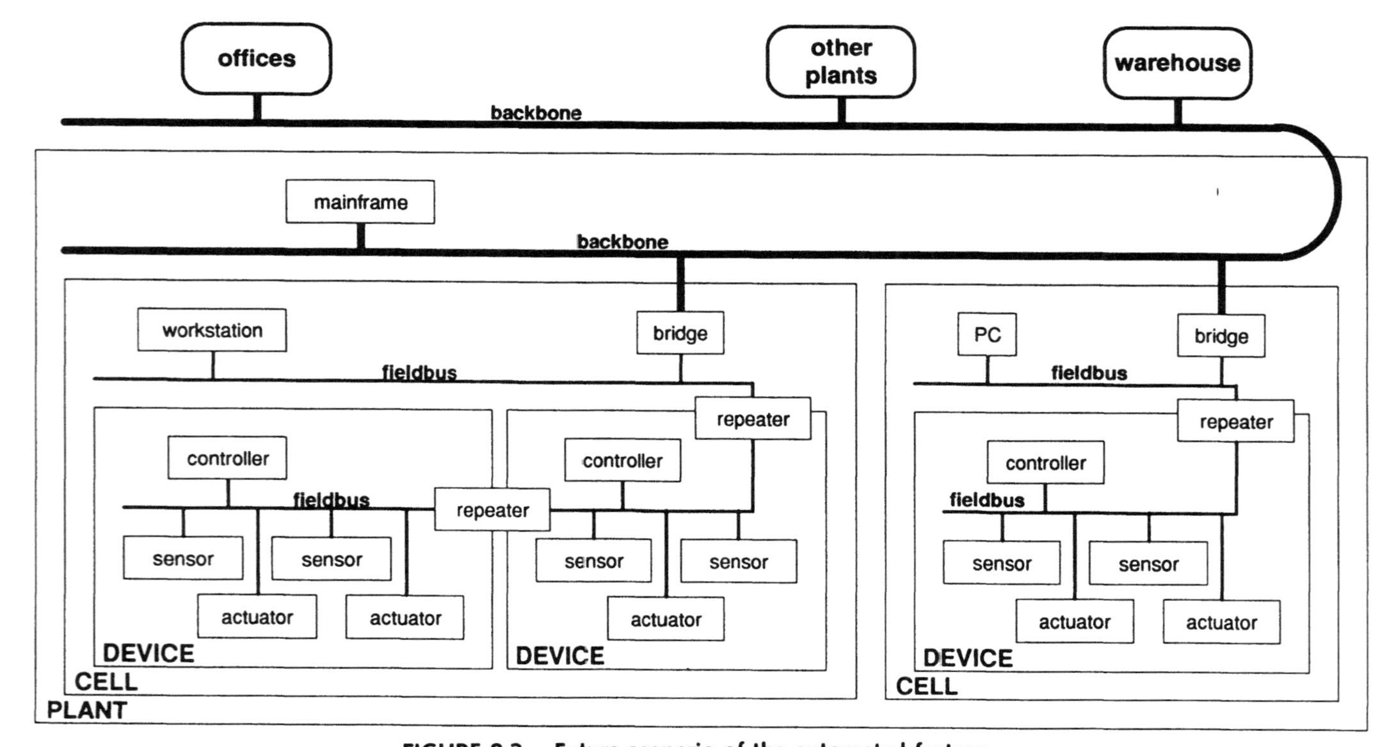

FIGURE 8.3. Future scenario of the automated factory.

- It is easy to establish remote access to the diagnostic and calibration capabilities of a device; this kind of operation is not feasible when analog technology is used.
- There is a clear distinction between the acquisition of data and processing; this allows each single device or the controller to be replaced when necessary, without affecting other equipment.
- The use of digital transmission greatly enhances the quality of the transmitted signal compared to the analog links, and this leads to a much better immunity to environmental noise. It must be noted that analog signals are converted to their digital form directly inside the originating devices.
- It is simple to check the integrity of the physical medium, thus the reliability of the overall system is improved.
- A certain amount of processing can be done directly within the field devices; thus control functions can be distributed in a better way.
- Each datum can virtually be made available to all the nodes in the network with none or minimal effort; this avoids the need of multiple cables when information must be shared among different sites.
- Usually it is not necessary to add cables or to modify the topology of the network when new devices have to be introduced into the system.

The use of fieldbuses brings some advantages at the cell level too. In fact, fieldbuses offer their users (i.e., application processes accessing the communication facilities) a simple interface and a powerful set of functions which are well suited to controlling equipment and field devices. Due to the adoption of a reduced protocol profile, better performance can also be expected. Moreover real-time communications can be supported.

The main drawback of fieldbus technology is that a single transmission medium is shared by all devices, and this can create a bottleneck and a point of failure for the overall system (when the cable breaks it is common to have many devices malfunctioning). This disadvantage can be reduced or eliminated by adopting a hierarchical network structure in which simpler fieldbus subnetworks are interconnected through bridges, and by introducing redundancy, when needed, of the physical media in order to achieve a higher degree of fault tolerance.

3. FIELDBUS REQUIREMENTS

As mentioned in the previous sections, fieldbuses can be considered LANs that are particularly suited for communication between equipment operating in automated manufacturing environments. Thus, unlike conventional networks, fieldbuses have to deal with the operation of time-critical devices, where the time involved in exchanging data and the safety of the system is of fundamental importance. The main requirements that an industrial field network must satisfy are introduced below, together with a description of the solutions adopted in FIP and PROFIBUS.

3.1. Communication Needs

Industrial communication needs can be put into three broad classes, depending on the nature of the information exchanged and on the related timing requirements. The first class deals with the cyclic operations involved in controlling simple devices, such as actuators and sensors. Operations on modern digital controllers rely on the periodic polling of a set of sensors so as to get the value of some measured parameters (i.e., position, speed, pressure, and so on). These values are used to generate (according to some predefined model) suitable control signals, used as commands for the actuators acting on the controlled system. A fundamental timing constraint that must be considered for this kind of communication is the maximum jitter that can be tolerated.

The second class takes into account events that can happen at unpredictable times, such as, for example, alarms and/or synchronization signals. Usually there is no need to carry a significant amount of information associated to these signals. Instead, there are severe constraints on the time taken to deliver them to the appropriate destination.

The third class concerns sophisticated operations such as remote diagnostics or the download of part programs for numerically controlled devices. In this case it must be possible to exchange an arbitrarily large amount of information at suitable data rates.

When fieldbuses are considered, the above-mentioned communication needs can be satisfied by three different kinds of data transfers known as cyclic data exchange, asynchronous urgent traffic, and high-level messaging services. A good fieldbus specification must contain mechanisms for each of these communication requirements.

3.1.1. Cyclic Data Exchange

A major requirement in the field environment is the ability to periodically query a number of simple field devices for sampled data represented by numeric values. This typical situation occurs when a controller polls a number of actuators and sensors to read out some measures or to set some parameters. Fieldbus protocols introduce the concept of cyclic data exchange of variables. By using a cyclic data exchange, the user can enforce the underlying network to carry out an automated periodic polling of a specified set of devices. The advantage is that, in this case, the network is responsible for managing the periodic inquiries regarding the connected devices.

The network must also provide for a certain degree of decoupling between the production of the values (the action of making a value available on the network) and their consumption (when the value of a variable is needed so as to do some kind of processing and must be read from the network). In other words the network must keep a memory in which information can be stored and retrieved in an asynchronous way by the application processes running on the devices, thus making it possible for the controllers to act as if conventional analog point-to-point links were being used. Cyclic data exchange services provide the user with a simple and powerful mechanism to deal with sampled systems. Irrespective of the kind of mechanisms used for the implementation, the main fieldbus architectures available today enable such kinds of data transfer to take place. The effectiveness of the cyclic data exchange is severely affected by jitter, that is the timing drift between successive sampling or between successive cyclic exchanges. It is clear that the jitter involved in a cyclic exchange of data must be as small as possible. In fact, a high jitter value can reduce the accuracy with which the system is controlled, leading to poor precision, or worse, to instability.

3.1.2. Asynchronous Urgent Traffic

In an automated environment where potentially dangerous situations can arise it is mandatory for a device to be able to notify its controller of alarm situations as soon as possible. Moreover it should be possible to synchronize a number of devices on a particular event, which, in this way, acts as a trigger. In both cases the communication system must ensure that the time which elapses between the transmission and the reception of the alarm or synchronization signal is bounded, the value of the limit depending on the controlled system. When digital networks are used, high-priority traffic, consisting of small-sized messages with tight time constraints, has to be provided for.

Some of the conventional techniques adopted by some very popular LANs (i.e., the Ethernet network) in order to access the physical medium are not deterministic, and, hence, cannot be accepted. However, even protocol profiles such as MAP, that were conceived for the factory environment, are not very suitable since they are not able to guarantee very short response times, and what is more important, the connection-oriented model used in MAP does not allow transmissions to several receiving entities at the same time, thus making it difficult to synchronize a number of devices. Fieldbuses are able to manage urgent messages, since they adopt deterministic (or nearly deterministic) medium access techniques. These techniques are designed to allow broadcasting and multicasting, and they foresee the use of priority in data transmission.

3.1.3. High-Level Messaging Service.

Fieldbuses must also provide for communication at the cell level. This has led to sophisticated communication facilities being offered which cover activities such as part program downloading or controlling the execution of a complex sequence of actions on a target device. For this reason, a set of services based on a quite complex model of the real device is usually offered to the user's programs. This kind of traffic has no particular timing constraints, but high throughput is often desirable. In this case, the most interesting aspect is perhaps the set of services offered to the user which builds up the programming interface for the application processes. At present, fieldbuses usually rely on a subset of the MMS (manufacturing message specification) application services since MMS is the only standard protocol defined for this purpose.

3.2. Environmental Requirements

Due to the environment in which fieldbuses have to operate, they have to tackle aspects such as intrinsic safety, remote powering, noise immunity, and medium redundancy that usually are not covered by other kinds of network.

Intrinsic safety is a topic of great importance when operating in dangerous environments, for example, in an explosive environment. In such situations fieldbus architectures turn to using fiber optics or low-voltage electrical signals to avoid potentially risky situations.

Remote powering is of importance when small and low-cost devices have to be connected to the network, since it is possible to adopt the same cable for transmitting data and also for supplying power. This advantage can be evaluated both in terms of reduced cabling complexity (each device need not be connected to the power supply by an additional cable) and in terms of the reduced cost and size of each device (which will not include a power supply section).

Since automation and field environment are usually noisy, great care has been taken to improve the noise immunity of fieldbuses. In particular, the use of fiber optics is suggested for particularly noisy environments. In general, digital transmissions are less error-sensitive and exhibit a higher immunity to noise even when traditional copper wires are used.

In order to improve the system's fault tolerance, fieldbuses can also be implemented with a redundant physical medium, that is to say a pair of identical physical cables are used. Special circuits inside each device are able to detect a possible failure on either of the two links and to switch the transmissions to the other (good) one. In this case the alternative link is automatically used. The mechanism is completely transparent to users and to communication software, both under normal operations and when a fault occurs.

3.3. Fieldbuses and OSI Services

A very important issue when dealing with an industrial communication system is the interface to the communication services being offered to the application processes running on the different devices connected to the network. Fieldbuses have been developed by taking into account the layered structure depicted in the

OSI (Open Systems Interconnection) reference model. However, since a full seven-layer architecture could hardly satisfy the real-time requirements of the automation environment, a three-layer reduced protocol has been adopted such as the one foreseen by miniMAP, and included in the extended performance architecture (EPA) of the MAP protocol. The reduced profile contains only the physical, the data link, and the application layers, while the network, transport, session, and presentation layers have been eliminated.

When the reduced architecture is adopted, there is a general simplification of the overall functions. Some of the features of the missing layers that are still needed have to be shifted into the data link and the application layer. For example, the management of subnetworks by using bridges has been shifted in the data link layer, while issues such as the connection management, some limited form of message segmentation, and the transfer syntax management have been included in the application layer.

3.3.1. Data Link Services.

In the existing fieldbus protocols, efforts have been made to hide from the data link users details of the mechanism to access the communication medium to provide the data link users with a clean interface. This interface must provide both the basic services which allow confirmed and unconfirmed transmissions and also those services which are peculiar to the particular protocol being considered. The first class of services is very similar to those provided by LAN's protocol. They offer a basic capability to exchange messages, and they ensure that fieldbus can be used to directly support the high-level protocols devoted to management functions. The second class instead deals with all those services that exploit the real-time communication features provided by the fieldbus, thus they give the user a means to effectively control operations in a field environment.

3.3.2. Application Services.

A fieldbus must basically provide the users of its application layer two kinds of services:

- A set of high-level messaging functions to accomplish tasks at the cell level, similar to those defined by the MMS standard
- A set of services to exploit the features related to the cyclic and real-time functions of the underlying data link layer.

At present, MMS is the only high-level standard explicitly defined for the manufacturing environment, and its role is now widely accepted. In particular it seems that the set of services provided by this protocol is well suited and sufficient to cover the communications needs in the context of the plant and of the cell.

The fieldbus protocols offer a set of services which are very similar to those described in the MMS specification [7, 8]. This is not surprising because one of the aims of the fieldbus standardization is to fully integrate the instrumentation subnetworks within the framework of the factory network architecture and, in particular, in the networks used at the cell and plant levels (type one and type two networks). Such integration can be achieved if the application processes on the different devices connected to the network share a common view of the underlying communication system or, in other words, if they can access a common set of objects and services to handle them.

The following services are usually provided for the application processes by every fieldbus protocol:

- Variable reading and writing
- Part program uploading/downloading
- Remote control of program execution
- Management of events such as alarms.

These services operate on a conceptual model of the field device, which is very similar to the virtual manufacturing device (VMD) object defined in MAP.

Unlike the MMS-like messaging services, the various fieldbus protocols are quite different with regard to the way they provide their users with cyclic and real-time services. In practice, two methods are used. The first obvious solution is to provide a complementary set of services to deal with these new forms of data exchange. Another solution is to instead modify the standard services or their parameters so as to get the same result. Both of these choices have advantages and disadvantages, and this is one of the reasons why an agreement on how the application layer of the standard IEC/ISA Fieldbus must be has not yet been reached.

4. FIELD INSTRUMENTATION PROTOCOL (FIP)

The FIP is a French national standard fieldbus developed in order to achieve a high rate in exchanging data between devices in a factory environment. It adopts a classical reduced protocol profile consisting of three communication layers and a network management block as shown in Figure 8.4. In a FIP network each data exchange is supervised by a special station called the *bus arbiter*, which is responsible for assigning the right to transmit data on the bus to data producers. When data information is broadcast on the bus, it can be read by each listening station. Thus, the number of information exchanges depends only on the quantity of data produced and not on the number of stations that consume the data.

The data traffic can be either real time or not. The real-time traffic consists of a periodic or cyclic exchange of objects (either variables or messages); while the non-real-time traffic consists of an aperiodic exchange of objects. Obviously,

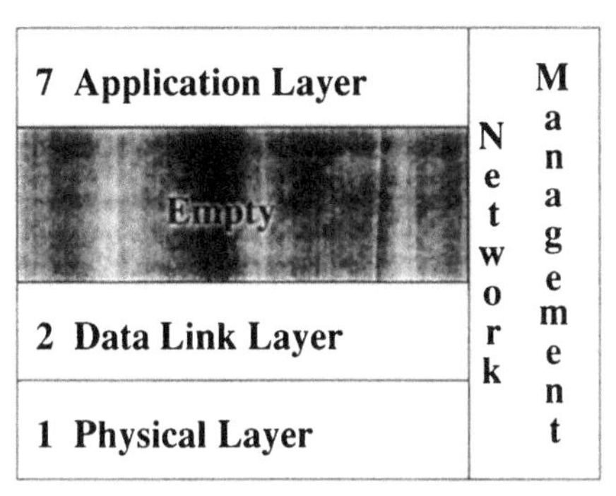

FIGURE 8.4. FIP protocol profile.

real-time traffic has a higher priority than aperiodic traffic. The amount of cyclic (real-time) data and their timing characteristics must be known in complete detail in advance so that the network can be set up in its configuration phase. The FIP protocol provides two different types of addressing: one for variables and one for messages. Each variable is assigned a logical identifier which is unique in the network and totally independent of the producer and the consumers stations, whereas messages contain the addresses of the transmitting and receiving stations. The protocol allows consumers of cyclic (real-time) data to check the spatial and temporal consistency of the variables. In this way each consumer of a given variable is allowed to know whether its own copy of the variable value is the last produced and is the same owned by all other consumers.

4.1. Physical Layer

The physical medium can be either a twisted pair or an optical fiber; the standard document also gives rules to handle mixed networks, i.e., networks where the two physical media coexist. Despite two different physical media being allowed, the set of services provided and the data coding techniques are the same; only the transmission speeds are slightly different.

4.1.1. Network Topology.

Figure 8.5 shows an example of network topology using a twisted pair. In the picture JB is a junction box acting as a passive multiple tap and provides two or more accesses for derivations. TAPs are used to provide connection points to the trunk cable, while repeaters bring two pairs of trunk cable together. DBs are diffusion boxes used to bring several terminal segments together on a trunk cable and DS and NDS are locally and nonlocally disconnectable subscribers respectively.

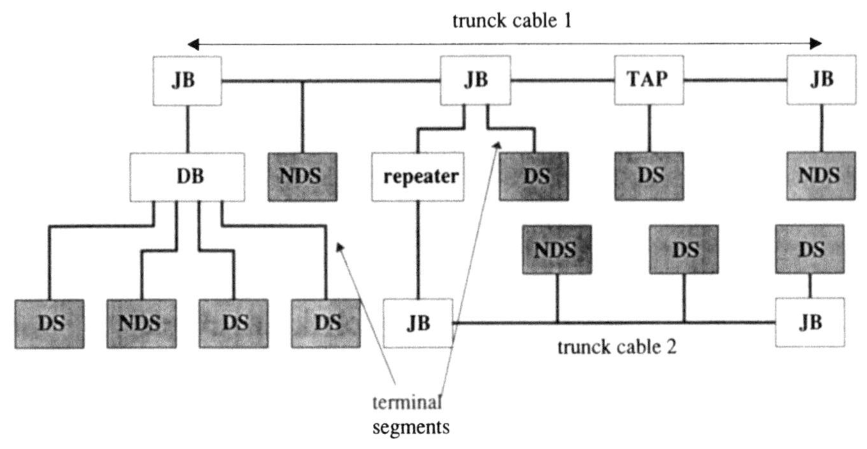

FIGURE 8.5. Network topology with twisted pair medium.

Figure 8.6 shows an example of FIP network topology based on optical fiber. In this case only point-to-point physical links are allowed, active optical star devices are therefore needed to provide connections.

4.1.2. Transmission Speeds and Data Encoding.

Transmission speeds allowed in FIP are 31.25Kb/s, 1Mb/s, 2.5Mb/s, and 5Mb/s. The last one is allowed for the fiber-optic medium only, while the other ones are supported by both physical media. The physical layer uses the Manchester encoding scheme (see Figure 8.7) to send the data obtained from the upper

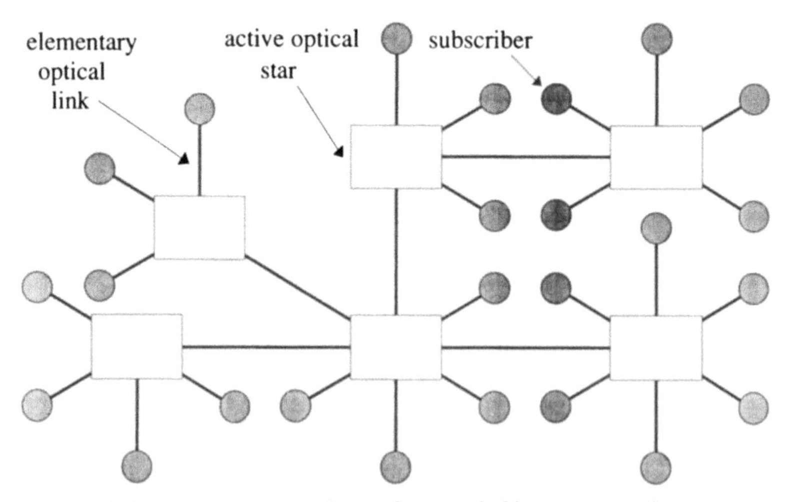

FIGURE 8.6. Network topology with fiber-optic medium.

layers on the network. As is well known, this coding technique allows the user to send data together with timing synchronization. Violation bits are used for frame synchronization and are also shown in Figure 8.7.

4.1.3. Service Primitives.

The physical layer offers the following three service primitives to the data link layer:

- `PHY_DATA.request( symbol_value )`
- `PHY_DATA.indication( symbol_value )`
- `PHY_DATA.confirm( status )`

where the `symbol_Avalue` can assume one out of eight values: `0, 1, EB+, EB-, V+, V-, silence, activity`. It should be noted that the `activity` and `silence` symbols are used by the data link layer to request the physical layer to start and stop a frame transmission, respectively, and are used by the physical layer to indicate the beginning and the end of a frame reception to the data link layer.

The `status` parameter can assume two logical values that are positive and negative and are used to notify the data link layer of the correctness or the failure of the previously transmitted symbol (bit).

The `PHY_DATA.request( symbol_value )` service is used by the data link layer to invoke the transmission of a single bit on the physical medium, while the `PHY_DATA.confirm( status )` is generated by the physical layer to inform the data link layer about the success/failure of the transmission of the last symbol issued in the form of a `PHY_DATA.request`. The `PHY_DATA.indication( symbol_value )` is generated by the physical layer after a symbol from has been received from the network.

Figure 8.8 shows interactions between data link layers using the lower layer's services.

4.1.4. Structure of Frame.

All FIP frames sent over the network by the physical layer have the structure shown in Figure 8.9. In practice only the control and data (CAD) field actually contains data while all the other fields are added by the data link layer to allow receivers to recognize the transmitted information correctly. The frame start sequence (FSS) is made up of a preamble (PRE), a frame start delimiter (FSD), and an equalization bit (EB+); the control and data (CAD) field contains useful data and the frame end sequence (FES) is made up of a frame end delimiter (FED) and two equalization bits. The PRE field allows the receivers to synchronize with the sender's clock; the FSD and the FED fields delimit the useful information in the CAD field.

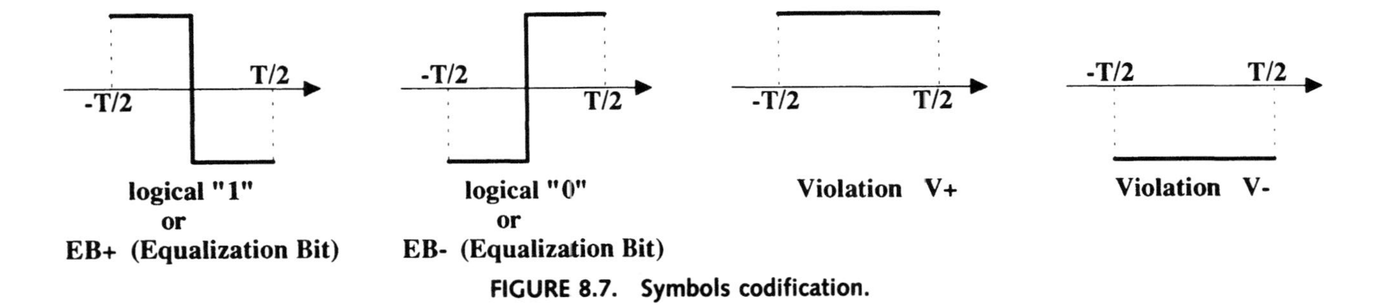

FIGURE 8.7. Symbols codification.

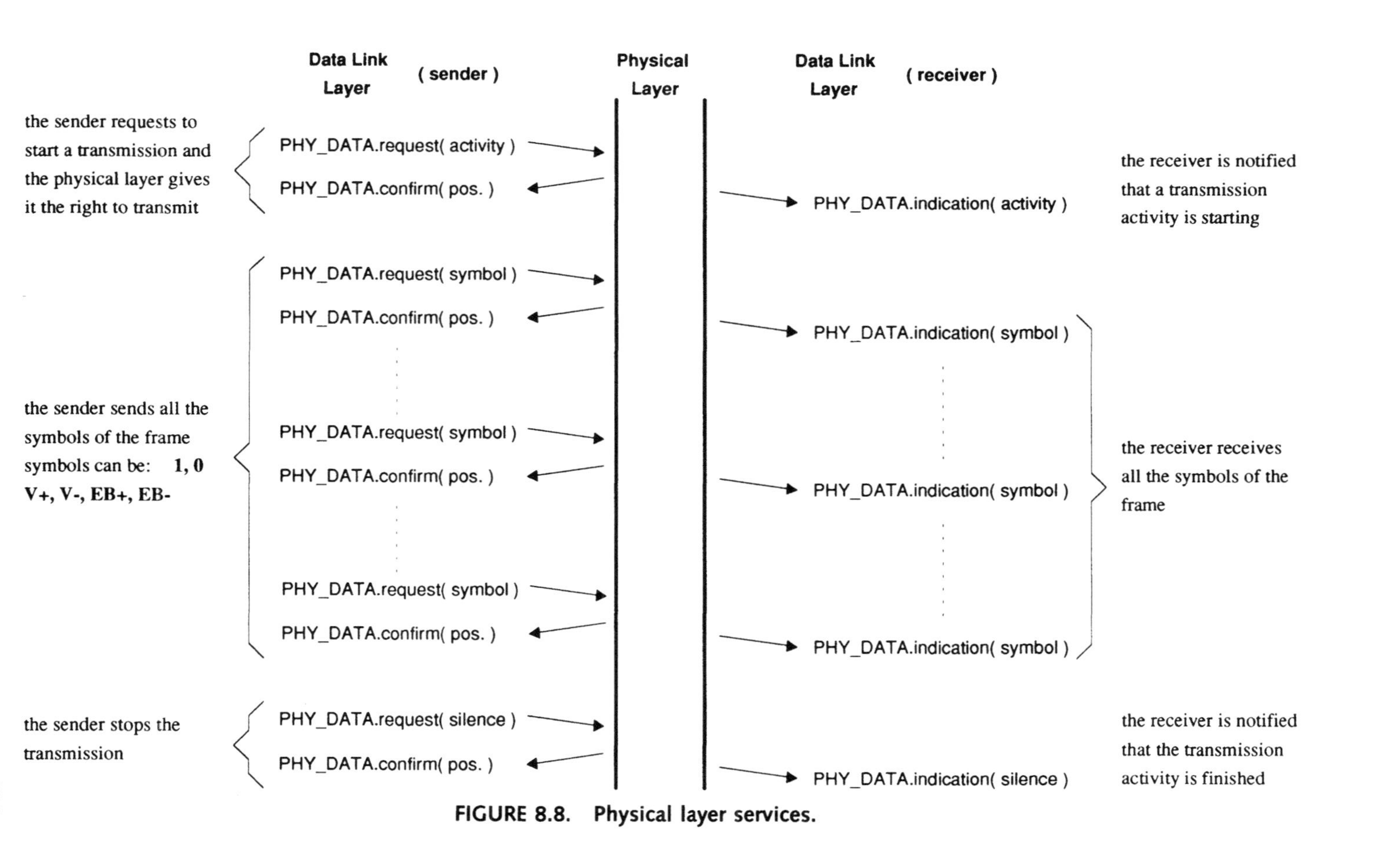

FIGURE 8.8. Physical layer services.

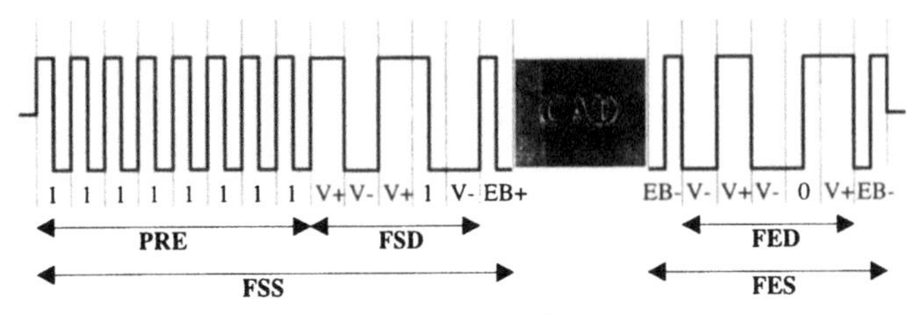

FIGURE 8.9. FIP frame.

4.2. Data Link Layer

The data link layer handles exchanges of variables and messages; these exchanges can be either cyclic or on-user-request (acyclic). Only those objects whose name and periodicity are set during the configuration phase of the system can be exchanged periodically. This kind of exchange takes place without any explicit user intervention. Explicit user requests cause the values of one or more objects to be exchanged on the bus in an asynchronous way.

All the object exchanges are driven by a bus arbiter (BA) located on a station attached to the bus. Only one bus arbiter can be active at any given time, and this function does not conflict with the other activities of the hosting station such as producing and consuming variables. The main task of the bus arbiter is to schedule the exchanges of cyclic objects among network entities at the correct rates and to satisfy explicit user requests if transmission time is not completely exhausted by the cyclic exchanges. The data link layer offer upper layer's users services to update and exchange cyclic and acyclic objects.

4.2.1. Object Addressing.

The data link layer has an internal buffer for each variable produced or consumed and an internal structure, called file, devoted to messages. Such structures are accessed only through a 16-bit-long identifier ID. Identifiers used for variables must differ from each other, but a variable identifier can also be assigned to messages. Variable identifiers have a global meaning in that they are used to identify variables at the producer and at the consumer(s) entities in a way which is completely independent from the physical location of the variable in the network. Message identifiers are used by the local entity and the bus arbiter and have a local meaning. This is due to the fact that when a message is broadcast on the network it contains the addresses (24 bits long) of the source and destination(s) entities.

4.2.2. Physical Medium Allocation.

During the network configuration phase the bus arbiter creates a scanning table containing the identifiers and the timing requirements of the objects that are to be exchanged cyclically. At the same time each station allocates suitable buffers for the identified variables together with files for cyclic messages. When the time to circulate a cyclic object on the bus is reached, the bus arbiter broadcasts on the bus a request frame containing the identifier of that object. Then the producer puts the current value of the object on the bus and all the consumers read it simultaneously. When the object is a message, an acknowledgment frame can be returned to the sender by the destination, and the transaction ends with a final frame sent by the message producer to the bus arbiter.

The cyclic exchange of objects is shown in Figure 8.10. Specifically, Figure 8.10a concerns cyclical variables (ID_A), while Figures 8.10b and 8.10c show the mechanisms adopted for messages (ID_J) where an acknowledgment is not/is required, respectively. Each one of the three transactions shown in Figure 8.10 is called an elementary sequence. When an elementary sequence ends, the arbiter starts the next one. It is worth noting that acknowledgments can be used only for point-to-point messages. Multipoint (broadcast or multicast) messages cannot be acknowledged.

When an ID_DAT or ID_MSG frame is recognized, a timer is started in each consumer. When the timeout is reached (the RP_DAT or RP_MSG_xx frame is lost or delayed), consumers simply return to a waiting state for another ID_DAT or ID_MSG.

The scanning table of the bus arbiter can contain identifiers of objects which are to be circulated on the bus with different periods. For example, assuming a set of cyclic objects as shown in Table 8.1, the bus arbiter will schedule them in the way shown in Figure 8.11. The variable *A* has the shortest periodicity and this time defines the duration of the elementary cycle. The time needed to carry out all the elementary sequences of each elementary cycle must be less than the duration of the elementary cycle itself. If there is some time left after the completion of all the elementary sequences in an elementary cycle, it can be used by the bus arbiter to satisfy aperiodic requests. Such a time interval is called the *aperiodic window.*

The pattern of elementary cycles which is repeated periodically is called a *macrocycle*. In the example of Figure 8.11 it is easy to see that the periodicity requirements are satisfied. The macrocycle is made up of 12 elementary cycles 5 ms long. In this period variable *A* is circulated 12 times while variable *B* is sent six times and message *J* and variable *C* are sent four and three times, respectively.

TABLE 8.1. Cyclic Objects and Their Periodicity.

Object	ID	Periodicity [ms]
variable	A	5
variable	B	10
message	J	15
variable	C	20

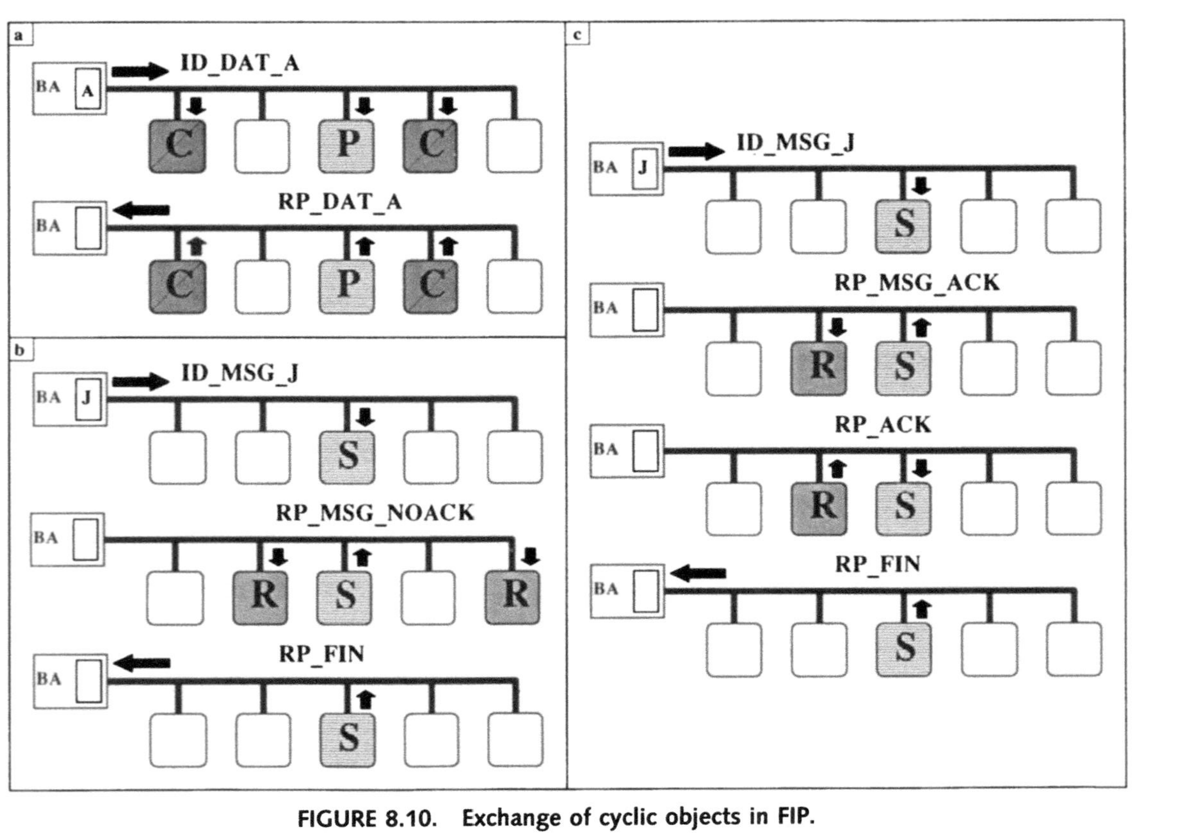

FIGURE 8.10. Exchange of cyclic objects in FIP.

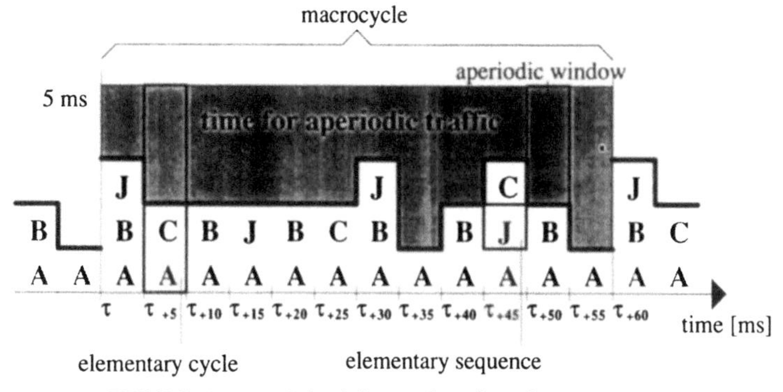

FIGURE 8.11. Scheduling of cyclic objects in FIP.

When a network entity requires one or more aperiodic transactions for variables, it sets a request bit in the RP_DAT frame of a cyclic transaction then, in an aperiodic window, the bus arbiter requests the list of identifiers of aperiodic variables to be exchanged with the network entity which is called, for this reason, *initiator (I)* of the aperiodic transaction(s). The initiator gives its answer in the same aperiodic window. Then, probably in a different aperiodic window, the arbiter performs a set of elementary sequences similar to those mentioned above which allow the aperiodic variables to circulate on the bus. Such a three-stage transaction is depicted in Figure 8.12. It should be noted that the FIP standard allows the initiator to be neither the producer nor the consumer of the aperiodic variables in the list; however, an acyclic transaction can be initiated only by the producer of a cyclic variable.

Messages are handled in almost the same way as acyclic variables. During a variable transaction the producer sets a bit in the RP_DAT frame in order to ask the bus arbiter for a message transaction and a sequence identical to that shown in Figure 8.10b (and 8.10c) is performed in an aperiodic window.

4.2.3. Format of Frames.

Figure 8.13 shows the format of the FIP frames. The FSS and FES fields are those shown in Figure 8.9, while the FCS field is a frame check sequence. The ID_DAT, ID_RQ, and ID_MSG frames share the same format; however, they differ in the control field since, by decoding this field, it is possible for each network entity to understand whether the bus arbiter is requesting a variable, a list of variable identifiers or a message. The value of the variable is contained in the data field of the RP_DAT frame. This value is encoded using the ASN.1 [9, 10] standard format. It should be noted that some bits in the control field can be set by the producer to request the bus arbiter to authorize an acyclic transaction. The ID_RQ frame is used by the bus arbiter to request the list of variable identifiers from the network entity which originated an aperiodic variables ex-

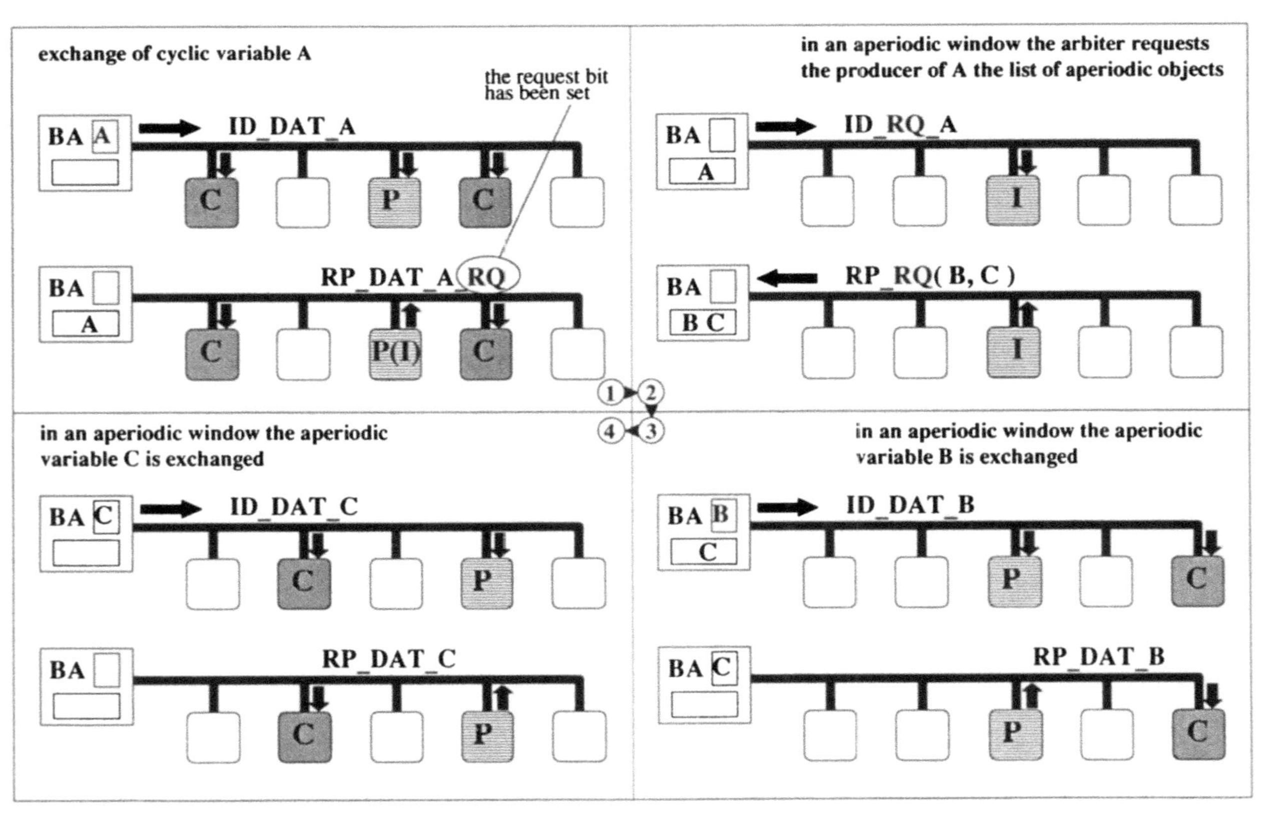

FIGURE 8.12. Exchange of acyclic variables in FIP.

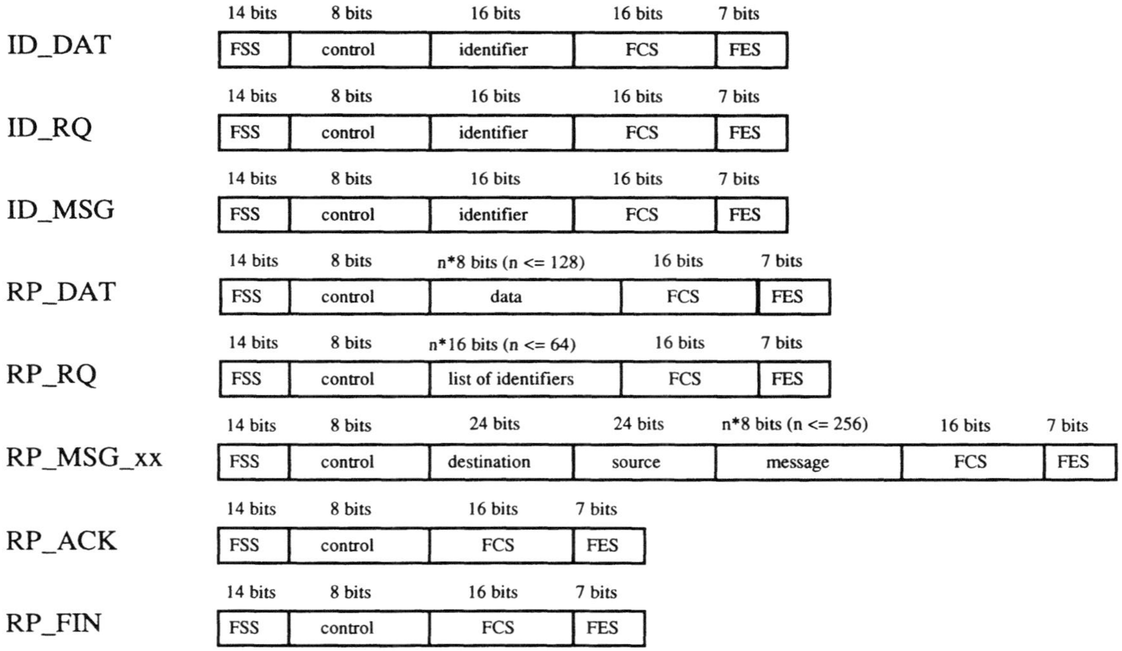

ID_DAT
14 bits | 8 bits | 16 bits | 16 bits | 7 bits
FSS | control | identifier | FCS | FES

ID_RQ
14 bits | 8 bits | 16 bits | 16 bits | 7 bits
FSS | control | identifier | FCS | FES

ID_MSG
14 bits | 8 bits | 16 bits | 16 bits | 7 bits
FSS | control | identifier | FCS | FES

RP_DAT
14 bits | 8 bits | n*8 bits (n <= 128) | 16 bits | 7 bits
FSS | control | data | FCS | FES

RP_RQ
14 bits | 8 bits | n*16 bits (n <= 64) | 16 bits | 7 bits
FSS | control | list of identifiers | FCS | FES

RP_MSG_xx
14 bits | 8 bits | 24 bits | 24 bits | n*8 bits (n <= 256) | 16 bits | 7 bits
FSS | control | destination | source | message | FCS | FES

RP_ACK
14 bits | 8 bits | 16 bits | 7 bits
FSS | control | FCS | FES

RP_FIN
14 bits | 8 bits | 16 bits | 7 bits
FSS | control | FCS | FES

FIGURE 8.13. Format of FIP frames.

change. Finally the control field of the RP_MSG contains a bit that can be set by the message producer to invoke an RP_ACK frame from the receiver.

4.2.4. Service Primitives.

Service primitives offered to the FIP application layer can be divided into two subsets: services which handle variables and services which handle messages. In the first subset the following service types can be found:

- Primitives to write buffer:

```
–L_PUT.request( id, value )
–L_PUT.confirm( id, status )
```

- Primitives to read buffer:

```
–L_GET.request( id )
–L_GET.confirm( id, status, value )
```

- Primitives to transfer buffer:

```
–L_SENT.indication( id )
–L_RECEIVED.indication( id )
```

where `id` is the variable identifier; `value` is the value of the variable; and `status` indicates whether the previous request was terminated correctly or not.

Figure 8.14 shows an example of the use of the FIP data link service primitives: the cyclic variables *X* and *Y* are respectively produced and consumed by the same network entity. The network entity is polled cyclically by the bus arbiter in order to broadcast the value of *X* on the network and to read the new value of *Y* from the network. Each time the data link layer broadcasts the value of *X* on the network, it passes an `L_SENT.indication(X)` to the application layer, and each time the data link layer receives the value of *Y*, it gives an `L_RECEIVED.indication(Y)` to the application layer. Then the `L_SENT.indication` and `L_RECEIVED.indication` primitives involve buffer transfers which are driven by the bus arbiter. Instead the L_PUT and L_GET primitives are used by the application layer to write and read a buffer and these actions do not initiate any bus activity.

Acyclic variable exchanges can be invoked by using ad hoc primitives and in particular:

- Primitives for specified explicit buffer requests:

```
–L_SPEC_UPDATE.request( id_spec, id_list )
–L_SPEC_UPDATE.confirm( id_spec, status )
```

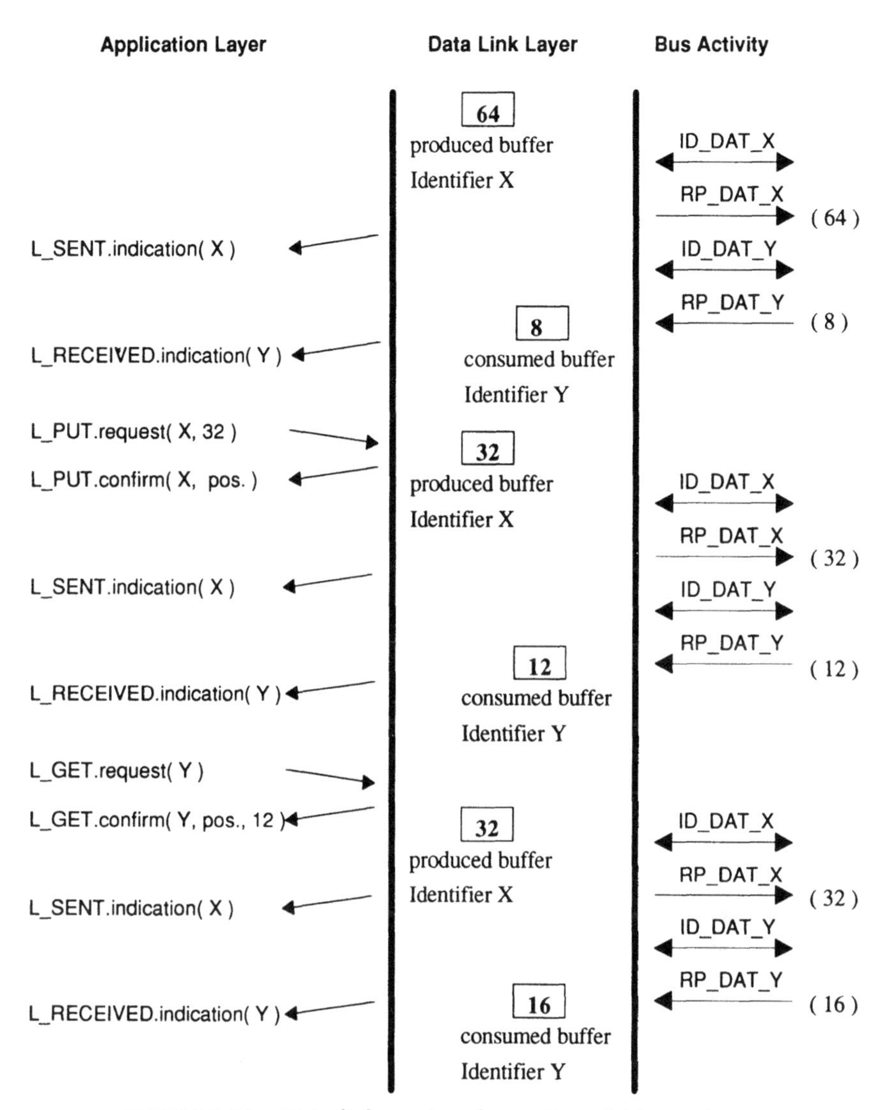

FIGURE 8.14. Data link services for cyclic variables exchange.

- Primitives for free explicit buffer request:

  –`L_FREE_UPDATE.request( id_list, priority )`
  –`L_FREE_UPDATE.confirm( status )`

The primitives L_SPEC_UPDATE and L_FREE_UPDATE are both used to request the transfer of acyclic variables. The first type allows the user to specify the cyclic variable to which the request must be appended (id_spec). These primitives are used only by internal functions of the application layer to manage the temporal consistency of data. The second type appends the request to the RP_DAT frame of the first cyclic variable produced after the request has been received from the upper layer. The parameter id_list contains the list of the identified variables to be exchanged acyclically; the priority value can be either normal or urgent: in an aperiodic window urgent requests are served first.

Figure 8.15 shows how the system behaves when acyclic transfers of variables are requested by using L_FREE_UPDATE primitives. Arrowed arcs originating from and terminating in the central section of Figure 8.15 indicate frames originating in the bus arbiter and addressed to the bus arbiter respectively. The second subset of data link services is dedicated to message transfers and contains:

- Primitives for connection establishment:

  –L_CONNECT.request( )
  –L_CONNECT.confirm(id)

- Primitives for disconnection:

  –L_DISCONNECT.request(id)
  –L_DISCONNECT.confirm(id, status)

These primitives are used by the application layer to request an identifier to be associated with a message. The identifier must be one of those of cyclic variables produced by the network entity and it must not be associated with any other message queue. Obviously these primitives must be used for acyclic messages only; in the case of cyclic ones the identifier is assigned statically during the configuration phase.

The same identifier can be used until it is released by using the primitives L_DISCONNECT and then several messages can be sent by using the same assigned identifier.

- Primitives to transfer messages without acknowledgment:

  –L_MESSAGE.request(id, dest_addr, source_addr. msg)
  –L_RAMESSAGE.indication(dest_addr, source_addr, msg)
  –L_MESSAGE.confirm(id, dest_addr, source_addr, status)

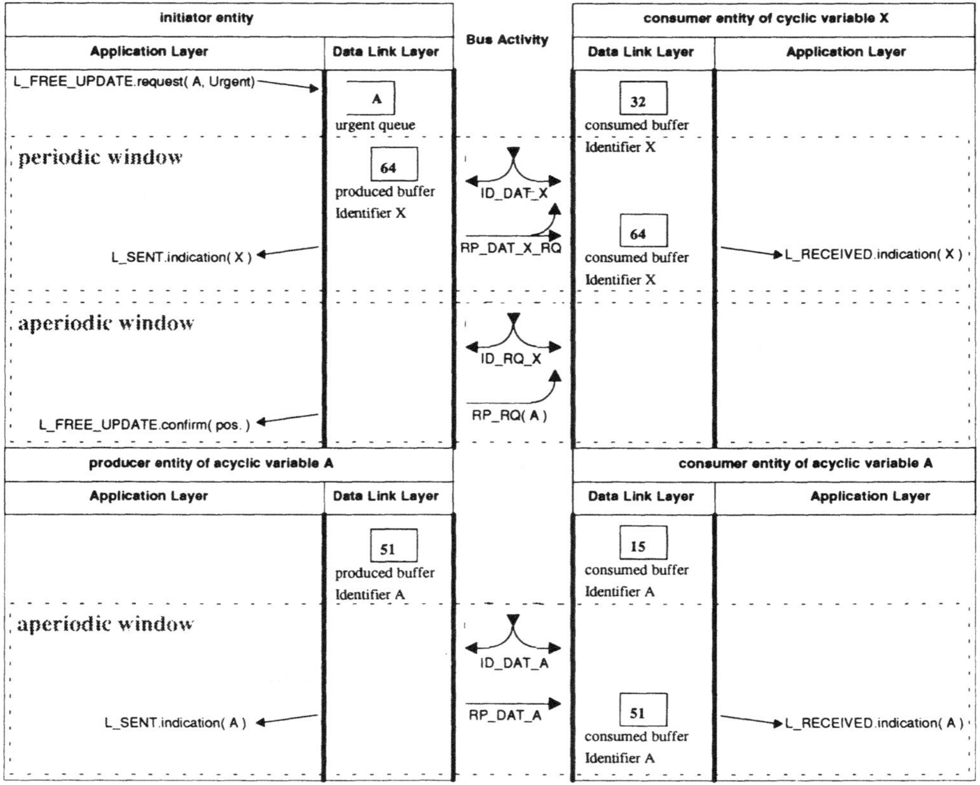

FIGURE 8.15. Data link services for acyclic variables.

- Primitives to transfer messages with acknowledgment:

  –L_MESSAGE_ACK.request(id, dest_addr, source_addr, msg)
  –L_MESSAGE_ACK.indication(dest_addr, source_addr, msg)
  –L_MESSAGE_ACK.confirm(id, dest_addr, source_addr, status)

It is necessary to remember that messages without acknowledgment can be either point-to-point or multipoint, while the acknowledged messages can only be point-to-point. Figure 8.16 shows how a cyclic message exchange with acknowledgment is carried out in FIP. The bus arbiter schedules the message queue with periodicity; if the queue is empty, the network entity responds with an RP_FIN to the ID_MSG_X frame. For acyclic messages, the application layer must issue an L_CONNECT_REQUEST to the data link layer and, once the identifier has been obtained, messages with that identifier can be exchanged. When the data link layer receives an L_MESSAGE_REQUEST_xx, it waits until the cyclic variable with the same identifier is due to be sent and it sets the bits for message request in the control field of the appropriate RP_DAT. Then in an aperiodic window the arbiter will schedule the message in the same way shown in Figure 8.16.

4.3. Application Layer

The FIP application layer services can be divided into two classes known as SubMMS and MPS: SubMMS is a subset of the messaging MMS (manufacturing message specification) services, while MPS defines manufacturing periodical/aperiodical services. This kind of organization is shown schematically in Figure 8.17. The SubMMS module offers the application process a subset of MMS (manufacturing message specification) enriched by the possibility of broadcasting, multicasting, and prenegotiating associations. These services are based on messaging services supported by the data link layer and can be used for the following control activities:

- Setup and configuration of network entities
- Downloading of programs and remote management performance

or for the following supervision activities:

- Variable monitoring
- Alarm management.

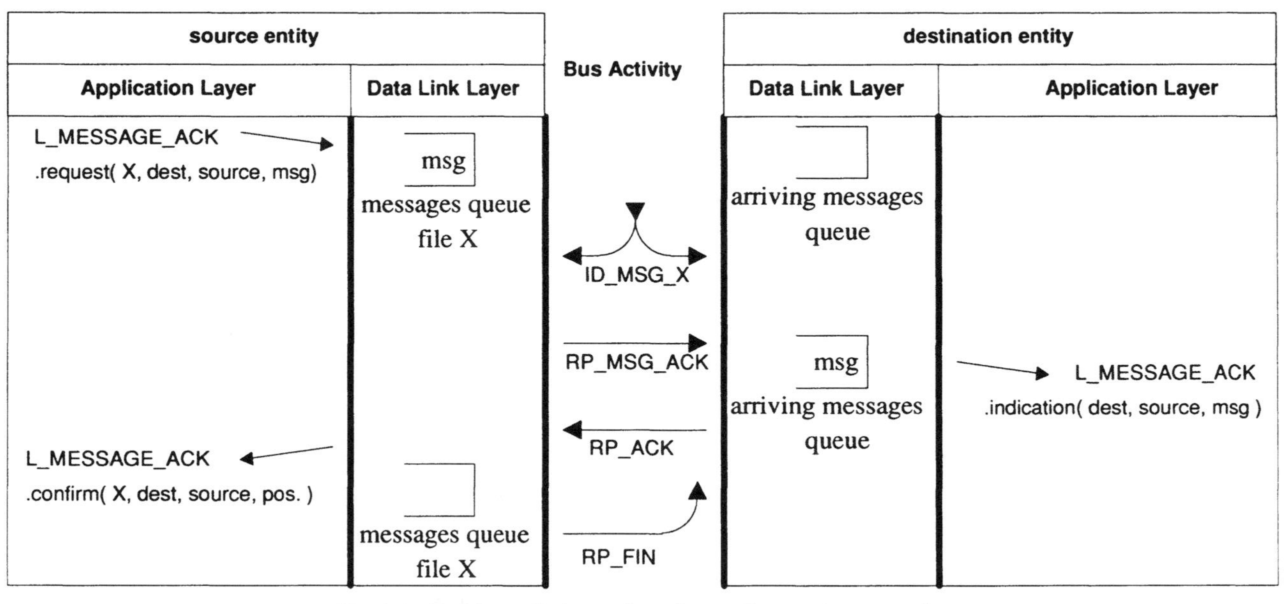

FIGURE 8.16. Data link services for cyclic messages exchange.

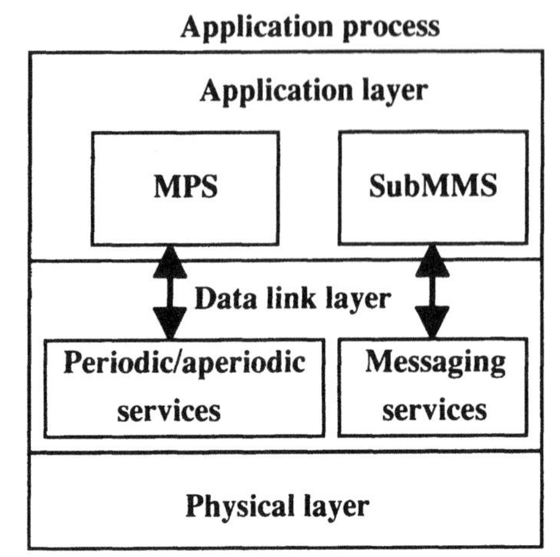

FIGURE 8.17. Application layer.

The MPS is the most important module of the application layer and the remaining part of this section mainly is dedicated to it.

4.3.1. MPS.

The MPS module manages different but related activities such as:

- Resynchronization of produced and consumed variables
- Management of information on the freshness and on the temporal and spatial consistency of data
- Reading and writing either local or remote variables, indication of reception and transmission.

The variable resynchronization mechanism guarantees that asynchronous and synchronous application processes can cooperate in a synchronous distributed application, and this is achieved by using a double memorization technique. In fact, for each variable concerned, the application layer handles two buffers: a *private buffer* and a *public buffer* as shown in Figure 8.18. The application process is allowed to access only the private buffer, whereas network functions can operate only on the public buffer.

The resynchronization mechanism can be used either for produced or for consumed variables. In the first case, the application process writes the new values of the variable in the private buffer in a completely asynchronous way while the network functions can read data from the public buffer. In the second case, data coming from the network are written into the public buffer and the

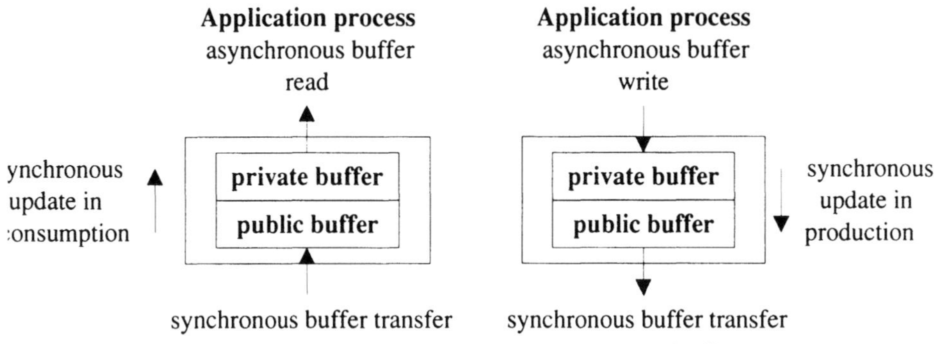

FIGURE 8.18. Resynchronized variables double buffering.

application process reads data from the private buffer. The update operation is triggered by the reception of a *synchronization variable,* whose period is called *synchronization period.* Figure 8.19 shows the contents of the private and public buffers for a produced resynchronized variable. The dotted line shows changes in the content of the private buffer, thick lines mean that the content of the public buffer has been updated where *s* is the reception time of the synchronization variable. A variable can be declared as resynchronized, if necessary, during the network configuration phase. In this case the synchronization variable must be specified too.

The application layer offers the application processes mechanisms with which to control the promptness and refreshment of variables. A status flag associated

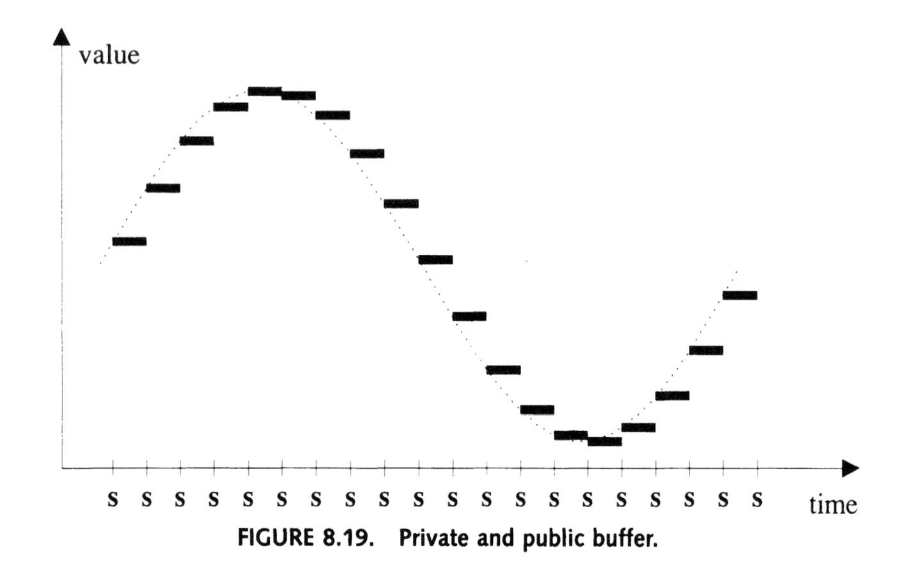

FIGURE 8.19. Private and public buffer.

with each variable gives application processes information about the promptness for the variables consumed and about the refreshment for the variables produced. These mechanisms are optional and must be selected during the network configuration phase. It is worth noting that both promptness and refreshment can be asynchronous, synchronous, or punctual.

In asynchronous promptness the status information tells the application process whether the bus arbiter has scheduled the variable correctly and/or the network and data link layers work properly. The status information is computed on the basis of a time interval set by the consumer. This time interval should reflect the minimum period between two accesses to the variable made by the consumer. The timer is triggered by the variable reception and the status becomes true until the time interval has elapsed. Figure 8.20 shows the timing behavior of the asynchronous promptness status versus the consumed variable updates. In Figure 8.20 the consumed variable is received at the time *c*.

The asynchronous refreshment status is computed by a variable producer and the mechanism is the same as the one for asynchronous promptness. The application layer of the producer handles a timer set to a value which should be equal to the production period of the variable itself. Each time the variable is produced, the timer is started and the status becomes true. The variable is sent on the network together with the status information so that all the consumers can be informed about the correctness of producer's operations. A false value of the status means that the timer has expired before a new value has been produced, i.e., the producer does not satisfy its production period requirement. The word *asynchronous* means that the true or false value of the status is related to the production or to the consumption interval of the application process and it is completely independent of the network operations.

In order to synchronize promptness and refreshment status information with network operations, the mechanisms discussed above must be modified slightly. The timer in this case is triggered by the indication of the reception of a synchronization variable and the status becomes true at the production or consumption time and does not change until a new synchronization variable is received (or

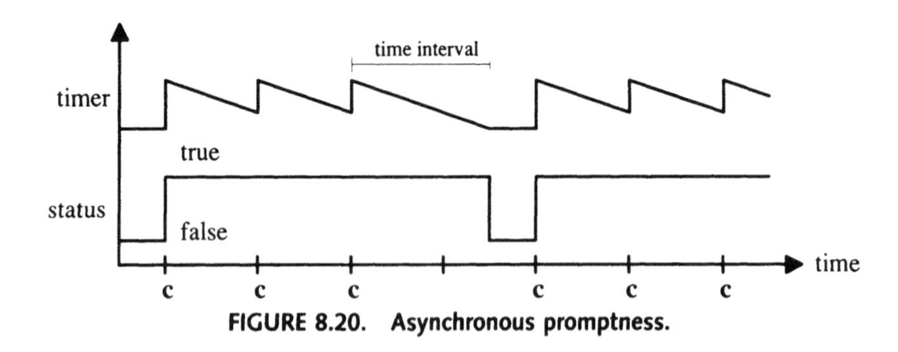

FIGURE 8.20. Asynchronous promptness.

until the time interval has elapsed). Figure 8.21 shows timing diagrams for synchronous refreshment (the synchronous promptness is slightly different) where s is in the reception time of the synchronization variable and p is the production time of the variable.

The timing diagrams for promptness and refreshment are very similar; only consumption time is substituted by production time. The conceptual difference is that the promptness status is computed in the application layer and is used by a local application process, while the refreshment status is computed in the application layer and is used by remote application processes. Punctual promptness and refreshment give status information which is true if the variable has been received/produced within a timing window started on the reception of a synchronization variable.

These concepts can be extended to lists of consumed or produced variables. The temporal consistency of a list of variables is satisfied if all the variables in the list have a true value of the promptness or refreshment status. The spatial consistency of a list of variables is information which allows a consumer of such a list to compute status information whose semantic properties satisfy its application process. For example, this status must be true if and only if all consumers of such a list share the same values of the variables of the list at the same time. This mechanism requires each consumer to periodically broadcast an additional consistency variable to all the other consumers. If some variables of a list exhibits a false promptness status, it is possible to request a further transmission of the invalid variables. This mechanism is implemented by the application layer in a transparent way to the application processes and uses the L_SPEC_UPDATE services offered by the data link layer.

The application layer offers the application processes services reading and writing of variables and indicates whether a variable has been sent or has been received. The services A_READLOC and A_WRITELOC allow the application process to read or write a variable in the local communication entity. These primitives do not involve network activity.

A_UPDATE is a service used to update a certain variable which can either be produced or consumed, or can either be not produced or not consumed by the

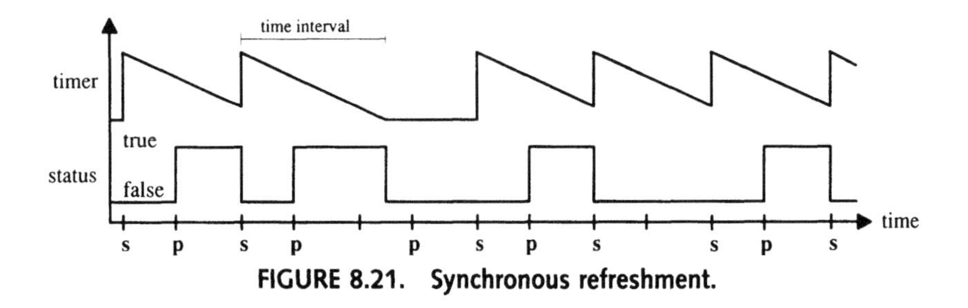

FIGURE 8.21. Synchronous refreshment.

application process which has issued the request. This service involves network activity and uses the L_FREE_UPDATE primitive of the data link layer; it can therefore be used to start acyclic exchanges. The A_READFAR allows the application process to start an acyclic read to update the local copy of the variable and, subsequently, to read the local copy. The A_READFAR uses the L_FREE_UPDATE primitives to carry out the acyclic exchange. If the variable is not resynchronized, primitives L_GET are used to read the value from the buffer in the data link layer. If the variable is resynchronized, the private copy of the buffer in the application layer is read and, in this case, no data link layer service primitives are needed. A_WRITEFAR behaves in a symmetric way. If the variable is not resynchronized, a new value is written in the local copy of the variable by using L_PUT primitives, and subsequently an acyclic transfer request is issued. If the variable is resynchronized the private copy of the buffer is updated. Then an acyclic transfer is started by means of the L_FREE_UPDATAE primitives. Both A_WRITEFAR and A_READFAR allow the user to specify either `urgent` or `normal` priority for the acyclic exchange.

Services A_SENT.`indication` and A_RECEIVED.`indication` are used by the application layer to inform the application process that a produced variable has been sent on the network and a consumed variable has been updated by the network respectively. The A_SENT.`indication` primitive is optional. There are two other services: A_READ and A_WRITE. They enable the application process to operate without knowing whether the variable is local or far. These services read attributes of the local copy of the variable and then automatically initiate a local or far operation.

4.4. Components

FIP needs ad hoc components either for the physical interface or the protocol high layers. A number of chips have appeared on the market during the last few years. For example, Telemecanique has developed the FIPART chip, which covers that part of the physical layer which is independent of the communication section (data encoding, error detection, etc.) and some services of the data link layer. Telemecanique has also developed the FIPIU chip, which enriches FIPART with all services of the data link layer. The French company CEGELEC has developed the FULLFIP component, which implements all of the FIP protocol stack. At the moment it is the most complete chip available for use in FIP networks.

Components with different levels of functionality are needed because of the different needs that network stations have: A simple sensor needs less functionality than a controller and so on. Together with the above-mentioned chips, line tools for twisted pair and optical fiber have also been developed. Figure 8.22 shows a simplified layout for an FIP board interfacing in a personal computer

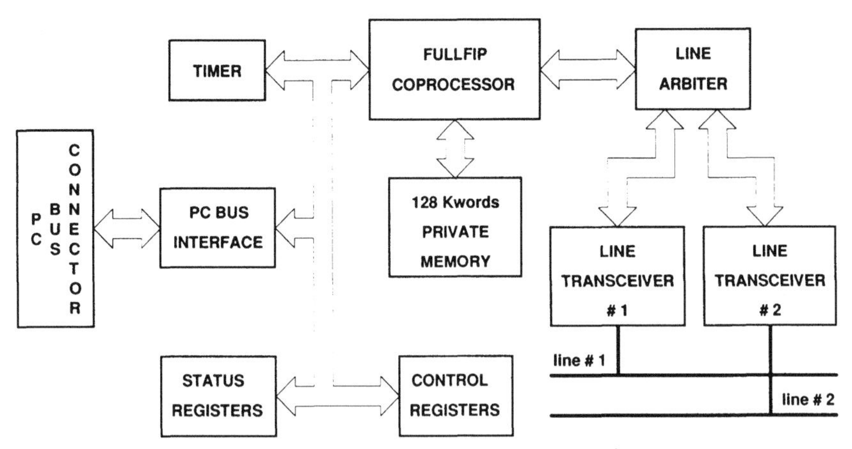

FIGURE 8.22. FIP board with PC interface.

environment based on the FULLFIP component. The *line arbiter* block is used to detect whether a line is broken and then to switch to the alternate one. The *timer* block is used to control line activity and to provide the clock of the FULLFIP chip.

5. PROFIBUS

PROFIBUS, which is a German national standard promoted by the German standard organization (DIN), is able to interconnect both low-cost devices, such as sensors and actuators, and more powerful machines such as PLCs (Programmable Logic Controller) and NCs (Numeric Controller). The network architecture is based on a reduced protocol profile and is quite similar to miniMAP. The network topology is based on a common bus and the access technique adopted is a slightly different version of the token-bus mechanism, thus guaranteeing bounded response times. A simple bit serial asynchronous transmission protocol has been adopted so that components available commercially can be used.

A PROFIBUS network can include two kinds of stations: masters and slaves. Masters take part in the circulation of the token while slaves play a passive role and can only be polled by the master stations. This hierarchy is typical in field control systems, where sensors and actuators are slave stations while PLCs and NCs are master nodes.

The PROFIBUS medium access control includes the services set out in the IEEE 802.4 MAC standard [11]; however, an additional service has been introduced called cyclic send and request data with reply (CSRD) to deal with cyclical polling of the slave stations. In this way a number of slaves can be specified in a list and then are polled automatically by a master at each token reception without any user intervention. The set of services provided to the application users by PROFIBUS is very similar to that provided by the manufacturing message specification (MMS). This leads to a high degree of compatibility between this kind of field network and the more complex factory automation networks.

PROFIBUS adopts a three-layer reduced protocol stack, as shown in Figure 8.23 (compared to the ISO/OSI seven-layer model). The functionalities of the missing layers, where necessary, are provided by the data link layer and the lower portion of the application layer.

5.1. Physical Layer

In order to meet the various requirements that can be found in different factory environments, the PROFIBUS architecture has been designed to support more than one physical layer. Here, version one of the PROFIBUS system is considered. It is a solution which has been conceived to facilitate cheap and easy implementations by making use only of commercial components like UARTs (Universal Asynchronous Receiver Transmitter) and low-cost line couplers. This choice implies that the physical layer implementations are slightly less efficient

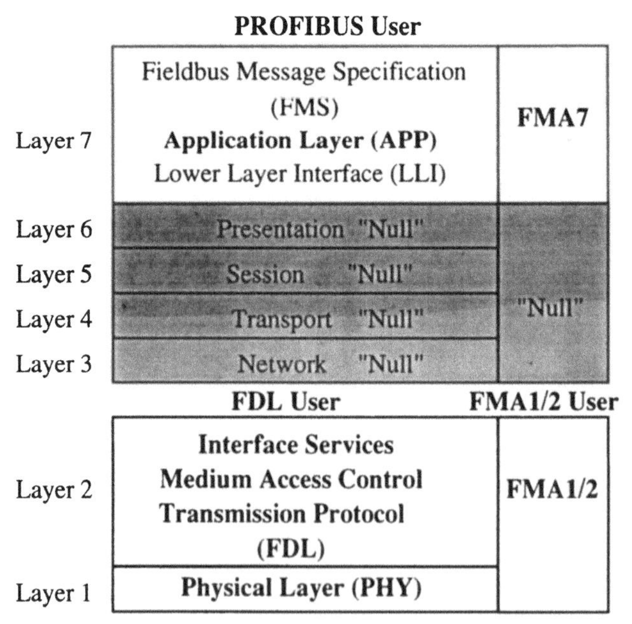

FIGURE 8.23. PROFIBUS protocol profile.

than those of other protocols which make use of components designed and developed ad hoc. Additional features included in the standard document are the remote powering of field devices connected to the network and the use of a redundant physical medium (a dual-communication bus architecture) to improve the reliability of the system. Future versions of the standard will allow longer segments, the use of fiber optics to reduce electromagnetic interference, and will provide intrinsic safety when operating in an explosive environment.

5.1.1. Network Topology.

The basic system segment is a linear bus, terminated at both the ends of the twisted pair cable. The maximum line length allowed is 1200m and up to 32 stations or repeaters can be connected to a single line. It is possible to increase the total network length and the number of stations connected to the network by means of repeaters. No more than three repeaters can be placed between any pair of stations in the network. This leads to two different network shapes: the linear bus (with repeaters) and the tree. The linear bus topology allows a maximum of 122 connected stations and a maximum distance between the end stations of 4.8Km (see Figure 8.24a). In the tree topology (using active stars) a larger area can be covered, and up to 127 stations can be connected (Figure 8.24b). In the pictures, M/S is a generic station (master or slave), REP is a repeater, and Rt is a bus terminator.

5.1.2. Transmission Speeds and Data Encoding.

Version one makes use of the nonreturn to zero (NRZ) encoding scheme, combined with the EIA RS-485 signaling technique, while the physical medium adopted in this case is the shielded twisted pair. A number of data signaling rates have been taken into account in the physical layer of version one so as to deal with different topologies and network sizes. Data rates ranging from 9.6Kb/s up to 500Kb/s are defined, depending on the segment length.

5.1.3. PROFIBUS Controllers.

Implementations of the PROFIBUS protocol profile rely on the availability of a so-called PROFIBUS controller (PBC). The purpose of a PBC is to connect a field device or a field controller to the physical medium. A PBC consists of three components: the line transceivers, UART, and a microprocessor used to implement the functionalities of the data link and the application layers and to interface the PBC to the user processes, as shown in Figure 8.25. Optionally electrical or optical couplers can be inserted between the transmission medium and the transceivers.

To obtain a satisfactory trade-off between performance and cost, three types of implementation are envisaged:

a) Repeater in a Linear Bus Topology

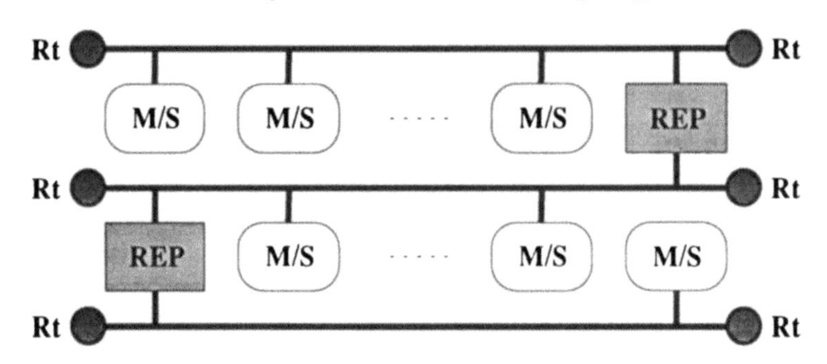

b) Repeater in Tree Topology

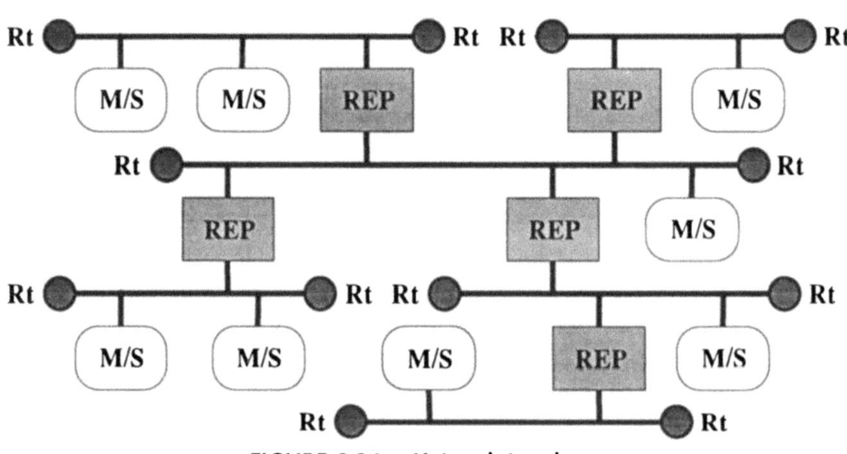

FIGURE 8.24. Network topology.

1. For high transmission rates, a dedicated CPU, that is a commercial microprocessor, should be used in order to carry out the data link and the application layer functions, while a UART is needed to send and receive characters on the network (PBC A in Figure 8.25).
2. For medium transmission rates, a single microcontroller chip with a built-in UART device can be used, thus reducing the PBC costs (PBC B in Figure 8.25).
3. For low transmission rates, the PROFIBUS protocol operations can be assigned to the processor of the device connected to the PBC, which is responsible for interfacing the UART (PBC C in Figure 8.25).

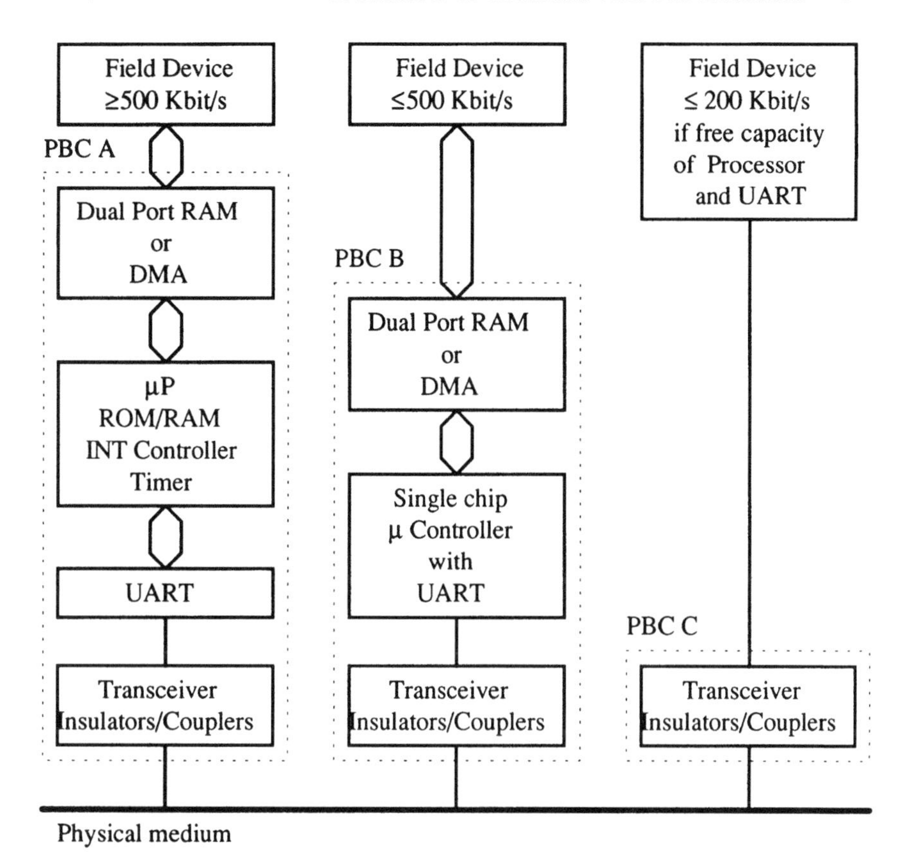

FIGURE 8.25. PBC controller.

5.2. Data Link Layer

The PROFIBUS data link (Fieldbus data link, FDL) must offer a well-defined set of services to the upper layer independent of the underlying physical layer. FDL is tailored to satisfy the typical requirements met in field environments, both in the method used to access to the communication medium (and to support real-time communications), and in the offered services (oriented to the controller/devices communication model).

5.2.1. Medium Access Method.

The medium access method is based on a mix of the decentralized token passing technique and centralized master-slave polling. The right to transmit on the bus is managed in a way very similar to the well-known IEEE 802.4 token bus protocol [11]. The token is circulated only among a subset of the stations connected to the physical medium, called master stations. When a master station becomes the token owner, it gets the right to transmit frames on the transmission medium. Slave stations instead play a passive role, in that they are not involved in the medium arbitration mechanism. A slave station can only transmit when it is requested to do so (that is, when it is polled) by a master station.

Each transaction on the bus is initiated by the master station holding the token. The token is passed among the master stations around a logical ring that is configured in the network startup phase and then is maintained dynamically according to a particular algorithm. This algorithm, which controls the correct circulation of the token and the maintenance of the consistence of the logical ring, requires each station to know its predecessor (previous station) PS and successor (next station) NS in the logical ring (Figure 8.26).

The medium access control technique used by a PROFIBUS protocol deals with (and is able to manage) the following error conditions and exceptions:

- Multiple token
- Lost token
- Error in token passing
- Duplicate station addresses
- Stations with faulty transmitter/receiver.

Moreover it is possible to add and remove stations dynamically during normal network operations, and to support any combination of master and slave stations.

Usually every exchange of messages takes place in two phases. First an action frame is sent by the master station holding the token to the designated destination (master or slave station), then the latter replies with a response or an acknowledgment message. Both the request and response frames can carry user data. Exceptions to this rule mainly consist of unacknowledged messages (for example,

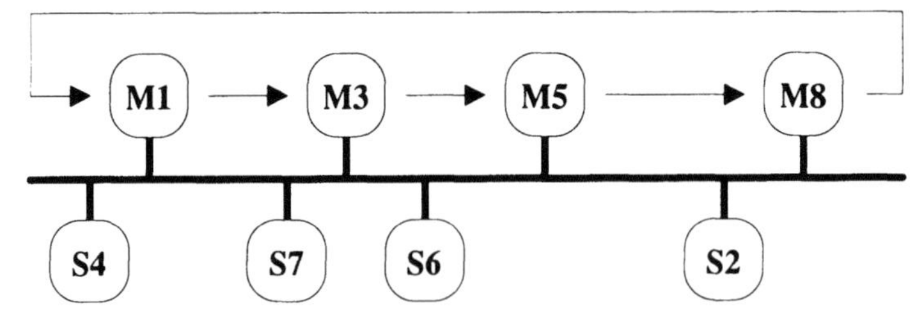

FIGURE 8.26. Token circulation in the logical ring.

broadcast messages are necessarily unacknowledged) and in token transmissions. Every time a master station sends a frame which requires a response, it starts a timeout counter. If the response (or acknowledgment) is not received within a certain amount of time, called slot time, the sender tries to retransmit the frame for a predetermined number of times. If there is still no response, the target device is marked nonoperational by the sending station. This prevents the sender from making further (multiple) retries by addressing the nonresponding device.

The PROFIBUS access method is able to enforce a policy which allows two priority levels in exchanging FDL data. Specifically, each protocol data unit (PDU) can be sent as a low-priority frame or a high-priority frame, and high-priority frames are always given precedence over low-priority ones. The choice of how to use the priority mechanism is left to the FDL user. This does not lead to any fairness problem, in that the medium access method guarantees that a station which misuses the priority mechanism (for example, by transmitting all its PDUs at the high-priority level) does not affect the high-priority traffic on the other stations.

The evaluation of the time taken by the token to travel around the logical ring is fundamental for medium access arbitration. Every time a master station receives the token, it reads and restarts a timer which measures the real token rotation time T_{RR}. The measured value is then compared with a target token rotation time T_{TR} selected for all the stations when the network is configured. If a nonzero token holding time T_{TH} is obtained ($T_{TH} = T_{TR} - T_{RR}$), message exchange cycles can be performed until the timer expires. High-priority messages have to be sent before any low-priority messages. When a transaction to exchange data is started it is taken to completion anyway (including any possible retries), irrespective of the actual token holding time measured by the initiating station. Even if there is no token holding time available, a station is allowed to perform at least one high-priority message cycle on the receipt of each token. This technique guarantees that reaction time, defined as the worst-case time interval between two high-priority exchanges on the same station, is bounded.

Besides normal data exchange cycles special mechanisms must be provided to build up and maintain the logical ring. Each master station keeps track of the other master stations in the logical ring by maintaining a data structure known as a list of active stations (LAS). The token is circulated among the stations listed in the LAS by increasing address order. It is up to each master station to dynamically update its LAS so as to reflect the actual situation of the logical ring. The insertion and removal of stations in the logical ring is determined by each master station as each master station is responsible of those stations which lie in the range of addressed beginning with the current station and ending with the next master station in the LAS. This range is called GAP, and is checked on the receipt of each token after all the messages queued in the token holder station have been sent.

5.2.2. Data Link Frame Format.

Frames are constituted by a sequence of characters, each one of which is encoded as a start-stop or UART symbol for asynchronous transmission. Figure 8.27 shows the character encoding scheme which is made up of one start bit (always binary 0), 8 bits representing the byte to send, one parity bit (even parity is adopted), and one stop bit (always binary 1). In this way, bit synchronization is achieved by looking for the falling edge of the start bit. Figure 8.28 shows the format of the data link frames. Frame synchronization is obtained by means of a start/stop delimiter (SD/ED) pair. Moreover, each action frame must be preceded by at least 33 idle bits (bits at the binary 1 value). There are four kinds of start delimiters (SDi) and they identify: fixed-length data frames with no data field (SD1); fixed-length frames with data field (SD3); frames with a variable-length data field (SD2); and the token frame (SD4). Each frame includes an eight-bit check sequence (FCS) provided for error detection purposes and computed as the arithmetic sum of all the characters in the frame, excluding the delimiters.

The destination (DA) and the source address (SA) fields which appear in each frame are one byte long. Since the most significant bit of this byte is used as an extension bit (EXT), valid addresses for the master and slave stations are in the range of 0–126, thus giving a maximum number of 127 stations that can be connected to a single PROFIBUS network. The address 127 in the DA field is reserved for the broadcast and multicast transmissions. If the EXT bit is set in the DA or SA field, an additional address extension for a link service access point (LSAP), or a region/segment address (to be used in hierarchical bus systems), is specified in the DATA UNIT field of the frame. LSAPs are used so that more than one service access point on each station can be made available to the FDL users. Each LSAP on each station can be addressed directly, thus making it possible for several higher-layer connections to coexist.

5.2.3. Data Link Services.

The fieldbus data link layer (FDL) offers four kinds of services to its users:

- Send data with acknowledgment (SDA)
- Send data with no acknowledgment (SDN)

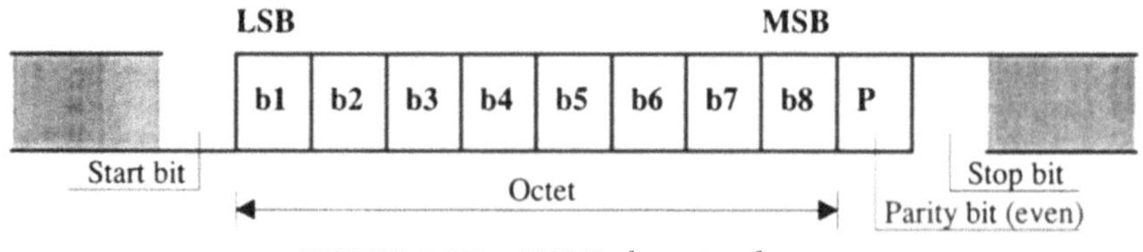

FIGURE 8.27. UART character frame.

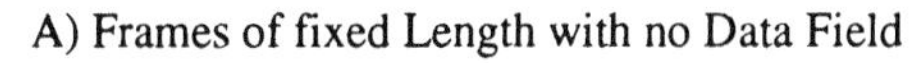

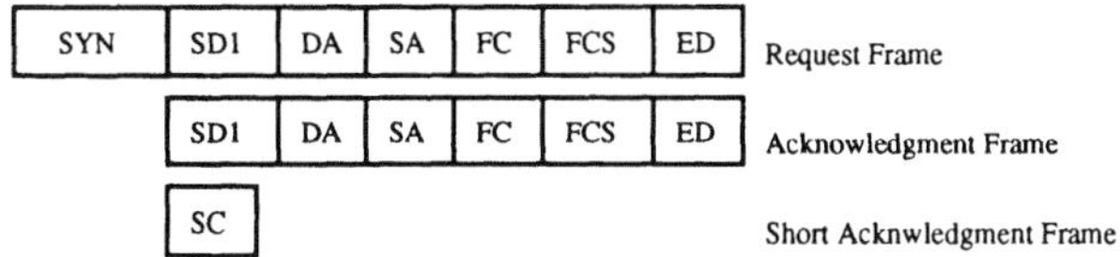

B) Frames of fixed Length with Data Field

SYN	SD3	DA	SA	FC	DATA UNIT	FCS	ED	Send/Request Frame
	SD3	DA	SA	FC	DATA UNIT	FCS	ED	Response Frame

C) Frames with variable Data Field Length

SYN	SD2	LE	LEr	SD2	DA	SA	FC	DATA UNIT	FCS	ED	Send/Request Frame
	SD2	LE	LEr	SD2	DA	SA	FC	DATA UNIT	FCS	ED	Response Frame

D) Token Frame

SYN	SD4	DA	SA

where:

SYN	Synchronization Period, a minimum of 33 idle bits
SD1	Start Delimiter 1
SD2	Start Delimiter 2
SD3	Start Delimiter 3
SD4	Start Delimiter 4
DA	Destination Address
SA	Source Address
FC	Frame Control
LE	Octet Length, allowed values 4 to 249
LEr	Octet Length repeated
DATA UNIT	Data Field, variable Length (L-3), max. 246 octets (variable) or 8 octets (fixed)
FCS	Frame Check Sequence
ED	End Delimiter

FIGURE 8.28. Data link frame format.

- Send and request data with reply (SRD)
- Cyclic send and request data with reply (CSRD).

SDA and SDN are very common transfer services and are provided by nearly all the industrial data link layers, while SRD has been derived from the IEC PROWAY C standard [12] and the IEEE 802.2 type 3 services [13]. The SDA service allows a user on a master station to send data to a single remote user in a confirmed way, that is to say, the remote station has to confirm the correct reception of the transmitted data (see Figure 8.29a). If an error occurs during the transmission, the sending station makes a certain number of retries. Then, if an acknowledgment is not received, the user is notified of the transmission failure.

a) SDA service

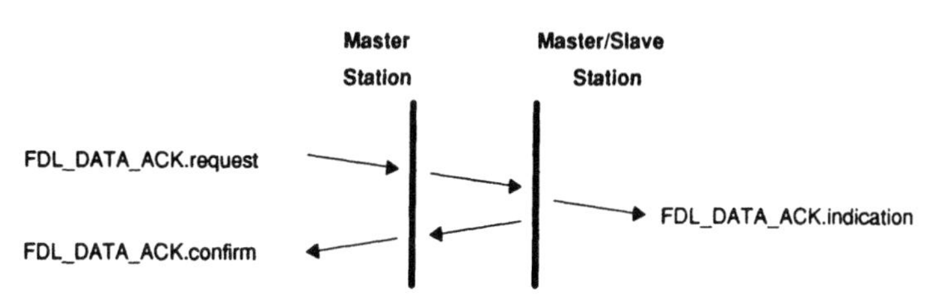

b) SDN service

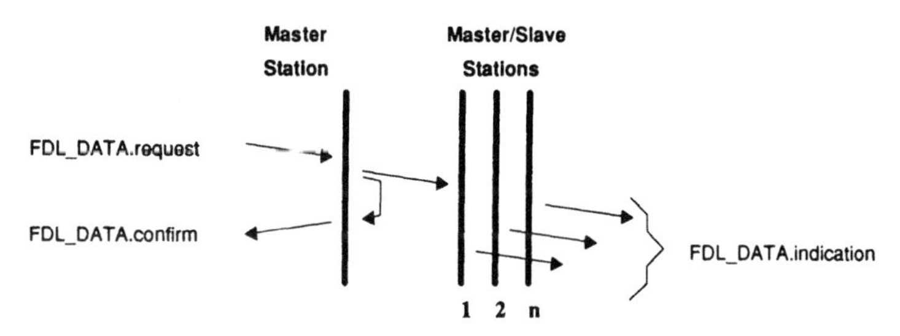

c) SRD service

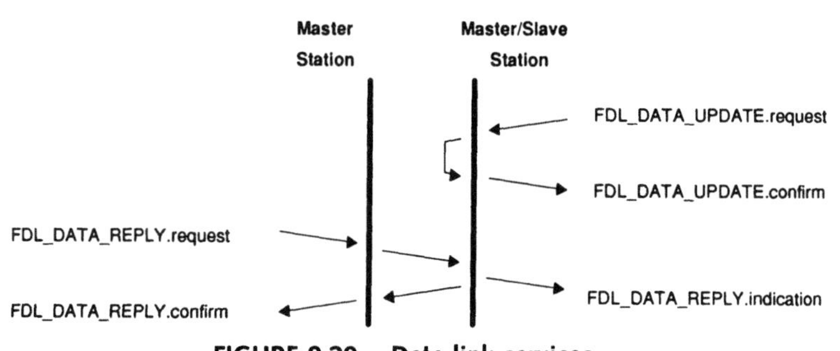

FIGURE 8.29. Data link services.

The following set of primitives, related to the SDA service, is for sending a message with acknowledgment:

- `FDL_DATA_ACK.request()`
- `FDL_DATA_ACK.indication()`
- `FDL_DATA_ACK.confirm()`

The `FDL_DATA_ACK.request()` primitive is used to actually send data. The receiver is notified of an incoming message by means of the

FDL_DATA_ACK.indication( ) primitive, and the acknowledge signal is passed back to the sender through the FDL_DATA_ACK.confirm( ) primitive.

The SDN service is used to send data without acknowledgment from the receiving station. This means that the sender has no confirmation of the correct reception from the remote user. Unlike the previous service, SDN can be used to send data either to individual remote users, to groups of remote users (multicast addresses), or to all the stations in the network (broadcast address), as shown in Figure 8.29b. Three service primitives, related to the SDN service, are for sending a message without acknowledgment:

- FDL_DATA.request( )
- FDL_DATA.indication( )
- FDL_DATA.confirm( )

For the SDN services the FDL_DATA.confirm( ) primitive is generated locally.

The SRD service is similar to SDA, but in this case the acknowledgment which comes from the remote station can contain data that must have been previously made available by the remote (polled) user (Figure 8.29c). This service is therefore used to poll remote stations in an asynchronous way. Two sets of service primitives are provided:

- Primitives for sending a message with reply:

    –FDL_DATA_REPLY.request( )
    –FDL_DATA_REPLY.indication( )
    –FDL_DATA_REPLY.confirm( )

- Primitives to update the buffer on the receiver side:

    –FDL_REPLY_UPDATE.request( )
    –FDL_REPLY_UPDATE.confirm( )

The FDL_REPLY_UPDATE( ) primitives are used by the receiving station to store data that will be returned to the polling master when it subsequently invokes an FDL_DATA_REPLY.request( ) primitive.

CSRD is peculiar to the PROFIBUS protocol, and is particularly suited to the field environment. The CSRD service is a way to make the FDL automatically perform a cyclic query on a set of destinations listed in a poll list (specified by the user) whenever the station gets the token. The following primitives are specified:

- Primitives to update the buffer on the sender side:

    –FDL_SEND_UPDATE.request( )
    –FDL_SEND_UPDATE.confirm( )

- Primitives to set the poll list:

  –FDL_CYC_DATA_REPLY.request()
  –FDL_CYC_DATA_REPLY.confirm()

- Primitives to enable single entries in the poll list:

  –FDL_CYC_ENTRY.request()
  –FDL_CYC_ENTRY.confirm()

- Primitives to deactivate cyclic polling:

  –FDL_CYC_DEACT.request()
  –FDL_CYC_DEACT.confirm()

The user at the master side can specify a poll list through the FDL_CYC_DATA_REPLY() service primitives, which also start polling. Each entry in the poll list can be individually enabled or disabled by means of the FDL_CYC_ENTRY() primitives, and the polling can be stopped by using the FDL_CYC_DEACT() primitives. Data exchange is similar to the SRD service, but, since this service is performed in a totally asynchronous way with respect to the user operations, data to be sent must be previously stored by the transmitting user in suitable buffers through the FDL_SEND_UPDATE.request() service. Both the master and the slaves are notified of each cyclic exchange initiated in the network by means of the FDL_CYC_DATA_REPLY.confirm() and the FDL_DATA_REPLY.indication() primitives, respectively. This kind of data exchange is depicted in detail in Figure 8.30. In this figure the initial phase in which polling is started, a generic poll cycle, and finally the poll deactivation phase are shown.

5.3. Application Layer

The PROFIBUS application layer has been designed to meet the needs found in real-time factory environments and specifically to support communications among application processes, even though they can represent simple devices such as actuators and sensors. Moreover, application services make it possible to easily integrate a PROFIBUS subnetwork into a hierarchical communication system such as that described by the manufacturing automation protocol (MAP) specifications. The PROFIBUS application layer consists of two logical entities, namely the fieldbus message specification (FMS) and the lower-layer interface (LLI).

CSRD service

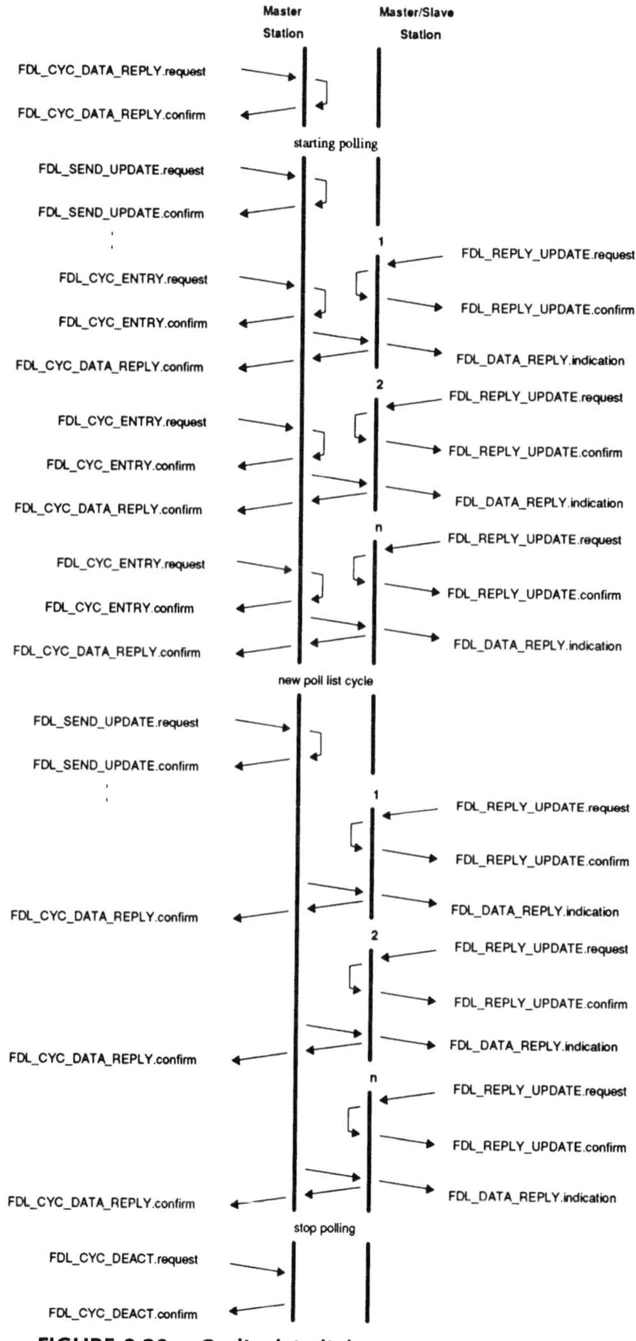

FIGURE 8.30. Cyclic data link services (CSRD).

5.3.1. Lower-Layer Interface.

The LLI provides interface functions between the application layer and the lower FDL layer. It maps the functionalities of the FMS on the services provided by the FDL, by managing the establishment and the release of connections.

5.3.2. LLI Services.

There are five kinds of services provided by the LLI to the FMS:

- Associate (ASS), used to establish a connection
- Data transfer confirmed (DTC), used for the transmission of messages for confirmed services on a previously established virtual connection
- Abort (ABT), used to release a connection
- Data transfer acknowledged (DTA), used for the transmission of messages for unconfirmed services
- Data transfer unconfirmed (DTU), used for the transmission of messages for unconfirmed services without establishing any virtual connection.

The last two services are optional, and may also be omitted from certain network stations.

5.3.3. Connection Relationships.

Application processes access the communication functionalities offered by the application layer of the PROFIBUS protocol through so-called communication end points (CEP). A communication between two application processes takes place on a communication relationship (CR), which logically connects a unique pair of CEPs. There are two kinds of CRs, i.e., connection-oriented CRs and connectionless CRs. The destination address is used to distinguish among one-to-one CRs, one-to-many CRs (multicast), and one-to-all CRs (broadcast). In PROFIBUS one-to-one CRs are connection-oriented, while the other two, broadcast and multicast CRs, are necessarily connectionless since they involve more than two dialoguing entities.

For the connection-oriented CRs, a connection between the two partners must be established before any data can be exchanged. Therefore, a connection establishment phase must precede any data transfer phase. When a connection is no longer needed it can be released by the user. Two kinds of connection-oriented CRs are possible. The first set contains the master-master acyclic CRs, while the second includes the master-slave CRs. Master-slave CRs can be further subdivided into cyclic and acyclic data transfers. In this case, the slave can be given the initiative to issue an unconfirmed service request to the master.

Connectionless CRs are used for one-to-many (multicast) or one-to-all (broadcast) connections. Since the connection need not be established nor released, these CRs are always in the data transfer state. They are useful for operations involving a community of users; in particular, they can be used for synchronization of remote stations, for global alarms, or for triggering some action such as starting application processes on a number of remote stations.

5.3.4. *Fieldbus Message Specification.*

The aim of this upper entity of the application layer is to provide a common view of the distributed factory environment to the application processes which are in it and reside on the different devices connected to the network. Usually, application processes are not aware of the actual communications that take place among the stations but deal with an abstract model which describes field devices. The interface to this model is enforced by the FMS entity.

The logical model mentioned above is named *virtual fieldbus device* (VFD), and closely resembles the virtual manufacturing device (VMD) structure described in the manufacturing message specification (MMS). The VFD model is based on the VFD object, which contains all the communication objects related to the VFD. Descriptions of the various objects which make up the VFD are stored in a suitable object dictionary (OD).

Interactions between the actors operating in a PROFIBUS environment occur according to the client-server model, that is to say, a process (the client) requests a service from another process (the server). Two kinds of service requests are defined: confirmed services and unconfirmed services. Confirmed services have to be acknowledged by the server, while this is not the case for the unconfirmed requests.

5.3.5. *FMS Services.*

A VFD is made of a number of constituting objects, each of which can be manipulated by means of an appropriate set of services. FMS services can be roughly divided into seven classes.

5.3.5.1. VFD Support Services. The VFD object is characterized by four attributes (vendor name, model name, revision, and profile number). These attributes describe some static properties of the VFD. The logical and physical status of the associated device are also stored in the VFD object. The two services `Identify` and `Status` let a remote user read the identification attributes and the status of the device, respectively. The physical status gives a rough indication of the current state of the real device, that can be operational, partially operational, not operational, or needing maintenance. The logical status instead keeps track of the communication functionalities of the device. Typical values of the logical status are: ready for communication, limited number of services, OD-LOADING-NON-INTERACTING and OD-LOADING-INTERACTING. The two last conditions occur when the object dictionary is being loaded.

5.3.5.2. Object Dictionary Management. The object dictionary (OD) is used to keep the description of the communication objects defined in the VFD. Object descriptions are identified with a unique numeric index. Some of them can also be identified optionally with a visible string. Two kinds of OD services are defined, one for retrieving the descriptions of objects and the other for writing one or several objects in the OD, respectively.

5.3.5.3. Context Management. The context management services allow the user to establish and release logical connections (or communication relationships). The basic object type used in modeling the communication relationships of a FMS entity is the Communication Relationship List (CRL) object. This kind of object contains a list of all the active CRs of the device, together with a set of attributes characterizing each CR. Services are provided to establish a new CR and to release an existing one. A third service is used by the FMS to reject improper protocol data units received from a remote station.

5.3.5.4. Domain Management. A domain object is defined to describe memory areas containing data or programs. Domains are viewed as sequences of uninterpreted bytes, each one having a given maximum length. Two sets of services are defined on domain objects: downloading and uploading the domain contents. Downloading consists of loading data from the client into a domain placed on the server, while uploading is used to transfer data from the server to the client. Since the uploading and downloading operations may require an arbitrarily large amount of data to be exchanged and since the size of data carried in each PDU is bounded, an upload and a download control scheme have been defined so that the correct transfer sequences can be carried out.

5.3.5.5. Program Invocation Management. The program invocation object type provides the means to describe the concept of the program and its interactions with the associated domains. A program invocation object can be created and deleted. Such an object specifies the list of domains the program will use and a number of options to define the access rights. Moreover, the execution of a program invocation can be started, stopped, resumed, reset, and killed by the application process.

5.3.5.6. Variable Access. One of the most important classes of objects defined by FMS is the variable object class. There are five kinds of variable objects: simple variables, arrays, records, variable lists, and physical accesses. Simple variables are the conventional primitive variables such as, for example, those defining boolean, integer, or floating point values. An array is a collection of simple variables, all of which have the same type. A record, instead, is made of a collection of data, possibly of different types. Both arrays and records can be accessed either as a whole or on an element-by-element basis. A variable list object is made up of a list of arrays, records, and simple variables. Two services are defined to create and to destroy variable list objects. Physical access objects are used to make raw accesses to portions of the device's memory (which are viewed as byte strings). They cannot be redefined or deleted, and accesses are specified in terms of the initial address and length of the byte string. Services are defined so that application processes can read and write the value of the variables in a VFD. It is also possible to transmit the value of a variable to a broadcast or multicast address. This (unconfirmed) service is useful when it is necessary to update several copies of the same variable (located on different devices) at the same time.

5.3.5.7. Event Management. Event objects are used to send important messages, such as alarms, from one device to another or to several other devices. Events are notified by using the event notification (unconfirmed) service. It is also possible to lock or unlock remote generations of events. When an event object is locked, further event notifications are disabled.

6. IEC 1158 FIELDBUS

The IEC 1158 Fieldbus is a standard proposal currently being considered by the International Electrotechnical Committee. This standard proposal started from a joint effort made by the IEC and the Instrument Society of America (ISA) and is also referred to as the ISA/SP 50 proposal. To facilitate progress and keep the ISA and IEC standards in step, combined IEC/ISA meetings were held. This protocol is also known as Fieldbus. The standardization activity concerning the physical layer has been completed while the data link layer is quite stable even though it still has to be approved as an international standard. The application layer is still under development and this activity seems to be far from being completed.

The IEC 1158 standard has a reduced profile stack with only three layers (Figure 8.4) and the data link services offered to the IEC Fieldbus application layer are supposed to be compatible with those offered to the network layer in the OSI environment, as shown in Figure 8.31. The main characteristics of the first two layers will be introduced below. The medium bandwidth allocation mechanism is centralized with a station which arbitrates the right to ask for and send data between the network stations. Real-time requirements and dynamic network configuration are handled by means of dynamic bandwidth allocation.

6.1. Physical Layer

The physical layer has been approved completely and it concerns three different media: copper wire, optical fiber, and radio. By using copper wire, remote powering is enabled, i.e., the power signal can be transmitted on the same medium as that used for the data signal. An IEC network can be made by several subnetworks connected together by means of bridges. Moreover, a bridge can be used to connect to the network a station which cannot satisfy the timing requirements imposed by communication rules. In such a case

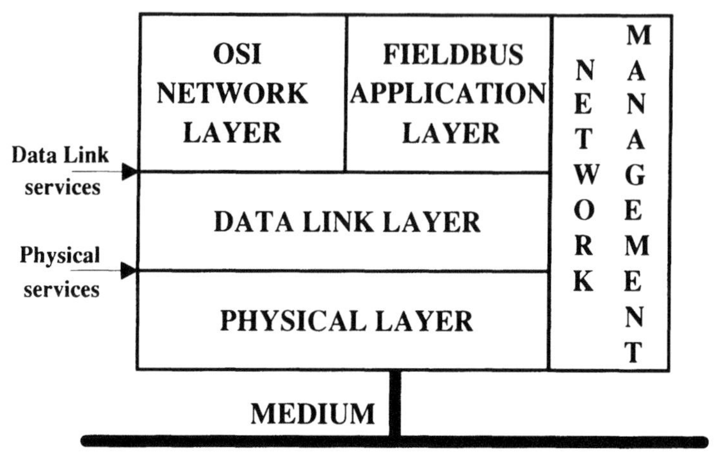

FIGURE 8.31. IEC 1158 stack vs OSI environment.

the bridge acts as an interface toward the network. The physical layer offers the data link layer a set of services which are very similar to the physical layer services of the other protocols examined above. There is only one new service: `Ph_CHARACTERISTICS.indication` which is used at the startup of the network by the physical layer to give the data link layer information about the physical data rate and the number of bits added to each frame before it is sent.

6.2. Data Link Layer

In the IEC Fieldbus, network stations must require and obtain system bandwidth in order to receive and send data. The mechanism is controlled by a network station called link active scheduler (LAS), which gives stations the right to communicate. This right is represented by a token.

There are three kinds of tokens circulating on the network:

- The scheduler token is owned by the station which acts as LAS. This token can be sent to another link master (LM), which becomes the new LAS. Only one LAS can be active in each subnetwork at any given time. Link masters are network stations which are capable of behaving as an LAS.
- The delegated token is sent by the LAS to those network entities that either make an explicit request for it or must be scheduled with a predefined periodicity. A network entity can hold the delegated token only for a predefined period of time.
- The reply token is used by the owner of the delegated token to request an immediate answer from another station without having to wait for the addressed station to be scheduled by the LAS.

When a network entity requests the delegated token from the LAS, it has to specify how long the token should be, the priority of the communication, the kind of scheduling and the timing requirements. The priority parameter can have three possible values: urgent, normal, and time-available. Different priorities involve different upper bounds on the maximum size of the message. For instance, only requests for data shorter than 64 bytes can be handled as urgent, while normal requests can involve data sizes up to 128 bytes, and time-available, up to 256 bytes. The scheduling policy can be one-time, periodic, and triggered by an external event. One-time scheduling means that the LAS deletes the request after it has been successfully satisfied. Periodic scheduling is used for requests which must be satisfied with a given periodicity specified in the requests themselves. Note that a periodic request is satisfied cyclically without any further explicit invocation. Finally a service request can specify that the LAS sends the delegated token only after a certain event has occurred, so that the event triggers the scheduling activity for that request, i.e., the reception of a certain frame. Timing requirement parameters are used to specify the maximum time that may elapse before the request is satisfied, the maximum acceptable jitter, and the periodicity of the scheduling for a periodic request.

In order to carry out the complex activities mentioned above, the network entities must be able to compute in advance the time needed to transmit a certain amount of data over the network. Furthermore they also have to be synchronized. For this reason at station startup the physical link layer notifies the data link layer of the physical data rate and the number of bits added to each frame before transmission. In such a way, when the application layer requests a certain service to the data link layer, the latter is able to compute the number of messages and their sizes and the time needed to send and/or receive each them. These operations allow the data link layer to know in advance how much time is needed to complete a certain service.

In the IEC Fieldbus, each station has its own clock which can be periodically synchronized with the one of the LAS. This is done by means of suitable synchronization frames. The priority, the kind of scheduling, and the time requirements define a set representing the quality of service (QOS) required to the data link by the application layer when the service is invoked. If the QOS parameters are in conflict, the data link notifies the upper application layer of the problem immediately, otherwise a request is sent to LAS containing all the constraints. In order to verify whether the LAS will satisfy the requirements, the data link layer starts a number of timers. If a timeout occurs, it means that the LAS is not able to allocate enough bandwidth to handle the request and the requesting data link layer returns a suitable indication to the upper layer.

When the application layer requests a service, it can specify different data structures to be used for storing the exchanged information. In practice these structures can be either buffers or FIFO queues (see Figure 8.32) and they have to take into account the way the communication is handled by the application layer.

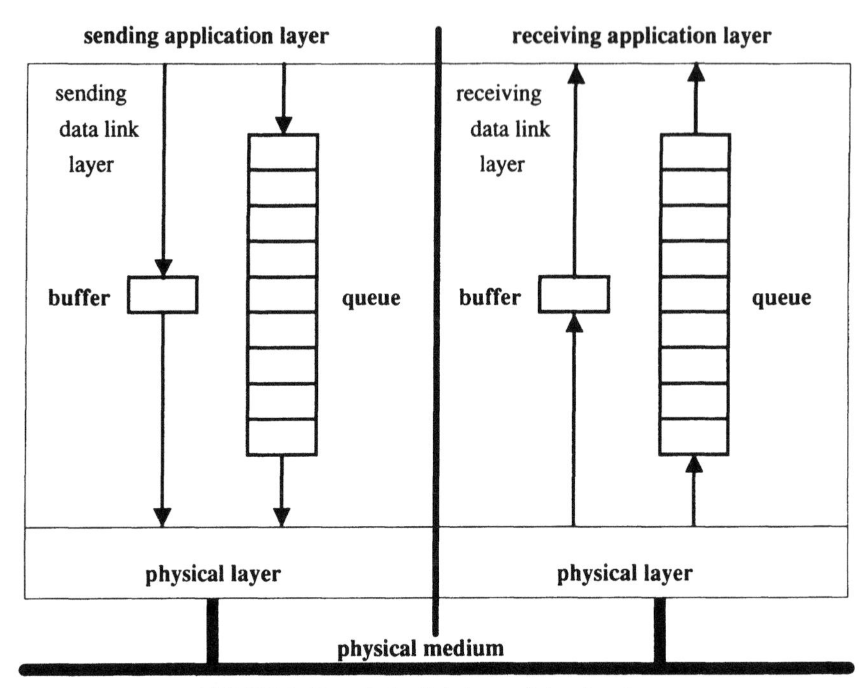

FIGURE 8.32. Data link layer data structures.

When buffers are used, their contents are overwritten whenever new data become available (either in transmission or in reception), while when FIFO queues are used, data are appended and dequeued without overwriting. In addition the application layer is allowed to specify whether an indication must be generated whenever a datum is received/transmitted from/to the network. For example, if the application layer has to send data and wants to obtain an indication on each transmission, a buffer can be used to contain the data to be sent and its content can be refreshed on each indication, without causing any loss of data.

The application layer can invoke both connection-oriented and connectionless services and require a particular QOS to handle the flow control for the connected services. The QOS values in this case are classical, disordered, ordered, and unordered. The classical QOS requires that data are delivered without any loss, duplication, and misordering. The disordered QOS does not tolerate any loss or any duplication, but the order of the messages received can differ from the sending order. The ordered QOS imposes the requirement that there is no duplication or misordering but data loss. Finally, the unordered QOS allows possible data losses, misordering, and duplications.

In order to handle the above flow control policies correctly, the data link layer has to implement suitable mechanisms based on data structures and counters. The scheduling activity of the LAS can be divided into three windows with decreasing priority: a cyclic window in which periodical and time-critical requests are served, an acyclic window in which time-available requests are served, and a third window used by the LAS to test whether new stations want to enter the network.

7. A BRIEF COMPARISON OF STANDARD FIELDBUSES

In this section the fieldbuses FIP and PROFIBUS and the standard IEC Fieldbus (still under study) will be compared, taking into account the requirements they have to meet as was discussed in Section 3.

7.1. Timing Requirements

As outlined earlier in this chapter a fieldbus specification has to satisfy a number of time constraints. In particular, a bounded response time must be guaranteed in order to provide urgent communications such as alarms, while the maximum jitter must be kept as low as possible in cyclic exchanges. The FIP protocol is deterministic by nature, so that all the transfer times are deterministic too. Since the transmission cycle of a certain cyclic variable is periodically repeated by the network arbiter, the delay experienced in exchanging this single datum is always bounded to the transmission cycle time itself.

Jitter is usually kept at very low values because a centralized polling scheme is adopted. In FIP no confirmation is returned to the producer of a variable value to indicate the success or the failure of the transmission. If an error occurs during a variable exchange, the receiver does not get the right value until the next cycle. FIP users, however, can request new values in place of the corrupted ones by using the window devoted to aperiodic data transfers in the FIP elementary cycle. The responsibility for such an error recovery action is left totally to the user.

PROFIBUS, just as in the well-known token bus approach, grants an upper bound to the time needed to access the channel and hence to the time required to send a high-priority message containing data values. PROFIBUS allows a master station to send at least one (high-priority) frame each time it gets the token. The transmission of low-priority messages and the polling of slaves can be carried out only if the measured real token rotation time is lower than an established target token rotation time. In this way the maximum delay that high-priority frames can experience is upper bounded, and the bound depends heavily on the target token rotation time [14]. Since a distributed mechanism is used to access the bus, PROFIBUS usually allows for a better utilization of the transmissive medium, but cyclic data exchanges can suffer from a nonnegligible jitter.

In the IEC Fieldbus all the decisions about scheduling are taken by the LAS. The mechanism adopted is very similar to the one used by FIP, but increased flexibility is expected since the scheduling operations are dynamically tuned at run time. Since all the scheduling information is managed by a single entity, better results can be expected both in terms of response time for the urgent asynchronous requests and of jitter for the cyclic exchanges. Finally, it is worth noting that the acyclic data transfer requests in FIP, the delivery requests of low-priority frames in PROFIBUS, and normal and time-available traffic in the IEC Fieldbus are queued on the transmitter side and are carried out by the network only when the urgent traffic allows it to do so, so that a bounded response time is not granted for such kind of services.

Fieldbuses are particularly oriented toward those applications which require a periodic exchange of small-sized data, even though the exchange rate can be significantly high. The term *efficiency* is used here to mean the ratio between the amount of useful user data (such as variable values) moved through the network during a time cycle and the number of bits needed to transmit such information. Fieldbus efficiency is largely affected by the mechanisms introduced in the medium access control (MAC) sublayer. Because of the different medium access techniques adopted, each network exhibits better efficiency under certain conditions. FIP, for instance, is very efficient when each produced variable is consumed by several stations. In this way only a single elementary sequence is needed to supply all the consumers of a variable with its current value simultaneously. As mentioned in Section 4, FIP variables can have different cycle times. This allows the network load to be kept as small as possible when some variables have to be updated at a rate which is slower than the elementary cycle.

The IEC Fieldbus behaves in a very similar way to FIP, but the increased flexibility usually has to be paid for in terms of efficiency since more bandwidth is used for transferring scheduling information. In fact, FIP cyclic data exchanges are defined statically in the network configuration phase, while in the IEC Fieldbus they can be dynamically scheduled, thus requiring a suitable exchange of control information. In PROFIBUS, when a master-slave cyclic connection is established, a real data transfer on the network takes place only when new data values become available on the producer side. Consumers (that is, receivers) have a copy of each datum involved in the transfer stored in a local buffer. This copy is updated whenever a new value is received from the producer. If the value on the producer side remains unchanged, the producer (slave) replies to a poll command from the consumer (master) with a very short acknowledgment since no new datum has to be returned in this case. This mechanism is provided to save system bandwidth so that the transmission of redundant information can be avoided and thus messages traveling on the network can be kept as short as possible. To further enhance the efficiency, all the above fieldbuses make use of compact data encodings and adopt particularly streamlined transfer syntaxes.

7.2. Data Exchange Priority

The mechanisms adopted to access the bus allow the medium to be shared out fairly among all the users connected to the network. Each device is instead responsible for the correct use of its own slice of the system bandwidth. In order to meet urgent traffic needs, it is necessary for each user to be able to specify a priority level associated with each outgoing message. Network policies will ensure that high-priority messages are delivered before any queued lower-priority message.

FIP cyclic data exchanges are ideally performed at the highest priority or, in other words, cyclic data transfers are always carried out in a deterministic way. Acyclic transfer requests are assigned to two priority levels, that is to say urgent and normal. Urgent requests are honored first by the bus arbiter that enforces the priority mechanism with a suitable scheduling of the servicing requests. No priority is associated with exchanged messages.

PROFIBUS provides a two-level priority scheme. Each PROFIBUS master station is allowed to send at least one high-priority frame when it gets the token; moreover, high-priority frames always have precedence over the low-priority transmissions. The high-priority option can be used only with the acyclic connections; in this way the cyclic polling of the slave stations could be delayed when the arrival rate of urgent frames is significantly high. The standard recommends that only very important events make use of the high-priority transmission services.

The IEC Fieldbus further extends the FIP and PROFIBUS mechanisms by adopting a scheme based on three priority levels: urgent, normal, and time-available. The substantial difference with respect to the priority scheme provided by FIP and PROFIBUS is that for each priority level a different upper limit of message length is also specified. This limit is lower for urgent messages and higher for messages which can be sent when there is enough available time to do so. Fairness and priority policies are handled in PROFIBUS by the medium access mechanism, while in FIP and in the IEC Fieldbus they are based on the operations of the bus arbiter and the LAS respectively.

7.3. Fault Tolerance

Operations in a FIP environment rely entirely on the availability of the bus arbiter. If the active bus arbiter becomes faulty, a method based on timeouts lets other stations configured as potential bus arbiters to activate a new arbiter election mechanism. FIP station faults do not prevent the whole network from working, even though some produced variables can become unavailable.

The PROFIBUS token passing mechanism is protected against token losses, corruptions, and/or duplications. Stations are free to enter or leave the logical ring at any time and the fault-tolerance characteristics are the same as those of IEEE 802.4 networks. The fault tolerance characteristics of the IEC Fieldbus are very similar to those of the FIP protocol, in that if the current LAS is faulty, a new LAS is selected from a set of link masters. Since scheduling is dynamic, restarting the new LAS is a more complex task than it is with FIP. For this reason LMs have to check the LAS operations more closely.

7.4. Services of the Data Link Layer

The data link layers of FIP, PROFIBUS, and IEC Fieldbus offer services to send messages in a connectionless fashion, with or without acknowledgment of the correct reception by the receiver, in accordance with the PROWAY C specification [12]. A transmission error occurring during an acknowledged transmission is detected at the sender's side when the receiver's acknowledgment is not received within a predefined period of time; in this case a number of retries is carried out by the transmitting station without releasing the bus mastership.

PROFIBUS also provides a service to ask for a reply containing data from the receiving station. Data values returned in this case must be previously made available to the responding station by invoking a reply update service. FIP supports a specific set of services aimed at managing the network distributed variable database. In this case exchanged objects are not messages but variables which have a global network identifier. Services are provided to read and write the local value of a consumed or produced variable and to invoke an acyclic update of the variable value.

The IEC Fieldbus offers the most comprehensive set of services among the fieldbuses described here. Other than virtually allowing every kind of data exchange possible with the other two protocols, it introduces a means to keep a common sense of the time in the overall network. Unlike FIP and PROFIBUS, the data link of the IEC Fieldbus also allows data exchanges to take place in a connection-oriented way, thus making it possible to introduce flow and error control mechanisms that avoid losses, duplications, and misordering of the exchanged messages.

7.5. Services of the Application Layer

At the application level, FIP adopts a subset of the MMS services, while PROFIBUS adopts a slightly modified version known as fieldbus message specification (FMS). FIP and PROFIBUS use different techniques to the map fieldbus functionalities (such as the automated cyclic exchange of a number of variables with short response times) on the application layer services.

The FIP application layer is structured in two distinct functional units called MMS and MPS. The former is responsible for the implementation of a subset of the MMS messaging services, while the latter allows the user to access the FIP-distributed variable database supported by the underlying data link layer. The application model offered to the user consists of a set of variables, described in terms of type, attributes, and value, and identified by character string names. Application services are provided to read and write local and remote variables. It is also possible to refresh a variable value and to obtain information about the transmission/reception of a variable.

In PROFIBUS, by contrast, variables can be accessed by the user of the application layer only through the conventional MMS services based on a message exchange support. Even though this method seems to be less efficient or simple than using a set of ad hoc protocol mechanisms, these MMS-like services offer, in practice, a suitable method to access shared variables. In fact, as mentioned in Section 5, PROFIBUS supports cyclical data exchanges by means of a special set of services (CSRD). This allows a master station to initiate an automatic poll operation on a set of slaves. Once the polling has been started it is periodically repeated each time the master becomes the token owner, without any other service invocation by the application processes. It is possible to access this mechanism by means of a special kind of virtual connection called cyclic connection. Application processes can exchange variables (both simple and structured) either on a conventional acyclic connection or on a cyclic connection. In the latter case, data is polled automatically from the slave stations by means of the CSRD mechanism in a way that is totally transparent to the user, thus leading the system behaving in a similar way to FIP where variable values are cyclically refreshed without any explicit user intervention. The application layer of the IEC Fieldbus is not yet standardized or stable; however, recent works on this topic are oriented toward the inclusion of those functionalities specified in FIP and PROFIBUS.

8. CONCLUSIONS

There is no doubt that communications will play an increasingly important role in the factory automation process in the near future. Fieldbuses, a kind of network conceived to satisfy communication needs in an automated factory environment, were first developed in the early 1980s and now appear to be the most suitable solution with which to interconnect intelligent devices at the shop floor and cell level in the automated manufacturing scenario. In this chapter the kinds and the typical communication requirements of industrial environments have been analyzed. Some significant fieldbus proposals have been examined, their advantages over the commonly used point-to-point links have been pointed out, and it has been explained how they can affect the evolution of the next generations of automation systems.

Unfortunately, a widely accepted solution of a unified standard fieldbus (the

IEC 1158/ISA SP 50 Fieldbus) is not available at present; thus, it was necessary to take into account two of the most important proposals for fieldbuses that have been promoted as national standards in France (FIP) and in Germany (PROFIBUS). Devices and products operating according to the specifications of both these two fieldbuses are already available on the market and are currently being used in several automated (pilot) plants. Moreover, many of the concepts introduced in these two protocol profiles have been used as a working model for the definition of the international standard fieldbus.

The main characteristics of FIP and PROFIBUS have been presented together with some aspects already defined in the IEC fieldbus standard draft, and an attempt has been made to compare their main features, paying special attention to the mechanisms adopted to enable real-time communication and to the set of services which the user is offered.

REFERENCES

1. International Electrotechnical Committee, "International Standard Fieldbus", preliminary version, IEC 1158, 65C(secr)120, 105, 106, 121 and 122, July 1993, Switzerland, Central Office of the IEC, 3 rue de Varembè, Geneva.
2. French Association for Standardization, "FIP Bus for Exchange of Information Between Transmitters, Actuators and Programmable Controllers", NF C46 601–607, March 1990, Union Technique de l'Eletricitè, Immeuble Lavoisier 4, place des Vosges, La Dèfense 5, Courbevoie.
3. German Institute of Normalization, "PROFIBUS standard part 1 and 2", DIN 19 245, April 1991, PROFIBUS Nutzerorganization e.V. Herseler Str.31, 5047 Wesseling, Germany.
4. McLean C., Mitchell M., and Barkmeyer E., "A Computer Architecture for Small-Batch Manufacturing", IEEE Spectrum, Vol. 20, no. 5, May 1983, pp. 59–64.
5. A. Valenzano, C. Demartini, and L. Ciminiera, "MAP and TOP Communications—Standards and Applications", Addison-Wesley Publishing Co., 1992.
6. General Motors, "Manufacturing Automation Protocol Specification Version 3.0", August 1988, North American MAP/TOP Users Group, ITRC, P.O.BOX 1157, Ann Arbor, MI 48106.
7. International Organization for Standardization, "Manufacturing Message Specification —Service Definition", ISO DIS 9506-1 TC 184/SC 5/WG 2, February 1990. Fachbereich Industrielle Automation (FB-IA) des NAM im DIN, D 60498 Frankfurt.
8. International Organization for Standardization, "Manufacturing Message Specification —Protocol Specification", ISO DIS 9506-2 TC 184/SC 5/WG 2, February 1990. Fachbereich Industrielle Automation (FB-IA) des NAM im DIN, D 60498 Frankfurt.
9. International Organization for Standardization, "Information processing—Open Systems Interconnection—Specification of Abstract Syntax Notation One (ASN.1)", ISO DIS 8824.2, August 1986.

10. International Organization for Standardization, "Information processing—Open Systems Interconnection—Specification of basic encoding rules for Abstract Syntax Notation One (ASN.1)", ISO DIS 8825.2, August 1986.
11. ANSI/IEEE Std 802.4-1985, ISO DIS 8802/4 "IEEE Standards for Local Area Networks: Token Bus Access Method", December 1984, The Institute of Electrical and Electronics Engineers, Inc, 345 East 47th Street, New York, NY 10017, USA.
12. International Electrotechnical Committee, "Proway C Specifications", IEC Std. 955, 1989.
13. ANSI/IEEE Std 802.2-1985, ISO DS 880/2 "IEEE standards for local area networks: Logical link control—Acknowledged connectionless service—Draft addendum," December 1984. The Institute of Electrical and Electronics Engineers, Inc, 345 East 47th Street, New York, NY 10017, USA.
14. P. Montuschi, L. Ciminiera, and A. Valenzano, "Time characteristics of IEEE 802.4 token bus," in IEE Proceedings Part E, Vol. 139, No. 1, Jan. 1992, pp. 81–87.

9
Triangle-Based Surface Models[1]

Leila De Floriani, Paola Magillo
Information and Computer Science Department
University of Genova
Via Dodecaneso, 35, 16146 Genova (Italy)

Silvia Bussi
Elsag Bailey
R & D Department
Via Puccini, 2, 16154 Genova (Italy)

ABSTRACT

The problem of reconstructing a digital model of a surface from a finite set of sampled points is a basic issue in many different application domains, including computer graphics, geographic data processing, computer vision and computer aided design. A triangulated surface model is often used because of the possibility of including surface features and of the simplicity of the topological structure.

The definition of a triangle-based surface model relies on the concept of triangulation. In this paper, we discuss the basic properties of triangulations, Delaunay triangulations, constrained, and conforming triangulations. We present a survey of algorithms for building these kinds of triangulations that represent the first step in the construction of a surface model.

Special attention is given to the surface reconstruction problem in $2^1/_2$ dimensions that is connected to digital terrain modeling in geographic information systems. The more general problem of reconstructing the bounding surface of a solid object from three dimensional scattered data is also considered, and a brief survey of the main approaches proposed in the literature is presented.

---

[1] This work has been supported by the Strategic Project "Knowledge through Images: on Application to Cultural Heritage" of the Italian National Research Council through contract N. 94.04221.ST74.

1. INTRODUCTION

Surface modeling plays an important role in many different application domains including computer graphics, geographic data processing, computer vision, and computer aided design.

The general problem can be stated as reconstructing a model of a surface by interpolating a finite set of points in space belonging to it. Defining a convenient relational model for encoding the neighborhood relations among the data points is necessary to build an effective surface representation.

The surface reconstruction problem in $2^1/_2$ dimensions is strongly related to applications to digital terrain modeling in geographic information systems. In this case, the problem can be mathematically stated as that of interpolating a bivariate function when its values are given either at a uniformly spaced grid or at a set of irregularily distributed points. The surface to be reconstructed is the graph of a bivariate function, and thus the problem reduces to that of computing a 2D tesselation of the domain of such function.

A 2D tesselation of the function domain provides a relational model for encoding the neighborhood relations among the projections of the data points on the $x - y$ plane. A triangle-based representation is a very flexible surface model; it approximates the surface by means of a network of planar, non-overlapping and irregularily shaped planar facets. Even when the surface is sampled at regular intervals in both coordinates, using a triangle-based surface model turns out to be advantageous because such models can adapt to the changes in the surface and include a set of "surface-specific" points and lines, that characterize the surface independently of the data sampling. The construction of a triangle-based approximation reduces to the problem of computing a straight-line plane graph, termed plane triangulation, wherein the data projections are joined by straight-line segments intersecting only at their endpoints and the projections of the given straight-line segments are included in a subset of its edges.

In general, an arbitrary triangulation may not represent an acceptable solution for numerical interpolation because of the elongated shape of its triangles. Intuitively, a "good" triangulation is one wherein triangles are as much equiangular as possible, so as to avoid thin and elongated triangular facets. A Delaunay triangulation is optimal with respect to such requirement, and thus has been extensively used as a basis for surface models.

This article describes the basic properties of standard Delaunay triangulations and of Delaunay triangulations constrained by, and conforming to, a given set of straight-line segments. It also contains a survey of algorithms for building these kinds of triangulations.

When a large number of sampled points is available, a triangulation joining all the data can be highly inefficient in storage and for search and retrieval operations. Approximated models based on triangular grids have been used in the past. Such models are built on the basis of a restricted subset of the data, chosen in such a way to provide a representation of the surface within a certain error tolerance. Approximated surface models are a good data compression mechanism, but give an approximation at a predefined level of accuracy. On the contrary, multiresolution surface models encode a set of increasingly finer surface representations, thereby allowing efficient access and manipulation of surface data at different levels of abstraction. A brief survey on multiresolution surface models is included in this paper.

In general, the problem of reconstructing a boundary model of a solid object from a finite set of data in 3D space finds applications in computer graphics, computer aided design, computer vision and robotics. As in the $2^1/_2$ case, input data may consist of points or straight-line segments lying on the object boundary. In addition, a set of plane contours, obtained by intersecting the object with a collection of planes, may also be given.

Many existing algorithm face the boundary reconstruction problem in an indirect way, building first a 3D representation of the object volume by using a tetrahedralization of the input data. We briefly discuss Delaunay tetrahedralization, possibly constrained or conforming, that represent an extension to three-dimensions of (constrained or conforming) Delaunay triangulations. More general geometric structures (such as the γ-neighborhood graph, that extends the concept of Delaunay triangulation) are also introduced.

This paper is organized as follows. In Section 2, the basic properties of standard Delaunay triangulation and of Delaunay triangulation constrained by, and conforming to, a given set of straight-line segments, that are the basis for digital models of two-and-half dimensional surfaces are described. In Section 3, a survey of algorithms is provided for building the above three kinds of triangulations. In Section 4, multiresolution models of surfaces in $2^1/_2$D are introduced and in Section 5, the problem of reconstructing the bounding surface of a solid object is considered, and a survey of existing approaches to its solution as well as a description of geometric structures used in such algorithms is provided.

2. TWO-AND-HALF DIMENSIONAL SURFACES

A two-and-half dimensional surface is the graph of a bivariate real function $z = \phi(x, y)$ defined over a connected subset D of the $x - y$ plane. The surface reconstruction problem faced in this survey consists of computing a *Digital Surface Model* (DSM), a discrete approximation of a surface, built by interpolation based on a finite set of data points sampled on it. Because of the one-to-one correspondence between points of the surface and points of the two-dimensional domain D, building a DSM reduces to computing a plane subdivision of D into regions, and to establishing an analytic expression for the correspondence between each region and the face defined over it.

More formally, a *Digital Surface Model* (DSM), built on a finite set $\mathcal{S}$ of sampled points, consists of a pair $\mathcal{D} \equiv (\Sigma, \Phi)$, where

1. Σ is a plane subdivision [1], having the set of points $\{(x, y) \mid (x, y, z) \in \mathcal{S}\}$ as its vertices;
2. Φ is a family of bivariate continuous functions, such that every function $\phi_i \in \Phi$ is defined on a closed region f_i of Σ, and interpolates the elevations of the data points whose projections are the vertices of f_i;
3. for every pair of adjacent regions f_i and f_j in Σ, functions ϕ_i and ϕ_j set the same values on the edge shared by f_i and f_j.

Surface patches, defined over the regions of Σ, are called *faces* of the DSM. Condition (3) can be necessary to ensure the continuity of the surface.

Since a plane subdivision with n vertices is composed of $O(n)$ edges and regions, the spatial complexity of a DSM with n vertices is a linear function of n.

Polyhedral Surface Models (PSMs) are DSMs characterized by linear interpolating functions (and, thus, have plane faces). PSMs used in practice have triangular faces and are built based on a triangulations of the set of data points. For a polyhedral surface, the continuity is ensured by the interpolation condition in (2), and thus condition (3) is redundant.

Points forming set $\mathcal{S}$, on which a DSM or PSM is built, can be either distributed on a regular grid, or can be irregularily sampled. *Triangulated Irregular Networks* (TINs), that are triangulated PSMs based on irregularily distributed data, can better adapt to surface features, because they have the capability of including special points (minima, maxima, saddle points) and, through the notion of constrained and conforming triangulation, also special lines (ridges, valleys), that characterize the surface. Thus, TINs succeed in representing a surface at a certain level of accuracy, using a smaller amount of data.

In many practical applications (for instance, in geographical applications), a large number of sampled values is available. Building a DSM based on the whole set of data points would be too expensive in terms of storage. Thus, a subset of the original dataset is selected according to some appropriate mechanism, and an approximate DSM is built based on them. An *approximate* digital surface model built on a data set $\mathcal{S}$ is a digital surface model built on a subset $\mathcal{S}'$ of $\mathcal{S}$. Different norms can be used to evaluate the approximation error of an approximate DSM. The approximation error at a point $P \equiv (x, y, z) \in \mathcal{S} - \mathcal{S}'$ is often evaluated as the infinite norm, that is, the absolute value of the difference between the interpolated and the measured elevation values. The error on a face is defined as the maximum error of points in $\mathcal{S} - \mathcal{S}'$ whose vertical projections lie inside or on the boundary of the face. A DSM is said to approximate a surface *at level* ϵ *of accuracy* if, for every face, the approximation error is less or equal to ϵ.

Even if the points of the original dataset $\mathcal{S}$ were distributed on a regular grid, the selected subset $\mathcal{S}'$, on which an approximate DSM is built, can have an irregular distribution; often, data points are incrementally inserted into $\mathcal{S}'$ until the approximation error becomes less than a predefined tolerance value, thus adding several points in rough areas, and fewer in flat ones, to reflect the different complexity of the surface. Thus, TINs are used to deal with such irregularly distributed data.

Building a TIN reduces to the problem of computing a triangulation of the domain with vertices at the projections of the data points on the $x - y$ plane. Often, a Delaunay triangulation [1] is used as domain subdivision for a TIN, because of its good behavior in numerical interpolation (intuitively, a Delaunay triangulation avoids long and thin triangles). As shown in [2], among all possible triangulations of a fixed data set, the Delaunay triangulation is the one that minimizes the roughness of the approximating surface.

3. TRIANGULATIONS

This Section deals with triangulations, Delaunay triangulations, and constrained and conforming Delaunay triangulations. A survey of related construction algorithms is also provided.

Given a finite set V of points in the plane, a *triangulation* of V is a maximal straight-line plane graph having V as its set of vertices. Thus, in a triangulation, every region, except for the external region, is a triangle.

The problem of finding an optimal triangulation of a given set of points has been considered for many different applications. In surface approximation problems, a criterion related to the size of the angles of triangles is used. When a triangulation is taken as the basis for a digital surface model, the approximated elevation of a point P, internal to a triangle, is obtained as a function of the elevations of the vertices of that triangle. A better approximation is obtained when the three vertices of the triangle lie as closely as possible to P. Intuitively, a *Delaunay triangulation* of a set V of points, is, among all the possible triangulations of V, the one wherein triangles are as much equiangular as possible (see Figure 9.1).

The Delaunay triangulation of a set V of points in the plane is usually defined in terms of another geometric structure, the *Voronoi diagram,* that describes the proximity relationship among the points of V. The Voronoi diagram of a set V of n points is a subdivision of the plane into n convex polygonal regions, called *Voronoi regions,* each associated with a point P_i of V. The Voronoi region of P_i is the set of points of the plane that lie closer to P_i than to any other point in V. Two points P_i and P_j of V are said to be *Voronoi neighbors* when the corresponding Voronoi regions are adjacent.

The *geometric dual graph* of the Voronoi diagram is a plane graph $T \equiv (V, E)$, called the *Delaunay graph* of V, whose edges join pair of points P_i, P_j $(i \neq j)$ of

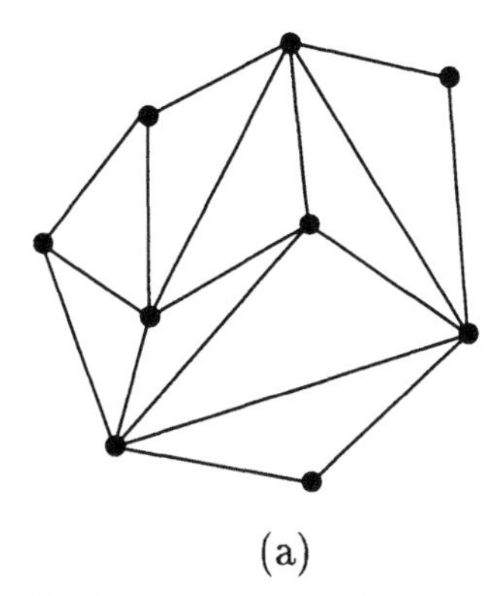

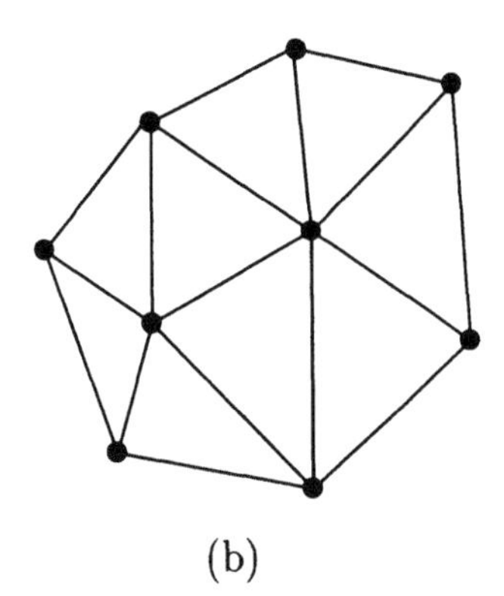

FIGURE 9.1. (a) An arbitrary triangulation of a point set, and (b) a Delaunay triangulation of the same set.

V, such that P_i and P_j are Voronoi neighbors (see Figure 9.2). The Delaunay graph explicitly represents the Voronoi neighborhood relation induced by the Voronoi diagram over set V. If every face of T is a triangle, then T is the same as the Delaunay triangulation of V. Otherwise, T can always be completed to a Delaunay triangulation by decomposing each of its nontriangular faces. Because nontriangular faces can be decomposed in different ways, there are many possible Delaunay triangulations generated by the same Delaunay graph. The Delaunay triangulation of a set V is unique (i.e., the Delaunay graph is a triangula-

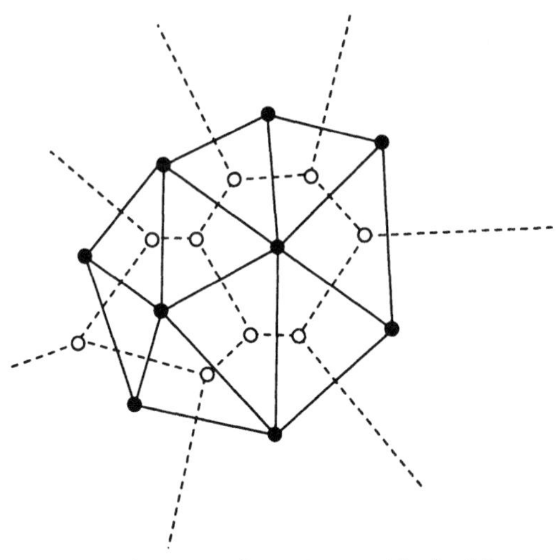

FIGURE 9.2. The Voronoi diagram of a point set (dashed lines), and its dual graph, the Delaunay triangulation (plain lines).

tion) if and only if, for any four points of V, such points do not belong to the same circle.

An alternative characterization of the Delaunay triangulation is given by the so called *empty circle property*. Let τ be a triangulation of a set V of points. A triangle t of τ is said to satisfy the *empty circle property* if and only if the circle circumscribing t does not contain any point of V in its interior. A triangulation τ of V is a Delaunay triangulation if and only if every triangle of τ satisfies the empty circle property (see Figure 9.3).

A Delaunay triangulation satisfies also the *max–min angle property*, that is used operatively by several construction algorithms. Let τ be a triangulation of V, let e be an edge of τ, and Q be the quadrilateral formed by the two triangles of τ adjacent to e. Edge e is said to satisfy the *max–min angle property* if and only if either Q is not strictly convex, or replacing e with the opposite diagonal of Q does not increase the minimum of the six internal angles of the resulting triangulation of Q. An edge e, that satisfies the max–min angle property, is also called a *locally optimal edge* (see Figure 9.4). A triangulation τ of V is a Delaunay triangulation if and only if every edge of τ is locally optimal.

The equivalence of the max–min angle property and the empty circle property has been shown by Lawson [3] by considering the triangulation of a strictly convex quadrilateral. It is important to point out that a locally optimal edge does not necessarily belong to a Delaunay triangulation. If the repeated application of the max–min angle criterion to the edges of a triangulation τ does not cause any diagonal flipping, then τ is a Delaunay triangulation.

Another characterization of the Delaunay triangulation of a set V of points in d dimensions is provided by its relation with the convex hull of a transformed set V' of points in $d + 1$ dimensions. When $d = 2$, then V' is obtained from V by

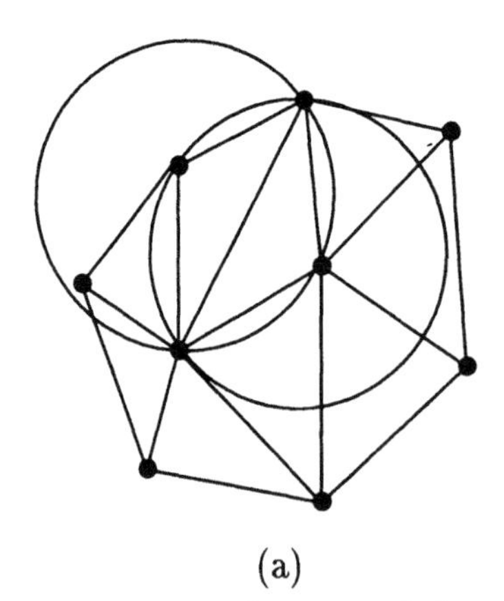

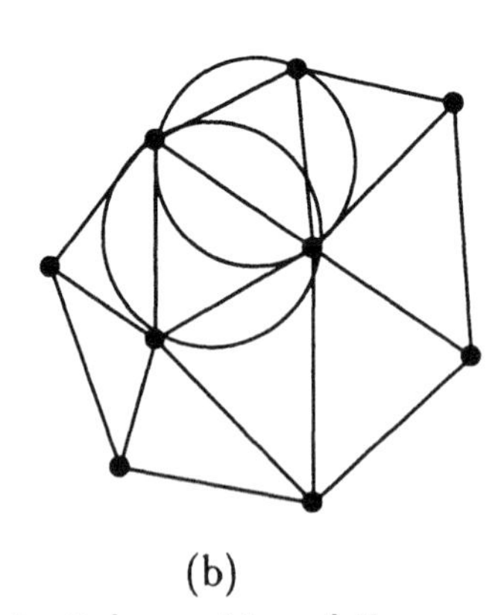

FIGURE 9.3. The empty circle property: (a) is not a Delaunay triangulation, (b) is a Delaunay triangulation.

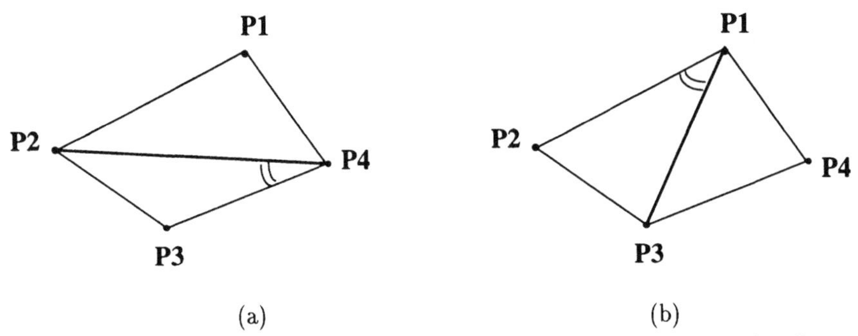

FIGURE 9.4. The max-min angle property: the thick edge in (a) is not locally optimal, the one in (b) is locally optimal.

transforming every point $P_i \equiv (x_i, y_i) \in V$ into the point $P' \equiv (x_i, y_i, x_i^2 + y_i^2)$ on the paraboloid $z = x^2 + y^2$. It can be shown that the Delaunay graph of V is the vertical projection of the lower portion of the convex hull of V'.

3.1. Constrained and Conforming Delaunay Triangulation

In practice, a triangulation must sometimes be forced to include a given set of segments among its edges. The need for including a constraining set of segments in a triangulation arises, for instance, when reconstructing an object from a set of three-dimensional stereo segments lying on the object boundary and obtained from several viewpoints [4]. In terrain modeling, the possibility of including some a priori known lineal features of the terrain, such as ridges, rivers, country boundaries [5], has a fundamental importance in order to obtain a representation that adapts to the natural characteristics of the surface. Moreover, it is important to combine the possibility of including special lines with the advantages of a Delaunay triangulation (i.e., a good behavior in numerical approximations).

The problem of including segments in a Delaunay triangulation has been faced in the literature by two different perspectives, leading to the definitions of *constrained* and *conforming Delaunay triangulation.* The constrained Delaunay triangulation is the best approximation of the Delaunay triangulation containing the set of given segments among its edges [6]. The conforming Delaunay triangulation is a proper Delaunay triangulation of an augmented set of vertices, such that each constraint segment is the union of edges of the triangulation [7] (see Figure 9.5).

Given a set V of points in the plane and a set L of straight-line segments having their endpoints in V, and such that they do not intersect each other except at their endpoints, the pair $G \equiv (V, E)$ defines a *constraint graph* (the segments in L are called *constraints*). A *triangulation of V constrained by L* is a triangulation of V containing the constraint graph as a subgraph.

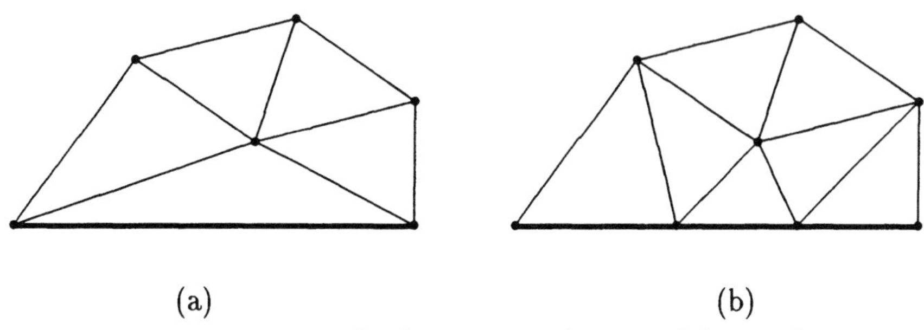

FIGURE 9.5. (a) A constrained Delaunay triangulation, and (b) a conforming one (there is only one constraint segment, that is drawn in thick lines).

An altcrnativc charactcrization of constrained triangulation can be given through the notion of visibility. Two points P_i and P_j in V are called mutually *visible* with respect to L if and only if they can be joined by a straight-line segment without intersecting the interior of any constraint segment in L. We call *visibility graph* the graph $G_v \equiv (V, E_v)$, where edges in E_v join pairs of mutually visible points in V. Note that the constraint graph is a subgraph of the visibility graph. A *triangulation of V constrained by L* is a maximal straight-line plane subgraph of the visibility graph, containing the constraint graph as a subgraph (see Figure 9.6).

A *Constrained Delaunay Triangulation* (CDT) of a point set V with respect to a set L of constraints is a constrained triangulation τ of V with respect to L, satisfying thc following constrained vertion of the *cmpty circle property.* The circumcircle of each triangle of τ does not contain in its interior any point of V that is visible from all the three vertices of the triangle (see Figure 9.7). Note that, if the set L of constraints is empty, then we have a standard Delaunay triangulation.

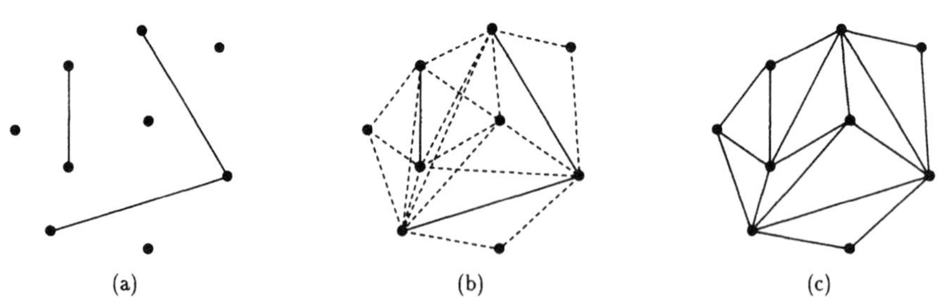

FIGURE 9.6. (a) A constraint graph; (b) the corresponding visibility graph, and (c) a constrained triangulation.

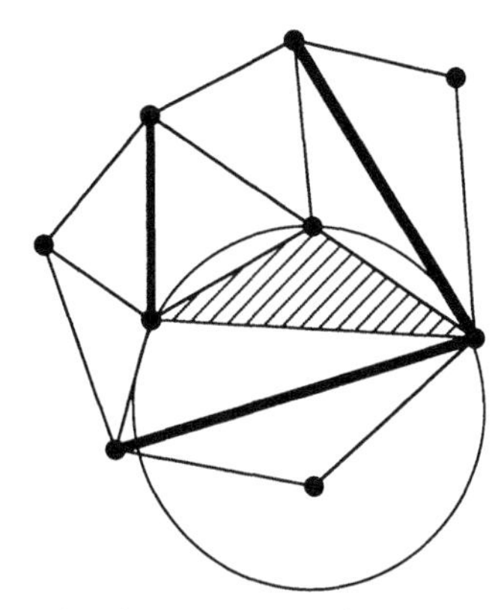

FIGURE 9.7. A constrained Delaunay triangulation (constraint segments are drawn in thick lines). Note that the dashed triangle satisfies the constrained version of the empty circle criterion.

The properties of a standard Delaunay triangulation can be extended to a CDT with some modifications due to the visibility constraints defined by the segments in L. The *max–min angle property* can be reformulated for a CDT by simply considering only those convex quadrilaterals formed by two adjacent triangles that do not share a constraint segment.

One problem with using a constrained Delaunay triangulation in surface approximation is that the forced inclusion of constraint edges may produce thin triangles. An alternative approach is provided by the notion of conforming Delaunay triangulation. Intuitively, the idea is splitting the constraint segments into shorter segments by adding new vertices (the elevations of such new points are inferred by linear interpolation from the ones of the two constraint endpoints) in such a way that the CDT of the resulting set of short segments will be a Delaunay triangulation of the augmented vertex set (see Figure 9.5 again).

Let $G \equiv (V, E)$ be a constraint graph, and let $W \supseteq V$ be a set of points. A triangulation τ of W *conforms* to G if every constraint in E is the union of edges of τ.

A *Conforming Delaunay triangulation* is a Delaunay triangulation of W that conforms to G (see Figure 9.8). Let us call the closed portion of an edge of G between two contiguous points of W on this edge an *interval*. A Delaunay triangulation τ of W conforms to G if every interval defined by G and W satisfies the empty circle property with respect to W. For every constraint graph $G \equiv (V, E)$, there exists a set $W \supseteq V$ such that the Delaunay triangulation of W conforms to G. In particular, it has been proven [8] that there alway exists a set W, whose Delaunay triangulation conforms to G, such that all the extra vertices in $W - V$ lie on the edges of G (most existing approaches to the construction of a conforming Delaunay triangulation are based on this latter result).

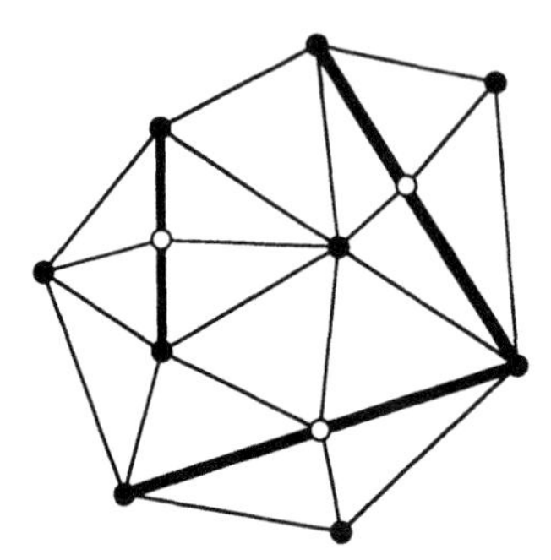

FIGURE 9.8. A conforming Delaunay triangulation.

3.2. Algorithms for Computing a Delaunay Triangulation

Existing algorithms for building a Delaunay triangulation or, equivalently, its dual graph (the Voronoi diagram) can be classified into the following five categories:

- *two-step algorithms*, that first compute an arbitrary triangulation, and then optimize it to a Delaunay triangulation by iteratively applying either the empty circle or the max–min angle criteria.
- *incremental algorithms* [9, 10, 11], that construct a Delaunay triangulation by stepwise insertion of the data points, while maintaining a Delaunay triangulation at each step.
- *divide and conquer algorithms* [12], that compute a Delaunay triangulation by recursively splitting the point set into two halves, and merging the computed partial solutions.
- *sweep-line methods* [13], that compute the Voronoi diagram of a set of points by first transforming it in such a way that the Voronoi region of a point P_i is considered only when P_i is intersected by the sweep-line.
- *three-dimensional algorithms*, that compute the convex hull in 3D, and then project the lower portion on the $x - y$ plane.

Three-dimensional algorithms will not be considered here. See [14] for a treatment of the topic.

The first Delaunay triangulation algorithms were based on a *two-step strategy*. An arbitrary triangulation of the given set V of points can be obtained through the following three steps:

- sort the points of V by increasing x-coordinate;
- form a triangle with the first three noncollinear points in the sorted sequence;
- iteratively add the next point P_i by connecting P_i to all the vertices of the

existing triangulation that are visible from P_i (i.e., they can be connected to P_i without intersecting existing edges).

The optimization step iteratively applies the max–min angle (or the empty circle) criterion to any internal edge of the current triangulation, such that its two adjacent triangles form a strictly convex quadrilateral until no more edge swapping occurs. Constructing the initial arbitrary triangulation requires $O(n \log n)$ operations in the worst case. Lawson [3] shows that the optimization process terminates and has a $O(n^2)$ worst-case time complexity.

Incremental algorithms can be further classified into *static* and *on-line* algorithms. *Static algorithms* usually start by sorting all the points according to their euclidean distance from a fixed origin and then build the triangulation in such a way that each created triangle belongs to the final tesselation [3, 15].

On-line algorithms are based on the incremental insertion of the internal points in an initial Delaunay triangulation of the domain. The initial triangulation of the domain can be obtained, for instance, by creating a triangle enclosing all the data points that will be removed together with all the edges incident in its vertices at the end of the process.

The update of the current Delaunay triangulation at the insertion of a new internal point P_i can be performed in two different ways. One approach [9] builds first an arbitrary triangulation by connecting P_i to the three vertices of the triangle t of the existing triangulation that contains P_i. The triangulation is then optimized by iteratively applying the max–min angle criterion until no more edge swapping occurs (see Figure 9.9).

Another approach, [10], locates first the triangle t of the current triangulation

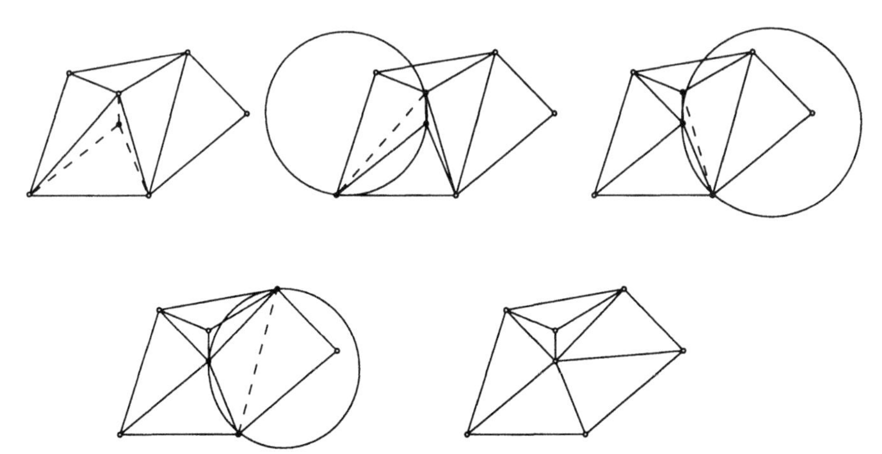

FIGURE 9.9. A working example of optimization process in Guibas and Stolfi's algorithm.

that contains point P_i. Starting from t, it deletes all the triangles of the current triangulation whose circumcircle contains P_i. Such triangles form a star-shaped polygon called the *influence polygon* of P_i, containing P_i in its kernel. The new Delaunay triangulation is simply obtained by connecting P_i with all the vertices of the influence polygon (see Figure 9.10).

In both algorithms, the update of the triangulation when a new point is inserted requires $O(n)$ steps in the worst case, and thus these algorithm have an $O(n^2)$ time complexity. On the contrary, the insertion of a new point requires only a constant number of steps in the average case [16].

In summary, both two-steps and incremental techniques are suboptimal because they have an $O(n^2)$ time complexity in the worst case, but they have a good expected time behavior: a randomized version of Guibas and Stolfi's algorithm runs in expected $O(n \log n)$ time [11]. On-line algorithms are especially appropriate when some of the data points are not known in advance. Such methods are used for the construction of models approximating a surface at a certain level of accuracy, through repeated insertion of data points until the required precision degree is met. They are also used for building multiresolution models that are collections of approximate models of a surface based on increasingly larger sets of data (see Section 4).

Compared with incremental techniques, *divide-and-conquer algorithms* are more complicated and require more storage space, but are, in general, computationally more efficient. An asyntotically optimal Delaunay triangulation algorithm, working in $O(n \log n)$ time, is due to Lee and Schacter [12].

Similar to the Voronoi diagram computation algorithm proposed by Shamos [1], the method of Lee and Schacter is based on a recursive splitting of the set of data points into two almost equally sized subsets, and on the pairwise merging of the Delaunay triangulations separately computed. The algorithm performs the following four steps:

- the points of V are preliminarily sorted from left to right (if two points have the same x-coordinate, then the y-coordinate is considered);
- set V is split into two subsets V_L and V_R, where V_L contains the leftmost half of the points of V, and V_R, the rightmost half;

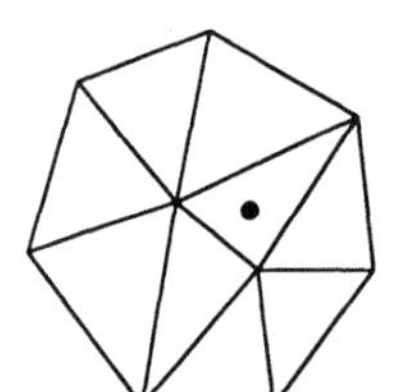
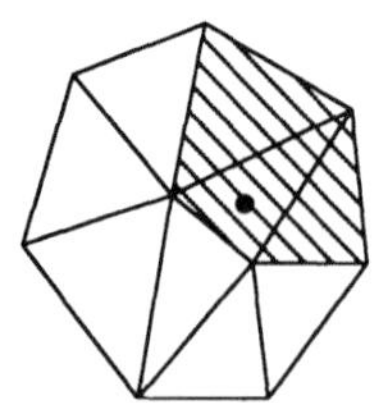
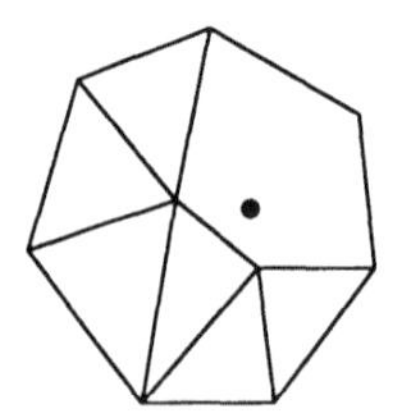
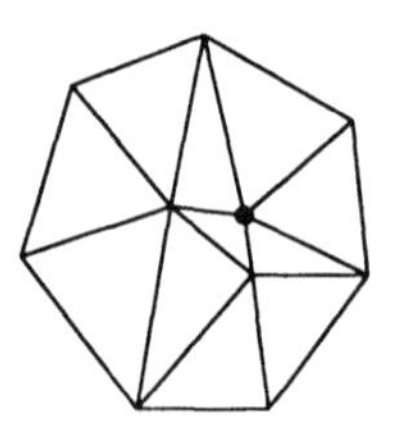

FIGURE 9.10. Illustration of the steps of Watson's algorithm.

- the Delaunay triangulations of V_L and V_R are recursively constructed and then merged together to form the Delaunay triangulation of V.

The merging step of the triangulations of V_L and V_R starts with the computation of the convex hull of $V = V_L \cup V_R$, that is the domain of the Delaunay triangulation of V. This reduces to determining the lower and upper common tangent of the convex hulls of V_L and V_R (domains of the corresponding triangulations). Then we move from the lower segment to the upper one by deleting the edges that are not in the final Delaunay triangulation of V, and by adding the new edges (see Figure 9.11). It can be shown that the common tangents can be found in $O(n)$ time and the whole merging phase has a linear time complexity as well. By inserting a preliminar sorting of the points of V from left to right, the split of V into V_L and V_R requires only a linear time at each iteration. Hence, the worst-case time complexity of this algorithm is globally equal to $O(n \log n)$ time, that is worst-case optimal.

The *sweep-line algorithm* proposed by Fortune [13] for Voronoi diagram computation is competitive in simplicity with incremental algorithms. Because it avoids the merging step, it is simpler than divide-and-conquer ones. The sweep-line technique conceptually sweeps a horizontal line across the plane, noting the regions intersected by the line as the line moves. Computing the Voronoi diagram directly with a sweep-line technique is difficult because the Voronoi region of a point may be intersected by the sweep-line before the point itself is intersected. Thus, the algorithm computes a geometric transformation of the Voronoi diagram that has the property that the lowest point of the transformed region of a point appears at the point itself, and, thus, the Voronoi region of a point is considered only when the point itself is intersected by the sweep line.

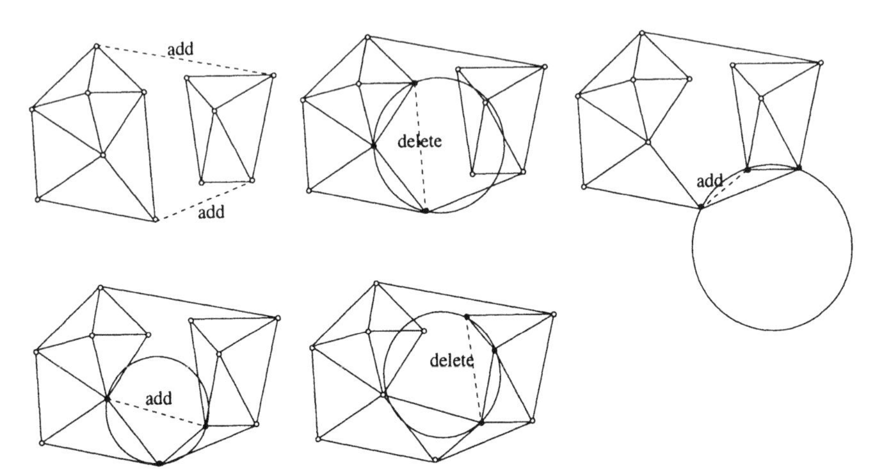

FIGURE 9.11. Merge step in the divide-and-conquer algorithm.

3.3. Algorithms for Building a Constrained Delaunay Triangulation

The problem of building a constrained triangulation has been first considered for the special case of an arbitrary triangulation of a simple polygon, in connection with applications to polygon decomposition. This case is not interesting for applications to surface modeling, thus here we focus on algorithms for the general case of arbitrary constraints.

Algorithms for solving the general constrained triangulation problem can be classified into two major categories:

- algorithms called *direct algorithms*, that compute the CDT by starting from the whole set of points and constraint segments [6, 4, 17];
- algorithms called *difference algorithms*, that modify a standard Delaunay triangulation (or a Voronoi diagram) of the set of data points by insertion of the constraint segments [18].

The algorithm by Lee and Lin [6] is based on the preliminary computation of the visibility graph, and on the successive iterative elimination of those edges that are not in the CDT. In a CDT, an edge $\overline{P_iP_j}$ is a Delaunay edge if and only if P_i and P_j are mutually visible and there exists a circle passing through P_i and P_j that does not contain any vertex visible from both P_i and P_j. For every vertex P_i of the visibility graph, the algorithm scans the list of vertices visible from P_i in counterclockwise order. At the beginning, the vertex P_j in such list, corresponding to the shortest edge around P_i, is found. Because of the previous result, $\overline{P_iP_j}$ must be a Delaunay edge. For every three consecutive vertices P_1, P_2, P_3 around P_i, the quadrilateral formed by P_i and P_1, P_2, P_3 is considered to decide whether P_iP_2 is a Delaunay edge. The visibility graph can be computed in $O(n^2)$ steps (where n is the number of data points in V), by applying the method described in [19], whereas the detection of Delaunay edges requires $O(n^2)$ steps in the worst case.

The application of the divide-and-conquer paradigm to the design of an efficient algorithm for constrained Delaunay triangulation seems to be difficult because of the problem related to the linear separability of the set of constraint segments. The algorithms by Chew [18] and by Joe and Wang [20] extend the divide-and-conquer algorithm of Lee and Schacter for computing a standard Delaunay triangulation. The worst-case time complexity is $O(n \log n)$ for both algorithms. In [18], the constraint graph $G \equiv (V, L)$ is assumed to be contained into a rectangle that is subdivided into vertical strips in such a way that there is exactly one vertex in each strip. According to the divide-and-conquer strategy, the CDT is computed for each strip and adjacent strips are merged together to form new strips containing twice as many vertices. The recursive partition of the problem can be represented by a tree of depth $O(\log n)$. The main difficulty in this process is due to cross edges, that is, edges of the constraint graph that cross a strip without having an endpoint in it. It can be easily seen that $O(l)$ constraint segments (where l is the cardinality of L) can cross a strip, giving a total of $O(n^2)$ cross edges for all the strips. Chew's algorithm only stores those cross edges that are visible from at least one vertex in the strip that restricts the total number of stored edges to $O(l)$. Thus, the complexity of all the subproblems at the same level of the tree is $O(n + l)$, where n is the number of points in V and l the number of constraints in L. Chew shows that two adjacent strips can be merged in linear time. Thus, the algorithm builds the resulting CDT in $O(n \log n)$ time.

An on-line incremental algorithm, based on the modification of an existing CDT by iteratively inserting new points and constraint segments has been proposed by De Floriani and Puppo [21]. This algorithm is an extension of Watson's algorithm for computing a standard Delaunay triangulation with the difference that, at each step, either a new point or a new constraint segment can be inserted.

The insertion of a new point P is performed as in Watson's algorithm, the difference now being that the "constrained" version of the empty circle criterion is adopted to build the influence polygon Q_P of P. A new constraint segment $s \equiv \overline{P_1P_2}$ can be inserted only when the two endpoints P_1 and P_2 of s are already vertices of the current triangulation. Segment insertion also involves the determination and retriangulation of an influence polygon. The *influence polygon* Q_s of s is formed by the union of all the triangles of the old triangulation that are intersected by s. Q_s is a simple polygon and has segment s as a diagonal. It has been shown that the final updated CDT can be obtained by removing all the triangle edges that lie inside the influence polygon Q_s, splitting Q_s into two subpolygons, separated by diagonal s, and computing a Delaunay triangulation inside the two polygons separately (see Figure 9.12). Intersecting segment s with

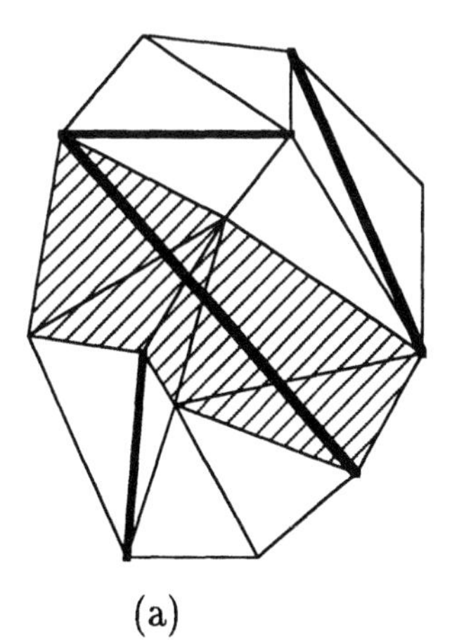

(a)

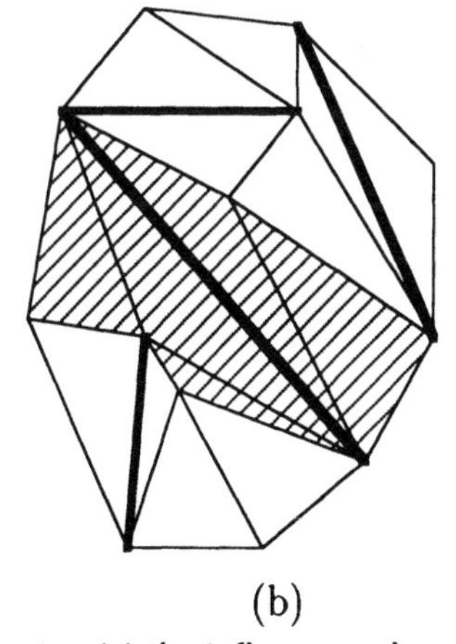

(b)

FIGURE 9.12. Insertion of a constraint segment *s*: (a) the influence polygon of *s*, (b) the updated CDT obtained by retriangulating the polygon.

the existing triangulation can be done in time linear in the number of intersected triangles, that is, $O(n)$ in the worst case. The CDT of a polygon can be computed in $O(k \log k)$ time, where k is the size of the polygon.

Because the algorithm is based on the incremental modification of a CDT by the insertion of a single point or segment at a time, it necessarily exhibits a higher worst-case time complexity than nonincremental ones. On the other hand, the approach is particularly interesting because it makes it possible to add data points and constraint segments at run time. The insertion of a point has an $O(n)$ complexity, whereas inserting a segment may require $O(n \log n)$ operations.

In [22], algorithms are described for incremental construction and dynamic maintenance of a constrained Delaunay triangulation that perform point insertion, point deletion, segment insertion and segment deletion. In particular, randomized algorithms are proposed that compute a CDT of n points and l constraint segments in $O(n \log n \log l)$ expected time and perform l segment deletions in $O(n \log l)$ expected time.

3.4. Algorithms for Building a Conforming Delaunay Triangulation

Constructing a conforming Delaunay triangulation with respect to a conflict graph $G \equiv (V, E)$ basically reduces to the problem of finding a set V' of points such that the Delaunay triangulation of $W \equiv V \cup V'$ conforms to G. This is generally a much harder problem than computing a constrained triangulation of G because a large number of extra points may be necessary in order to achieve conformity. The lower bound for the size of V' is equal to $\Omega(nl)$, where n and l are, respectively, the number of vertices and of constraints in G. An $O(l^2 n)$ upper bound for the number of points to be added has been proven by Edelsbrunner and Tan [7].

Proposed approaches [4, 8, 23, 24] are based on the idea of placing a sufficient number of points on the edges of the constraint graph in such a way that the Delaunay triangulation of the augmented point set is guaranteed conform to G. The problem is that there is generally no function $f(n)$ that can bound the number of inserted points. In particular, the number of points added grows as the constraint segments move closer to each other.

The algorithm by Boissonnat et al., [4], is based on the remark that a subsegment e of a constraint segment is a Delaunay edge if the circle having e as diameter does not intersect any other constraint segment. Thus, if the circle associated with a constraint segment e intersects some other segment, then e is split into a finite number of subsegments such that none of their circles intersects any constraint. When two constraint segments intersect at an endpoint, one new point is inserted on both segments in such a way that the circumcircle of the triangle defined by the common endpoint, and by the two new points, does not intersect any constraint segment.

The method proposed by Saalfeld [8] is based on an iterative process that, at each step, computes an ordinary Delaunay triangulation of the current vertex set (initially, set V): if the triangulation does not conform to G, then the intersection points between the edges of the triangulation and the constraint segments are added to the set and the process is iterated.

It should be said that the approaches just illustrated present a time complexity that is not necessarily polynomial because there is no upper bound to the number of inserted points.

The algorithm by Edelsbrunner and Tan [7] has an $O(l^2n + n^2)$ worst-case complexity and is optimal with respect to the number of extra points it inserts (at most $O(l^2n)$). This algorithm is based on a different approach that inserts new points lying not necessarily on constraint segments. The algorithm consists of two phases. The first phase constructs $O(n)$ circles with disjoint interiors (they are allowed to be tangent) such that their union is connected and contains V. The intersections between these circles and the edges of G are added together with the possible tangent points between pairs of circles. At the end of this phase, every constraint segment has been split into *protected* intervals (i.e., intervals contained into a circle), and *unprotected* intervals. Unprotected intervals will be further subdivided during the second phase.

4. MULTIRESOLUTION SURFACE MODELS

Organizing surface representations at different levels of detail is a recent research issue. *Multiresolution surface models* provide a data compression mechanism as well as a variable resolution method for representing a surface at different levels of abstraction. A multilevel organization allows an easy implementation of searching and other geometric operations, such as finding surface intersection, or zooming when visualizing the surface. Moreover, it makes real-time simulation and visualization possible for those applications in which describing less important areas with fewer details is a relevant issue.

Multiresolution models consist of collections of DSMs built based on increasingly large subsets of the given data set S. The structure wherein such individual DSMs are connected induces a classification of multiresolution surface models into *hierarchical* and *pyramidal* models (see [25] for a survey).

4.1. Hierarchical Models

The concept of the hierarchical surface model is based on that of hierarchical subdivision of the domain. Intuitively, given a subdivision Σ, a region f of Σ can be seen as an individual entity and refined into a subdivision Σ_f, whose domain covers f. The refinement of f is performed by adding new vertices either inside f

or on its sides. Recursive application of the refinement process leads to a hierarchy of subdivisions.

More formally, a hierarchical subdivision is defined based on an ordered collection $\{\Sigma_0, \ldots, \Sigma_m\}$ of subdivisions such that, for every $j > 0$, Σ_j has more than one region, and there exists exactly one $i < j$ such that the domain of Σ_j is a region of Σ_i. A *hierarchical subdivision* $\mathcal{H}$ is a tree with labelled arcs where the nodes are the subdivisions $\Sigma_0, \ldots, \Sigma_m$. Σ_0 is the root. Every subdivision Σ_j is linked as a child to the unique subdivision Σ_i (with $j < i$) containing the region f that is the domain of Σ_j. Arc (Σ_i, Σ_j) is labeled with region f. An example of a hierarchical subdivision is shown in Figure 9.13.

A region that is refined in the hierarchy (i.e., that appears as the label of some edge) is called a *macroregion*. A region is called *simple* otherwise. The total number of regions in a hierarchical subdivision is linear in the number of simple regions, that, in turn, linearly depends on the total number of inserted data points [26].

A *Hierarchical Surface Model* (HSM) is built on a hierarchical subdivision by associating with each subdivision Σ_i in the hierarchy a DSM $\mathcal{D}_i \equiv (\Sigma_i, \Phi_i)$. Any

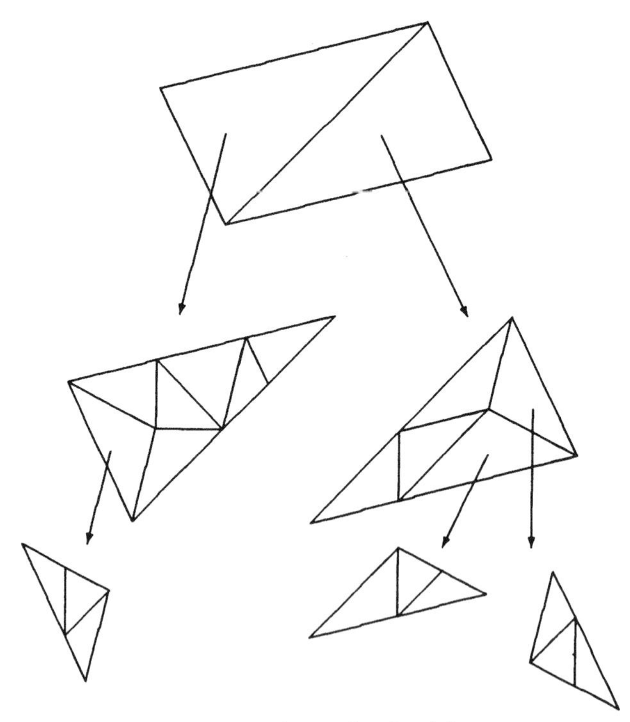

FIGURE 9.13. A hierarchical subdivision.

vertex $P_i \equiv (x_i, y_i)$ that is common to more than one subdivision Σ_j has the same elevation value z_i in each corresponding surface model $\mathcal{D}_j$.

In a hierarchical subdivision $\mathcal{H}$, the refinement of a macroregion can split one of its edges into a chain of edges through insertion of new vertices. Inconsistent refinement of an internal edge e may have undesirable effects on the surface model supported by $\mathcal{H}$, because the continuity of the surfaces on adjacent patches could not be guaranteed. In order to avoid vertical discontinuities in the overlying surface, a special *matching rule* must be satisfied in $\mathcal{H}$. This rule requires that, for any two adjacent regions f_1 and f_2 in a subdivision Σ_i in the hierarchy, the same set of vertices is inserted on the common edge of f_1 and f_2 in the subdivisions refining f_1 and f_2 at the next level.

According to the refinement criterion, existing HSMs (see Figure 9.14) can be classified into *quadtree-based models* and *hierarchical triangulated models*. Quadtree-based models, such as *quadtrees* [27, 28, 29] and *quaternary triangulations* [30, 31], are HSMs built on gridded data, wherein the refinement process follows a regular geometrical pattern with rectangles and equilateral triangles, respectively, as basic elements.

Triangle-based HSMs built on irregularily distributed data provide a more flexible surface description because they can better adapt to the roughness of the surface and include surface-specific points and lines that characterize the surface independently from the data sampling. Such models are based on a hierarchical decomposition that is not geometrically fixed, wherein faces are arbitrary triangles. *Ternary triangulations* [26, 32] represent a first attempt in this direction, using a decomposition pattern with fixed topology and variable geometry. However, they have the disadvantage of producing long and thin triangles, and thus a less efficient surface approximation. *Hierarchical Triangulated Irregular Networks* (HTINs) are hierarchical models based on a hierarchy of triangulations wherein the recursive refinement process is driven by an accuracy based crite-

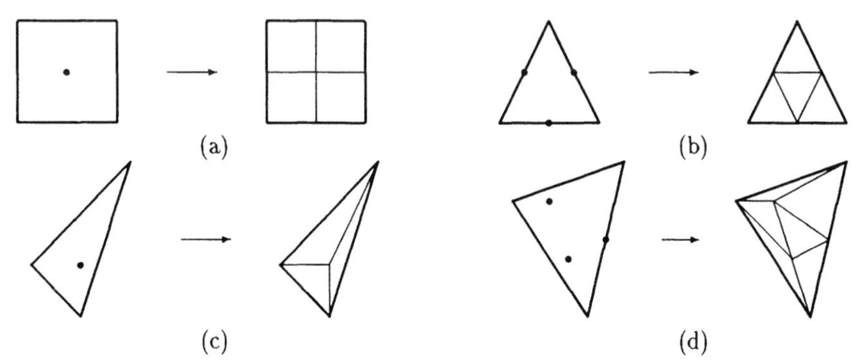

FIGURE 9.14. Different refinement rules in existing HSMs: (a) quadtree, (b) quaternary triangulation, (c) ternary triangulation, (d) hierarchical TIN.

rion. Given a sequence $\epsilon_0, \ldots, \epsilon_m$ of decreasing tolerance values, each level in the hierarchy represents the surface at one of the predefined values ϵ_i. The top level corresponds to the coarsest description, satisfying the largest tolerance ϵ_0. Passing from a level i to level $i + 1$, every triangle t is expanded in a finer triangulation by iteratively inserting points in its interior or on its sides, until the accuracy ϵ_{i+1} is reached. The number of points in the refinement of a triangle is not predefined.

The rule used for selecting and inserting points in a triangle characterizes the different types of HTINs (in all cases, the criterion is based on an error evaluation and ensures the matching rule to be satisfied). In *Adaptive Hierarchical Triangulations* (AHTs) [33], a triangle t is refined by iteratively applying a *splitting rule*. Four maximum-error points related to the interior and to the three edges of t are computed and, among them, the ones corresponding to an error greater than ϵ_{i+1} are selected for insertion. A predefined decomposition pattern is then applied depending on the configuration of selected points. This process is repeated until no more points are selected for insertion.

In *Hierarchical Delaunay Triangulations* (HDTs) [26, 34], are based on a hierarchy where each element is a Delaunay triangulation. Notice that, although the subdivision inside every macrotriangle is locally a Delaunay triangulation, the global expanded subdivision of the whole domain generally is not. The refinement of a triangle t is performed by an iterative application of the *Delaunay selector,* that, at each step, updates the current Delaunay triangulation by inserting the point having the maximum error. The basis of the construction algorithm for an HDT must be an on-line method that incrementally builds a Delaunay triangulation through iterative insertion of points, as those described in Section 3.2.

4.2. Pyramidal Models

A strictly hierarchical structure cannot be imposed on a Delaunay triangulation because the insertion of a new point might cause a modification that can involve the whole tesselation. Even in HDTs, any description of the surface, corresponding to an intermediate level in the tree describing the hierarchical model, does not correspond to a Delaunay triangulation of the domain. Pyramidal surface models have been developed as alternative multiresolution methods to hierarchical ones; such models allow global properties (such as the Delaunayhood of a triangulation) to be satisfied with respect to the whole domain. Again, the concept of pyramidal surface model relies on that of pyramidal subdivision.

A pyramidal subdivision is defined on the basis of a sequence $\{\Sigma_0, \ldots, \Sigma_m\}$ of plane subdivisions of the domain D, each representing an increasingly finer description of D. Two regions belonging to different subdivisions may overlap. The structure of a pyramidal subdivision is stratified rather than hierarchical. A *pyramidal subdivision* $\mathcal{P}$ is represented by a labelled multigraph (i.e., a graph with parallel arcs), having $\{\Sigma_0, \ldots, \Sigma_m\}$ as its set of nodes. There is an arc (Σ_i, Σ_{i+1}), joining two consecutive subdivisions in $\mathcal{P}$, for every pair of faces $f_i \in \Sigma_i$ and $f_j \in \Sigma_{i+1}$, such that f_i and f_j have a non-empty intersections, and $f_i \neq f_j$. Such arc is labeled with the pair (f_i, f_j). An example of a pyramidal subdivision is shown in Figure 9.15. Several arcs (Σ_i, Σ_{i+1}), labeled with different pairs of faces (f_k, f_l), can join the same two subdivisions Σ_i and Σ_{i+1}. In the worst case, every region at level i can be linked to every region of the subdivision at level $i + 1$, thus leading to an $O(n^2)$ worst-case space complexity of the structure (where n denotes the number of vertices in the most refined level of the pyramid).

The *Delaunay Pyramid* [35] is a multiresolution surface model composed of a sequence of Delaunay triangulation that represent the surface at increasingly finer levels of detail over the whole domain. Every triangulation is obtained from the previous one by iteratively inserting the data point corresponding to the maximum error until the current Delaunay triangulation satisfies the required

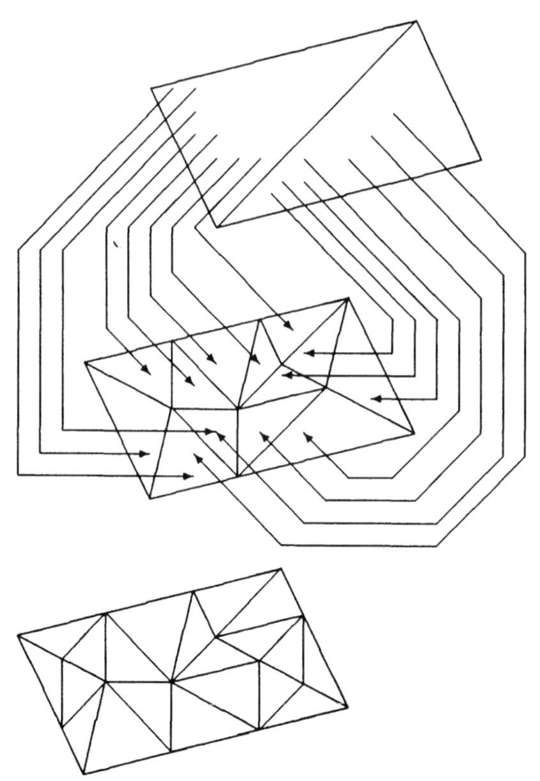

FIGURE 9.15. A pyramidal subdivision (only links between the first two levels are shown).

accuracy. Unlike a hierarchical Delaunay triangulation, the Delaunay pyramid guarantees the equiangularity property to be globally satisfied. This advantage is paid in terms of an increased space complexity. Moreover, a pyramidal structure does not allow easy local refinements in areas of interest.

In many applications, the different surface approximations are defined by increasingly larger sets of both points and segments (corresponding to lineal features that must be included in the representation). The definition of *constrained Delaunay pyramid* is analogous to the one of standard Delaunay pyramid, with the difference that every level in the structure consists of a constrained Delaunay triangulation. The construction of a constrained Delaunay pyramid is based on the on-line method for incrementally computing a CDT described in Section 3.3.

5. SURFACES IN THREE DIMENSIONS

The problem of reconstructing the bounding surface of a three-dimensional object can be formally stated as follows: given a set of entities in space known to lie on the boundary of the object, find a polyhedral surface that contains all input data, thus approximating the object boundary.

Input data may consist either of a set of points, of a set of straight line segments, or of a set of polygonal cross sections lying on parallel planes intersecting the object. The format of the input depends on the type of data source. If it is not known how data have been obtained, or if a single reconstruction method is to be used for data from various source types, then no structural relation between the input points can be assumed, except that all of them lie on the boundary of an object. Input data consisting of straight-line segments lying on the boundary of an object are typically acquired by means of a stereo process and mainly occur in robotics applications. Surface reconstruction from a set of parallel cross sections is an important problem in clinical medicine, anatomic research, and computer graphics.

The simplest boundary approximation is the one consisting of planar triangles, thus producing a three-dimensional simple closed polyhedron with triangular faces, that is, a triangulation of the surface.

In this section, we briefly review some of the existing methods proposed in the literature for triangulating the boundary of *3D* objects. The interested reader is referred to the various papers mentioned herein for a more thorough treatment of this problem. The different approaches can be classified, according to the specific dataset they are suited for, into:

- methods suitable for a set of irregularily distributed points,
- methods suitable for a set of straight-line segments, and
- methods dealing with a set of polygonal contours.

Variable resolution representations of *3D* surfaces have also been developed mainly for applications in computer vision and robotics: in [32], a hierarchical model, the *prismtree,* for describing the boundary of a *3D* object is described together with algorithms for surface intersection and neighbor finding based on such representation.

5.1. Reconstruction from Scattered Points

The reconstruction problem from scattered data points can be stated as follows: given a set V of vertices in *3D,* find a simple closed polyhedron passing through all vertices of V.

The problem just described can be addressed in two ways:

- through the *direct* construction of a triangulation of the surface defined by the data [36, 37].
- through the intermediate construction of some 3D structure, obtained by filling the interior of the object with tetrahedra, and, then, deriving a triangulated representation of the boundary of the object from such auxiliary structure [38, 39].

Direct methods have been proposed in [36, 37]. O'Rourke [36] suggests the use of minimal surface area polyhedra: the convex hull is used as a starting point of the reconstruction algorithm that modifies it systematically in order to include all data points internal to the hull in the boundary. In [37], Boissonnat reduces the problem of building a triangular-faced polyhedron to that of finding several convex hulls on the Gaussian sphere. The important property of such a spherical surface representation is that the description of sufficiently small regions with no change in the sign of the curvature is one-to-one, thus ensuring that a triangulation on the unit sphere corresponds to a triangulation on the surface.

Indirect approaches have been proposed by Boissonnat [38] and Veltkamp [39] that are based on two different three-dimensional structures. The approach proposed by Boissonnat [38] uses a tetrahedralization of the input data as an auxiliary structure, and extracts from it a representation of the object boundary as the collection of external faces.

A Delaunay tetrahedralization generalizes the concept of Delaunay triangulation introduced in Section 3, to three dimensions. The *Voronoi diagram* of a set $V = \{P_1, \ldots, P_n\}$ of n points in the 3D space is defined as a collection of n convex polyhedra, each associated with a point in V, whose union covers $\mathbb{R}^3$. The polyhedron associated to P_i is the locus of points of the space that lie closer to P_i than to any other point in V, and is called the *Voronoi polyhedron* of P_i. The geometrical dual of the Voronoi diagram, obtained by linking pairs of points of V whose Voronoi polyhedra are face-adjacent, is called the *Delaunay graph* associated with V. If no five points are cospherical, the $3D$ cells of the Delaunay graph are tetrahedra; otherwise, they can still be decomposed into tetrahedra. The resulting tetrahedralization is called a *Delaunay tetrahedralization* of set P.

The Delaunay tetrahedralization fills the interior of the convex hull of the points in V with tetrahedra. Thus, if all the points of V lie on the convex hull, the Delaunay tetrahedralization is a volumetric polyhedral representation of the object. Otherwise, some tetrahedra are eliminated in such a way that all points of V are on the boundary of the resulting polyhedral shape. This "sculpture" of the convex hull is done by sequentially eliminating one tetrahedron at a time, according to some heuristic criteria, until all points of V are on the boundary.

In order to ensure that, at each step, the boundary of the current three dimensional shape is a polyhedron, topological constraints, summarized in the following three rules, must be respected (see Figure 9.16):

1. any tetrahedron with exactly three vertices on the current boundary can be eliminated,
2. any tetrahedron with exactly five edges on the current boundary can be eliminated, and
3. any tetrahedron not satisfying the previous rules cannot be eliminated.

The heuristic criteria try to minimize the variation in the curvature of the boundary by eliminating less regular tetrahedra first. At the end of the sculpturing process, the set of remaining tetrahedra provides a volumetric representation of the object, and the triangular faces of these tetrahedra, that lie on the external boundary, correspond to a polyhedral approximation of the object surface.

Boissonnat suggests an implementation of the method with a worst case complexity equal to $O(n^2 \log n)$, where n is the number of input data points. A drawback to this algorithm is that the sculpturing process can get locked before all the innermost vertices have been included into the boundary, even if a closed

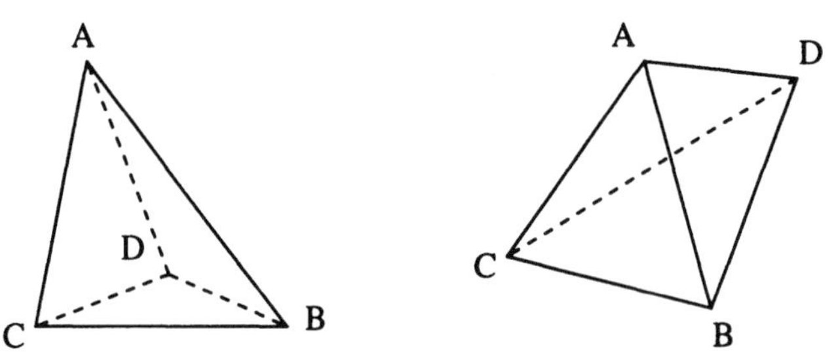

FIGURE 9.16. Tetrahedron *ABCD* (left) can be eliminated if triangle *ABC* is on the external boundary and vertex *D* is in the interior; Tetrahedron *ABCD* (right) can be eliminated if triangles *ABC* and *ABD* are on the external boundary and edge *CD* is in the interior.

polyhedral boundary through all vertices exists. Moreover, it is not known if every Delaunay tetrahedralization contains a polyhedron through all its vertices.

The approach presented in [39] is based on a different geometric structure, called the *γ-neighborhood graph.* The γ-neighborhood graph is a parametric graph that unifies a number of graphs, such as the convex hull and the Delaunay graph, into a continuous spectrum of graphs that range from the Delaunay tetrahedralization to the complete graph.

Let V be the set of data points, and t be a triangle having three points P_1, P_2, $P_3 \in V$ as its vertices. Let r be the radius of the smallest possible sphere touching the three vertices, that is, the one such that P_1, P_2, P_3 lie on its diametrical circle. The radius of any sphere touching P_1, P_2, P_3 can be expressed as $r/(1 - c)$, where c is a parameter in the range $[0, 1]$. Note that the case of a sphere with an infinite radius (i.e., a halfspace) is included. A negative sign may be attributed to parameter c, depending on whether the centers of the two spheres lie on the same or on opposite sides with respect to t; thus, the radius must be expressed as $r/(1 - |c|)$, where $-1 \leq c \leq 1$. The γ-graph $\gamma([-1, 1], [d, 1])$, for some parameter $d \in [-1, 0]$, includes all triangles t wherein there exist two values $c_0 \in [-1, 1]$ and $c_1 \in [d, 0]$ such that two spheres touching the three vertices of t, with radii $r/(1 - |c_0|)$ and $r/(1 - |c_1|)$, respectively, do not contain any point of V in their intersection. A smaller value of d corresponds to a smaller volume of the intersection, and, thus, more triangles are included in the γ-graph (see Figure 9.17 for some two-dimensional examples of a γ-graph). If $d = -1$, then both spheres are allowed to be two halfspaces, and the γ-graph contains all possible triangles defined by ternes of vertices in V. If $d = 0$, then the γ-graph is the same as the Delaunay tetrahedralization of V because the Delaunay tetrahedralization can be characterized as the collection of all triangles t having vertices at three points P_1, P_2, $P_3 \in V$, such that there exists an empty sphere touching P_1, P_2 and P_3.

For any $d \in [-1, 0]$, graph $\gamma([-1, 1], [d, 1])$ has the convex hull of set V as its boundary. Triangles lying on the boundary of the γ-graph are thus called *hull triangles*. The polyhedron corresponding to the boundary of the given object is obtained by constricting the graph $\gamma([-1, 1], [d, 1])$ through iterative deletion of hull triangles. A γ-graph where triangles have been deleted is called a *pruned* γ-graph. The constriction process is continued until all data points lie on the pruned graph hull.

Four triangles of the graph may implicitly form a tetrahedron; a tetrahedron formed by at least one hull triangle is called a *boundary tetrahedron*. The selection of the triangle $t \equiv P_1, P_2, P_3$ to be deleted at a generic step is based on the observation that the opposite vertex P_4 of the tetrahedron P_1, P_2, P_3, P_4 that has the largest solid angle φ has the largest probability to be sensed from outside the boundary. Intuitively, the solid angle φ at P_4 depends on how close P_4 lies to t relative to the size of the tetrahedron, and on the shape of t. This notion is formalized by introducing a γ-indicator, associated with t with respect to P_4. If r is the radius of the smallest sphere touching P_1, P_2, P_3, and R is the radius of the

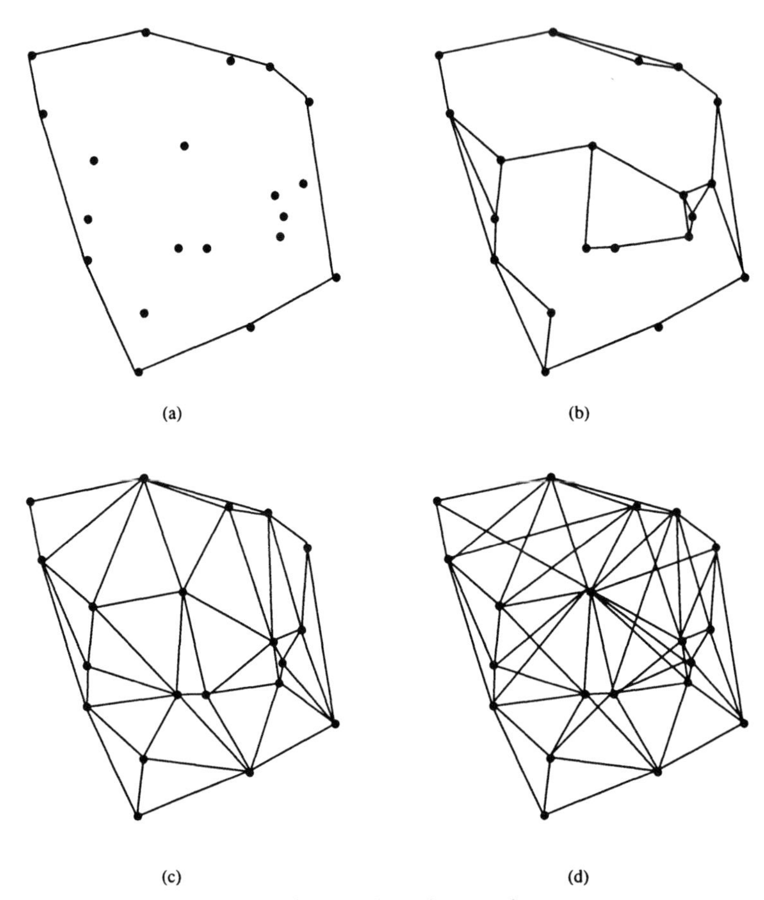

FIGURE 9.17. Two-dimensional examples of γ-graphs containing increasingly larger sets of edges: (a) γ([−1, 1], [1, 1] (convex hull); (b) γ([−1, 1], [1/4, 1]; (c) γ([−1, 1], [0, 1] (Delaunay triangulation); (d) γ([−1, 1], [−1/4, 1].

sphere touching P_1, P_2, P_3, P_4, then the γ-indicator is defined as the value of c for which $R = r/(1 - |c|)$, where the sign of c discriminates on the side where the center of the sphere lies with respect to t.

Deleting a hull triangle $[P_1, P_2, P_3]$, due to the γ-indicator with respect to a vertex P_4, intuitively means that tetrahedron $[P_1, P_2, P_3, P_4]$ is merged into the "empty space". Note that the interior of a tetrahedron formed by triangles of the γ-graph can be intersected by other triangles of the graph. In order to leave the hull of the graph properly defined, any triangle crossing P_1, P_2, P_3, P_4 must also removed. In addition, the boundary polyhedron must remain connected. This is achieved by introducing topological constraints, similar to the ones mentioned for the algorithm by Boissonnat that forbid merging a tetrahedron into the free space in some cases.

Any $\gamma([-1, 1], [d, 1])$ can be used for constriction, but the existence of a solution can be guaranteed only for d "small enough". Unfortunately, the threshold value is a priori unknown. On the other hand, a small value of d implies a larger graph and, thus, a higher computational complexity. It is possible, though, to start the constriction process from the graph $\gamma([-1, 1], [0, 1])$, (i.e., the Delaunay tetrahedralization), and adaptively add triangles of a $\gamma([-1, 1], [d, 1])$, $-1 \leq d < 0$, when necessary, so that the process never gets stuck.

The three dimensional $\gamma([-1, 1], [d, 1])$ graph can be constructed in $\Theta(n^2)$ time for $d = 0$, and $O(n^3)$ time for $d < 0$, where n is the number of points. Once the γ-neighborhood graph has been built, the boundary constriction algorithm takes $\Theta(t \log t)$ time, with t the number of triangles in the γ-graph.

5.2. Reconstruction from Scattered Segments

Reconstructing the boundary of an object from a set $\mathcal{S}$ of straight-line segments belonging to such boundary implies finding a simple closed polyhedron including the segments of $\mathcal{S}$ as a subset of its edges. Such edges are obtained through a stereo process applied from several viewpoints. The approach used by existing algorithms consists of constructing first a Delaunay tetrahedralization of the input, including the given set of segments. Then, a suitable subset of tetrahedra is eliminated in such a way that all given segments lie on the boundary of the resulting shape. As in the two-dimensional case, the problem of including a given set of segments into a tetrahedralization can be faced from two perspectives, thus leading to the concepts of *constrained* and of *conforming Delaunay tetrahedralization*.

The problem of computing a constrained Delaunay tetrahedralization in 3D has been shown to be NP-complete [40] and, thus, not suitable for developing algorithms based on it. A reconstruction algorithm based on the use of a conforming Delaunay tetrahedralization has been proposed by Boissonnat et al., [4]. Their algorithm for building a conforming Delaunay tetrahedralization, already described in Section 3.4 for the two-dimensional case, replaces each data segment with a set of points (that generates a set of points on the object boundary); then, a Delaunay tetrahedralization of the set of segments endpoints augmented with such additional set is computed. Unfortunately, there is no bound to the number of points that must be added to guarantee that all given segments are included in the resulting tetrahedralization.

The conforming Delaunay tetrahedralization, obtained through the method just mentioned, is used as a starting point for reconstructing the boundary of the object. The basic idea is that triangles linking stereo segments to the viewpoints represent optical rays and, thus, cannot intersect the object (see Figure 9.18).

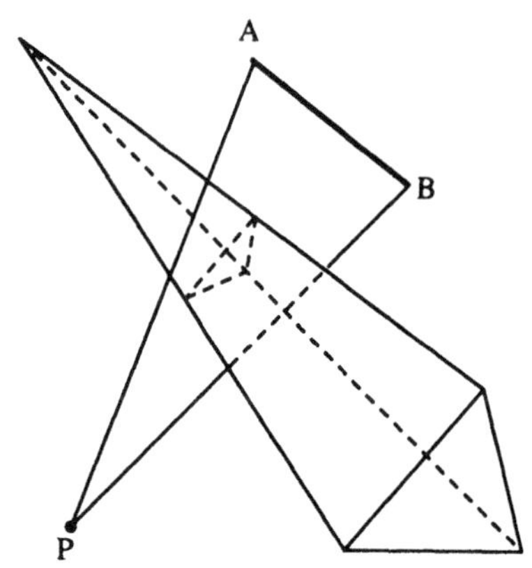

FIGURE 9.18. A tetrahedron that cannot be part of the object since it is intersected by a triangle connecting a stereo segment $\overline{AB}$ to the corresponding camera center P.

Thus, for each triangle t having an edge in one of the segments and the vertex opposite to the edge in one of the camera optical center, all tetrahedra intersected by t are marked as empty. In general, a better result can be obtained by keeping, for each tetrahedra, a count of the number of times it has been crossed by a triangle, and marking only tetrahedra that have been intersected by a certain number of triangles.

Eliminating tetrahedra marked to be part of free space is not enough to obtain a valid representation of the object. In fact, some tetrahedra that belong to the free space may not be intersected by any triangle joining a camera center with a measured segment, and thus remain in the final tetrahedralization. This can create singularities on the object surface. A valid representation of the object is obtained by removing tetrahedra that create such singularities.

5.3. Reconstruction from Planar Cross Sections

The problem of reconstructing a surface from a sequence of planar cross-sections can be formulated as follows. Let $C_1, \ldots, C_n$ be a set of contours obtained by intersecting a three-dimensional object by n cutting planes $\Pi_1, \ldots, \Pi_n$. Each contour C_i consists of a collection of simple polygons (some possibly lying inside others). The aim is finding a simple closed polyhedron such that, for every $i = 1, \ldots, n$, the intersection of the polyhedron with plane Π_i is the same as contour C_i.

In existing approaches, the problem of constructing an object over the n cross section is reduced to the problem of constructing a sequence of $n - 1$ partial shapes, each of them connecting two cross sections C_i and C_{i+1}, lying on adjacent planes. Methods proposed in the literature are based on two main techniques:

- reducing the problem to a path searching in a toroidal graph [41, 42];
- using Delaunay triangulations [43].

By assuming that the resulting polyhedron has vertices on the given cross sections and consists of triangular faces only, it has been shown that the remaining problem of finding the edges of the polyhedron that spans the space between two consecutive contours reduces to that of finding a minimum cost cycle in a toroidal graph [42]. Keppel [41] was the first to reduce the problem of connecting vertices to a search in a toroidal graph. He used a maximum volume heuristic to select a path. Fuchs, Kedem, and Uselton [42] presented the first efficient algorithm. Their method generalizes the results of Keppel and can be used with various optimization criteria. They proposed a minimal surface heuristic, but other optimization criteria may be used.

Boissonnat [43] fills the slice of the 3D object lying between two adjacent planes Π_i and Π_{i+1} by starting from a Delaunay tetrahedralization and successively deleting empty tetrahedra from it. Let M_j be the set of vertices of contour C_j, for a generic $j \in [0, n]$. The method computes first the Delaunay triangulations of the sets of points M_i and M_{i+1} of C_i and C_{i+1}. The three-dimensional Delaunay tetrahedralization of the points M_i and M_{i+1} is constructed from the two-dimensional ones. Note that both slice $(i - 1, i)$ and slice $(i, i + 1)$ intersect the cutting plane Π_i along the Delaunay triangulation of M_i, thus ensuring the coherence between the two slices. Finally, the algorithm eliminates those tetrahedra having an edge in Π_i or in Π_{i+1} outside the contours, and those

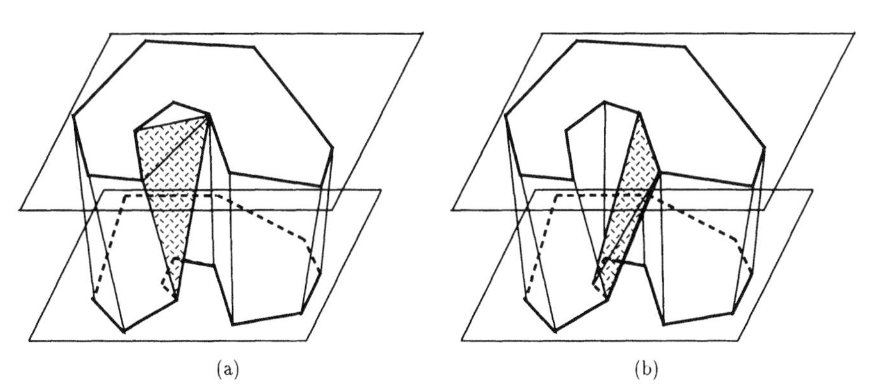

FIGURE 9.19. (a) A tetrahedron that must be eliminated because it is composed by edges not belonging to the cross section. (b) A tetrahedron that must be eliminated because it is connected to the remainder of the structure only by two edges.

contributing to nonsolid connections (see Figure 9.19). This procedure is repeated for all the pairs of adjacent cross-sections.

The complexity of the reconstruction algorithm is $O(t)$ if t is the maximum number of tetrahedra in the Delaunay triangulation between two adjacent planes Π_i and Π_{i+1}, containing $m = m_i + m_{i+1}$ points. In the worst case, t is $O(m^2)$, but when the contours in two adjacent cross sections are not too different, it is expected to be $O(m)$.

The triangle-based approach to surface reconstruction from cross-sections is not the only possible one. Sloan and Hrechanyc [44] address the surface reconstruction problem as that of generating a sweeping rule for a generalized cylinder representation giving rise to the contours. This approach is expecially well suited to deal with sparse data, and provides acceptable reconstructions from cross sections that are too sparse for the smoothness assumption common to the triangulation algorithms just mentioned.

6. CONCLUSIONS

We have presented a survey of methods for triangle-based description of surfaces defined by sets of points and straight-line segments in 3D space.

We have focused our attention on functional surfaces, that is, surfaces defined by a function $z \equiv \phi(x, y)$ of two independent variables. The first problem in defining a triangle-based surface representation is the choice of an appropriate topological structure to connect the projections of the data points and of the constraint segments. Delaunay triangulation is the most common choice for this purpose. We have presented a classification of the different approaches to computing a Delaunay triangulation in both the standard and the constrained and conforming cases.

The reconstruction of the boundary of a three-dimensional object from scattered data has also been addressed. For this problem, input data may consist of a set of points, of a set of segments and of a sequence of parallel cross-sections. An appropriate topological structure connecting the data belonging to the surface to be reconstructed can be derived from a tetrahedralization of the 3D volume of the object. Again, a Delaunay tetrahedralization is usually chosen for such purposes. We have presented an overview of the major known approaches in literature and described geometric structures (Delaunay triangulations, possibly constrained or conforming, and extensions) used by such algorithms.

REFERENCES

1. F.P. Preparata and M.I. Shamos, *Computational Geometry: an Introduction*, Springer Verlag, 1985.

2. S. Rippa, Minimal roughness property of the Delaunay triangulation, *Computer Aided Geometric Design*, 7, 7, 1990, pp. 489–497.
3. C.L. Lawson, Software for C^1 surface interpolation, *Mathematical Software III*, edited by J.R. Rice, Academic Press Inc., 1977, pp. 161–164.
4. J.D. Boissonnat, O.D. Faugeras and E. Le Bras-Mehlman, Representing Stereo Data with the Delaunay Triangulation, *Artificial Intelligence*, 44, pp. 41–89, 1990.
5. T.K. Peuker and D.H. Douglas, Detection of surface-specific points by local parallel processing of discrete terrain elevation data, *Computer Graphics and Image Processing*, 4, 1975, pp. 375–387.
6. D.T. Lee and A.K. Lin, Generalized Delaunay triangulation for planar graphs, *Discrete and Computational geometry*, 1, 1986, pp. 201–217.
7. H. Edelsbrunner and T.S. Tan, An Upper Bound for Conforming Delaunay Triangulations, *Discrete and Computational Geometry*, 10, 1993, pp. 197–213.
8. A. Saalfeld, Delaunay edge refinements, *Proceedings 3rd Canadian Conference on Computational Geometry*, 1991, pp. 33–36.
9. L.J. Guibas and J. Stolfi, Primitives for the manipulation of general subdivisions and the computation of Voronoi diagrams, *ACM Transactions on Graphics*, 4, 1985, pp. 74–123.
10. D.F. Watson, Computing the n-dimensional Delaunay tessellation with application to Voronoi polytopes, *The Computer Journal*, 24, 1981, pp. 728–746.
11. D.E. Knuth, L.J. Guibas and M. Sharir, Randomized incremental construction of Delaunay and Voronoi diagrams, *Algorithmica*, 7(4), 1992, pp. 381–413.
12. D.T. Lee and B.J. Schacter, Two algorithms for constructing a Delaunay triangulation, *International Journal of Computer and Information Sciences*, 9(3), 1980, pp. 219–242.
13. S. Fortune, A sweep-line algorithm for Voronoi diagrams, *Algorithmica*, 2(2), 1987, pp. 153–174.
14. H. Edelsbrunner, *Algorithms in Combinatorial Geometry*, Springer Verlag, 1987.
15. D.H. Mc Lain, Two-dimensional interpolation from random data, *The Computer Journal*, 19(2), 1976, pp. 178–181.
16. R. Sibson, Locally equiangular triangulations, *The Computer Journal*, 21, 1878, pp. 243–245.
17. L.P. Chew, Constrained Delaunay triangulations, *Algorithmica*, 4, 1989, pp. 97–108.
18. L. Schubert and C.A. Wang, An optimal algorithm for constructing the Delaunay triangulation of a set of line segments, *Proceedings 3rd ACM Symposium on Computational Geometry*, Waterloo, June 1987, pp. 223–232.
19. T. Asano, T. Asano, L. Guibas, J. Hershberger and H. Imei, Visibility of disjoint polygons, *Algorithmica*, 1, 1986, pp. 45–63.
20. B. Joe and C.A. Wang, Duality and construction of constrained Voronoi diagrams and Delaunay triangulations, *Technical Report*, Department of Computer Science, University of Alberta.
21. L. De Floriani and E. Puppo, An on-line algorithm for constrained Delaunay triangulation, *CVGIP, Graphical Models and Image Processing*, 54 (3) pp. 290–300, Academic Press, Orlando, FL, 1992.
22. T.C. Kao and D.M. Mount, Incremental construction and dynamic maintenance of

constrained Delaunay triangulations, *Proceedings 3rd Canadian Conference on Computational Geometry,* 1991, pp. 170–175.
23. L.R. Nackman and V. Srinivasan, Point placement for Delaunay triangulation of polygonal domains, *Proceedings 3rd Canadian Conference on Computational Geometry,* 1991, pp. 37–40.
24. A.A. Oloufa, Triangulation applications in volume calculation, *Journal Computer Civil Engineering,* 5, 1991, pp. 103–119.
25. L. De Floriani, P. Marzano and E. Puppo, Hierarchical Terrain Models: Survey and Formalization, *Proceedings ACM Symposium on Applied Computing* (SAC'94), Phoenix, Arizona, March 1994.
26. L. De Floriani and E. Puppo, A Hierarchical Triangle-based Model for Terrain Description, *Theories and Methods of Spatio-Temporal Reasoning in Geographic Space,* 1992, pp. 236–251.
27. Z.T. Chen and W.R. Tobler, Quadtree representation of digital terrain, *Proceedings Autocarto 86,* London, 1986, pp. 475–484.
28. H. Samet and R. Sivan, Algorithms for constructing quadtree surface maps. *Proceedings 5th International Symposium on Spatial Data Handling,* Charleston, 1992, pp. 361–370.
29. B. Von Herzen and A.H. Barr, Accurate triangulations of deformed, intersecting surfaces, *Computer Graphics (Proceedings SIGGRAPH 87),* 21(4), July 1987, pp. 103–110.
30. R. Barrera and A.M. Vaquez, A hierarchical method for representing relief, *Proceedings Pecora IX Symposium on Spatial Information Technologies for Remote Sensing Today and Tomorrow,* Sioux Falls, South Dakota, October 1984, pp. 87–92.
31. D. Gomez and A. Guzman, Digital model for three-dimensional surface representation, *Geo-processing,* 1, 1979, pp. 53–70.
32. J. Ponce and O. Faugeras, An object centered hierarchical representation for 3D objects: the prism tree, *Computer vision, Graphics and Image Processing,* 38(1), 1987, pp. 1–28.
33. L.L. Scarlatos and T. Pavlidis, Hierarchical triangulation using terrain features, *Proceedings IEEE Conference on Visualization,* San Francisco, CA, October 1990.
34. L. De Floriani, D. Mirra and E. Puppo, Extracting Contour Lines from a Hierarchical Surface Model *Computer Graphics Forum (Proceedings Eurographics 93),* 12(3), 1993, pp. 249–260.
35. L. De Floriani, A pyramidal data structure for triangle-based surface description, *IEEE Computer Graphics and Applications,* 8, March 1989, pp. 67–78.
36. J. O'Rourke, Triangulation of minimal area as 3D object models, *Proceedings 7th International Conference on Artificial Intelligence,* 1981, pp. 664–666.
37. J.D. Boissonnat, Representation of objects by triangulating points in 3-D space, *Proceedings 6th International Conference on Pattern Recognition,* Munich, 1982, pp. 830–832.
38. J.D. Boissonnat, Geometric structures for three-dimensional shape representation, *ACM Transactions on Graphics,* 3(4), pp. 266–286, 1984.
39. R.C. Veltkamp, 3D Computational Morphology, *EUROGRAPHICS '93,* pp. 115–127, 1993.
40. J. Ruppert and R. Seidel, On the difficulty of tetrahedralizing 3-dimensional non-

convex polyhedra, *Proceedings 5th ACM Symposium on Computational Geometry*, 1989, pp. 380–393.

41. E. Keppel, Approximating complex surfaces by triangulation of contour lines, *IBM Journal of Research and Development*, 19(1), 1975, pp. 2–11.
42. H. Fuchs, S.P. Usenton and Z. Zedem, Optimal surface reconstruction from planar contours, *Communications ACM*, 20(10), 1977, pp. 693–702.
43. J.D. Boissonnat, Shape Reconstruction from Planar Cross-Sections, *Computer Vision, Graphics and Image Processing*, 44, pp. 1–29, 1988.
44. K.R. Sloan and L.D. Hrechanyk, Surface reconstruction from sparse data, *Proceedings Conference on Pattern Recognition and Image Processing*, 1981, pp. 45–48.

10

Representation and Conversion Issues in Solid Modeling

Leila De Floriani
Dipartimento di Informatica e Scienze dell'Informazione
Università di Genova
Via Dodecaneso, 35 16146 Genova, Italy

Enrico Puppo
Istituto per la Matematica Applicata
Consiglio Nazionale delle Ricerche
Via De Marini, 6 (Torre di Francia), 16149 Genova, Italy

1. INTRODUCTION

The problem in giving a description of geometric objects within computer systems has become more important in the last few years. The significant development of computer application fields dealing with solid objects, like CAD/CAM and robotics, motivates a research for effective and efficient solutions. Solid modeling is an important discipline whose goal is to be able to express the entire nature of three-dimensional objects [50], and to make computer systems capable of answering geometric questions algorithmically [39]. Key points in designing a solid model are the capability of satisfying some general requirements, like completeness, integrity, regularity and so on, while giving a representation that is expressive and efficient enough to be used in practical applications.

Several object representation schemes have been developed in the past for different applications; interesting classifications and discussions of such schemes are given in [39, 43, 50]. Some object models can be considered *special-purpose,* in the sense that they are devoted to represent special classes of objects. Among this class of models, we can mention sweep representations, generalized cylinders and cones, blob models, primitive instancing, prism-trees, and some $2^1/_2$-dimensional representations (ternary and quaternary decompositions, quadtree decompositions, cone trees, 2D triangulations). *General-purpose* models are characterized by a better descriptive power on a wide class of solid objects, and are often identified with the models that have been developed for direct support to modern CAD/CAM systems. In this chapter, we will focus on this class of models and on the possibilities and ways of converting from one model to another.

CAD/CAM systems must allow users to perform different operations on the objects they represent. Two nontrivial, yet basic, operations are the creation (that involves an interaction between the designer and the machine) and the display and drawing of an object. Other important operations involve computations of integral properties of an object (e.g., volume or mass computation). Furthermore, modern CAD/CAM system should allow an efficient interaction between different objects (i.e., performing Boolean and interference operations), supporting graphic interactive modifications, and simulating and driving machining and assembly operations.

No representation scheme, among the ones known at the present state of the art, is the best one for all operations. Most commercial solid modelers developed in the past ten years are based on either *Constructive Solid Geometry* (CSG) [64] or *boundary representation* (BRep) (tables listing modeling systems can be found in [7, 44]). CSG describes an object as a Boolean combination of primitive components or halfspaces. In a BRep, objects are defined by their enclosing surfaces. CSG schemes are especially well suited for the object creation task (interactive input), whereas a BRep is possibly the best model for display purposes and allows an efficient computation of the integral properties (only in the case of objects with planar faces), and the explicit representation of tolerance and surface finish information.

Spatial enumerations and hierarchical space decomposition models (namely the *Octree* [40] and its variants) have been proposed mainly as secondary schemes in solid modelers. Although such models give only an approximate description of an object (depending on their level of resolution), they are especially well suited for computing integral properties and Boolean operations, and can be easily obtained from both CSG and BRep descriptions.

Since the early 1980s, the need for systems that are able to support hybrid representations had been highlighted [44]. As a compromise between the two major traditional schemes, many systems have been built that accept their input in a CSG form, and automatically convert it to a boundary representation that is used as a primary scheme for all other system operations. Even these systems suffer from some disadvantages, such as inefficiency in evaluating Boolean operations, and a heavy drawback in the complexity of the conversion algorithms [39, 45].

More recently, some hybrid models have been proposed that should be able to incorporate or easily interface previously existing models. The *PM-Octree* [10, 13, 24, 50] can be interpreted as a combination of Octree and BRep, and can be easily obtained from a CSG description and converted from, and to, a BRep. Such structure has been proposed mainly as a bridge between CSG and BRep as well as being a support structure that allows efficient computations as standard Octrees. A different combination of models is the *PM-CSG* [73], that is essentially an octree structure whose leaves refer to CSG primitives. *Modular Boundary Models* (MBMs) [15, 17], on the other hand, try to incorporate the advantages of CSG and BRep models while encoding information useful for Boolean and interference computation. Such models seem especially well suited for feature-based design, and for describing machining and assembly operations [16].

Hybrid modelers clearly require conversion algorithms that are able to translate data among the different schemes. Conversion algorithms have to satisfy both theoretical requirements and practical applicability. In other words, an algorithm must guarantee that the output model is always correct and consistent with the input model. At the same time, the algorithm must be efficient enough to be used in a practical system. Consistency is an important issue if one wants to keep different models of the same object (as is desirable in hybrid systems), and perform an automatic update of all the models when one of them is (interactively) modified.

The general problem of converting between different representations has been discussed by Requicha [43] and Mäntylä [39]. Requicha makes a distinction between *exact* and *approximate* conversions. Approximate conversions are typical when passing from an exact model, like CSG or BRep, to any spatial enumeration (included the Octree), because the latter are intrinsically approximate models. Of course, information is lost when performing an approximate conversion. Consistency in these cases must be intended in a loose meaning, that is, an algorithm gives a result that is consistent if it generates the best possible approximation of the input model with respect to the resolution of the output scheme.

No information is lost in exact conversions, thus giving *theoretically invertible* descriptions, provided that the models are unique and unambiguous. Such conversions are, in principle, the only ones that can guarantee the full consistency of two models. Unfortunately, as pointed out by Mäntylä, the total consistency is possible only if the modeling system is limited to support such operations on solids that can be mapped on each of the representations. As a consequence, if the total consistency is guaranteed, the more representation schemes we include in a system, the more limitations we get on the objects we can represent and on the operations we can perform on them.

In practice, the total consistency is hardly ensured and the invertibility is often impossible or still unsolved. On the other hand, approximate conversions are useful and utilized within existing systems. Most important conversion algorithms developed in the literature can be summarized as follows:

- approximate algorithms from BRep to Octree or variants;
- approximate algorithms from CSG to Octree or variants;
- exact algorithms from BRep to PM-Octree and vice versa;
- exact algorithms from CSG to PM-Octree;
- exact algorithms from CSG to PM-CSG tree;
- exact algorithms from CSG to BRep (boundary evaluation).

Notice that the only invertible conversion is the one between BRep and PM-Octree. Conversions from the Octree to exact models give poor results (stair-faceted solids) unless geometrical information unavailable in the Octree model could be somehow inferred. This topic is more typical of computer vision and robotics, and will not be addressed here. As conversions to Octrees cannot be inverted in a meaningful way, they make sense mainly "on the fly", to solve certain kind of problems using representations that are discarded afterwards [39].

Maybe the most crucial problem in conversions is the inverse boundary evaluation, that is, the creation of a CSG model from a boundary representation. CSG models are not unique, and thus prevent a priori invertible conversions. Hence, it is still impossible to convert data from CSG to a boundary representation, modify them (with any tool/algorithm working on the BRep) and give them back (algorithmically) to the designer for further interactive intervention.

Another aspect of conversion is when one wants to build a representation of a solid object from another representation without needing to convert back to the original model after the second one has been modified. This is the case when an object model suitable for manufacturing purposes must be constructed from the model produced through the design phase. A representation for manufacturing purpose should describe the object in terms of the so called form features [67] (e.g., through holes, pockets, slots, chamfers) that are related to specific machining and assembly processes. The algorithms that construct a CSG model or an MBM from a BRep through a feature recognition and organization process are examples of such irreversible conversions.

In the chapter we will survey the solid models mentioned earlier, describing their structure and outlining their characteristics in relation with application to CAD/CAM. We will then classify and analyze existing conversion algorithms, discussing their efficiency and applicability. Furthermore, we will address open problems in conversion.

The rest of the chapter is organized as follows. Section 2 addresses general issues and gives a theoretical basis for solid modeling. Section 3 describes volumetric schemes, classifying them into decomposition schemes (3.1), like tetrahedralization and Octrees, and constructive schemes (3.2) like CSG. Sections 4 and 5 describe boundary and hybrid schemes, respectively. Section 6 shows, in detail, algorithms that convert data to Octree or its variants. Similarly, Section 7 presents conversion algorithms towards PM-Octrees and PM-CSG trees, whereas section 8 describes conversions to boundary representations. Section 9 discusses the problem of converting data to CSG. Finally, section 10 discusses some concluding remarks.

2. OBJECT REPRESENTATION SCHEMES

The first problem in defining a computer representation of a solid object is to establish a mathematical model for solid objects. The most general mathematical abstraction of a real solid object is a subset of the three-dimensional Euclidean space $\mathbb{E}^3$. Very few subsets of $\mathbb{E}^3$ are adequate models of solid objects. The mathematical model of a solid object should capture properties that correspond to our intuitive notion of a solid [43] and are informally expressed as follows:

1. a solid object must be a bounded, closed subset of $\mathbb{E}^3$ (*finiteness*)
2. a solid object must be invariant under rigid transformation (*rigidity*)
3. a solid object must have an interior, and its boundary cannot have isolated or "dangling" parts (*regularity*)
4. the boundary of a solid object must determine unambiguously what is "inside" (and "outside") the object (*boundary determinism*)
5. a solid object must be representable by a finite number of entities (*finite describability*).

Mathematical models of solids satisfying properties (i)–(v) are subsets of $\mathbb{E}^3$ that are bounded, closed, and regular, and are called *r-sets* [43]. Under the conventional Boolean set operations, r-sets are not closed (see Figure 10.1), but they are closed under the so called regularized set operations [60].

Another way of characterizing solid objects is achieved by looking at the surfaces enclosing them. Intuitively, a surface can be regarded as a subset of $\mathbb{E}^3$ that is essentially two-dimensional. We define a 2-manifold as a topological space Σ where every point has a neighborhood topologically equivalent to an open disk of $\mathbb{E}^2$ [39]. If S is an r-set, $b(S)$ its boundary and Σ a 2-manifold, and Σ and $b(S)$ are topologically equivalent, then S is a realization of Σ in $\mathbb{E}^3$. Not all r-sets are realizations of 2-manifolds (see Figure 10.2). The mathematical models of solid objects we want to represent are r-sets bounded by 2-manifold surfaces.

We can now consider the problem of defining representations of the solid objects belonging to the mathematical modeling space M defined above. A solid representation is a finite collection of symbols (of a finite alphabet) describing a solid. The collection of all syntactically correct representations is denoted as R. Thus, a representation scheme is a relation $s : M \rightarrow R$. The domain of S is denoted by D and its image under s by V. Any representation in V is called *valid*. A valid representation r is *unamabiguous* (or *complete*) if it corresponds to a single object, it is *unique* if its corresponding object in M do not admit another representation than r.

Intuitively, a representation scheme is a relation between (abstract) solids and representations. A representation is invalid if it does not correspond to any solid. A valid representation is ambiguous if it does correspond to several solids. A solid has nonunique representations if it can be represented in several ways in the scheme.

The properties of a representation scheme can be summarized as follows (see [39] and [43] for a more thorough discussion):

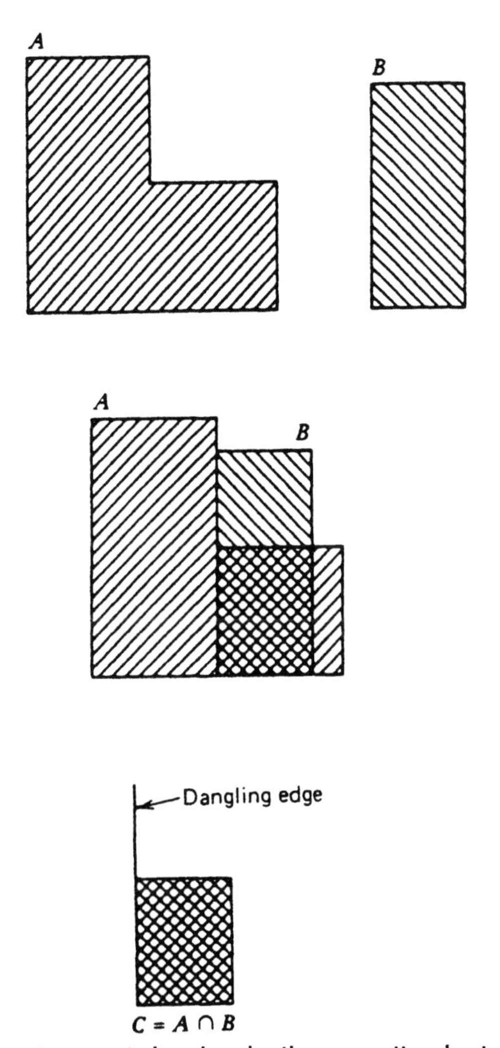

FIGURE 10.1. r-sets are not closed under the conventional set operations [43].

1. *Descriptive power:* the domain of a representation scheme characterizes its descriptive power, that is, the set of objects in M representable in the scheme.
2. *Validity:* the range V of a representation scheme is the set of representations that are valid. It is highly desirable to have a representation scheme wherein all admissible representations are valid (i.e., $V = R$), otherwise algorithms must be provided to check the validity of representations after they have

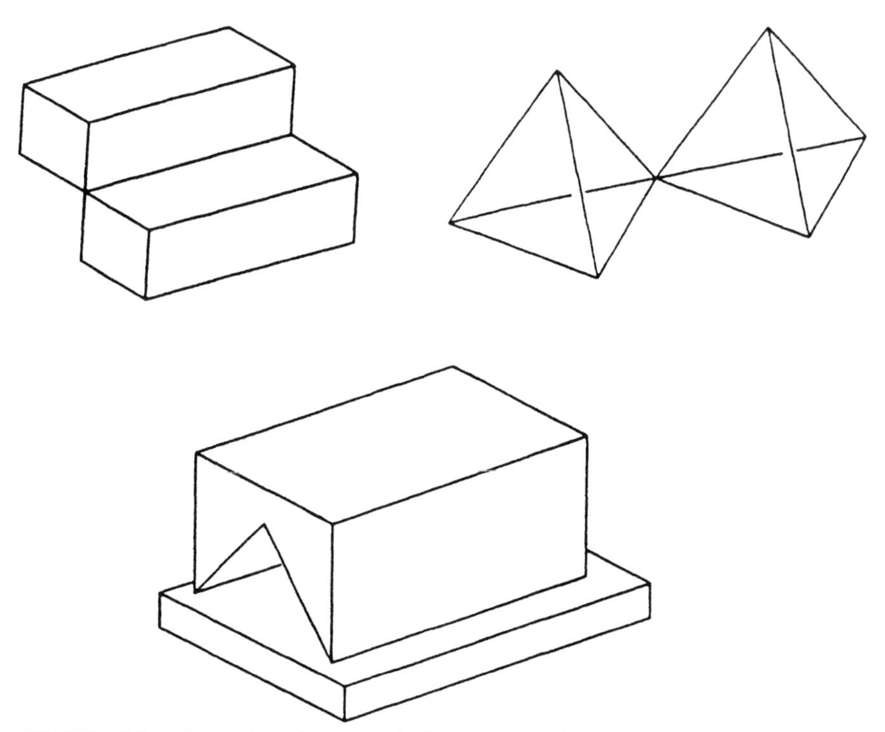

FIGURE 10.2. Examples of r-sets which are not realization of 2-manifolds [39].

been constructed, or validity constraints must be embedded in the procedures that construct the representations.

3. *Completeness and uniqueness:* a scheme is complete when all its valid representations are complete; it is unique when all its valid representations are unique. Representation schemes that are complete and unique would be extremely valuable because there exists a one-to-one mapping between objects and their representations.

Other properties, like conciseness (size of a representation scheme), ease of creation (by a human user), applicability, computational ease, closure of operations (do manipulation operations preserve validity?), are discussed in [39] and [43].

Since no representation scheme is good for all operations that we might want to perform on it (visualization, Boolean operations, computation of integral properties), solid modeling systems sometimes make use of several representation schemes. The presence of multiple representations raises the important issue of consistency. *Consistency* means that the various representations of an object in different schemes must not be contradictory. More formally, two representations $r \in R$ and $r' \in R'$ are said to be *consistent* if there exists at least an object $m \in M$ having representations r and r'. A stronger concept is that of equivalence: two representations are *equivalent* if they describe the same set of objects. Thus, equivalent representation schemes have the same domain.

Because we are focusing on solid object representations for applications in CAD/CAM and in computer graphics, a broad classification of representation schemes is into *volumetric, boundary* and *hybrid* schemes. Volumetric schemes describe an object in terms of solid primitives covering its volume, whereas boundary schemes describe an object in terms of the surfaces enclosing it. Hybrid schemes, like the PM-Octree and the modular boundary representations, are combinations of the two approaches. Representations belonging to the three groups are described in detail in the next sections.

Other schemes have been developed for applications that are not included in the previous classification, whereas, on the other hand, there exist representations that will be described in the following sections that have been applied also in computer vision, or robotics (for instance, boundary representations or octrees).

Among representations used in other applications we can mention are *sweep* representations [53, 54]. In a sweep representation, an object S is described by a two-dimensional set moving along a curve in the 3D space. We can have translational and rotational sweeping when the two-dimensional set is moved along a straight-line segment perpendicular to the plane where it lies, or when the set is rotated around an axis [43]. Usually, these methods are used to generate boundary descriptions; many boundary modelers have sweeping primitives (defined by plane curves and not two-dimensional sets) to generate object representations [39]. A "generalized" form of sweep representation is extensively used in computer vision: the so-called *generalized cylinders* (or *cones*). A generalized cylinder describes an object in term of a principal axis (that is a space curve, called *spine*) and a cross section varying with regularity along the spine. Generalized cylinders are quite useful for describing shapes that can be decomposed into a few elongated simpler shapes such as airplanes, animals, or the human body.

3. VOLUMETRIC SCHEMES

Volumetric representation schemes describe an object in terms of volumetric primitives. They can be classified into *decomposition models* that describe an object as a collection of simple primitive objects combined with a single "gluing" operation, and into *constructive* models, that describe an object as the Boolean combination of primitive point sets. Examples of decomposition schemes are cell decompositions, spatial enumerations, space-based adaptive subdivision schemes. Examples of constructive models are Constructive Solid Geometry trees and their variations. In the following two subsections, these two classes of schemes are described in detail.

3.1. Decomposition Schemes

Decomposition schemes can be classified into *object-based* and *space-based* schemes. The former describe an object as the combination of pairwise quasi-disjoint elementary 3D cells whose union covers the object. The latter decompose the 3D space into elementary volumes (usually cubes), and represent the object as the combination of the volume elements belonging to it.

Object-based decomposition schemes are known as cell decompositions. Cells may be any objects topologically equivalent to a sphere. Any two cells must be either completely disjointed, or meet at one vertex, or along one line or face. Usually, tetrahedra are used as basic cells. In this case, a cellular decomposition is a *tetrahedralization*. Figure 10.3 shows an example of a tetrahedralization. Cell decompositions are extensively used in finite elements and also in computer vision for object reconstruction [8, 12]. Cell decompositions are unambiguous schemes, but they are not unique. They have several important properties—invariance through rigid transformations, ease of update, and computational efficiency. They are mainly used as auxiliary representations for spe-

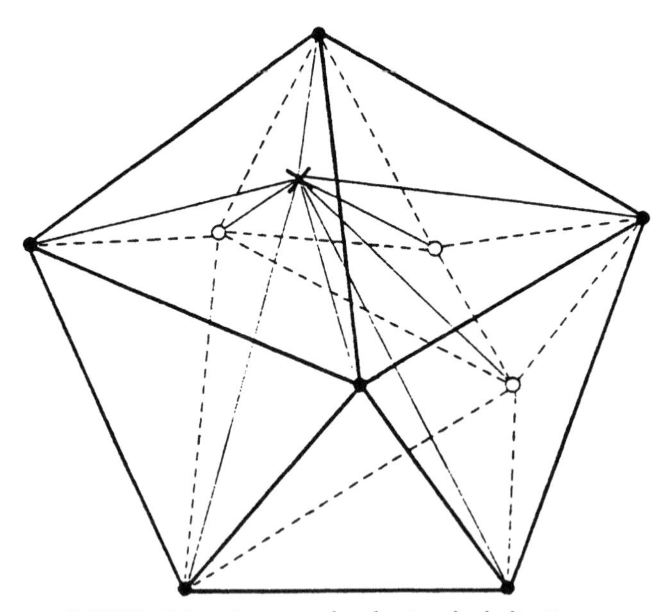

FIGURE 10.3. An example of a tetrahedralization.

cific computations and are generated either by starting from sparse sets of points, or, more often, by starting from another solid model (in finite element applications [22]).

Space-based schemes describe objects in terms of the 3D space they occupy by subdividing the space into regular volume elements, called *voxels*. Thus, the object representation will depend on the position in the space occupied by the object, that is, such schemes are not invariant under translations and rotations. Furthermore, they give an approximate representation of the object that depends on the resolution of the spatial subdivision into voxels (i.e., on the size of the voxel).

The first space-based scheme, a generalization of the pixel-based schemes used in image processing, is the spatial enumeration scheme. In such a scheme, a representation of an object S is a list of spatial cells occupied by S. The cells (voxels) are cubes of fixed size and lie in a fixed spatial grid, thus forming a regular subdivision of the space. Each voxel is described by a single vertex (because of the regularity of the subdivision), and the list of voxels having a nonempty intersection with the object is described in the form of a three-dimensional binary array. Spatial enumeration schemes are unambiguous, unique (except for positional nonuniqueness) and are easy to validate (because we have to pay attention only to connectivity requirements). They tend to be quite verbose and storage costs dramatically increase with increased resolution. A great advantage of such representations is the conceptual simplicity of Boolean operations on them. Also, even if such operations may be inefficient on serial computers (because of the size of the representation), they are extremely well suited for implementations on parallel machines. The main disadvantage of such schemes is the fact that they produce only approximate object descriptions, where the quality of the approximation is determined by the voxel size.

Spatial enumeration schemes have large storage requirements and a poor feasible resolution. To overcome these problems, adaptive subdivision schemes have been developed [40]. Such schemes basically combine together neighboring voxels that are completely internal (full) or completely external to the object. They have the fundamental property that the number of cells needed for the representation of a solid is proportional to its surface area. Examples of adaptive subdivision schemes are the Octree and the Bintree.

The Octree [40, 50] is the three-dimensional generalization of the quadtree used for two-dimensional entities. The Octree uses a recursive regular subdivision of a cubic universe into eight octants. An octant is not subdivided when it is completely inside or outside the object, or its size is equal to the resolution (voxel level). This recursive subdivision is described by a tree of degree 8 where the root represents the universe, and each nonroot node is a cube obtained from the subdivision of its parent. Leaves describe portions of space for which no further subdivision is necessary or possible (voxels). An attribute is associated with the nodes of an octree: the leaves are said to be *full* or *void* depending on whether their corresponding cubes are entirely within or outside the object, respectively. Leaves corresponding to voxels that lie on the boundary of the object (i.e., partially occupied) are considered full or void depending on the portion of full space the voxels contain. All internal nodes are said to be partially occupied (*partial*). Figure 10.4 shows an example of an octree.

As spatial enumeration schemes, octrees are approximate representations; the quality of the approximation is determined by a fixed resolution. If no special connectivity requirement is imposed, octree schemes are valid. Up to the limits of the resolution, octrees are unambiguous representations. They are also unique, except for positional nonuniqueness: all space-based representations vary under

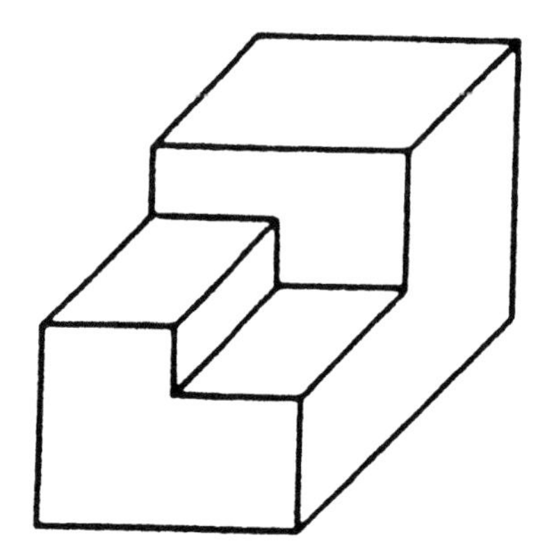

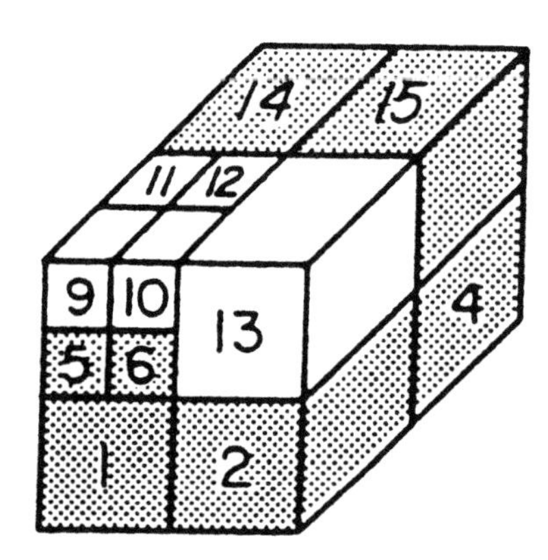

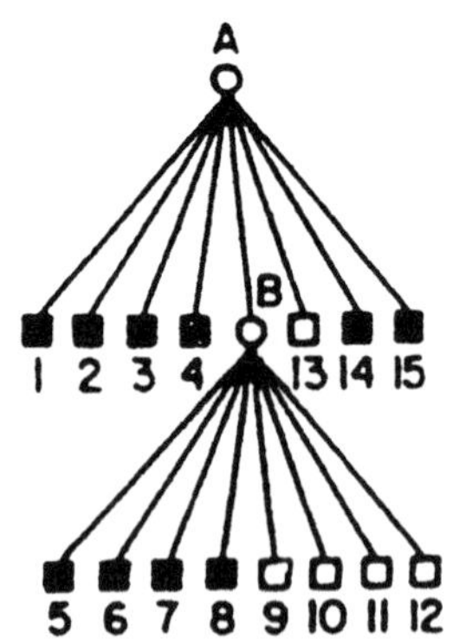

FIGURE 10.4. Example of an object (a), its Octree decomposition (b) and its tree representation (c) [50].

rigid transformations. Efficient algorithms based on a single tree traversal have been developed to perform Boolean operations on octrees and to compute volume and integral properties efficiently.

Some variants of the Octree scheme have been proposed in the literature. The MX-Octree, the natural 3D extension of the MX-quadtree, encodes (an approximation of) the boundary of the object [50]. Leaves in the tree can be of the types interior, boundary, and exterior, depending on their position with respect to the object. An alternative adaptive subdivision scheme is the *Bintree* [57], that is based on a binary recursive subdivision of the space. In a bintree, each non-homogeneous node (partially occupied) is subdivided into two halves. The subdivision is performed in order, in the x, y and z directions. The bintree structure requires less space in comparison with octrees, and it is also extendable to spaces of dimension higher than three. The Octree and bintree data structures can be converted into one another straightforwardly.

A general disadvantage of octree-like representations is the amount of storage they require. Although the number of nodes in an octree describing an object is proportional to the surface area of the object, octrees are still quite large. Thus, many researchers have designed and implemented alternative representations of octrees that replace the explicit tree structure with a pointerless linear description [50]. The so called *linear octrees* are representations of octrees as collections of leaf nodes. There are a number of different linear representations that can be grouped into two categories: those based on a locational code and those based on a preorder traversal of the octree.

The basic idea underlying the locational code method is that each leaf is encoded by a base 8 number, the locational code corresponding to a sequence of digits that locate the leaf along a path from the root of the octree. Usually, the collection of leaf nodes is kept as a list sorted according to the values of the locational codes of the nodes, or as tree structure, in order to speed up search and retrieval operations. The second class of pointerless representations are in the form of a preorder tree traversal of the nodes of the octree. The result is a string consisting of the symbols 'F', 'V', 'P' corresponding to full, void, and partially occupied nodes. This representation, credited to Kawaguchi and Endo [32], is called a *DF-expression*. Linear representations are sufficient for many algorithms. Boolean set operations on linear octrees consists of merging two character strings. Algorithms for nearest neighbor finding for connected component labeling and performing integral computations have been developed for such representations [51].

Linear representations have been developed for bintrees as well; they are more efficient than linear octrees because there are less leaf nodes in a bintree and some improvements can be made to the linear encoding leading to a representation requiring approximately one bit per node (compared to two bits per node for an octree) [55].

3.2. Constructive Schemes

Constructive Solid Geometry (CSG) defines a family of schemes for representing solids as Boolean combinations of primitive components. The most natural way to represent a CSG model is the so called *CSG tree*. A CSG tree is a binary tree wherein internal nodes represent operators that can be either rigid motions or regularized union, or intersection or difference. Terminal nodes, however, are either primitive leaves that represent subsets of $\mathbb{E}^3$ or transformation leaves that contain the defining arguments of rigid motions. Each subtree that is not a transformation leaf represents a set resulting from applying the indicated operators to the sets represented by primitive leaves.

We have two kinds of CSG schemes depending on whether the primitives are *r-sets* (called *CSG*), or are halfspaces (*CSG-based on general halfspaces*). The former is the most common, because for human users it is easier to operate on bounded primitives than with unbounded halfspaces. Figure 10.5 shows an example of a CSG tree (with bounded primitives). On the other hand, CSG schemes based on general halfspaces are a suitable input for conversion algorithms (see sections 6.2, 7.2, 8.2) and can be obtained straightforwardly from the ones based on r-sets if primitive components are objects with planar faces.

CSG schemes are unambiguous but clearly not unique; different CSG trees may describe the same object. The domain of a CSG scheme depends on the set of primitive solids and on the transformation and Boolean operators available. If the primitives of a CSG scheme are bounded, then any CSG tree is a valid representation if the primitive leaves are valid. This means that the validity of a CSG tree can be checked at a purely syntactical level. CSG trees based on halfspaces may represent unbounded sets and thus be invalid: checking if a set is bounded can be a computationally expensive process [43]. In principle, CSG schemes are concise but, in practice, the storage costs tend to grow because additional information for making graphical operations efficient is attached to the model.

The main disadvantages of CSG schemes are the difficulties in the generation of images with hidden lines removed, and in computing integral properties directly from the model. For visualization, it is necessary to perform a boundary evaluation of the object from its CSG representation that consists of performing Boolean operations by traversing the CSG tree. Some geometric modelers based on CSG (for instance, PADL2) perform an incremental boundary evaluation by updating the boundary description after each change in the CSG tree. On the other hand, there exist efficient algorithms for rendering CSG models by ray tracing (see [47]). The approximate evaluation of integral properties of CSG representations is usually based on conversion algorithms that implicitly construct a decomposition model approximating the object [36, 37].

Wyvill and Kunii [73] propose a variant of the CSG tree (with bounded primitives) that we call a *CSG-DAG*. This scheme describes an object as the combina-

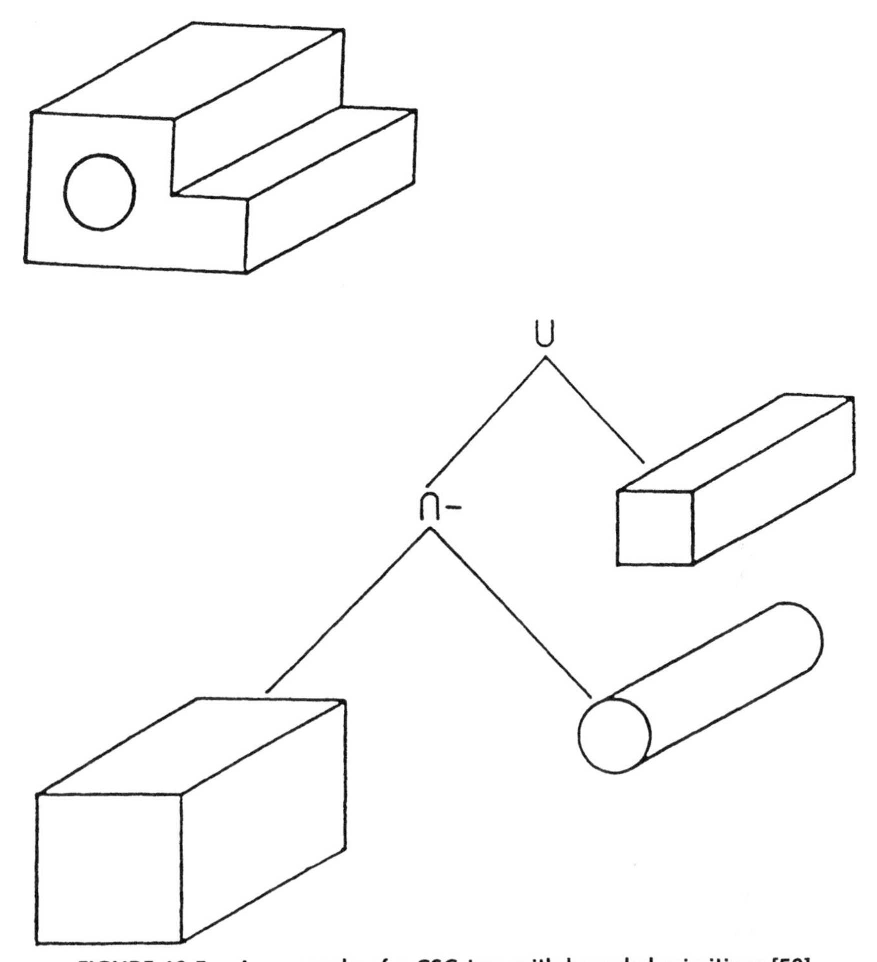

FIGURE 10.5. An example of a CSG tree with bounded primitives [50].

tion of previously defined objects, or primitive objects (e.g., cylinders, blocks). The Boolean operations allowed are set addition and set subtraction. The CSG-DAG is a directed acyclic graph (dag) wherein the internal nodes are disjoint union or set subtraction, and the leaf nodes correspond to objects that can be primitive or nonprimitive. If a leaf node is a nonprimitive object, then the node contains a pointer to the CSG-DAG describing it. A directed acyclic graph is used instead of the tree because the same subobject (primitive or nonprimitive) can be used several times in the same structure without replicating its description. Each occurrence of a subobject in the CSG-DAG also contains two matrices describing the relative spatial location and shape of the instance of the subobject that is being referred by this node. The first matrix is a transformation matrix (describing translation, rotation, and scaling) that is used to define the object instance. An object is properly instantiated by traversing the dag and applying matrix multiplication operators to the initial position. The second matrix is an inverse matrix that facilitates operations on the model.

The CSG-DAG has the same properties of classical CSG. This representation has been defined to facilitate the construction of a spatial index (called a *PM-CSG* tree) on top of the representation in order to perform solid modeling operations efficiently on objects described by a constructive model.

4. BOUNDARY SCHEMES

In a boundary representation (BRep) scheme, solids are defined by their enclosing surfaces. Boundary representations can be viewed as a generalization of wireframe schemes. In a wireframe scheme, an object is described as a collection of curved edges. Wireframe schemes are ambiguous even for the domain of plane-faced solids. Nevertheless, many of the early 3D modeling systems were based on these schemes.

A boundary representation of an object is a geometric and topological description of its boundary. The object boundary is segmented into a finite number of bounded subsets called *faces*. Each face is, in turn, represented by its bounding *edges* and *vertices*. Thus, a BRep consists of three primitive topological entities: *faces, edges* and *vertices*. Faces are contiguous portions of the surface bounding the object. Edges are the elements that, when joined together, form the face boundaries. Vertices are those points on the surface where edges join.

In order to describe objects with multiple connected faces or internal cavities, two derived topological entities are defined, the *loop* and the *shell*. A loop on a face f is a closed chain of edges bounding f. A shell S is defined as any maximal connected set of object faces.

In a BRep, a clear separation is made between the topological and geometric information. Topological information is concerned with the adjacency relations between pairs of individual topological entities (e.g., edges and vertices). There are 25 different adjacency relations. Each consists of a unique ordered pair from the five topological entities (shells, faces, loops, edges, and vertices) [65]. Figure 10.6 shows an example of a BRep; Figure 10.7 shows the adjacency relations among faces.

In a data structure for describing a BRep, the five topological elements must be stored together with a subset of the 25 adjacency relations. The idea is to store a subset of the 25 relations that allows a complete reconstruction of the nonstored relations without errors or ambiguities [66]. Also, each nonstored relation should be retrieved from the data structure in a time proportional to the number of elements involved in the relation itself. For example, the edges incident on a

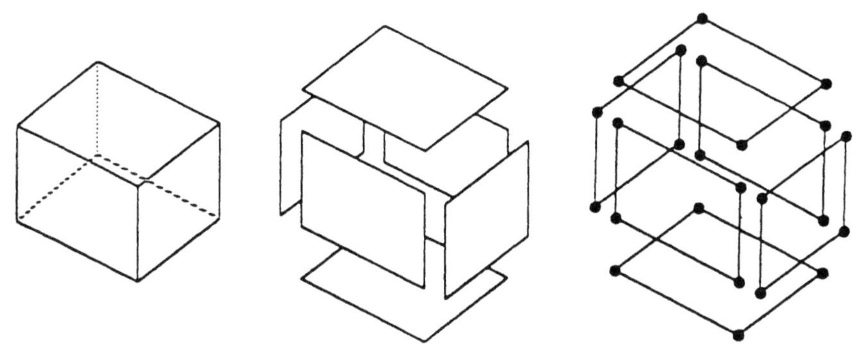

FIGURE 10.6. The boundary representation of a parallelepiped.

given vertex must be retrieved in time proportional to their number. Several data structures have been proposed in the literature to encode a BRep that differ in the number and kinds of relations they store: the winged-edge structure [5], the symmetric structure [69], the face adjacency hypergraph [2].

The geometric description consists of the shape and location in space of each of the primitive topological entities. For example, the geometry of a face may be described by a surface equation, the geometry of an edge by a three-dimensional spline curve, or the geometry of a vertex by its cartesian coordinates. A BRep must contain complete topological and geometric descriptions that, on the other hand, are not independent. A change in the geometry of a topological entity may render the object invalid or introduce new topological entities as a side effect.

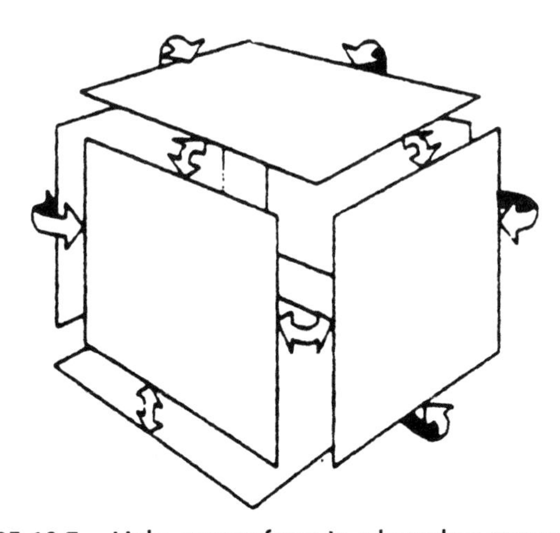

FIGURE 10.7. Links among faces in a boundary representation.

The modeling space of boundary schemes depends on the surfaces that can be used. Thus, boundary schemes can represent a wide variety of solid objects at arbitrary levels of detail. Boundary schemes are unambiguous (if faces are represented unambiguously), but generally they are not unique.

Validity on a BRep is, in general, quite difficult to establish. Validity criteria can be subdivided into topological constraints and geometric constraints. A topologically valid BRep can be constructed by the use of a limited set of primitive functions called *Euler operators* [2, 9, 21, 38, 39]. Euler operators ensure that the topological integrity expressed by Euler-Poincaré formula is satisfied at each update. Beside the low level Euler operators, more complex operations, like glue, rotational, or translational sweeps are available in modelers based on BRep. Such operators are defined in terms of basic Euler operators [2, 39].

Boundary representations are usually quite large, especially when curved-faced objects are approximated by polyhedral models. Boundary representations are especially useful to generate graphical output because of the availability of boundary information, and for describing tolerances (that are information attached to the boundary entities of an object). Integral properties can be easily and efficiently computed from a BRep when operating in a planar-faced environment. Boolean set operations are costly and tedious to implement on such representations.

5. HYBRID SCHEMES

Hybrid representation schemes have been defined as combinations of the approaches discussed in the previous sections. In this section, we present two major classes of hybrid models, PM-Octrees and PM-CSG trees, that combine a boundary representation or a CSG model with an octree, and Modular Boundary Models, that combine a BRep with a restricted set of Boolean operations.

5.1. PM Structures

The major drawback of adaptive space-based decomposition models like the Octree or the Bintree, is that they provide an approximate description of a solid object. To overcome this problem, several authors have proposed schemes that combine the octree with a boundary representation [3, 13, 24, 50]. These schemes have different names but their underlying ideas are very similar. Ayala et al., [3] call the resulting model an *extended octree,* Carlbom et al., [13] call it a *polytree,* Fujimura and Kunii [24] talk about a *hierarchical space indexing scheme.* Samet [50] discusses these structures and introduces the name *PM-Octree,* that is used here.

Like an Octree, a PM-Octree is based on the recursive regular subdivision of a finite cubic universe containing the object into octants. The root of the PM-Octree describes the universe whereas internal nodes are octants. Each nonterminal node is divided into eight equally sized octants. Terminal nodes can be *full* or *void* as, in the octree, or can be *face, edge* or *vertex* nodes. Face nodes are crossed by a single object face, edge nodes contain exactly two edge-adjacent faces together with a part of the common edge, vertex nodes contain exactly one vertex and portions of all the edges and faces incident on it (see Figure 10.8).

The subdivision criteria that a PM-Octree is based on can be stated as follows [50]:

1. At most, one vertex can lie in a volume represented by an octree leaf node
2. If a leaf node's volume contains a vertex, it cannot contain any edge or face that is not incident on that vertex
3. If a leaf node's volume contains no vertices, it can contain, at most, one edge.
4. If a leaf node's volume contains no vertices and contains one edge, it can contain no face that is not incident on that edge
5. If a leaf node's volume contains no edges, it can contain, at most, one face
6. Each leaf node's volume is maximal.

For the proper execution of many operations, the fact that a node is of type vertex, edge, or face is not sufficient to characterize the object completely. Carlbom et al., [13] store, with each node, the polygons determined by the boundary of the object that intersects the node. This requires a considerable amount of extra information. Navazo [42] uses a more efficient representation by including sufficient information to allow the classification of any point with respect to that node (i.e., if a point inside the node is inside, outside, or on the boundary of the object). For instance, in the case of a face node, in addition to the equation of the plane of the face, the direction of the object relative to the

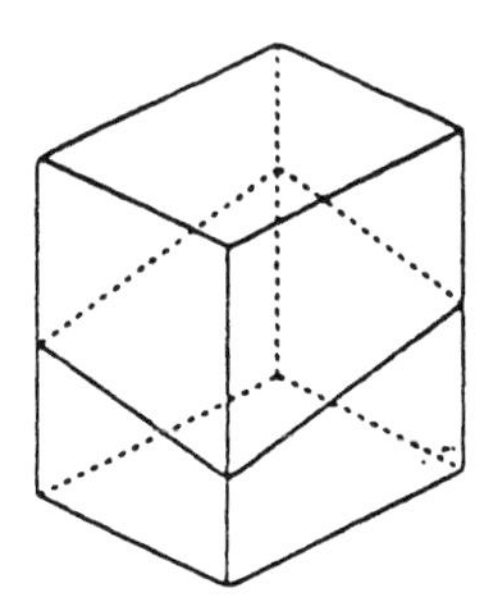
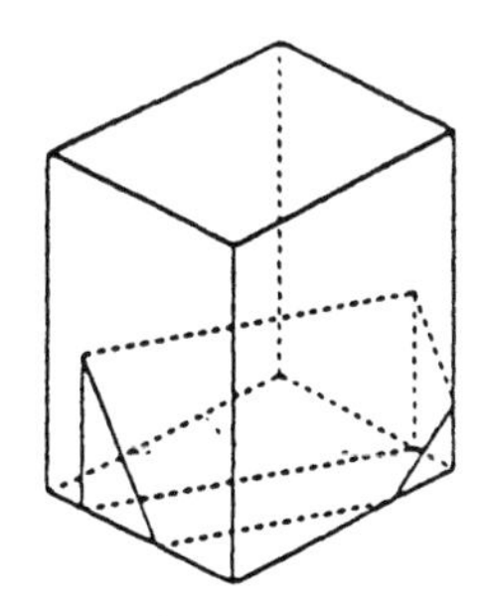
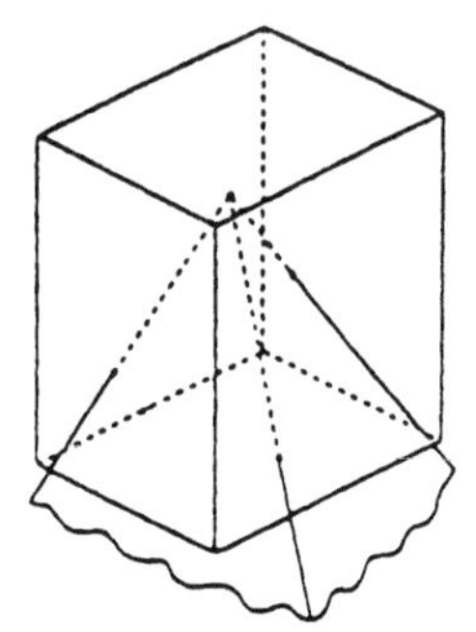

FIGURE 10.8. Face, edge and vertex nodes in a PM-Octree [42].

face is stored. In the case of an edge node, the direction of the object relative to the planes of the two faces sharing the edge are stored (this tells if the edge is convex or concave).

The domain of PM-Octrees is not restricted to planar-faced objects. They have been extended to deal with faces described by biquadratic patches. The difficulty in organizing curved surface patches by using octrees lies in devising efficient methods of calculating the intersection between a patch and an octree node. PM-Octree schemes are unambiguous but not unique, because of positional non-uniqueness (like octrees) and of nonuniqueness of boundary schemes. The main advantage of PM-Octrees is the simplicity of the algorithms that implement Boolean operations. These algorithms are based on a parallel traversal of the two input trees and are conceptually quite similar to algorithms for performing Boolean operations on octrees. Visualization and computation of integral properties can be performed quite efficiently on PM-Octrees as well [11, 42, 50].

In some special cases, PM-Octrees would require "infinite" subdivisions. This can happen when a vertex is very close to the clipping plane between two octants and the edges and faces connected to this vertex cross the border at a small angle (see Figure 10.9). As a maximum possible resolution (subdivision level) is fixed for PM-Octrees too, some leaves could still result as partial nodes. These cases seldom occur. However, a variation to PM-Octrees has been proposed by Dürst and Kunii [20] that overcomes this disadvantage. The model they define is called an *extended polytree,* and it is an attempt to improve the PM-Octree by encoding more kinds of nodes. In an extended polytree we can have two kinds of edge nodes and three kinds of vertex nodes. A node of type edge can contain either an edge with two faces adjacent on it, or two faces that do not intersect inside the node but are connected to an edge outside. A node of type

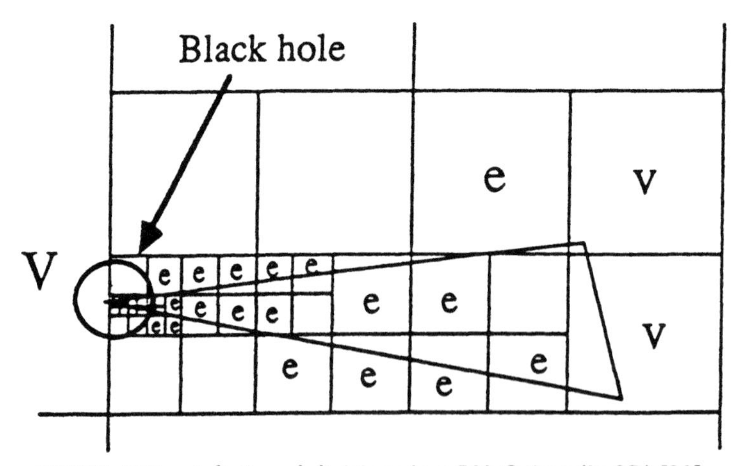

FIGURE 10.9. Infinite subdivisions in a PM-Octree (in 2D) [20].

vertex may contain either a vertex and all edges and faces incident in it, or edges and faces all incident in the same vertex, or faces all incident in the same vertex (in the latter two cases the vertex is outside the node). Apart from the more sophisticated information encoding, extended polytrees have the same property as PM-Octrees.

Wyvill and Kunii [73] define a different model, called the *PM-CSG tree,* that is a combination of an Octree with a CSG. The model is useful to perform solid modeling operations on objects defined by a CSG-DAG. In essence, the underlying concept is similar to the one that the PM-Octree is based on: they vary the definition of a leaf node so that it refers to a primitive object instead of a vertex, edge or face, as in the case of the PM-Octree. The decomposition criterion is such that only one primitive object is allowed to occupy each cell. In a PM-CSG tree we can have five types of leaf nodes:

1. a *full* node, that is entirely inside a primitive object;
2. an *empty* node, that is entirely outside the object;
3. a *positive boundary* node, that describes the boundary between empty space and exactly one positive object;
4. a *negative boundary* node, that contains the boundary between a primitive object S_1, and another primitive object S_2, where S_2 is subtracted from S_1, in such a way that the space corresponding to S_2 is actually empty;
5. a *nasty* node, that is a node at the finest level of resolution (that, thus, cannot be further decomposed) and cannot be classified in the previous types.

Nasty nodes can contain more than one primitive object and usually occur along edges where primitive (positive) objects meet. The model cannot be considered completely exact because nasty cells occur wherever two different primitive objects meet. In practical applications [73], however, nasty cells can be ignored because the volume of such cells at a sufficiently high level of resolution is very small. Figure 10.10 shows an example of a PM-CSG tree.

The decomposition criterion is realizable only because, in the CSG-DAG, primitive objects (or subparts) can be combined together only by disjoint union or set subtraction. A pure CSG representation would not allow a single primitive object to occupy each cell in the PM-CSG tree. In [4], Badouel and Hégron introduce a model very similar to the PM-CSG tree. They call such a model a *Boolean octree,* derive it from a standard CSG and use it for computing Boolean operations.

5.2. Modular Boundary Models

Modular Boundary Models (MBMs) describe the boundary of a solid object as the combination of face-abutting object parts, called components, each of these

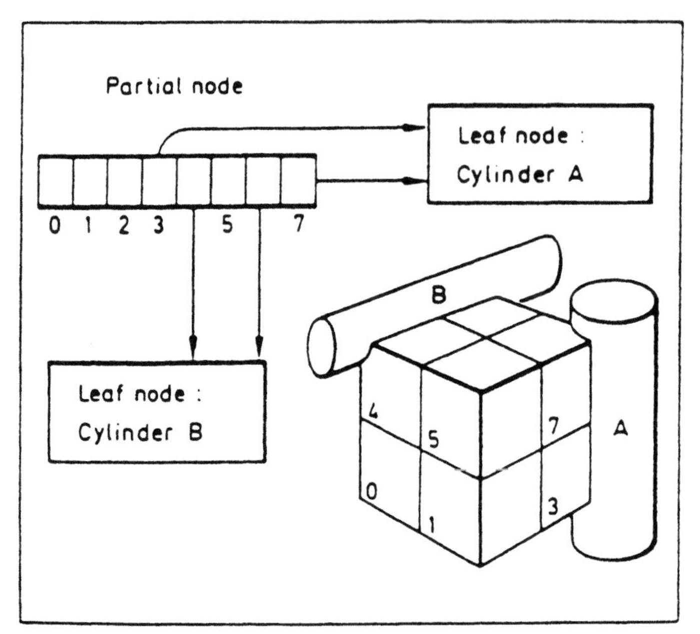

FIGURE 10.10. An example of PM-CSG octree [73].

is described by a boundary scheme. The information required to combine the boundaries of the various parts is stored as a graph [15, 17]. The purpose of MBMs is to combine the ease of design of CSG schemes with ease of display and surface representation of conventional boundary schemes.

There are two kinds of components in an MBM: *positive* and *negative* components. A positive component is a valid solid wherein the face normals are directed towards the outside. A negative component is an unbounded component obtained from a positive component by reversing the directions of its face normals. Two face-adjacent components (i.e., components intersecting at some faces) can be combined to form a composite component by gluing them at their common faces, called connection faces. This corresponds to a disjoint union or an intersection operation on the two original components. The combination rule in an MBM is more restrictive than in CSG schemes because compositions of parts that do not share portions of the boundary are not allowed.

First MBMs represented the face-to-face connection information in the form of a graph, the Hierarchical Face Adjacency Hypergraph described in [15], while further developments of the work have led to the design of a more complete description of an MBM, the Face-to-Face Composition Model (FFC) [17]. The novelty of the FFC model lies in the explicit encoding of information about abutting and intersecting component faces. The FFC model consists of two graphs, those being the *connection* graph and the *interference* digraph. The nodes of both the connection and the interference graphs describe the components (each component is represented in a boundary form). An arc in the connection graph joins two face-adjacent components and describes the set of connection faces shared by the two components. A (directed) arc in the interference graph joins two components (C1, C2) having a spatial interference, and describes the set of subfaces of the first component C1 that are internal to C2. Figure 10.11 shows an object, its decomposition into components, and the connection and interference graphs describing such decomposition. Updates of the FFC model and validity checks when adding new components are greatly simplified by the explicit encoding of interference information.

MBMs are quite general in the sense that any object that can be represented in a boundary scheme can be described by an MBM as well. Current implementa-

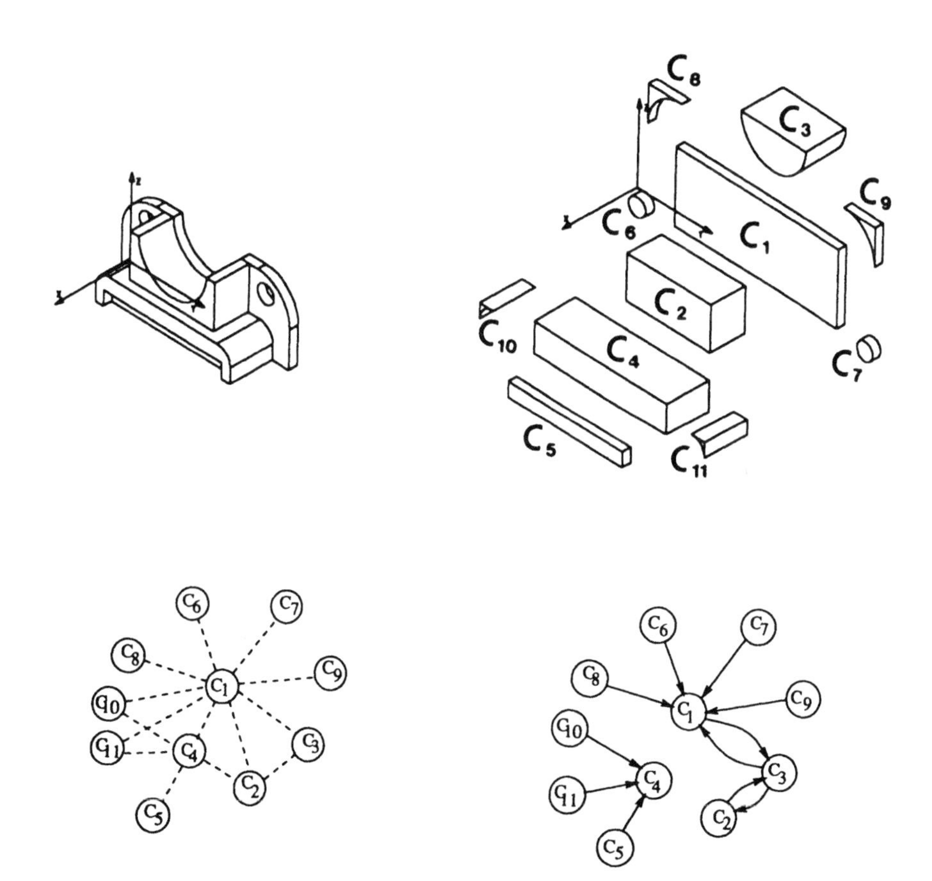

FIGURE 10.11. An object (a), its decomposition into face butting components (b), its connection (c) and interference graph (d) [16].

tions of MBMs, however, are based on polyhedral approximations. These schemes are unambiguous, but are generally not unique. The validity of an MBM can be tested in a similar way as for object-based decomposition models. The topological validity of each single component is enforced by the use of Euler operators for its generation. The geometric integrity problems for the MBM components are the same as for conventional boundary representations.

Compared with boundary and CSG schemes, MBMs offer advantages such as locality of manipulation and update (because of the modular approach), efficient access to geometric and topological information, the possibility of attaching tolerances to the various object parts and the capability of representing the so-called *form features* [67]. Form features, such as pockets, slots, chamfers, protrusions, and through holes are usually attached to object faces and, thus, can be effectively described as components of an MBM [17, 23]. Thus, MBMs are successfully used as the basic object representation scheme in feature-based solid modelers. Disadvantages of MBMs are the verbosity of the representation because of the large number of connection and interference information encoded. Boolean operations and integral computations can be performed more efficiently on an MBM than on a BRep because of the locality of manipulation and the availability of interference information.

6. ALGORITHMS FOR CONVERSION TO DECOMPOSITION SCHEMES

As we pointed out in the introduction, decomposition models give approximate descriptions and generally have a secondary role in modeling systems. Nevertheless, algorithms that convert from exact models to decomposition schemes are well set in the literature. We introduce them first because they give an interesting view about the different information encoded in the various schemes. Moreover, basic conversions to Octree and its variants create the basis for algorithms that convert to hybrid structures (i.e., PM-Octree, PM-CSG), as it will be discussed in the next section.

Spatial enumeration is historically the first decomposition scheme. Conversions to such models are possible through algorithms that compute a point membership classification function [61]: every voxel in the space is independently classified as internal or external to the object. A voxel is usually identified with its center. The crucial part of the algorithms is in discriminating the position of a point in the space with respect to other geometric entities (point-in-polygon and point-in-polyhedron tests). Point discrimination tests are usually computationally intensive and numerically unstable [31], yet they cannot be avoided in many geometric problems. However, such conversions are not very interesting because the output scheme is quite impractical due to the large amount of storage it requires.

Conversions to *hierarchical decomposition schemes* are more interesting. In the practice, the bintree structure is usually preferred to the Octree because of its smaller storage cost. As a general paradigm, the conversion of any model to Octree-like space decompositions can be interpreted as a task driven by the structure of the output. The universe that contains the object is recursively subdivided into subuniverses; for every subuniverse the input information that is *active* is selected among the information that was active in the upper level. In such a way, the input data to be tested is limited to the minimum at each recursion level. When no input information is active within a subuniverse, it is possible to decide if the subuniverse itself is full or void. Special tests must be generally performed when a subuniverse is still active at the voxel level.

The following two subsections present conversion algorithms from a boundary representation and from a CSG, respectively. Different approaches to the conversion problem are discussed and the crucial problems are highlighted.

Approximate decomposition models obtained from exact models do not allow inverse conversions because basic information is lost. Algorithms that convert from the Octree to a boundary representation exist in the literature. Such algorithms are exact in the sense that they maintain the information contained in the decomposition model. In other words, they allow the evaluation of the boundary of the portion of full space in the Octree representation [33]: a stair-faceted solid is obtained as the output unless the original object has only planar faces parallel to the cartesian planes. Therefore, if the Octree had been generated from an exact scheme with an approximated algorithm, the result of the inverse algorithm is a different object that is often useless for practical applications.

6.1. Conversion from Boundary Representation to Bintree

An algorithm that has to convert data from a BRep to a bintree must deal basically with two kinds of entities: the faces of the input model and the volumetric blocks of the output model. The faces are specified by their surface equation and by the edges that form their boundary. The blocks (parallelepipeds) are specified by their eight vertices.

There exist different algorithms for this conversion problem that can be classified roughly into three classes. The first (direct) approach is called *block classification* and consists of deciding whether a block is internal or external to the object, or it intersects its boundary; this approach is followed in [1, 56]. A second major approach is based on *connectivity,* and subdivides the conversion task into two phases. First, it determines all blocks that intersect the boundary of the object, and produces an approximation (MX-bintree) of the boundary itself. Then, connectivity issues are used to label all the blocks that are completely internal to the object. Such approach is applied in [55]. A third alternative approach consists of slicing the input polyhedron with parallel planes, solving the corresponding 2D problem of converting polygons to quadtrees, and integrating the results obtained on adjacent slices. Such solution is adopted in [58].

Algorithms based on block classification directly follow the general conversion paradigm. At the beginning of computation there exists just one big block (the universe) containing the whole object, that is, all its faces. At the generic step, given a face and a block, the algorithm must test whether the two entities intersect or not. If no face intersects the block, the algorithm must classify it as interior (full) or exterior (void) to the object, and make it a leaf node in the tree; otherwise the block is considered partial, is split, and its sons are processed in the recursive step.

The key points of an algorithm based on this approach are the following: a) to decide the relevance of a face with respect to a block; b) to subdivide input data in such a way that only relevant faces are tested at each step; c) to decide the color (i.e., the "fullness") of a leaf block. Altenhofen and Diehl propose an algorithm that uses a so called *modified ray-tracing* technique [1]. The initial universe is supposed to be large enough so that all its vertices lie outside the object(s), that is, they are *void*. At each recursion step, the algorithm computes the color of the new vertices of the spatial subdivision. The current universe is split into two blocks separated by a clipping plane and the sets of planes and edges relevant to the left and right blocks are computed; the partition can be done with different methods that require to evaluate point-in-polygon or point-in-polyhedron tests. Planes and faces that are relevant to both blocks intersect the clipping plane. Four finite rays are then traced. The first ray moves from a vertex of the universe to a vertex of the clipping plane along an edge of the universe, whereas the other ones follow around the clipping plane (see Figure 10.12). Intersections between each ray and the faces that are relevant to the clipping plane are computed. At each ray tracing, the color of the starting vertex and the number of intersections detected give the color of the new vertex. If no plane is active in the current universe, its color is the one of its vertices; if planes are still active at the voxel level, the color of the voxel is the one of the majority of its vertices.

The algorithm is theoretically linear in the number of output nodes. In practice, the tests performed to subdivide the active faces at each step are computationally expensive and make the execution time quite high.

The connectivity approach seems to adapt better to the nature of input data and allows the algorithm to exploit efficient Boolean operations on bintrees to carry on the conversion. The algorithm presented by Tamminen and Samet in [55] seems to be the most efficient among the ones appeared in the literature until now. In the first phase, the algorithm separately converts each face to a bintree approximating it; such trees all have full nodes at the voxel level because they encode two-dimensional objects. The trees corresponding to the various faces are then recursively overlayed in pairs to obtain a bintree of the whole boundary, that is, a MX-bintree of the object (see Figure 10.13).

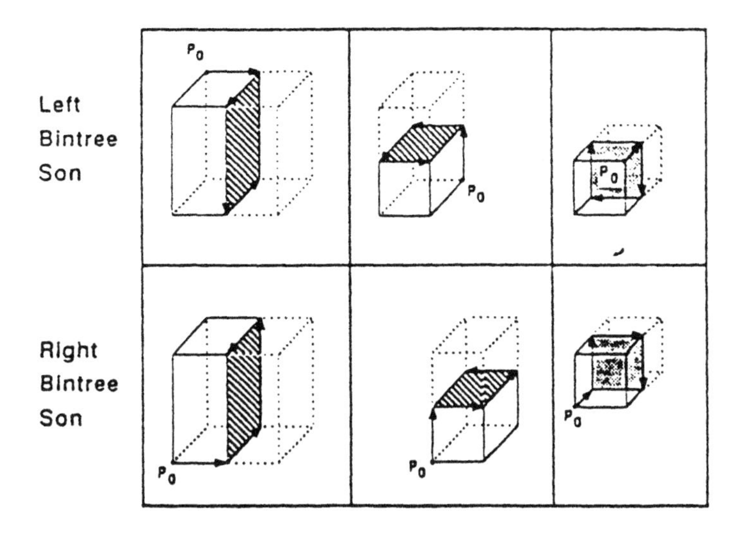

FIGURE 10.12. Ray tracing technique to compute the color at the vertices of the clipping plane [1].

In the second phase, the bintree is traversed and a connected component labeling is performed. Its void components that are not connected to the outside of the image are labeled as full, that is, the interior of the object is filled in order to give a standard bintree representation. As a requirement, the universe must be

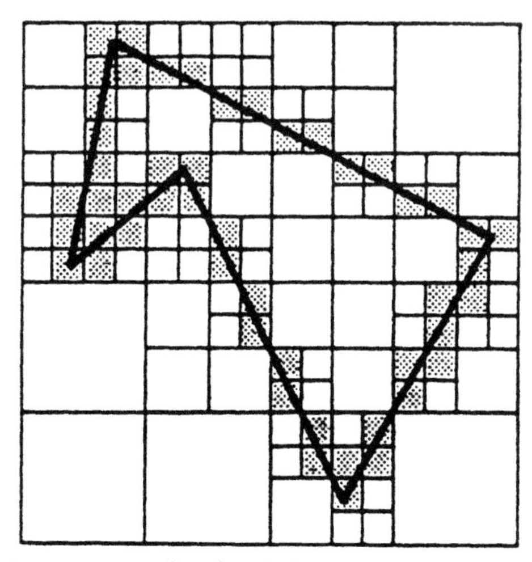

FIGURE 10.13. An example of MX-Octree in 2D (MX-quadtree) [26].

large enough to loosely contain all modeled objects, and the objects themselves must be placed in such a way that the empty space is connected. The algorithm supports multiple solids, but works only on objects without interior cavities (for instance, it would not work on a solid made with a void cube completely contained in a full cube). It could be extended to such objects through an accurate analysis of the adjacency graph of the connected components of the MX model. We present, in detail, the first phase of the algorithm. The phase of connected component labeling is related more to general Octree algorithms than to the specific conversion process, and is analogous to the algorithm for connected component labeling of quadtrees presented in [51].

The conversion from the boundary representation to the MX-bintree is performed by a recursive procedure COMBINE3 that is passed the whole list of faces of the boundary model at the first call. At each recursion step, procedure COMBINE3 computes two bintrees that correspond to the two halves of the input list of faces, then merges the two representations by calling procedure OVERLAY. The conversion of a single face at the lowest level of recursion is computed by procedure CONVERT3. The algorithm returns the MX-bintree of the object(s) in a linear DF-expression (see Section 3.1).

```
Procedure COMBINE3(n1, n2, T);
// The procedure handles data in a global array faces containing the
   complete list of faces; n1 and n2 are array indices which limit the
   sublist the procedure must consider; T is the DF-expression of the
   output tree. //
begin
  if n2 - n1 > 1 then
    COMBINE3(n1, (n1 + n2/2), T1);
    COMBINE3(n1 + n2/2 + 1), n2, T2):
    OVERLAY(T1, T2, T);
  else if n2 - n1 = 1 then
    CONVERT3(faces[n1], T1);
    CONVERT3(faces[n2], T2);
    OVERLAY(T1, T2, T);
  else
    CONVERT 3(faces[n1], T);
  end if
end COMBINE3;
```

Procedure OVERLAY is a basic manipulation primitive that computes the Boolean union of two bintrees [50]. The procedure traverses two linear trees synchronously, and generates the output tree according to the following rules:

1. if either of the nodes is full, the resulting node is full, and the other subtree is skipped;

2. if either of the nodes is void, the other subtree is copied to the result;
3. if both nodes are partial, the result is also partial;
4. the output tree is pruned by replacing recursively PVV by V and PFF by F (i.e., adjacent leaves with the same color at the same level are merged).

Procedure CONVERT3 can be somehow considered the core of the algorithm because it performs the conversion from a face to a bintree. Tamminen and Samet present the solution for planar faces, expressed by an equation

$$P(x, y, z) = ax + by + cz + d = 0.$$

A generalization to curved surfaces can also be obtained. Tamminen and Samet use a projection of the face on one of the coordinate planes to simplify the problem and avoid performing any expensive point-in-polygon test. They apply the following criterion—A block of the bintree intersects the face if and only if:

1. it intersects the infinite plane defined by the equation of the face;
2. its projection on the given coordinate plane intersects the projection of the face.

The main steps of procedure CONVERT3 are the following:

1. Choose a projection plane (say xy) that maximizes the remaining coefficient (c), and compute the projections of the edges bounding the input face on such a plane; a boundary description of a polygon in 2D is then obtained.

2. Compute the quadtree corresponding to the polygon on the projection plane. Notice that this problem corresponds to a conversion from a boundary representation to a hierarchical space decomposition in 2D. The quadtree can be obtained either by a standard algorithm [14, 26], or computed through a reduced version of the 3D algorithm, that is: compute the trees of the edges (the intersection test between a quadrant and an edge can be computed simply), overlay them recursively to obtain a MX-quadtree, then fill it by connectivity labeling.

3. Recursively halve the universe by planes alternatingly perpendicular to the x, y, and z axes. At each subdivision, keep track of the minimum and maximum values *minP* and *maxP* of $P(x, y, z)$ in each block (i.e., in its vertices). To each block B in the 3D subdivision, a corresponding node N is present in the quadtree (namely, the node that refers to the projection of B). Then, the following tests are applied:

- if N is void, then B is void;
- if not $minP(B) \leq 0 \leq maxP(B)$, then B is void (it does not intersect the infinite plane);
- if B is at the voxel level and N is full, then B is full;
- otherwise B is partial and must be split.

The time complexity of the algorithm depends linearly on different factors including the area of the object (for computing the MX-Octree of single faces), the perimeter of the projection of the faces (for computing the MX-quadtree in 2D), and the number of nodes of the MX-Octrees (for overlaying the trees). The connected component labeling is almost linear. In practice, both the approach followed and all optimization criteria adopted make the algorithm very efficient.

The algorithm presented by Tang and Lu [58] is not as efficient as the one previously presented because it computes everything at the voxel level. The algorithm generates an Octree, offering an interesting example of an alternative approach that uses sweep techniques. A plane parallel to a cartesian plane (say, xy) is swept along the third axis (z), in such a way that slices of the input object (i.e., layers of voxels) are considered at each step. The vertices of the input model are ordered along the z axis, and sweep techniques are applied to compute efficiently the intersection of each plane with the object. Each slice is viewed as a digital image (i.e., a grid of pixels). A line parallel to one coordinate axis (say y) is swept along the other axis (x) and a sweep technique is applied again in order to compute the intersections of the sweep line with the portion of the object in the slice. At each step, pixels on the sweep line that are interior to the object are set to "full". A fast technique to compute the index of a voxel in the Octree from its cartesian coordinates is used, and the output is stored in a linear octree using the locational code (see section 3.1). Layers are considered in pairs, first at the voxel level, and then up in the hierarchy. Groups of eight adjacent cells with the same color are condensed in a cell at the higher level. Just two layers must be stored in the main memory while processing the whole structure.

6.2. Conversion from CSG to Bintree

Set operations, as well as analysis and display of solid objects, are hard to perform directly on a CSG scheme. In order to facilitate such operations, some work has been done on evaluating CSG models through conversion to octree-like structures. Lee and Requicha [37] convert a CSG representation to a cellular decomposition based on variable size blocks to compute integral properties on a CSG efficiently. They do not construct the output structure, but add the contribution of each block belonging to the object to the computed integral property. Woodwark et al., [70, 71, 72] use a recursive subdivision technique that is similar to the one wherein the octree is based, for picture generation from CSG models based on hyperplanes. Their approach is similar to Woodwark's hidden surface elimination algorithm, the difference being that Woodwark applies it in the object space. The recursive subdivision of the input CSG model terminates when the submodel in a block meets a criterion of simplicity. Hidden surface [72] or hyperplane generation [71] techniques are applied to reduced "submodels".

To generate a hierarchical space indexing structure by converting from a CSG is conceptually more complicated than to convert from a BRep. When taking a BRep as input, a structure that is relational and intrinsically flat is transformed to a hierarchical structure. On the contrary, in the case of a CSG, two distinct and deeply different hierarchical structures must be manipulated: the CSG tree in input, and the space indexing tree in output.

On the other hand, basic operations (e.g., point discriminations) are both conceptually and computationally simpler in the case of CSG because primitive entities are expressed by mathematical functions; all such operations can be performed by computing the value of such functions at the points under consideration.

The general approach to conversion is a recursive subdivision of the universe combined with a selection of the active information at each subdivision step. The selection task involves a visit and eventually a modification (pruning) of the CSG hierarchy. Elementary components (either primitives or halfspaces) that are not active with respect to the current subuniverse must be eliminated from the CSG tree and Boolean operations on them must be evaluated.

In [48, 49] Samet and Tamminen described an algorithm for converting from a CSG representation to a bintree that is further detailed and discussed in [50]. The algorithm works on a CSG wherein internal nodes correspond to Boolean set operations (union and intersection) as leaves correspond to planar halfspaces. The CSG tree can also contain leaves that correspond to completely *full* or completely empty (*void*) space. These types of nodes, that are not present in the input tree, are generated at intermediate steps by the algorithm when pruning the CSG tree to a subuniverse corresponding to a bintree block. The output of the algorithm is a DF-expression (see section 3.1) encoding the bintree. Of course, any CSG scheme based on primitives with planar faces could be transformed into a CSG based on planar halfspaces by expanding the description of the primitives themselves before applying the conversion algorithm.

The algorithm traverses the universe in depth first order, and evaluates each successive subuniverse S against the CSG tree, that is it selects the nodes of the CSG tree that are *active* with respect to S. A leaf node in the CSG tree is active in S if the plane bounding the corresponding halfspace crosses S. Each inactive halfspace is substituted in the tree with a *full* or a *void* node (depending on S being inside or outside the halfspace itself); full and void nodes are considered inactive as well. An internal node is active in S if both its sons are active in S; otherwise, the Boolean operation corresponding to such a node is evaluated, and the CSG tree is pruned as a consequence.

Thus, the basic operations that must be performed are the following:

- evaluating a halfspace with respect to a block (subuniverse);
- pruning the CSG tree by evaluating Boolean operations at inactive nodes.

In order to decide if a leaf node is active, it is sufficient to evaluate the function defining the halfspace at the eight vertices of the current subuniverse. If all values obtained have the same sign, then the node is not active, and the sign itself indicates if either the subuniverse is full or empty with respect to such a halfspace; otherwise the node is still active. Samet and Tamminen keep track (on the CSG tree nodes) of the maximum and minimum values of the subuniverse with respect to each active halfspace. When subdividing the current block into two subuniverses, just four new vertices (corresponding to the clipping plane) must be evaluated, and, for each son, either the maximum or the minimum (never both) changes. This fact allows the algorithm to efficiently discriminate a halfspace with respect to a subuniverse. Figure 10.14 shows the (2-dimensional) bintree of a halfspace.

Boolean operations on the CSG tree are computed only when at least one of the operands (sons) is either a full or a void leaf. Thus, given an internal node B in the CSG tree corresponding to a Boolean operation, $B \in \{\cup, \cap\}$, such that its operands are a subtree T and either a void node V or a full node F, the following rules apply:

1. $F \cup T = F$
2. $V \cup T = T$
3. $F \cap T = T$
4. $V \cap T = V$

The algorithm starts computing, for each halfspace in the CSG tree, the minimum and maximum values in the whole cubic universe. Then, a recursive

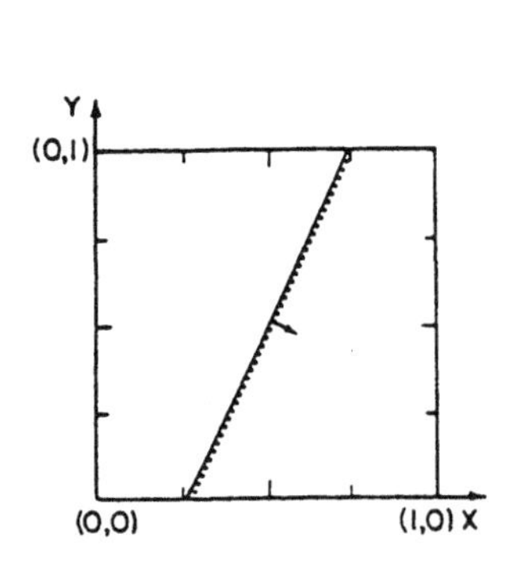

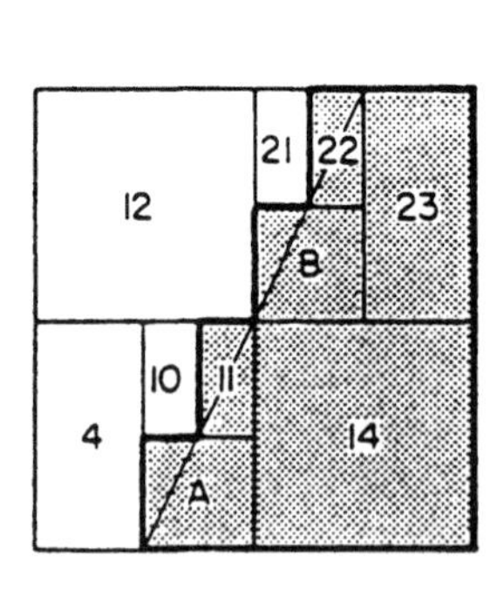

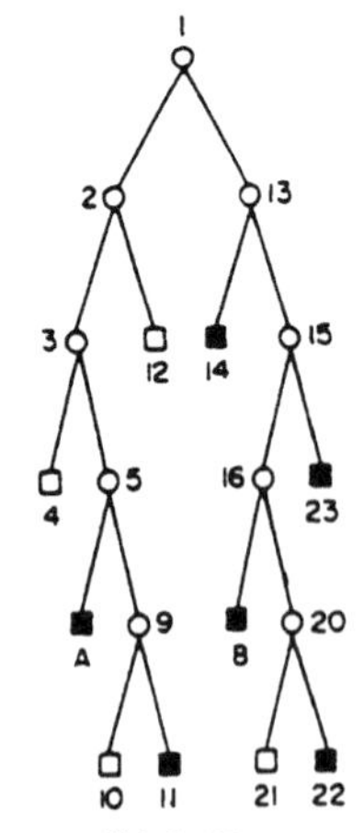

FIGURE 10.14. A half-space (a) and its corresponding bintree (b) [50].

procedure CSG_TRAVERSE is invoked that is passed the universe and the entire CSG tree at the first call. At each recursive call, the procedure will be passed a subuniverse of reduced size, and a copy of the CSG tree pruned to such a subuniverse. For the sake of efficiency, all input data are scaled to be contained in the unit cube (the initial universe) before processing. A subuniverse is simply identified in the algorithm by its level in the bintree hierarchy and by the width of its smallest side; this is not a complete characterization, but is sufficient for computing purposes, as the maximum and minimum values of halfspaces in a subuniverse are stored in the corresponding CSG tree nodes (see procedure HSP_EVAL). A procedure PUTOUT is used to update the global output (DF-expression), with the three types of nodes P (*partial*), F (*full*), and V (*void*). The pseudocode of procedure CSG_TRAVERSE follows.

```
procedure CSG_TRAVERSE (T,lev,width);
// T is the current CSG tree.
   lev is the level of the current block, and width is the width of its
   smallest side. //
begin
  if (TYPE(T) = "F") or (TYPE(T) = "V"' then
    PUTOUT(TYPE(T))
  else if lev = VOXEL_LEVEL then
    PUTOUT(TYPE(CLASSIFY_VOXEL(T)))
  else //subdivide and prune the CSG trees //
    PUTOUT("P");
    if lev mod 3 = 0 then width ← width/2;
    PRUNE(T,lev + 1,width,"LEFT",T_l);
    CSG_TRAVERSE(T_l,lev + 1, width);
    PRUNE(T,lev + 1,width,"RIGHT",T_r);
    CSG_TRAVERSE(T_r,lev + 1, width);
  end if
end CSG_TRAVERSE;
```

Procedure PRUNE reduces the size of the CSG tree relevant to the current bintree block. PRUNE traverses the CSG tree in depth-first order and removes inactive CSG nodes by applying the rules specified above. The evaluation of each halfspace with respect to the current block is computed by procedure HSP_EVAL. Such procedure takes the maximum and minimum values of a halfspace with respect to the father of the current block (such values are stored on the current CSG tree at the node corresponding to the halfspace itself), and updates those values to the current block. In order to do that, the algorithm computes a *delta* value on the basis of the halfspace function and of the width of the block; *delta* is then either added or subtracted to either the minimum or the maximum, according to its sign and to the position of the current block with respect to the father. The pseudocode of procedure PRUNE follows.

```
procedure PRUNE(T,lev,width,dir,T1);
// T is the current CSG tree;
   lev is the level of the current block, and width is the width of its
   smallest side.
   dir is either "LEFT" or "RIGHT" and indicates the position of
   current block wrt its father. //
begin
  if TYPE(T) = "HALFSPACE" then
    HSP_EVAL(T,lev,width,dir,T1)
  else
    PRUNE(LEFT(T),lev,width,dir,Tl);
    if ((TYPE(T) = "∪") and (TYPE(Tl) = "F")) or
        ((TYPE(T) = "∩") and (TYPE(Tl) = "V")) then
      T1 ← Tl //fast application of pruning rules 1 and 4 //
    else
      PRUNE(RIGHT(T),lev,width,dir,Tr);
      if ((TYPE(T) = "∪") and (TYPE(Tr) = "F")) or
          ((TYPE(T) = "∩") and (TYPE(Tr) = "V")) then
        T1 ← Tr //fast application of pruning rules 1 and 4//
      else if ((TYPE(T) = "∪") and (TYPE(Tl) = "V")) or
          ((TYPE(T) = "∩") and (TYPE(Tl) = "F")) then
        T1 ← Tr //apply rule 2 or 3//
      else if ((TYPE(T) = "∪") and (TYPE(Tr) = "V")) or
          ((TYPE(T) = "∩") and (TYPE(Tr) = "F")) then
        T1 ← Tl //apply rule 2 or 3//
      else //both subtrees are active: T cannot be evaluated //
        T1 ← MAKE_CSG_TREE(ROOT(T),Tl,Tr);
      end if
    end if
  end if
end PRUNE;
```

Function CLASSIFY_VOXEL invoked by procedure CSG_TRAVERSE is used to decide the color (either *full* or *void*) of a voxel when a CSG tree is still active in it. In the simplest case of just one halfspace active, the value of the halfspace equation at the center of the voxel is computed, and the color is set according to the sign of the result. If more than one halfspace is active, the procedure evaluates recursively the subtrees of the CSG tree, and applies the same rules as procedure PRUNE to combine the results.

Samet [50] gives a thorough analysis of the complexity of the algorithm. The main results of such analysis are the following:

- The "practical" complexity of the CSG tree evaluation is $O(2^{2n})$, where n is the resolution of the bintree.
- The average number of active CSG tree nodes in a bintree block approaches one asymptotically, as the resolution is increased.

- The computational complexity of converting a CSG tree approximation of a (curved) object to a bintree is asymptotically independent of the number of halfspaces used in the approximation.

Thus, the linear approximation of curved halfspaces can be practical even if it would lead to a considerable increase in the size of the CSG tree [50].

7. ALGORITHMS FOR CONVERSION TO PM-OCTREE AND PM-CSG

Exact octree structures have been developed on the wave of the corresponding approximated models to obtain structures that are both exact and computationally efficient. In the same way, algorithms that perform conversions to such structures have been derived, at least at a high conceptual level, from the ones described in the previous section.

Differences in the algorithms that compute exact models are mainly in saving geometric information that is otherwise lost in the approximate models. These differences substantially affect more of the data structure than the code of an algorithm. Geometric entities must be encoded using appropriate structures (like lists of plane equations for the PM-Octree, or CSG primitive descriptions for the PM-CSG) and suitably linked to the Octree structure.

Exact trees are usually much smaller in terms of number of nodes than the corresponding approximate models (at a reasonable level of resolution). Indeed, unlike in approximate models, a leaf node in a PM-Octree [a PM-CSG] can be partially occupied, provided that there is not too much information in it. Then, branch tests in conversion algorithms will be more sophisticated in measuring the information within a block, instead of simply checking it.

In the following subsections, we will discuss in detail some algorithms for converting from a BRep to a PM-Octree [10, 13], a CSG tree to a PM-Octree [41], and from a CSG-DAG to a PM-CSG [73].

7.1. Conversion from Boundary Representation to PM-Octree

In [10], Brunet and Navazo propose a conversion algorithm that takes a boundary representation of an object with planar faces and converts it to a PM-Octree. The algorithm follows the classical block classification paradigm wherein the space is recursively subdivided into octants. At each recursion step, the list of active faces for each octant are selected until such information can be encoded in a leaf node. The authors use a linear DF-expression for the output tree, where six different types of nodes are used, namely: B (full or black), W (empty or white), F (face), E (edge), V (vertex), and G (partial or grey).

Of course, nodes of type F, E and V must contain a pointer to a structure describing the face, edge or vertex contained in the node respectively. Given the list of plane equations of all the object faces (from the boundary model), it is sufficient to store pointers to such list: one face is described by one plane, one edge by two planes, and one vertex by three or more planes intersecting at the vertex itself. If planes are encoded in the list in such a way that the normal vectors of the corresponding faces are always directed towards the interior [the exterior] of the object, such normal vectors define the *configuration* of the block, that is, the portion of the block that is void and the one that is full (see figure 10.15).

The main procedure that performs the recursive subdivision and outputs the PM-Octree follows. The output structure is considered global and is updated by procedure PUTOUT that adds the new nodes to the DF-expression and sets links to the list of planes. At its first call, the procedure is passed the whole universe and the complete list of faces from the boundary representation.

```
Procedure BUILD_OCTREE(x,y,z,scale,facelist);
// x, y, z and scale define the block that is analyzed (its origin and
   the length of its edge);
   facelist is the list of active faces in the block. //
begin
  sq ← scale/2;
  for i ← 1 to 8 do
    compute the origin xi, yi, zi of the ith octant;
    CLIPPING (xi,yi,zi,s,facelist,facelist1);
    if facelist1 is empty then
      if the node is interior then
        PUTOUT("B")
      else
        PUTOUT("W")
      end if
    else if only one face in facelist1 then
      PUTOUT("F",facelist1)
    else if only two faces meeting at an edge in facelist1 then
      PUTOUT("E",facelist1)
    else if all faces in facelist1 meet at a point inside the block
  then
      PUTOUT("V",facelist1)
    else if sq > minscale then
      PUTOUT("G");
      BUILD_OCTREE(xi,yi,zi,sq,facelist1)
    else
      PUTOUT("G",facelist1)
    end if
end BUILD_OCTREE;
```

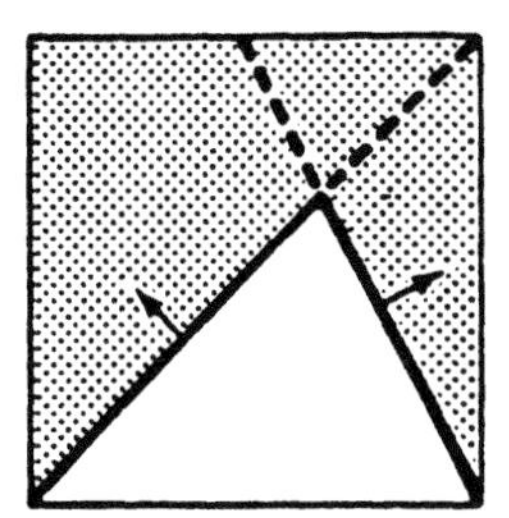
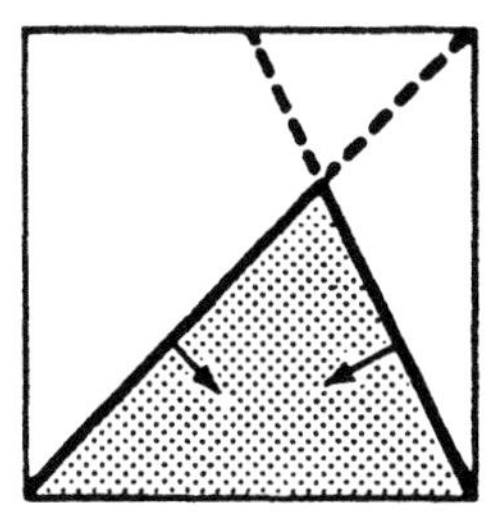

FIGURE 10.15. Two possible configurations for an edge node in a PM-Octree [50].

In their implementation, the authors also maintain a list of active vertices within the block that is used and recomputed by procedure CLIPPING, and is consulted by the branch tests to improve efficiency. Procedure CLIPPING must select the list of faces [of vertices] contained in the current block from the list of faces [of vertices] active in the parent block. Such task a priori involves computing cumbersome and numerically unstable intersection algorithms, as outlined in the previous sections. The authors propose to apply, for each face, the following cascade of heuristic tests to avoid, if possible, such computations:

- detect if any vertex of the face lies inside the block (to accept the face)
- use a max-min test between the block and the parallelepiped containing the face (to discard the face)
- see if the plane of the face intersects the block (to discard the face)
- try to intersect the edges of the face with the faces of the block (to accept the face)

If all the previous tests do not give sufficient information to accept or discard the face, an explicit intersection between the face and the block is computed.

When the condition for a full or void block holds, the color of the output node can be decided efficiently on the base of its adjacent face, edge, and vertex nodes. The color can be left undefined first, and it can be set to B or W by visiting the output tree after all "boundary" nodes are detected. Alternatively, a connected component algorithm can be applied to label black nodes, like in the algorithm for converting to bintrees by Tamminen and Samet discussed in section 6.1 [55].

Some tests in the **if** cascade inside procedure BUILD_OCTREE require computing intersection of planes, and eventually the point or edge location with respect to the current block. Special cases, like a block containing two planes that do not intersect inside it, must be treated.

The complexity of the algorithm is linear in the number of nodes of the output tree, but each step requires more computation than the corresponding block classification algorithms for computing an Octree. On the other hand, Brunet and Navazo show a comparison table between the resulting Octree and PM-Octree of the same solid object: the byte length of the output is reduced of a factor 1000 and the number of leaf nodes even of a factor 10,000 in the case of the PM-Octree.

The algorithm proposed by Carlbom et al., [13] differs from the previous one in using clipped polygons instead of complete faces as active information at each recursion step. The main effort in the subdivision task is devoted to the clipping algorithm; when subdividing a block into octants, three orthogonal clipping planes are activated in a sequence. Each polygon is traversed sequentially along its edges, and intersections with the current clipping plane are computed. The vertices of the polygon are parted into three classes, one each for the vertices on either side of the plane, and the third for the vertices on the plane (note the analogy with the approach of Altenhofen and Diehl described in section 6.1). The subdivision step (computing clipped polygons) seems more complex with respect to the one of the previous algorithm. On the other hand, when the leaf level is reached, the information related to each block can be stored explicitly (i.e., through vertices and edges) without any further computation whereas Brunet and Navazo store it implicitly (i.e., through plane equations).

Like in the previous algorithm, in the first step, only the boundary of the object (i.e., face, edge and vertex nodes) is built. Carlbom et al., propose to decide the color of void and full blocks using a ray-tracing technique like the one described in [47].

Finally, Dürst and Kunii [20], propose another algorithm that although following the classical subdivision paradigm, tries to improve its performance by exploiting all the geometrical information available in the BRep, namely vertices, edges, and faces. At each subdivision step, active vertices are checked first against the current block. If more than one vertex is still active, the block is recursively split without any further computation. If just one vertex is active, a vertex leaf is returned. Similarly, the algorithm checks in an **if** cascade if more than one, one, or none edge, more than one, one, or none face is still active, in order to perform branches. The algorithm is slightly more complex than the previous ones because the output structure is more sophisticated.

7.2. Conversion from CSG to PM-Octree and from CSG-DAG to PM-CSG

As in the case of boundary representations, algorithms that convert CSG schemes to exact octree models are quite similar to the algorithms for converting from CSG to approximate octree models. Navazo, Fontdecaba and Brunet propose an algorithm [41] that follows the same approach as the algorithm by Samet and Tamminen presented in section 6.2.

The algorithm takes a CSG based on planar halfspaces (that the authors call an *expanded CSG tree*) as input. First the model is scaled and translated into a cubic universe with an edge of length 2^n. The algorithm traverses the CSG tree and builds the PM-Octree recursively by checking at each recursion step the information active within the current block (subuniverse).

The original input is a CSG tree where internal nodes are intersection and union operations and leaves are halfspaces. As in the algorithm for conversion to the Octree, *full* or *void* leaves can appear when pruning the CSG tree to restrict it to a subuniverse. Again, the subdivision process stops when the current block can be encoded as a PM-Octree node, or if the block is at the voxel level.

The main recursive procedure is passed initially the entire universe, the complete CSG tree, and the list of all halfspaces in the tree. The output format is a linear octree, as in the algorithm presented by Brunet and Navazo in the previous section, and is again updated by procedure PUTOUT. Such procedure also sets the pointers to the appropriate plane equations and computes the node configurations for nodes of type face, edge, and vertex. The pseudocode description of the main procedure follows. The names of procedures and functions are self-explanatory.

```
procedure BUILD_OCTREE(x,y,z,scale,CSGT,spacelist);
// x, y, z and scale define the block that is analyzed (its origin and
   the length of its edge);
   CSGT is the CSG tree related to the current block.
   spacelist is the list of active halfspaces wrt the current block.
   Local symbols π1, π2 are used to denote halfspaces.
   Symbol [BO] denotes a generic Boolean operation. //
begin
  if CSGT = "B" then
    PUTOUT("B")
  else if CSGT = "W" then
    PUTOUT("W")
  else if CSGT = π1 then
    PUTOUT"F",CSGT)
  else if CSGT = π1[BO]π2 then
    if EDGE_IN_CUBE(x,y,z,scale,spacelist) then
      PUTOUT("E",CSGT,spacelist)
    else
      SUBDIVIDE(x,y,z,scale,CSGT,spacelist)
    end if
  else if ALL_PLANES_THROUGH_A_POINT(spacelist,V) and
                                   POINT_IN_CUBE(V,x,y,z,scale) then
    PUTOUT("V",CSGT,spacelist)
```

```
      else
         newsc ← scale;
         if newsc = minscale then
            PUTOUT("G",CSGT,spacelist)
         else
            PUTOUT("G");
            for i ← 1 to 8 do
               COMPUTE_COORD(i,x,y,z,newsc,xn,yn,zn);
               CLIPPING(xn,yn,zn,newsc,CSGT,spacelist,NCSGT,nspacelist);
               BUILD_OCTREE(xn,yn,zn,newsc,NCSGT,nspacelist);
            end for
         end if
      end if
   end BUILD_OCTREE
```

The PM-Octree produced by the previous procedure can be pruned by traversing it and condensing each group of eight adjacent nodes at the same level that are similar into a unique node at the higher level. Navazo, Fontdecaba and Brunet propose to prune the tree dynamically at each new insertion. The pruning procedure is called by procedure PUTOUT and checks the last nine elements of the output list; if there is one G node followed by eight similar sons, then they are condensed and the procedure is called recursively using the updated list. The concept of similarity among nodes is not straightforward when nodes of type E or V are present.

Procedure SUBDIVIDE handles the special case of a block intersecting just two planes that do not intersect inside the block itself. In this case, the block is recursively split until all leaf blocks obtained are blocks of type F, B or W.

Procedure CLIPPING determines then which halfspaces in the list are still active in passing from a node father to a node son. The procedure substitutes each halfspace that is no more active (i.e., the plane defining it does not intersect the block) with a B or a W node. Then, the CSG tree is pruned by a procedure analogous to procedure PRUNE described in Section 6.2. Boolean function ACTIVE uses a max-min test to decide if a halfspace is active; similarly, procedure CLIPPING uses the max [min] value on the block of the function defining the halfspace, to decide the color of the block itself.

```
Procedure
   CLIPPING(xn,yn,zn,newsc,CSGT,spacelist,NCSGT,nspacelist);
begin
   nlist ← ∅;
   NCSGT ← CSGT;
   for every halfspace π_k ∈ spacelist do
      if ACTIVE(π_k,xn,yn,zn,newsc) then
         ADD(nlist,p_k);
```

```
        else
          if MAX(πk,xn,yn,zn,newsc) > 0 then
            change node corresponding to πk in NCSGT to "W"
          else
            change node corresponding to πk in NCSGT to "B"
          end if
        end if
      end for
      PRUNE(NCSGT)
    end CLIPPING
```

An analysis of the complexity of this algorithm is essentially analogous to the one made for the algorithm by Brunet and Navazo presented in the previous Subsection. The theoretical complexity is the same as the algorithm for bintrees by Samet and Tamminen. Again, in practice, tests made inside the algorithm are somehow more complicated, and, on the other hand, the number of nodes and the byte length of the output is drastically reduced.

In [73], Wyvill and Kunii propose an algorithm for coverting from a CSG-DAG (see section 3.2) to a PM-CSG (see section 5.1). The algorithm follows a bottom-up approach—it is an interesting application that clarifies the data structures encoding both the CSG-DAG and the PM-CSG.

The first step of the algorithm is devoted to an independent evaluation of the positions of all instances of primitive objects in the input model. As the CSG-DAG is a structure whose leaves refer to primitive objects in a library, instances of such objects must be located in the three-dimensional space by traversing the DAG and evaluating the transformation matrices stored in the internal nodes. Once all primitives are located, an elementary PM-CSG tree is built for each one of them that consists of a single leaf-node (recall that to contain a single primitive is a sufficient condition to be a leaf node for a block in a PM-CSG tree). The leaf node will contain a pointer to all semantic information about the primitive (e.g., position, functional definition), and a flag indicating if the primitive is full (positive) or empty (negative).

The second step of the algorithm traverses again the CSG-DAG and recursively composes the PM-CSG trees. Both steps can be implemented by a unique recursive procedure that carries on both the location and the composition tasks; in fact, the recursion must expand first to the leaves of the DAG, (completing the location), and then rise up the hierarchy to compute the space division (i.e., the PM-CSG tree compositions). As all Boolean operations in a CSG-DAG are disjoint unions and differences, the space division is performed by algorithms that compute disjoint unions and differences of PM-CSG trees. We do not specify here the rules applied to perform such Boolean operations. Lists of the branch tests performed to compute each operation can be found in [73], though general paradigms followed by such procedures are just extensions of Boolean operations on quadtrees described in [51]. Of course, such operations involve recursive subdivisions of the space, and branch tests on blocks and primitive objects. The basic inclusion test of a primitive against a block is performed by computing the function(s) defining the object itself on the eight vertices of the block, transformed to the original location of the primitive model through inverse matrices. Such test returns three possible values, namely indicating if the block is completely inside, completely outside, or it crosses the primitive. In [74], Wyvill proposes a fast technique for computing such test that exploits a projection of the transformed block on a plane, as in the case of Tamminen and Samet (see section 6.1), geometric computations in 2D are simpler than in 3D, and simple considerations allow to turn 2D results back to the 3D case.

8. ALGORITHMS FOR CONVERSION TO BREP

The nature of the information encoded in a boundary representation is not hierarchical. A "minimal" description of a BRep could be given by three lists storing vertices, edges, and faces of the object respectively, whereas all topological relations could be inferred, in principle, by a geometrical analysis of the elements of such lists. On the other hand, a minimally complete description of the topological relations yields to a very complex network of pointers.

As pointed out by Requicha and Voelcker [45], there are two schools of thought concerning BReps. One school tends to use one single scheme system (the BRep) and to hold rich information in it, in terms of incidence and adjacency (i.e., topological relations—see, for instance [5, 39]). The other school that is frequently associated to the use of hybrid dual-scheme (e.g., BRep/CSG) systems, accepts BReps that are poor in topological information. The generation of a boundary representation should then involve at least the production of face, edge and vertex lists, and possibly the computation of topological relations sufficient to close the relation diagram (see section 4).

Algorithms for generating a BRep are known that convert from PM-Octrees and from CSG (refer to section 6 for considerations on converting from an approximate Octree structure to a BRep). The approaches and the structures of such algorithms are deeply different. In particular, the information encoded in a PM-Octree is somehow strictly related with the one present in a BRep; thus, this kind of conversion is simple and can produce a rich boundary representation. On the contrary, conversion from a CSG (*boundary evaluation*) is a very complicated task that has a high computational complexity and can require cumbersome numerical computations. Moreover, the boundary evaluation problem has been considered essentially in connection with CSG-based or hybrid systems in order to obtain data that are well suited for computing Boolean expressions and for rendering purposes. Thus, no attention in the literature has been paid so far to the production of "rich" BReps. All algorithms are mainly concerned with the evaluation of faces, vertices and edges that can remain somehow "mixed" without any explicit relation in the output model. Some authors even prefer to avoid a complete reconstruction of the BRep. Thibault and Naylor [59], for instance, propose conversion from CSG to an intermediate, but hierarchical structure, called a *Binary Space Partitioning Tree* [50], that is a binary tree representing a recursive partition of the 3D space by hyperplanes. This structure allows a fast evaluation of set theoretic Boolean expressions and the development of efficient ray-tracing algorithms.

8.1. Conversion from PM-Octree to Boundary Representation

The PM-Octree structure can be regarded as a superimposition of a spatial index on a boundary representation: all geometric information present on the BRep is implicitly maintained in the PM-Octree. For this reason, converting a PM-Octree structure to a BRep is a simple task.

In [10], Brunet and Navazo describe an algorithm that performs the conversion from a PM-Octree to a BRep in two steps. First, the PM-Octree is traversed, and only nodes of type vertex are considered. For every vertex, its coordinates are computed; moreover, for every face incident into the vertex, a pointer to the vertex itself and two pointers to the two neighboring faces are stored. This is simple because a list containing all faces of the object (or, more precisely, their equations) is part of the data structure encoding the PM-Octree, and pointers to the faces are stored in a cyclic order in a vertex node. After the first step, all vertices, faces, and relations face-vertex, vertex-face, and face-face are obtained.

```
procedure CREATE_LISTS(x,y,z,scale,facelist,vertexlist);
// x, y, z and scale define the block that is analyzed (its origin and
   the length of its edge);
   facelist is the list of faces of the object; vertexlist is the output list
   of vertices of the object.
   tree is a global array storing the linear PM-Octree.
   pa is a global index of the current position in the tree; pa is set to
   the root (i.e., the first position) at the first call. //
begin
  sq ← scale/2;
  for i ← 1 to 8 do
    pa ← pa + 1;
    if TYPE (tree[pa]) = "G" then
      CREATE_LISTS(x + ax_i * sq, y + ay_i * sq, z + az_i * sq)
    else if TYPE [tree[pa]) = "V" then
      compute the new vertex v of the BRep as intersection of its
          faces;
      insert v into vertexlist;
```

```
          for every face f ∈ facelist incident on v do
             store a pointer from f to v;
             store pointers from f to the two faces incident on v and
                  adjacent to f;
          end for
       end if
    end for
end CREATE_LISTS;
```

In a second step, the algorithm scans the list of faces, and uses the information stored in it to detect closed polygons bounding the faces. This corresponds to compute all the edges of the object, and the relations edge-vertex and face-edge, thus completing all topological information. A simplified pseudocode of the second step follows; the code of the procedure should be extended to handle faces with holes and to store relations.

```
procedure COMPUTE_POLYGONS(facelist,vertexlist,edgelist);
// facelist is the list of faces of the object; vertexlist is the list of
   vertices of the object;
   edgelist is the output list of edges of the object. //
begin
   for every face f ∈ facelist do
      let v be a vertex of f;
      let f1 be a face incident into v and adjacent to f;
      repeat
         let v1 be the other vertex belonging to both f and f1;
         if not e = vv1 is an edge of edgelist then
            insert e as a new edge into facelist;
         end if
         v ← v1;
         let f1 be the face incident into v1 and adjacent to f, other than
              f1;
      until v is the initial vertex
   end for
end COMPUTE_POLYGONS;
```

The time complexity of the first procedure is linear in the number of nodes of the PM-Octree, whereas the complexity of the second one is linear in the number of edges of the boundary model.

8.2. Conversion from CSG to Boundary Representation

The problem of converting from a CSG model to a BRep is well known in solid modeling under the name of *boundary evaluation*. As soon as first solid modelers based on CSG had been developed, the need for algorithms that could compute the boundary of an object was remarked. Boundary evaluation was initially seen as more of a way to produce data for displaying purposes and Boolean operations computation, than as a support tool for hybrid systems. Hereafter, the increasing popularity of combined CSG and BRep schemes made boundary evaluation algorithms become more and more important as a core tool of many solid modelers.

Despite the importance of the problem, the literature on this topic is surprisingly poor. Beginning in the 1970s and continuing through the 1980s several boundary evaluation algorithms were developed, both in academic and industrial environments. Nevertheless, for quite some time only few descriptions of such algorithms were made available and were found mainly in reports with limited circulation. Moreover, algorithmic descriptions were often given at a low level, and the attention was mainly focused on implementation details and efficiency heuristics.

A theory underlying the boundary evaluation (and other important problems in computational geometry) was presented in 1980 by Tilove [61], who developed the concept of *set membership classification.* Only in 1985, Requicha and Voelcker published a paper based on such a theory that addressed the problem in a general way and offered a high level description of the approach and the algorithms. Since then, few progresses have been made in designing efficient algorithms that convert from CSG to BRep; all efforts have been spent in finding ways to reduce the complexity in the edge/face detection task (see, for instance [46]), whereas no attention has been paid to the problem of reconstructing topological relations.

The boundary evaluation problem is inherently hard because two complex, yet deeply different structures must be handled and a great amount of information that is only implicitly encoded in the input model must be made explicit. Most algorithms work only on planar approximations of objects, that is, on CSG models whose primitives are all objects with planar faces. Extensions to curved objects are possible in principle through algebraic methods for computing intersections of 3D surfaces [52]. The analysis that follows is mainly based on the work of Requicha and Voelcker; some alternative or innovative issues by Mäntylä, Rossignac and Voelcker, and Beacon et al., will be briefly discussed as well.

Requicha and Voelcker propose a classification of the possible approaches into *nonincremental boundary evaluation* and *incremental boundary evaluation.* Algorithms that evaluate the boundary in a nonincremental way take the CSG description of a solid object S in input and produce the BRep of the same object without recording any intermediate result related to subparts of S. On the contrary, the incremental approach keeps track of all intermediate results and can be formulated as follows: given CSG and BRep descriptions for two objects A and B, find a BRep for object $S = A \otimes B$, where $\otimes$ denotes one of the regularized set operators [60]; in this formulation, the problem is also known as *boundary merging*. If BReps are given or can be computed for the basic components of the CSG, the incremental approach can be applied recursively bottom-up on the CSG tree to compute the BRep of a complex object. The latter approach is conceptually more complex and requires large storage for intermediate boundaries. Thus, nonincremental boundary evaluation is suitable whenever a stored "library" object is brought into a workspace [45]. On the other hand, if one wants to track evolving solids, and, mainly, if a hybrid system must be kept on-line consistent, incremental boundary evaluation is much better, and somehow realizes a cooperation between the two different models.

No matter what approach is followed, all known algorithms are based on *set membership classification*. The set membership classification was introduced by Requicha in the late 70s as a generalization of the "clipping" operation of computer graphics. Tilove formalized a complete theory in [61], that had an important impact on general geometric intersection problems (see, for instance [62]). The classification $M(X, S)$ of a *candidate set* X with respect to a *reference set* S is a segmentation of X into three subsets $XwrtS = (XinS, XonS, XoutS)$, where the subsets contain the parts of X that are inside, on the boundary of, or outside S respectively. Classifications can be combined, making set membership classification a powerful tool. The problem of computing a *combination* of two classifications through a regularized set operator is defined as follows: given classifications $XwrtA$ and $XwrtB$, find $XwrtS$, where $S = A \otimes B$. Combined classifications are obtained very simply if objects do not have overlapping boundaries (e.g., $XinS = XinA \cup XinB$, $XoutS = X -^* XinS$). In the case of overlapping boundaries, singularities can arise, and more information is needed in order to compute combinations: a classification can be augmented including neighborhood information for the elements of $XonS$. *Augmented classifications* are discussed both in [61] and in [45], and are fundamental in practical applications; on the other hand, neighborhood information is somehow complex to represent and essentially related to the treatment of geometric singularities; for the sake of brevity, we do not address this topic here.

Classification algorithms differ deeply according to the nature of the sets under consideration—the dimensionality of geometric entities determines the structure of an algorithm. Several algorithms for point/face, line/face, point/solid, line/solid classification, have been developed and discussed in brief by Requicha and Voelcker [45]. Sophisticated point/face and point/point algorithms working directly on BReps are also described in detail by Mäntylä [39]. All such algorithms can be part of a boundary evaluation (or a boundary merging) algorithm, depending on the strategy used to design the algorithm itself.

Generally speaking, set membership classification is applied to the boundary evaluation problem by exploiting the following fact: given an object S, if a superset X of the boundary ∂S of S is known, then $\partial S = XonS$. In the (conceptually) simplest case, a superset $\mathcal{F}$ of the faces of S is considered, in such a way that $\mathcal{F}onS$ will contain exactly the collection of the faces of S. In this case, a *Generate-and-Test* paradigm is applied, that follows:

```
paradigm FACE_GENERATE_AND_TEST(S,B);
// S is the input object; B is the output BRep //
  ⟨generate a sufficient set of tentative faces 𝓕 for S⟩;
  for every F ∈ 𝓕 do
    FwrtS ← CLASSIFY_FACE(F,S);
    ADD_TO_BREP(FonS,B);
  end for
end FACE_GENERATE_AND_TEST;
```

In order to understand how the above paradigm works, some concepts should be clarified:

- what a *sufficient set of tentative faces* is;
- how the elements of the boundary are extracted from the tentative set (procedure CLASSIFY_FACE);
- how faces are inserted into the boundary representation.

A set $\mathcal{F} = \{F_i \mid i = 1, \ldots, k\}$ is a set of tentative faces for an object S if $\cup_{i=1,\ldots,k} F_i \supset \partial S$, where ∂S denotes the boundary of S. In practice, the previous definition means that each face of S is some subset of some $F_j \in \mathcal{F}$. The generation of a sufficient set of tentative faces is somehow straightforward, and is based on the following fact (that is applied recursively): for every pair of solid objects A and B, $\partial(A \otimes B) \subset (\partial A \cup \partial B)$. Different choices are possible for tentative sets of faces, and characterize the approach of the algorithm. If either the faces of all primitives in the CSG (provided that they are available) or the infinite planes containing such faces (in the case of CSG based on halfspaces) are chosen, a nonincremental algorithm must be applied. On the contrary, if either the faces of only two solids A and B, or the corresponding infinite planes, are considered at a time, an incremental approach is implied.

The classification of geometric entities with respect to a solid object described by a CSG is a delicate subject. As we outlined in previous sections, the hierarchical structure of CSG often leads to a Divide-and-Conquer approach that is schematized by the following function (names of subfunctions are self-explanatory):

```
function CLASSIFY(X,S):classification;
// X is the candidate set; S is the reference set ancoded by CSG;
  return classification will be (XinS,XonS,XoutS). //
begin
  if IS_A_PRIMITIVE(S) then
    return(CLASSIFY_WRT_PRIMITIVE(X,S))
```

```
    else
      return(COMBINE(CLASSIFY(X,LEFT_SUBTREE(S)),
                     CLASSIFY(X,RIGHT_SUBTREE(S)),
                     OPERATOR(S)))
    end if
  end CLASSIFY;
```

In principle, face/solid classification could be implemented directly through the previously shown paradigm, in order to obtain procedure CLASSIFY_FACE used in the Generate-and-Test paradigm. In practice, explicit face/solid classification can be very complex to perform because Boolean operations on face subsets are involved; thus, such approach is seldom used.

Since the faces of a solid are bounded by closed loops of edges, an alternative edge-based approach is usually preferred, that follows an appropriated Generate-and-Test paradigm:

```
paradigm EDGE_GENERATE_AND_TEST(S,B);
// S is the input object; B is the output BRep //
  ⟨generate a sufficient set of tentative edges 𝓔 for S⟩;
  for every E ∈ 𝓔 do
    EwrtS ← CLASSIFY_EDGE(E,S);
    if ON_A_SINGLE_SURFACE(EonS,S) then
      DISCARD(E)
    else
      ADD_TO_BREP(EonS,B);
    end if
  end for
end EDGE_GENERATE_AND_TEST;
```

Function ON_A_SINGLE_SURFACE is aimed to detect if a tentative edge is on ∂S but lies on a unique surface, that is, it does not separate two different faces. The procedure manipulates neighborhood for its computations, and the corresponding **if** test deals with a special case that will not be discussed here. Edge generation is instead a fundamental task that must be further specified.

To generate a sufficient set of edges is less straightforward than to generate a sufficient set of faces. Again, the sufficient set is defined for the combination of two objects, and eventually extended to the whole CSG tree by recursion. Given objects A and B, the desired superset of edges for $S = A \otimes B$ is obtained as union of two different sets:

- *self edges* are all edges of A and all edges of B;
- *cross edges* are all edges obtained by intersecting faces of A with faces of B.

If BReps of A and B are available (like in the case of incremental algorithms), self edges are available in such a structure; in the case of a nonincremental approach applied to a CSG based on general halfspaces the set of self edges will be completely empty. Cross edges can be either computed by intersecting pairs of faces from A and B boundaries respectively, or by trimming oversized edges obtained by intersecting the planes containing such faces. The generation of cross edges can involve a two-dimensional line/face classification algorithm.

Based on the arguments discussed so far, we present in the following the pseudocode of a simplified version of the boundary merging edge-based procedure described by Requicha and Voelcker [45]. For the sake of simplicity, all parts concerning the treatment of neighborhoods (that are necessary for an actual implementation) have been suppressed here. The procedure is based on edge/solid classification and combination in 2D and 3D. Suitable procedures that perform the above tasks were developed in [45]; we do not describe them here in detail. Other procedures used here have been described already or are self-explanatory.

```
procedure BOUNDARY_MERGING(A,B,op,S);
// A and B are the BReps of the input objects; op is the Boolean
   operation which combines them;
   S is the output BRep corresponding to the combination of A and B
   through op. //
begin
  for every F face of A do
    for every E edge of F do   // classify self edges of A //
      EwrtA ← E;   // E is an edge of A //
      EwrtB ← CLASSIFY_EDGE(E,B);
      EwrtS ← COMBINE_EDGE(EwrtA,EwrtB,op);
      if not ON_A_SINGLE_SURFACE(EonS,S) then
        ADD_TO_BREP(EonS,S);
      end if
    end for;
    for every G face of B do   // classify cross edges //
      if not COPLANAR(G,F) then   // compute an oversized edge
  then trim it //
        C ← SURFACE_CONTAINING(G) ∩ SURFACE_CONTAINING(B);
        CwrtF ← CLASSIFY_EDGE_2D(C,F);
        CwrtG ← CLASSIFY_EDGE_2D(C,G);
        E ← CinF ∩* CinG;
        EwrtS ← E;   // EonS = E //
        if not ON_A_SINGLE_SURFACE(EonS,S) then
          ADD_TO_BRIEF(EonS,S);
        end if
      end if
    end for
  end for;
  for every G face of B do
    for every E edge of G do   // classify self edges of B //
      EwrtB ← E;   // E is an edge of B //
      EwrtA ← CLASSIFY_EDGE(E,A);
      EwrtS ← COMBINE_EDGE(EwrtA,EwrtB,op);
```

```
        if not ON_A_SINGLE_SURFACE(EonS,S) then
          ADD_TO_BREP(EonS,S);
        end if
      end for
    end for
  end BOUNDARY_MERGING;
```

The definition of an incremental boundary evaluation algorithm that uses recursively the procedure just described is straightforward by applying a divide-and-conquer paradigm, provided that BReps for primitives are available or can be computed easily (e.g., through intersection of enclosing planes).

```
algorithm BOUNDARY_EVALUATION(S,B);
// S is the input CSG; B is the output BRep. //
begin
  if IS_A_PRIMITIVE(S) then
    B ← EVALUATE_PRIMITIVE(S)
  else
    BOUNDARY_EVALUATION(LEFT_SUBTREE(S),B_l);
    BOUNDARY_EVALUATION(RIGHT_SUBTREE(S),B_r);
    BOUNDARY_MERGING(B_l,B_r,OPERATOR(S))
  end if
end BOUNDARY_EVALUATION;
```

Requicha and Voelcker do not give an explicit analysis of the computational complexity of the boundary evaluation, boundary merging, and set membership classification and combination algorithms discussed so far. All algorithms are polynomial with high exponentials, but worst-case complexity seems to be quite pessimistic with respect to experimental results. The average complexity has not been analyzed at all, and experiments show that such algorithms are still quite inefficient.

Several heuristic short cuts have been proposed to improve performance by trying to avoid useless and cumbersome computations. A formal approach was taken by Rossignac and Voelcker that address the problem of computing only on entities and over regions of space that can affect the desired final result [46]. Rossignac and Voelcker introduce and formalize the concept of *active zone* in CSG. Roughly speaking, the active zone Z of a node A in a CSG representation of a solid S, is the spatial region wherein changes to A affect S and hence ∂S. Active zones can be computed as combination of certain nodes of the CSG tree (depending on the position of A in S) and can be used for boundary evaluation as follows. If A is a subcomponent of object S, and $X \subset \partial A$ is a candidate set we want to classify with respect to S, then if Z is the active zone of A in S, the classical set membership classification can be substituted by an approach that trims X with respect to Z. The use of active zones allows on the average to early reject classified elements, thus avoiding redundant computations.

Problems related to the actual reconstruction of topological relations within the boundary representation have not been addressed in the literature. Algorithms developed for hybrid systems so far are usually expected to produce only "poor" BReps (e.g., just lists of faces, edges and vertices), whereas topological relations should be inferred somehow in a different stage (when required). On the contrary, in his book [39] Mäntylä refers to a modeling system based on BRep, where a "rich" representation is encoded (winged-edge data structure [5]). Mäntylä does not study, in detail, the boundary evaluation process, but he describes a sophisticated algorithm for the boundary merging of the BReps of two solid objects that takes care of reconstructing all the encoded relations. The algorithm is still based on set membership classification, but it computes vertex/face and vertex/vertex classifications by exploiting implicit vertex neighborhoods and explicit adjacency relations stored in the data structure. In [35] a complete description of an algorithm for set operations on polyhedral solids with convex polygonal faces is given: the algorithm first splits the faces in the two objects, so that no two polygons intersect, and then uses a set membership classification approach applied to the subdivided faces and to the two objects.

Finally, the extension of boundary evaluation to *free-form* (or *sculptured*) surfaces is addressed in a paper by Beacon et al., [6]. Their method (called the ISOS method) computes approximations of free-form surfaces by inner and outer bounding sets with planar faces, and performs face classifications on such sets.

9. CONVERTING TO CSG SCHEMES

The problem of converting any representation to a CSG description is quite hard, because of the intrinsic nonuniqueness of CSG schemes: the same object can be expressed as the Boolean combination of primitives in many different ways. Also, CSG is essentially a method for expressing the way an object is designed: using a CSG scheme makes sense only if the designer can interact with it.

As we pointed out in the introduction, conversion algorithms that produce CSG descriptions would be desirable in order to ensure the complete consistency of hybrid systems. On the other hand, converting another representation to CSG is useful only if the output structure is understandable and easily manageable by human beings.

Limited work has been done on algorithms for converting to CSG and all existing algorithms, however, produce descriptions that are useless for practical purposes. The existing papers dealing with such problems convert from a boundary representation or from a PM-Octree.

Woo [68] presents an algorithm based on a convex hull technique. The purpose is to produce from a boundary description of a solid given as input a volumetric description of the part for numerical control operations. The method finds the convex hull C of the input solid S, and then the difference between C and S, thus expressing S as C minus such difference. The process is applied recursively because each cavity of S can be expressed as the convex hull of the cavity plus its protrusions, and so on. The resulting CSG tree will be a Boolean combination (through disjoint union and subtraction only) of convex parts. A disadvantage to this method is that the resulting convex volumes are neither necessarily related to specific machining tasks, nor do they correspond to predefined primitive components. No information is produced for classifying these volumes as specific form features, and thus no guidance can be provided to choose the manufacturing method appropriate for each part.

Juan [29, 30], follows an analogous approach to Woo, but he produces a CSG based on half-spaces. The algorithm consider the set of halfspaces corresponding to the faces of the object, and it subdivides the object itself into *convex sets* and *deficiency sets*. At a given level of recursion, a convex set is a set of faces belonging to the convex hull, whereas a deficiency set is a maximally connected set of faces that do not belong to the convex hull. The subdivision is applied recursively on each deficiency set until only convex sets are obtained. Each convex set corresponds to an intersection of halfspaces (that bounds the convex volume containing the set), whereas each deficiency induces a union of halfspaces (that bounds the corresponding cavity). The resulting CSG tree is a binary tree whose leaves are halfspaces, and whose internal nodes are regularized unions and intersections (see Figure 10.16).

Both the previous algorithms have a complexity $O(n^2)$, where n is the number of faces in the boundary representation. An efficient algorithm for solving the conversion problem in two dimensions generating a CSG tree based on halfspaces has been presented by Dobkin et al., [19] that achieves a better time complexity. The algorithm can also be applied to certain kinds of polyhedra wherein the problem is equivalent to the two-dimensional one. The basic idea is the following: every simple polygon in the plane admits a representation expressed by a Boolean formula based on the halfplanes supporting its sides. Also, each of the supporting halfplane appears in the formula exactly once. Although a naive algorithm for computing such formula would require $O(n^2)$ operations, where n is the number of vertices of the input polygon, the method presented in [19] performs the conversion in $O(n \log n)$ steps.

An algorithm that converts from a PM-Octree to a CSG based on halfspaces has also been presented by Juan [30]. The algorithm essentially computes the CSG tree describing the portion of the object within each node in the PM-Octree (that is a simple task, because only a small number of possible configurations are allowed), then it tries to condense together CSG trees related to adjacent blocks, according to some *consensus* rules. The resulting CSG tree can be quite complex and somehow "non-natural" because the same halfspace can be replicated many times inside it.

A conversion from the PM-CSG model [73] to CSG-DAG has not been attempted yet. Yet, the PM-CSG seems to be better suited to this task than any other model. In fact, the way the PM-CSG description of an object relates to its

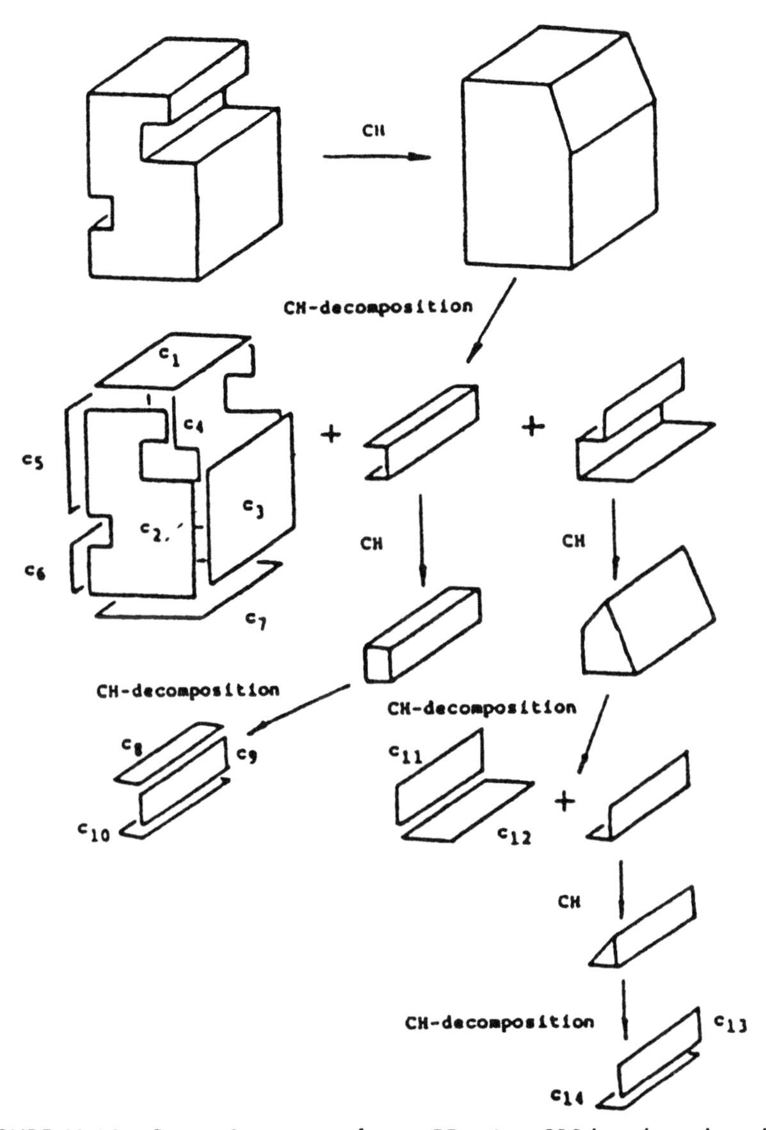

FIGURE 10.16. Conversion process from a BRep to s CSG based on planar half-spaces [29].

corresponding CSG-DAG is similar to the way the PM-Octree relates to the BRep. In other words, all information present in the CSG-DAG are maintained in the PM-CSG tree (mainly, all primitive objects are conserved), it is only somehow fragmented on the nodes. It would be interesting to investigate inverse transformation algorithms from PM-CSG. In particular, it would be important to understand if the application of such algorithms to objects, that have been modified while they are encoded by a PM-CSG, would produce the "right" modifications on the original CSG.

Among the algorithms described in the first part of this Section, the method proposed by Woo has possibly been the only attempt towards the use of CSG representations as solid models for manufacturing. This subject is quite wide and concerns not only the conversion problem, but also the general use of CSG models within a CAD/CAM system and the way the design and manufacturing techniques can be interfaced with computerized systems. Roughly speaking, an object should be designed with the best possible relation to a manufacturing sequence that produces it. Also, if a CSG is used to encode the design, the compositions of primitives within the model should also correspond to assembly or machining operations. Although some machining operations are essentially material removal and thus can be expressed as subtraction operations, assembly operations are more effectively expressed as face-to-face part combinations.

It might be useful to compute a primitive-based CSG from a BRep as a description for numerical control machines. In this sense, a CSG can be automatically derived from a BRep by methods for recognizing the so called *form features* of an object (holes, pockets, slots, etc.). Several methods have been proposed for automatic feature recognition. Kyprianou [34] uses a feature grammar to recognize features characterized by protrusions and depressions from a boundary description of the solid (see also [27]). Henderson [25] extracts cavities form a BRep by using expert system rules. Joshi and Chang [28] extract polyhedral features such as pockets, slots, steps, holes, from a boundary description, by using a graph matching approach. The features to be recognized are described by graphs representing the adjacency relations among the part faces. A method for feature recognition based on topological information is described in [17] that identifies a certain class of form features by using connectivity information extracted from a graph description of the object boundary. A conceptually similar, but simpler approach is described in [18]. A combination of this method with Kyprianou's approach is described in [23].

These latter methods, however, produce a description of the object where the form features are extracted in terms of a modular boundary model. An MBM gives a more flexible and powerful representation of form features because form features are attached to object faces and have tolerance information associated with them. Moreover, an MBM is better suited to describe assembly operations.

10. THE MBM AS A MODEL FOR CAD/CAM

One of the major limitations of classical CAD models (CSG, BRep, Octrees) is their lacking capability of describing form features and their relations. "Pure" solid modelers cannot be used for assembly and machining planning because they do not contain information (e.g., tolerances and dimensions, form features, materials) required for these tasks [23]. Modular boundary models are an attempt to fill the gap between the design and manufacturing phases because of their capability of representing form features as model components. Moreover, the MBM description of the object produced by the designer, where components represent design features, can be locally modified (because of the modular nature of the model) to produce an MBM description in terms of manufacturing features.

Conversion problems on an MBM have been considered only as far as conversion to and from a boundary representation is concerned [63]. The problem of converting from an MBM to a BRep is conceptually simple because it involves a traversal of the connection graph with an elimination of the connection facets among the various components of the MBM. An algorithm performing such conversion has been implemented as part of a solid modeler based on the Face-to-Face Composition Model [17]. The problem of converting from a boundary representation to an MBM involves form feature recognition. The algorithms mentioned in section 9 all produce a description of the form features extracted in terms of their boundary. The algorithms described in [17, 18, 23] also organize the features extracted in an MBM, they are methods for building an MBM from a BRep. The purpose of producing an MBM from a BRep is not to facilitate operations on the BRep, as in the case of conversions to Octrees or PM-Octrees, but to generate a solid model suitable for manufacturing and assembly.

REFERENCES

1. M. Altenhofen and R. Diehl, "Conversion of boundary representations to bintrees", *Proceedings of the EUROGRAPHICS '88 Conference,* D. Duce, Ed., pp. 117–127, North Holland, Amsterdam, 1988.
2. S. Ansaldi, L. De Floriani and B. Falcidieno, "Geometric modeling of solid objects by using a face abjacency graph representation", *Computer Graphics,* Vol. 19, No. 3, pp. 131–139, July 1985.
3. D. Ayala, P. Brunet, R. Juan and I. Navazo, "Object representation by means of nonminimal division quadtrees and octrees", *ACM Transaction on Graphics,* Vol. 4, No. 1, pp. 41–59, January 1985.
4. D. Badouel and G. Hégron, "Set operation evaluation using Boolean octree", *New Trends in Computer Graphics,* N. Magnegat-Thalmann, D. Thalmann, Eds., Springer–Verlag, 1988.
5. B.G. Baumgart, "Winged-edge polyhedron representation", *Technical Report* STAN-CS-320, Computer Science Department, Stanford University, Stanford, CA, 1974.
6. G.R. Beacon, J.R. Dodsworth, S.E. Howe, R.G. Oliver and A. Saia, "Boundary evaluation using inner and outer sets: the ISOS method", *IEEE Computer Graphics and Applications,* Vol. 9, No. 2, March 1989.

7. P.J. Besl and R.C. Jain, "Three-Dimensional object recognition", *ACM Computing Surveys,* Vol. 17, No. 1, pp. 75–145, March 1985.
8. J.D. Boissonnat, "Geometric Structures for Three-Dimensional Shape Representation", *ACM Transactions on Graphics,* Vol. 3, No. 4, pp. 266–286, April 1984.
9. I.C. Braid, R.C. Hillyard and I.A. Stroud, "Stepwise construction of polyhedra in geometrical modeling", in *Mathematical Models in Computer Graphics and Design,* K.W. Brodlie, Ed., Academic Press, New York, pp. 123–141, 1980.
10. P. Brunet and I. Navazo, "Geometric modeling using exact octree representation of polyhedral objects", *Proceedings EUROGRAPHICS '85,* pp. 159–169, 1985.
11. P. Brunet and I. Navazo, "Solid representation and operation using extended octrees", *ACM Transaction on Graphics,* Vol. 8, 1989.
12. E. Bruzzone, L. De Floriani and E. Puppo, "Reconstructing three-dimensional shapes through euler operators", *Proceedings 5th International Conference on Image Analysis and Processing,* Positano (Italy), September 20–22, 1989.
13. I. Carlbom, I. Chakravarty and D. Vanderschel, 'A hierarchical data structure for representing spatial decomposition of 3-D objects", *IEEE Computer Graphics and Applications,* Vol. 5, No. 4, pp. 24–31, April 1985.
14. T. Casciani, B. Falcidieno, G. Fasciolo and C. Pienovi, "An algorithm for constructing a quadtree from polygonal regions", *Computer Graphics Forum,* pp. 269–274, 1984.
15. L. De Floriani and B. Falcidieno, "A hierarchical boundary model for solid object representation", *ACM Transactions on Graphics,* Vol. 7, No. 1, pp. 42–60, 1988.
16. L. De Floriani, A. Maulik and G. Nagy, "Manipulating a modular boundary model with a face-based graph structure", *IFIP WG 5.2 Workshop on Geometric Modeling,* Rensselaer Ville, NY, 1989.
17. L. De Floriani, "Feature extraction from boundary models of three-dimensional objects", *IEEE Transactions on Pattern Analysis and Machine Intelligence,* Vol. 11, No. 8, pp. 785–798, August 1989.
18. L. De Floriani and E. Bruzzone, "Building a feature-based object description from a boundary model", *Computer Aided Design,* December 1989.
19. D. Dobkin, L. Guibas, J. Hershberger and J. Snoeyink, "An efficient algorithm for finding CSG representation of a simple polygon", *Technical Report* 50a, Digital–System Research Center, Palo Alto, CA, 1989.
20. M.J. Dürst and T.L. Kunii, "Integrated polytrees: a generalized model for the integration of spatial decomposition and boundary representation", *Theory and Practice of Geometric Modeling,* W. Straßer, H.P. Seidel, Eds., Springer-Verlag, pp. 329–348, 1989.
21. C.M. Eastman and K. Weiler, "Geometric modeling using Euler operators", *Proceedings of the First Conference on Computer Graphics and CAD/CAM Systems,* Cambridge, Ma, pp. 248–259, May 1979.
22. D.A. Field, "Implementing Watson's algorithm in three dimensions", *Proceedings Second Annual Simposium on Computational Geometry,* Yorktown Heights, NY, pp. 246–259, June 1986.
23. B. Falcidieno and F. Giannini, "Automatic Recognition and Representation of Shape-Based Features in a Geometric Modeling System", *Computer Vision, Graphics and Image Processing, 48, 1,* pp. 93–123, October 1989.

24. K. Fujimura and T.L. Kunii, "A hierarchical space indexing method", *Proceedings of Computer Graphics '85*, Tokyo, TI-4, pp. 1–14, 1985.
25. M.R. Henderson, "Extraction of feature information from three-dimensional CAD data", *PhD Thesis*, Purdue University, 1984.
26. G.M. Hunter and K. Steiglitz, "Operations on images using quadtrees", *IEEE Transactions on Pattern Analysis and Machine Intelligence*, Vol. 1, pp. 145–153, 1979.
27. G.E. Jared, "Shape features in geometric modeling", *Solid Modeling by Computers: from Theory to Applications*, Plenum, New York, 1984.
28. S. Joshi and T. Chary, "Graph-based heuristics for recognition of mechanical features from a 3D solid model", *Computer Aided Design*, Vol. 20, No. 2, 1988.
29. R. Juan, "Boundary to Constructive Solid Geometry: a step towards 3D conversion", *Proceedings EUROGRAPHICS'88 Conference*, D. Duce, Ed., Amsterdam, pp. 129–139, 1988.
30. R. Juan, "On Boundary to CSG and extended octree to CSG conversions", *Theory and Practice of Geometric Modeling*, W. Straßer, H.P. Seidel, Eds., Springer-Verlag, pp. 349–367, 1989.
31. M. Karasick, "On the representation and manipulation of rigid solids", *PhD thesis*, School of Computer Science, McGill University, Montreal, 1988.
32. E. Kawaguchi and T. Endo, "On a method of binary picture representation and its application to data compression", *IEEE Transactions on Pattern Analysis and Machine Intelligence*, Vol. 2, No. 1, pp. 27–35, January 1980.
33. T.L. Kunii, T. Satoh and K. Yamaguchi, "Generation of topological boundary representations from octree encoding", *IEEE Computer Graphics and Applications*, Vol. 5, No. 3, pp. 29–38, March 1985.
34. L.K. Kyprianou, "Shape classification in Computer-Aided-Design", *PhD Dissertation*, Computer laboratory, University of Cambridge, England, July 1980.
35. D.H. Laidlaw and J.F. Hughes, "Constructive Solid Geometry for polyhedral objects", *ACM Computer Graphics*, 20, 4, pp. 161–170, August 1986.
36. Y.T. Lee and A.A.G. Requicha, "Algorithms for computing the volume and other integral properties of solids. I. Known methods and open issues", *Communications of the ACM*, 25, 9, pp. 635–641, September 1982.
37. Y.T. Lee and A.A.G. Requicha, "Algorithms for computing the volume and other integral properties of solids. II. A family of algorithms based on representation conversion and cellular approximation", *Communications of the ACM*, 25, 9, pp. 642–650, September 1982.
38. M. Mäntylä and R. Sulonen, "GWB: a solid modeler with Euler operators", *IEEE Computer Graphics and Applications*, 2, 7, pp. 17–31, September 1982.
39. M. Mäntylä, *An introduction to Solid Modeling*, Computer Science Press, Rockville, MD, 1987.
40. D. Meager, "Octree encoding: a new technique for the representation, the manipulation, and display of arbitrary 3-d objects by computer", *Technical Report*, Electrical and System Engineering IPL-TR-80-111, Rensselaer Polytechnic Institute, Troy, NY, October 1980.
41. I. Navazo, J. Fontdecaba and P. Brunet, "Extended octrees, between CSG trees and boundary representations", *Proceedings EUROGRAPHICS'87*, North-Holland, pp. 239–247, 1987.

42. I. Navazo, "Extended octree representation of general solids with plane faces: model structure and algorithms", *Computers & Graphics,* 13, 1, pp. 5–16, 1989.
43. A.A.G. Requicha, "Representations of rigid solids: theory, methods, and systems", *ACM Computing Surveys,* 12, 4, pp. 437–464, December 1980.
44. A.A.G. Requicha and H.B. Voelcker, "Solid modeling: a historical summary and contemporary assessment", *IEEE Computer Graphics and Applications,* Vol. 2, No. 2, pp. 9–24, March 1982.
45. A.A.G. Requicha and H.B. Voelcker, "Boolean operation in solid modeling: boundary evaluation and merging algorithms", *Proceedings IEEE,* Vol. 73, 1, January 1985.
46. J.R. Rossignac and H.B. Voelcker, "Active zones in CSG for accelerating boundary evaluation, redundancy elimination, interference detection, and shading algorithms", *ACM Transaction on Graphics,* Vol. 8, No. 1, pp. 51–87, January 1988.
47. S.D. Roth, "Ray casting from modeling solids", *Computer Graphics and Image Processing,* Vol. 18, pp. 109–114, 1982.
48. H. Samet and M. Tamminen, "Bintrees, CSG trees, and time", *Computer Graphics,* Vol. 19, No. 3, pp. 121–130, July 1985.
49. H. Samet and M. Tamminen, "Approximating CSG trees of moving objects", *Technical Report,* CS-TR-1472, University of Maryland, College Park, MD, January 1985.
50. H. Samet, *The Design and Analysis of Spatial Data Structures,* Addison–Wesley, Reading, MA, 1990.
51. H. Samet, *Applications of Spatial Data Structures,* Addison-Wesley, Reading, MA, 1990.
52. R.F. Sarraga, "Algebraic methods for intersections of quadric surfaces in GMSolid", *Computer Graphics and Image Processing,* Vol. 22, pp. 222–238, 1983.
53. B.I. Soroka and R.K. Bajcsy, "A program for describing complex three-dimensional objects using generalized cylinders as primitives", *Proceedings of the Pattern Recognition and Image Processing Conference,* Chicago, Il, pp. 331–339, June 1978.
54. S.A. Shafer and T. Kanade, "The theory of straight homogeneous generalized cylinders", *Technical Report,* CMU-CS-83-105, Carnegie–Mellon University, Pittsburgh, Pa, January 1983.
55. M. Tamminen and H. Samet, "Efficient octree conversion by connectivity labeling", *Computer Graphics,* Vol. 18, No. 3, pp. 43–51, July 1984.
56. M. Tamminen, O. Karonen and M. Mäntylä, "Ray-casting and block model conversion using a spatial index", *CAD Journal,* Vol. 16, No. 4, 1984.
57. M. Tamminen, "Comment on quad- and octrees", *Communications of the ACM,* Vol. 27, No. 3, pp. 248–249, March 1984.
58. Z. Tang and S. Lu, "A new algorithm for converting boundary representation to octree", *Proceedings EUROGRAPHICS'88 Conference,* D. Duce, Ed., North–Holland, Amsterdam, pp. 105–116, March 1984.
59. W.C. Thibault and B.F. Naylor, "Set operations on polyhedra using space partitioning trees", *ACM Computer Graphics,* Vol. 21, No. 4, pp. 153–162, July 1987.
60. R.B. Tilove and A.A.G. Requicha, "Closure of Boolean operations on geometric entities", *Computer Aided Design,* Vol. 12, No. 5, pp. 219–220, September 1980.

61. R.B. Tilove, "Set membership classification: a unified approach to geometric intersection problems", *IEEE Transactions on Computers,* C-29, Vol. 10, pp. 874–883, October 1980.
62. R.B. Tilove, "A null-object detection algorithm for constructive solid geometry", *Communications of the ACM,* Vol. 27, No. 7, pp. 684–694, July 1984.
63. R. Valle, "Conversion algorithms on solid object representation", Dissertation, Università degli Studi di Genova, Italy, 1988, (in Italian).
64. H.B. Voelcker and A.A.G. Requicha, "Geometric modeling of mechanical parts and processes", *IEEE Computer,* Vol. 10, No. 12, pp. 48–57, December 1977.
65. K. Weiler, "Edge-based data structures for solid modeling in a curved-surface environment", *IEEE Computer Graphics and Applications,* Vol. 5, No. 1, pp. 21–40, January 1985.
66. K. Weiler, "Topological structures for geometric modeling", *Ph.D. dissertation,* Department of Computer and System Engineering, Rensselaer Polytechnic Institute, Troy, NY, August 1986.
67. P.W. Wilson and M. Pratt, "Requirements for support of form features in a solid modelling system", *Technical Report,* Geometric Modeling Project, CAM-I, 1985.
68. T.C. Woo, "Feature extraction by volume decomposition", *Proceedings Conference on CAD/CAM in Mechanical Engineering,* MIT, Cambridge, MA, March 1982.
69. T.C. Woo, "A combinatorial analysis of boundary data structure schemata," *IEEE Computer Graphics and Applications,* Vol. 5, No. 3, pp. 19–27, March 1985.
70. J.R. Woodwark and K.M. Quinlan, "The derivation of graphics from volume models by recursive subdivision of the object space", *Proceedings Computer Graphics 80 Conference,* Online Publishers, London, pp. 335–343, 1980.
71. J.R. Woodwark and K.M. Quinlan, "Reducing the effect of complexity on volume model evaluation", *Computer-aided Design,* Vol. 14, No. 2, pp. 89–95, March 1982.
72. J.R. Woodwark, "Generating wireframes from set-theoretic solid models by spatial division", *Computer-aided Design,* Vol. 18, No. 6, pp. 307–315, July/August 1986.
73. G. Wyvill and T.L. Kunii, "A functional model for constructive solid geometry" *Visual Computer,* Vol. 1, No. 1, pp. 3–14, July 1985.
74. G. Wyvill, "Geometry and modeling for CAD systems", *Tutorial Notes–CG International '90 Conference,* Singapore, June 1990.

Index

A

B

C

D

E

G

K

L

M

N

O

P

Q

R

S

U

V

W